MW00770330

Structural Equation Modeling

A Second Course
(2nd edition)

A volume in
*Quantitative Methods in Education and the Behavioral Sciences:
Issues, Research, and Teaching*
Ronald C. Serlin, *Series Editor*

Quantitative Methods in Education and the Behavioral Sciences:
Issues, Research, and Teaching

Series Editor
Ronald C. Serlin
University of Wisconsin–Madison

Multilevel Modeling of Educational Data (2008)
Edited by **Ann A. O'Connell** and **D. Betsy McCoach**

Real Data Analysis (2007)
Edited by **Shlomo S. Sawilowsky**

Structural Equation Modeling: A Second Course (2006)
Edited by **Gregory R. Hancock** and **Ralph O. Mueller**

Structural Equation Modeling

A Second Course
(2nd edition)

edited by

Gregory R. Hancock
University of Maryland

Ralph O. Mueller
University of Hartford

INFORMATION AGE PUBLISHING, INC.
Charlotte, NC • www.infoagepub.com

Library of Congress Cataloging-in-Publication Data

Structural equation modeling : a second course / edited by Gregory R. Hancock & Ralph O. Mueller. – 2nd ed.
    pages cm – (Quantitative methods in education and the behavioral sciences)
 ISBN 978-1-62396-244-9 (pbk.) – ISBN 978-1-62396-245-6 (hardcover) – ISBN 978-1-62396-246-3 (ebook) 1. Structural equation modeling. 2. Analysis of covariance. 3. Linear models (Statistics) 4. Multilevel models (Statistics) I. Hancock, Gregory R. II. Mueller, Ralph O.
 QA278.3.S77 2013
 519.5'3–dc23

                                                2013004599

Printed in the United States of America

To our students,
from whom we learn far more than we could ever teach.

—Greg and Ralph

CONTENTS

PART I

FOUNDATIONS

PART II
EXTENSIONS

PART III
ASSUMPTIONS

ACKNOWLEDGEMENTS

We are grateful to the American Educational Research Association's Special Interest Group for Educational Statisticians for their willingness to take a risk with the first edition, and for their continued support throughout the writing of this greatly expanded second edition. We also acknowledge the unwavering encouragement of George Johnson and his team at Information Age Publishing.

We further wish to recognize the members of this edition's "Breakfast Club" who, over many months, spent Tuesday mornings sharing breakfast, discussing and critiquing chapter drafts, and offering amazingly detailed technical and pedagogical feedback: Yoonjeong Kang, Yong Luo, Xiulin Mao, and Xiaoshu Zhu. Congratulations on earning your Mickey Mouse ears!

Finally, and most importantly, we give our deepest appreciation to Goldie, Sydney, Quinn, and Tate, for all their love, patience, and support throughout this project, and for perspective on what truly matters most.

Structural Equation Modeling: A Second Course (2nd ed.), page xi
Copyright © 2013 by Information Age Publishing
All rights of reproduction in any form reserved.

PREFACE

Let me be the first to welcome you to the second edition of *Structural Equation Modeling: A Second Course*. The first edition was the initial volume in the series, *Quantitative Methods in Education and the Behavioral Sciences: Issues, Research, and Teaching*, sponsored by the American Educational Research Association's Special Interest Group for Educational Statisticians. Structural equation modeling (SEM) remains a critical methodological technique for testing theoretical models. This is, in part, attested to by the existence of the second edition of this edited volume, with chapters written by many of the top researchers in the field. The second edition of any edited book is rather rare. And obviously second editions are not developed without a very successful first edition. Here I will briefly describe the content and importance of the revised chapters, as well as provide a few details on the new chapters. So let us have a look at what awaits you in this volume.

As in the first edition, the purpose of this volume remains to provide a user-friendly, applied, intermediate-level coverage of SEM. As before, the volume is divided into three parts, *Foundations*, *Extensions*, and *Assumptions*. The *Foundations* part consists of revisions of three chapters from the first edition, as well as an entirely new chapter. The chapter on equivalent models (Hershberger & Marcoulides) reminds us that there are typically several equivalent best-fitting models. Here we are given some advice as to how to deal with this issue. The next chapter is on reverse arrow dynamics (Kline) and deals with models that involve feedback loops and/or reciprocal relations (often a challenging issue for the modeler). This is followed by a new chapter on partial least squares (PLS) modeling (Rigdon). Although PLS modeling has been around for longer than most other forms of SEM, pio-

Structural Equation Modeling: A Second Course (2nd ed.), pages xiii–xv
Copyright © 2013 by Information Age Publishing

neered by Herman Wold in the 1960s, it has mostly been used in Europe until recently. Thus it is refreshing to see PLS becoming more of a mainstream technique as evidenced by this chapter. Rigdon frames PLS as an exploratory and component-based form of modeling. The final chapter in this section on power analysis (Hancock & French) further develops this important component of statistical analysis. Having 100 subjects does not necessarily guarantee a sufficient degree of power. Thus, determining the sample size for a desired amount of power is quite important.

The second part on *Extensions* moves us beyond the basics and into more advanced structural models. The first chapter in this part (Thompson & Green) considers between-group differences in the context of multiple sample models. This is followed by a new chapter on conditional process models (Hayes & Preacher). The possibilities of mediated and/or moderated models in SEM have existed for some time, but their use has exploded in recent years due to new developments. My students ask more about these types of models than just about any other advanced topic in SEM. So I am thrilled to see this chapter appear in the volume. The next chapter moves us into interaction and quadratic effects latent variable models (Marsh, Wen, Hau, & Nagengast). These types of models continue to increase in popularity. Next we see a chapter on latent growth models (Hancock, Harring, & Lawrence). Modeling longitudinal growth is not only popular, but presents numerous challenges to existing methods and software in SEM. Chapter 9 considers mixture modeling and mean structure modeling (Pastor & Gagné). It is quite frequent that we have mixtures of continuous and ordinal measures, and means structure models have become increasingly popular. The last chapter in this part is a new chapter on exploratory SEM (Morin, Marsh, & Nagengast). Although exploratory factor analysis has existed for over a century, exploratory SEM is rather new to the field. This will become an important topic as exploratory structural modeling develops in the coming years.

The last part of the volume is on *Assumptions*. Although often overlooked, assumptions represent a critical component of any inferential technique. Thus it is important to not only know what the assumptions are in SEM, but when to be concerned about violations and when robustness is of assistance. This part begins with a chapter on nonnormal and categorical data (Finney & DiStefano). As some estimation methods rely on meeting the multivariate normality assumption, more often than not our models involve measures that are more than slightly nonnormal and/or categorical in nature. Knowing what to do in this context is critical. The next chapter considers missing data in the SEM context (Enders). Since it is typical that there are some missing data on one or more measures, and since there are more missing data techniques than ever before, it is important to have a guide of what to do and when to do it. Chapter 13 involves a discussion of multilevel modeling (Stapleton). This is critical as quite often our observations are not independent due to some hierarchical data structure (e.g., classrooms). I consider this another topic that has greatly developed over the past decade or so (although much work is still needed). This is followed by a new chapter on Bayesian SEM (Levy & Choi). Although we have had Bayesian statistics almost since the advent of frequentist statistics, Bayesian methods in the SEM context is a rather new and potentially important development. The chapter describes many potential uses, such as using prior information in multilevel models as well as in missing data methods (e.g., MCMC). The final chapter in the volume is on Monte Carlo or simulation studies (Bandalos & Leite). This is one of the most important techniques in all of statistics, as this is where we find out where our methods work well and where they begin to break down. Since many doctoral students in quantitative methods (and even their professors) are quite interested in conducting simulation studies, this is a rather important chapter.

Altogether, the chapters in this volume represent important contributions to our knowledge of SEM beyond the basics. As the method continues to develop at a rather rapid pace, it is critical that we keep up with the knowledge base. I want to thank the editors and the authors for putting together this important volume for all SEMers.

—**Richard G. Lomax**
Department of Educational Studies
The Ohio State University

SERIES INTRODUCTION

Quantitative Methods in Education and the Behavioral Sciences: Issues, Research, and Teaching is a unique book series sponsored by the American Educational Research Association's Special Interest Group for Educational Statisticians. Motivated by the group's central purpose—to increase interaction among educational researchers interested in the theory, applications, and teaching of statistics in the social sciences—the series continues to be devoted to supporting didactically oriented presentations that introduce, extend, and clarify state-of-the-art quantitative methods for students and researchers in the social and behavioral sciences. The series' books present selected topics in a technically sophisticated yet didactically oriented format. This allows for the individual volumes to be used to enhance the teaching of quantitative methods in course sequences typically taught in graduate behavioral science programs. Although other series and journals exist that routinely publish new quantitative methods, the current series is dedicated to both the teaching and applied research perspectives of specific modern quantitative research methods through each volume's relevant and accessible topical treatments. In line with this educational commitment, royalties from the sale of series volumes will continue to be used in large part to support student participation in annual Special Interest Group conference activities.

—**Ronald C. Serlin**
University of Wisconsin–Madison
Series Editor

Structural Equation Modeling: A Second Course (2nd ed.), page xvii
Copyright © 2013 by Information Age Publishing

INTRODUCTION TO THE SECOND EDITION OF *STRUCTURAL EQUATION MODELING: A SECOND COURSE*

Gregory R. Hancock and Ralph O. Mueller

The origins of modern SEM are usually traced to biologist Sewall Wright's development of path analysis (e.g., Wright, 1921, 1934; see Wolfle, 1999, for an annotated bibliography of Wright's work). With respect to the social and behavioral sciences, path analysis lay largely dormant until the 1960s when Otis Duncan (1966) and others introduced the technique in sociology. Simultaneously, statistical developments by Karl Jöreskog (e.g., 1966, 1967) articulated a method for confirmatory factor analysis (CFA), an application of normal theory maximum likelihood estimation to factor models with specific a priori hypothesized theoretical latent structures. A milestone in the development of modern SEM was Jöreskog's provision for a formal χ^2-test comparing the observed pattern of relations among measured variables to that implied by an a priori specified factor model, thereby allowing for the disconfirmation (or tentative confirmation) of such an hypothesized model. Soon after, quite unceremoniously it seems, the fusion of Wright's measured variable path analysis and Jöreskog's CFA occurred and SEM was

Structural Equation Modeling: A Second Course (2nd ed.), pages xix–xxvii
Copyright © 2013 by Information Age Publishing

quietly born (see Wolfle, 2003, for an annotated bibliography of the introduction of SEM to the social sciences).

Despite its tremendous potential, SEM remained generally inaccessi-
ble to researchers in the social and behavioral sciences until well into the
1970s. Not only did it require access to a special statistical software package, LISREL,[1] but utilizing this package required knowledge of matrix algebra. However, by the 1980s and 1990s, examples of measured variable
path models, CFAs, and latent variable models started to become increasingly common, largely due to the vastly improved user-friendliness of SEM
software. EQS (Bentler, 1985) was the first program to offer non-matrix
based syntax, followed by the SIMPLIS command language of LISREL 8
(Jöreskog & Sörbom, 1993), and later Mplus (Muthén & Muthén, 1998).
Social scientists could now use rather intuitive commands that mirrored
the structural equations themselves. Versatile graphic interfaces were also
developed within some SEM programs to further simplify the modeling
process; in fact, a primarily graphics-based SEM program, AMOS, was created (see the AMOS 4 User's Guide, Arbuckle & Wothke, 1999). By the end
of the 1990s, SEM had ascended to the ranks of the most commonly used
multivariate techniques within the social sciences.[2]

Also responsible for SEM's growth in popularity was the increase in accessibility of the materials used to train greater numbers of social science
graduate students in SEM. In the 1970s and 1980s, a first wave of influential
SEM texts relied heavily on matrix formulations of SEM and/or emphasized
measured variable path analysis (e.g., Bollen, 1989; Byrne, 1989; Duncan,
1975; Hayduk, 1987; James, Mulaik, & Brett, 1982; Kenny, 1979). Authors
of a second wave of books starting in the 1990s, many of which are now in
subsequent editions, moved toward less mathematical and more accessible
treatments. Their focus was more applied in nature, trading matrix-based
explanations for coverage of a broader range of topics, tending to include
at a minimum such model types as measured variable path analysis, CFA, latent variable path analysis, and multi-group analyses, as well as applications
with SEM software packages (e.g., Byrne, 1998, 2006, 2010, 2012; Dunn,
Everitt, & Pickles, 1993; Kline, 2011; Loehlin, 2004; Maruyama, 1998; Mueller, 1996; Raykov & Marcoulides, 2006; Schumacker & Lomax, 2010).

With regard to more advanced SEM topics, a host of books are also now
available. Specific topics covered by such works include latent growth models (Bollen & Curran, 2006; Duncan, Duncan, & Strycker, 2006; Preacher,
Wichman, MacCallum, & Briggs, 2008), multilevel models (Heck & Thomas, 2008; Reise & Duan, 2003), latent variable interactions (Schumacker
& Marcoulides, 1998), in addition to other advanced topic compendia
(e.g., Cudeck, du Toit, & Sörbom, 2001; Hoyle, 2012; Marcoulides &
Moustaki, 2002; Marcoulides & Schumacker, 1996, 2001) and texts covering
both introductory and some more advanced topics (Kaplan, 2008; Mulaik,

2009). Most of these resources, however, serve researchers with specialized interests and tend to be more appropriate for a technically sophisticated audience of methodologists. Less common are texts addressing advanced topics while achieving the accessibility that is characteristic of the second wave of introductory texts. The present volume is intended to parallel the movement toward more accessible introductory texts, serving as a didactically-oriented resource covering a broad range of advanced topics that are often not discussed in an introductory course. Such topics are important in furthering the understanding of foundations and assumptions of SEM as well as in exploring SEM as a potential tool to address new types of research questions that might not have arisen during a first encounter with SEM.

THE PRESENT VOLUME:
GOALS, CONTENT, AND NOTATION

In a second course in SEM the treatment of new topics can, arguably, be approached in two ways: (1) emphasizing topics' mathematical and statistical underpinnings and/or (2) stressing their relevance, conceptual foundations, and appropriate applications. As much as we value the former, in keeping with the goal of maximizing accessibility to the widest audience possible we have tried to emphasize the latter to a greater degree. Accordingly, with regard to the selection of chapter authors, all have established technical expertise in their respective areas; however, the most salient professional attribute for our purposes was that their previous work was characterized by a notable didactic clarity. Drawing on these skills in particular, selected authors were asked to present a thorough, balanced, and illustrated treatment of the state-of-knowledge in the assigned topic area; they were asked *not* to set forth a specific agenda or to propose new methods. Where appropriate, authors focused on the clear explanation and application of topics rather than on analytical derivations, and provided syntax files and excerpts from popular software packages of their choosing (e.g., EQS, LIS-REL/SIMPLIS, Mplus) in the chapter or in supplemental materials available from the authors (e.g., online).

Regarding the topics themselves, we assembled chapters targeted toward graduate students and professionals who have had a solid introductory exposure to SEM. In doing so, we intended to provide the reader with three important things: (1) the necessary foundational knowledge to fill in possible gaps left during a first course on SEM, (2) an extension of the range of research questions that might be addressed with modern SEM techniques, and (3) ways to analyze data that violate traditional assumptions. As such, the chapters in the book—all of which are substantially updated since the first edition or are new to this second edition—have been organized into three sections. The first, *Foundations*, complements introductory texts, filling in some foundational gaps by addressing the areas of equivalent models (Hershberger & Marcoulides), feedback loops and formative measurement (Kline), partial least squares modeling (Rigdon; a new chapter for this edition), and power analysis (Hancock & French). The second section, *Extensions*, presents methods that allow researchers to address applied research questions that introductory SEM methods cannot satisfactorily address. These extensions include latent means models (Thompson & Green), conditional process models for addressing mediation and moderation (Hayes & Preacher; a new chapter for this edition), latent interaction and quadratic effects (Marsh, Wen, Hau, & Nagengast), latent growth models (Hancock, Harring, & Lawrence), mixture models (Pastor & Gagné), and exploratory SEM (Morin, Marsh, & Nagengast; a new chapter for this edition). The third and last section, *Assumptions*, acknowledges that real-world data rarely resemble the textbook ideals. Rather, they are replete with complexities often serving as obstacles to the real substantive questions of interest. The topics in this section are nonnormal and categorical data (Finney & DiStefano), missing data (Enders), multilevel models with complex sample data (Stapleton), Bayesian SEM (Levy & Choi; a new chapter for this edition), and Monte Carlo methods (Bandalos & Leite), with the last chapter providing researchers with a means to investigate for themselves the consequences of assumption violations in their models.

Finally, in order to maximize the utility of this text as a teaching tool, we strived to ensure as much homogeneity in chapter structure and notation as possible. As illustrated in Figure I.1, the Bentler-Weeks "VFED" system is used for labeling variables and residuals wherever possible in the chapters.[3] These designations, and their more traditional Jöreskog-Keesling-Wiley (JKW) notational analogs in parentheses, are shown below:

V = observed Variable (X, Y)
F = latent Factor (ξ, η)
E = Error/variable residual (δ, ε)
D = Disturbance/factor residual (ζ).

Chapter figures also generally follow the convention of showing measured variables (V) in squares/rectangles and latent variables (F) in circles/ellipses. Measured or latent residuals (E, D) are not typically enclosed in any shape (although the latent nature of residuals could warrant an ellipse). For mean structure models, the pseudovariable (1) is enclosed in a triangle.

As for the parameters relating the VFED elements, we use the parameter notation system that we call the "a-b-c" system, which was developed for the first edition of this volume (Hancock & Mueller, 2006). While the JKW system is very useful for mathematical representations of SEM, for didactic

Covariance structure model (not all parameters shown)

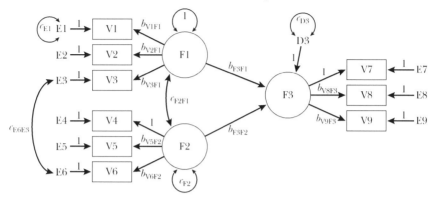

Mean structure model (not all parameters shown)

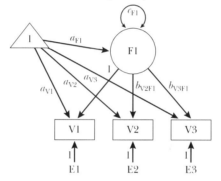

Figure I.1 Diagrammatic and notational conventions for this volume.

expositions we find the *a-b-c* system generally more intuitive. First, drawing from regression notation, in mean structure models the symbol *a* is used to designate an intercept (or mean) parameter, which is represented diagrammatically as a path from the pseudovariable (1) to a measured or latent variable. A subscript on *a* indicates the variable whose intercept (or mean) is being modeled, such as a_{V1} or a_{F1} shown in the bottom of Figure I.1. Analogs to *a* in the JKW system include subscripted τ (measured variable intercept), κ (exogenous factor mean), and α (endogenous factor intercept). Sample-based estimates of an *a* parameter in the *a-b-c* system have an additional hat designation, such as $\hat{a}_{F1}$.

Second, also drawing from regression notation, the symbol *b* is used to designate a structural (causal) relation hypothesized between any VFED elements; this relation is represented diagrammatically as a single-headed (usually straight) arrow. Subscripts on *b* follow the "*to–from*" convention, indicating first where the arrow is going *to* and second where the arrow is coming *from*; examples include the factor loading b_{V2F1} or the latent structural relation b_{F3F2} shown in the top of Figure I.1. In measured variable path models, where no latent variables exist, the subscripts may be shortened to include variable numbers only (e.g., b_{V4V2} becomes b_{42}). Analogs to b in the JKW system include subscripted λ (measurement model loading), γ (structural path from exogenous variable to endogenous variable), and β (structural path from endogenous variable to endogenous variable). Sample-based estimates of a b parameter in the *a-b-c* system have an additional hat designation, such as $\hat{b}_{F3F2}$.

Third, the symbol c is used to designate a covariance (or variance) parameter for any exogenous VFED elements. This relation is represented diagrammatically as a two-headed (usually curved) arrow; a covariance parameter is depicted by an arrow connecting two elements, while a variance parameter is depicted by a circular arrow connecting an element to itself (as a variance is a covariance of a variable with itself). Subscripts on c indicate the elements being connected, where two subscripts (in arbitrary order) indicate the covarying exogenous elements and a single subscript indicates the exogenous element whose variance is a parameter. Covariance examples in Figure I.1 include the c_{E6E3} and c_{F2F1}, while variance examples include c_{E1}, c_{F2}, and c_{D3}. In measured variable path models, where no latent variables exist, the subscripts may be shortened to include variable numbers only (e.g., c_{V1V2} becomes c_{12}, and c_{V3} becomes c_3). Analogs to c in the JKW system include subscripted ϕ (exogenous variable co/variance), θ (measured residual co/variance), and ψ (latent residual co/variance). Sample-based estimates of a c parameter in the *a-b-c* system have an additional hat designation, such as $\hat{c}_{F2F1}$ and $\hat{c}_{D3}$.

To summarize the *a-b-c* notational system:

> a = intercept or mean path from a pseudovariable (1) to V or F;
> b = structural path connecting any VFED;
> c = co/variance associated with exogenous VFED elements.

LOOKING FORWARD

Over the last 40 years or so, SEM has matured into a methodological framework that allows researchers to address theory-driven causal research questions at both the observed variable and latent variable levels. In our view, the integration of structure and measurement facilitated by this analytical paradigm stands as one of the most significant methodological developments of the 20th century. We now observe many institutions of higher education offering SEM instruction, as well as a need in the professional social/

behavioral science community for SEM tutelage. We conceived this book out of a mutual passion for teaching SEM, both to beginning and more experienced modelers, and indeed learned much about SEM ourselves from selecting and editing the chapters in the first and second editions of this volume. Our hope is for you to share in our excitement about SEM as you fill in some of the holes that might be in your foundational knowledge (Part I), extend the kind of research questions that you can answer with SEM (Part II), and learn what can be done when assumptions commonly made about data do not necessarily hold true in yours (Part III).

NOTES

1. According to Sörbom (2001), LISREL III became available in 1974.
2. Between 1994 and 2001, the number of both substantive and technical SEM journal publications more than doubled to over 340 articles; over 60% of the total number of journals published by the American Psychological Association during the same time period contained articles using SEM (Hershberger, 2003; for the 1985–1994 period, see Tremblay & Gardner, 1996).
3. Due to the more mathematical nature of some of the topics, adhering to this notational system was not possible in all of the chapters.

REFERENCES

Arbuckle, J. L., & Wothke, W. (1999). *Amos 4.0 User's Guide.* Chicago: SPSS.

Bentler, P. M. (1985). *Theory and implementation of EQS, A structural equations program.* Los Angeles: BMDP Statistical Software.

Bollen, K. A. (1989). *Structural equations with latent variables.* New York, NY: Wiley.

Bollen, K. A., & Curran, P. J. (2006). *Latent curve models.* Hoboken, NJ: Wiley.

Byrne, B. M. (1989). *A primer of LISREL: Basic applications and programming for confirmatory factor analytic models.* New York, NY: Springer-Verlag.

Byrne, B. M. (1994). *Structural equation modeling with EQS and EQS/Windows.* Thousand Oaks, CA: SAGE Publications.

Byrne, B. M. (1998). *Structural equation modeling with LISREL, PRELIS, and SIMPLIS.* Mahwah, NJ: Lawrence Erlbaum.

Byrne, B. M. (2006). *Structural equation modeling with EQS: Basic concepts, applications, and programming* (2nd ed.). Mahwah, NJ: Lawrence Erlbaum.

Byrne, B. M. (2010). *Structural equation modeling with AMOS* (2nd ed.). New York, NY: Routledge.

Byrne, B. M. (2012). *Structural equation modeling with Mplus.* New York, NY: Taylor & Francis.

Cudeck, R., du Toit, S., & Sörbom, D. (Eds.) (2001). *Structural equation modeling: Present and future—A Festschrift in honor of Karl Jöreskog.* Lincolnwood, IL: Scientific Software International, Inc.

Duncan, O. D. (1966). Path analysis: Sociological examples. *American Journal of Sociology, 72,* 1–16.

Duncan, O. D. (1975). *Introduction to structural equation models.* New York, NY: Academic Press.

Duncan, T. E., Duncan, S. C., & Strycker, L. A. (2006). *An introduction to latent variable growth curve modeling: Concepts, issues, and applications* (2nd ed.). Mahwah, NJ: Lawrence Erlbaum.

Dunn, G., Everitt, B., & Pickles, A. (1993). *Modelling covariances and latent variables using EQS.* London: Chapman & Hall.

Hancock, G. R., & Mueller, R. O. (Eds.). (2006). *Structural equation modeling: A second course.* Greenwich, CT: Information Age Publishing.

Hayduk, L. A. (1987). *Structural equation modeling with LISREL.* Baltimore, MD: The Johns Hopkins University Press.

Heck, R. H., & Thomas, S. L. (2008). *An introduction to multilevel modeling techniques* (2nd ed.). New York, NY: Routledge.

Hershberger, S. L. (2003). The growth of structural equation modeling: 1994–2001. *Structural Equation Modeling: A Multidisciplinary Journal, 10,* 35–46.

Hoyle, R. (Ed.) (2012). *Handbook of structural equation modeling.* New York, NY: Guilford Press.

James, L. R, Mulaik, S. A., & Brett, J. (1982). *Causal analysis: Models, assumptions, and data.* Beverly Hills: SAGE Publications.

Jöreskog, K. G. (1966). Testing a simple structure hypothesis in factor analysis. *Psychometrika, 31,* 165–178.

Jöreskog, K. G. (1967). Some contributions to maximum likelihood factor analysis. *Psychometrika, 32,* 443–482.

Jöreskog, K. G., & Sörbom, D. (1993). *LISREL 8: Structural equation modeling with the SIMPLIS command language.* Chicago: Scientific Software International.

Kaplan, D. (2008). *Structural equation modeling: Foundations and extensions* (2nd ed.). Thousand Oaks, CA: Sage.

Kenny, D. A. (1979). *Correlation and causation.* New York, NY: Wiley.

Kline, R. B. (2011). *Principles and practice of structural equation modeling* (3rd ed.). New York, NY: The Guilford Press.

Loehlin, J. C. (2004). *Latent variable models* (4th ed.). Hillsdale, NJ: Lawrence Erlbaum.

Marcoulides, G. A., & Moustaki, I. (Eds.) (2002). *Latent variable and latent structure models.* Mahwah, NJ: Lawrence Erlbaum.

Marcoulides, G. A., & Schumacker, R. E. (Eds.) (1996). *Advanced structural equation modeling.* Mahwah, NJ: Lawrence Erlbaum.

Marcoulides, G. A., & Schumacker, R. E. (Eds.) (2001). *New developments and techniques in structural equation modeling.* Mahwah, NJ: Lawrence Erlbaum.

Maruyama, G. M. (1998). *Basics of structural equation modeling.* Thousand Oaks, CA: SAGE Publications.

Mueller, R. O. (1996). *Basic principles of structural equation modeling: An introduction to LISREL and EQS.* New York, NY: Springer.

Mulaik, S. A. (2009). *Linear causal modeling with structural equations.* Boca Raton, FL: Chapman & Hall/CRC.

Muthén, L. K., & Muthén, B. (1998). *Mplus user's guide.* Los Angeles, CA: Muthén & Muthén.

Preacher, K. J., Wichman, A. L., MacCallum, R. C., & Briggs, N. E. (2008). *Latent growth curve modeling.* Thousand Oaks, CA: Sage Publications.

Raykov, T., & Marcoulides, G. A. (2006). *A first course in structural equation modeling* (2nd ed.). Mahwah, NJ: Lawrence Erlbaum.

Reise, S. P., & Duan, N. (Eds.) (2003). *Multilevel modeling: Methodological advances, issues, and applications.* Mahwah, NJ: Lawrence Erlbaum.

Schumacker, R. E., & Lomax, R. G. (2010). *A beginner's guide to structural equation modeling* (3rd ed.). New York, NY: Routledge.

Schumacker, R. E., & Marcoulides, G. A. (Eds.) (1998). *Interaction and nonlinear effects in structural equation modeling.* Mahwah, NJ: Lawrence Erlbaum.

Sörbom, D. (2001). Karl Jöreskog and LISREL: A personal story. In R. Cudeck, S. Du Toit, & D. Sörbom (Eds.), *Structural equation modeling: Present and Future* (pp. 3–10). Lincolnwood, IL: Scientific Software International.

Tremblay, P. F., & Gardner, R. C. (1996). On the growth of structural equation modeling in psychological journals. *Structural Equation Modeling: A Multidisciplinary Journal, 3*, 93–104.

Wolfle, L. M. (1999). Sewall Wright on the method of path coefficients: An annotated bibliography. *Structural Equation Modeling: A Multidisciplinary Journal, 6*, 280–291.

Wolfle, L. M. (2003). The introduction of path analysis to the social sciences, and some emergent themes: An annotated bibliography. *Structural Equation Modeling: A Multidisciplinary Journal, 10*, 1–34.

Wright, S. (1921). Correlation and causation. *Journal of Agricultural Research, 20*, 557–585.

Wright, S. (1934). The method of path coefficients. *Annals of Mathematical Statistics, 5*, 161–215.

PART I

FOUNDATIONS

CHAPTER 1

THE PROBLEM OF EQUIVALENT STRUCTURAL MODELS

Scott L. Hershberger and George A. Marcoulides

Structural equation modeling (SEM) is a powerful and flexible tool for the analysis of data that is currently enjoying widespread popularity. A major reason for its frequent use is that it allows researchers to posit complex multivariate relations among observed and latent variables whereby direct and indirect effects can be evaluated along with indexes of their estimation precision. Another major reason is the availability of easy-to-use computer programs (e.g., AMOS, EQS, LISREL, Mplus). Researchers conducting SEM analyses generally proceed in three stages: (i) a theoretical model is hypothesized, (ii) the overall fit of data to the model is determined, and (iii) the specific parameters of the proposed model are evaluated. A variety of problems might be encountered at each of these stages. This chapter provides a general review of the problem of *equivalent models* from an SEM perspective.

Broadly defined, equivalent models differ in causal structure and substantive interpretability but are identical in terms of data-model fit. Narrowly defined, they are a set of models that yield identical (a) implied co-

Structural Equation Modeling: A Second Course (2nd ed.), pages 3–39
Copyright © 2013 by Information Age Publishing

variance, correlation, and other observed variable moment matrices when analyzing the same data, (b) residuals and fitted moment matrices, and (c) fit function and related goodness-of-fit indices (e.g., chi-square values and p-values). Econometricians were among the first to formally investigate the properties of equivalent models (Koopmans, Rubin, & Leipnik, 1950) and the influence of their early work is still present in the many contemporary econometric investigations of equivalent models (e.g., Chernov, Gallant, Ghysels, & Tauchen, 2003). Within the field of SEM, interest in equivalent models dates back to Stelzl (1986). Since then, the problem of model equivalence as a source of difficulty for inferring causal relations and substantive interpretations from structural equation models has received considerable methodological attention (e.g., Breckler, 1990; Hershberger, 1994; Lee & Hershberger, 1990; Levy & Hancock, 2007; Luijben, 1991; MacCallum, Wegener, Uchino, & Fabrigar, 1993; Marcus, 2002; McDonald, 2002; Raykov, 1997; Raykov & Marcoulides, 2001, 2007; Raykov & Penev, 1999; Williams, Bozdogan, & Aiman-Smith, 1996). Additional developments in the study of model equivalence have also arisen from graph theorists' attempts to formulate a mathematical basis for causal inferences (Pearl, 2009). This is not surprising given the many shared characteristics of graph models and structural equation models. Other developments have also emerged from work in the fields of computing and data mining (Marcoulides & Ing, 2012). Yet, interest in model equivalence in the SEM field appears to be limited to only a few quantitatively-oriented researchers; for example, reviews of published studies find few researchers show any concern or even any awareness of the problem (e.g., Breckler, 1990; Roesch, 1999). In the sections that follow we identify different types of equivalent models and illustrate how their presence can be detected from the structure of a model.

A DEFINITION OF EQUIVALENT MODELS

An in-depth understanding of the idea of equivalent models first requires comprehending how data-model fit is evaluated in structural equation models. Let us consider the vector $\mathbf{x} = (V1, V2, \ldots, Vp)'$ to represent a set of p observed variables with a population covariance matrix Σ and sample covariance matrix $\mathbf{S}$, θ the vector of parameters of a proposed structural equation model M, and $\hat{\Sigma}(\theta)$ the population covariance matrix implied by the model M at θ. Under the null hypothesis $H_0 : \Sigma = \hat{\Sigma}(\theta)$ for some θ, the basis for most fit indices used in SEM is the discrepancy between the sample covariance matrix of the observed variables ($\mathbf{S}$) and the covariance matrix of these variables implied by the model (i.e., $\hat{\Sigma}(\theta)$, which is often simply written as $\hat{\Sigma}$). Using numerical algorithms, parameter estimates describing the structure of a model M are generated that produce an implied covariance matrix $\hat{\Sigma}$

with least discrepancy from the observed sample covariance matrix $\mathbf{S}$. In the special case that the implied covariance matrix is identical to the sample covariance matrix, the discrepancy between $\mathbf{S}$ and $\hat{\boldsymbol{\Sigma}}$ (i.e., $\mathbf{S} - \hat{\boldsymbol{\Sigma}}$) is a matrix of zeros. Such a case implies a perfect or exact fit of the data to the model (in this case, the model has zero residuals) and would lead one to retain the considered null hypothesis. Nevertheless, because the implied covariance matrix is bound by the restrictions imposed by the model M, and because of the inevitable presence of sampling error in the observed data, the sample and implied covariance matrices are almost never identical; that is, $\mathbf{S} - \hat{\boldsymbol{\Sigma}} \neq 0$ in most modeling situations. It is this (nonzero) value of the discrepancy yielded during parameter estimation that is used in the computation of most fit indices in SEM. For example, the contribution of $\mathbf{S} - \hat{\boldsymbol{\Sigma}}$ to the commonly used goodness-of-fit index (GFI, Jöreskog & Sörbom, 1996) is:

$$\text{GFI} = 1 - \frac{\text{tr}\left[(\mathbf{s} - \hat{\boldsymbol{\sigma}})'\mathbf{W}(\mathbf{s} - \hat{\boldsymbol{\sigma}})\right]}{\text{tr}\left[\mathbf{s}'\mathbf{W}\mathbf{s}\right]}$$

where tr is the trace operator, $\mathbf{s}$ and $\hat{\boldsymbol{\sigma}}$ refer to vectors of the nonduplicated elements of $\mathbf{S}$ and $\hat{\boldsymbol{\Sigma}}$, respectively, and $\mathbf{W}$ is a weight matrix that the method of parameter estimation defines (e.g., $\mathbf{W}$ is an identity matrix for unweighted least squares; $\mathbf{W}$ derives from $\mathbf{S}^{-1}$ for generalized least squares; and $\mathbf{W}$ derives from $\hat{\boldsymbol{\Sigma}}^{-1}$ for reweighted least squares, the asymptotic equivalent of maximum likelihood). The GFI is scaled to range between 0 and 1, with higher values interpreted as signifying better fit (although, this issue has also received considerable attention in the literature; for more details see Hu & Bentler, 1999; Marsh, Hau, & Wen, 2004). In those cases where the discrepancy value does not appear in the formula for a fit index, other terms are used that reflect the difference between the observed and model implied matrices.

Equivalent models were defined previously as a set of models that, regardless of the data, yield identical (a) implied covariance, correlation and other observed variable moment matrices when fit to the same data, (b) residuals and fitted moment matrices, and (c) fit function and related goodness-of-fit indices (e.g., chi-square and p values). Yet, one most frequently thinks of equivalent models as described in (a) above. Now let us consider two hypothesized alternative models, denoted as M1 and M2, each of which is associated with a set of estimated parameters (contained in the vectors $\hat{\boldsymbol{\theta}}_{M1}$ and $\hat{\boldsymbol{\theta}}_{M2}$) and a covariance matrix implied by those parameter estimates (denoted simply as $\hat{\boldsymbol{\Sigma}}_{M1}$ and $\hat{\boldsymbol{\Sigma}}_{M2}$). Models M1 and M2 are considered equivalent if, for any sample covariance matrix $\mathbf{S}$, the implied matrices are identical. Mathematically, this can be written as $\hat{\boldsymbol{\Sigma}}_{M1} = \hat{\boldsymbol{\Sigma}}_{M2}$, or alternatively, $(\mathbf{S} - \hat{\boldsymbol{\Sigma}}_{M1}) = (\mathbf{S} - \hat{\boldsymbol{\Sigma}}_{M2})$. That is to say, model equivalence is not defined by the data, but rather by an algebraic equivalence between model parameters. In turn, because of this model equivalence, the values of any considered statistical tests of global fit (e.g., test of exact fit, tests of close fit) will be identical, as will any goodness-of-fit indices that are based on the discrepancy between the sample covariance matrix and the model-implied covariance matrix. Thus, even when an hypothesized model fits well according to the examined fit indices, there may still be equivalent models with identical fit—even if the theoretical implications or substantive interpretations of those models are very different. Indeed, Raykov and Marcoulides (2001) presented an instance of an infinite series of equivalent models to an initially hypothesized one.

As an example, we define a model involving three observed variables, V1, V2, and V3. We note that this model contains three observed variables only for purposes of simplicity but could easily be expanded to more complicated models including ones with latent variables. This model is shown in Figure 1.1 as Model 1 (M1). The covariance matrix for the three variables (S), the set of parameter estimates ($\hat{\theta}_{M1}$) obtained from fitting the model to the covariance matrix using maximum likelihood estimation, and the covariance matrix implied by the parameter estimates ($\hat{\Sigma}_{M1}$) are as follows:

$$\mathbf{S} = \begin{bmatrix} 2.000 & & \\ 0.940 & 1.800 & \\ 1.030 & 0.650 & 1.500 \end{bmatrix}$$

$$\hat{\theta}_{M1} = \begin{bmatrix} \hat{b}_{V2V1} = 0.470 \\ \hat{b}_{V3V2} = 0.361 \\ \hat{c}_{V1} = 2.000 \\ \hat{c}_{E2} = 1.358 \\ \hat{c}_{E3} = 1.265 \end{bmatrix}, \text{ and}$$

$$\hat{\Sigma}_{M1} = \begin{bmatrix} 2.000 & & \\ 0.940 & 1.800 & \\ 0.339 & 0.650 & 1.500 \end{bmatrix}.$$

A chi-square goodness-of-fit test suggests that Model 1 is consistent with the data, but the process of model confirmation has not completed. For this particular case, there in fact exist two other models equivalent to Model 1: namely, Model 2 (M2) and Model 3 (M3), also shown in Figure 1.1. Fitting

Model 1:

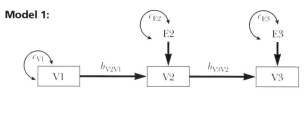

Model 2:

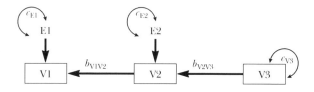

Model 3:

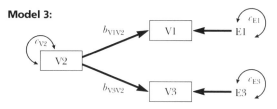

Figure 1.1 Three equivalent models.

Model 2 and Model 3 to the covariance matrix **S** would result in two new sets of maximum likelihood parameter estimates:

$$\hat{\theta}_{M2} = \begin{bmatrix} \hat{b}_{V1V2} = 0.522 \\ \hat{b}_{V2V3} = 0.433 \\ \hat{c}_{V3} = 1.500 \\ \hat{c}_{E1} = 1.509 \\ \hat{c}_{E2} = 1.518 \end{bmatrix}, \quad \hat{\theta}_{M3} = \begin{bmatrix} \hat{b}_{V1V2} = 0.522 \\ \hat{b}_{V3V2} = 0.361 \\ \hat{c}_{V2} = 1.800 \\ \hat{c}_{E1} = 1.509 \\ \hat{c}_{E3} = 1.265 \end{bmatrix},$$

it turns out, and despite the different parameterizations of Model 1, Model 2, and Model 3, they all have the exact same model-implied covariance matrix:

$$\hat{\Sigma}_{M1} = \hat{\Sigma}_{M2} = \hat{\Sigma}_{M3} = \begin{bmatrix} 2.000 \\ 0.940 & 1.800 \\ 0.339 & 0.650 & 1.500 \end{bmatrix}.$$

Therefore, because Model 1, Model 2, and Model 3 are equivalent, we observe for this specific example that $(\mathbf{S} - \hat{\Sigma}_{M1}) = (\mathbf{S} - \hat{\Sigma}_{M2}) = (\mathbf{S} - \hat{\Sigma}_{M3})$.

WHY EQUIVALENT MODELS ARE IMPORTANT TO IDENTIFY

Identifying equivalent models can be a very difficult and time-consuming task. But there is clearly a compelling reason for undergoing such an arduous activity: Model equivalence is methodologically significant because it dramatically exposes the limitations of SEM to test theories. The potential existence of equivalent models reminds us that we never test *a* model but rather a whole class of models from which the hypothesized model cannot be distinguished by any statistical means.

Although it is true that (over-identified) structural equation models have testable implications, those implications are only one part of what a model represents: a set of claims, assumptions, and implications. Equivalent models can cause damage by not only offering false causal relations; they might also lead one to accept their equally false implications. For example, Williams et al. (1996) discussed many hypothesized models of job satisfaction and its correlates; for each model, at least one model equivalent to it altered both the size and direction of the job satisfaction relations. The determination and subsequent interpretation about what factors contribute to job satisfaction very much depended on which equivalent model was selected.

Most researchers conducting SEM analyses would agree that alternative models should always be considered. So why is it that equivalent models should be considered a special threat to causal inference, much more so than alternative but nonequivalent models? After all, we could readily propose alternative but nonequivalent models that fit the data almost as well as the hypothesized model, alternative models so close in fit as to render trivial the choice among the models based solely on fit. The reason equivalent models are so special is that any one of them can *never* be supported without *all* of them being supported. On the other hand, alternative models that fit almost as well as an hypothesized model in one sample could be discarded if their fit decreases substantially in a new sample.

In the remainder of this chapter, different types of model equivalence are first discussed, and then strategies for identifying them are presented. Next we describe important statistical and conceptual similarities between model equivalence and model identification. The final section of the chapter offers some suggestions that have been proposed in the literature for selecting among equivalent models.

TYPES OF MODEL EQUIVALENCE

Observationally Equivalent/Covariance Equivalent

Model equivalence is a generic term that could refer to one of several specific situations of equivalence. Hsiao (1983), Luijben (1991), McDonald (2002), Pearl (2009), and Raykov and Penev (1999) have proposed a taxonomy that can be used to distinguish among several different types of model equivalence. Two models are considered *observationally equivalent* only if one model can generate every probability distribution that the other model can generate. Observational equivalence is model equivalence in the broadest sense and can be shown using data of any type. In contrast, models are considered *covariance equivalent* if every covariance matrix generated by one model can be generated by the other. Thus, observational equivalence encompasses covariance (model) equivalence; that is, observational equivalence requires the identity of individual data values, whereas covariance equivalence requires the identity of summary statistics such as covariances and variances. We note that observationally equivalent models are always going to be covariance equivalent, whereas covariance equivalent models might not necessarily be observationally equivalent. For the purposes of the current chapter, the term *equivalent* is used specifically to refer to models that are covariance equivalent, as well as *globally equivalent* as described next.

Globally Equivalent/Locally Equivalent

Another distinction can be drawn between equivalent models, namely, those that are *globally equivalent* models and those that are *locally equivalent* models. For two models to be globally equivalent, *a function* must exist that translates *every* parameter of one model into the parameters of another model. The result of this mapping of one parameter set into another is an identity between the implied covariance matrices of the two models. On the other hand, if only a *subset* of one model's parameter set is translatable into the parameter set of another model, the models are then considered *locally equivalent*. Local equivalence does not guarantee that the implied covariance matrices of the two models will be the same.

To further clarify this point, let us consider an illustrative example of globally equivalent models. This illustrative example involves a factor analysis model (M0) with just one common factor F1 for three observed variables, V1, V2, and V3, specified as

$$\begin{bmatrix} V1 \\ V2 \\ V3 \end{bmatrix} = \begin{bmatrix} b_{V1F1} \\ b_{V2F1} \\ b_{V3F1} \end{bmatrix} F1 + \begin{bmatrix} E1 \\ E2 \\ E3 \end{bmatrix}.$$

The parameters b_{V1F1}, b_{V2F1}, and b_{V3F1} are the factor loadings; E1, E2, and E3 are the error terms with variances c_{E1}, c_{E2}, and c_{E3}, and covariances c_{E1E2}, c_{E1E3}, and c_{E2E3}; and c_{F1} is the variance of F1. Let us assume that three constraints have been placed on the parameters of M0: (1) $c_{F1} = 1$, (2) $c_{E1} = c_{E2} = c_{E3} = c_E$, and (3) $c_{E2E1} = c_{E3E1} = c_{E3E2} = c_{EE}$. Based upon these constraints, the implied covariance matrix $\hat{\Sigma}_{M0}$ for M0 is

$$\begin{bmatrix} b_{V1F1}^2 + c_E & & \\ b_{V2F1}b_{V1F1} + c_{EE} & b_{V2F1}^2 + c_E & \\ b_{V3F1}b_{V1F1} + c_{EE} & b_{V3F1}b_{V2F1} + c_{EE} & b_{V3F1}^2 + c_E \end{bmatrix}.$$

Because of the imposed constraints, M0 is over-identified with one *df* (i.e., the difference between the six variances and covariances minus the five parameter estimates: b_{V1F1}, b_{V2F1}, b_{V3F1}, the common error variance c_E and the common error covariance c_{EE}).

Now let us consider two other models, M1 and M2, both derived from M0. In addition to the homogeneous error variances, in M1 the covariance c_{E3E2} is constrained to be zero while in M2 the covariance c_{E2E1} is constrained to be zero. Thus, the implied covariance matrix $\hat{\Sigma}_{M1}$ for M1 is

$$\begin{bmatrix} b_{V1F1}^2 + c_E & & \\ b_{V2F1}b_{V1F1} + c_{EE} & b_{V2F1}^2 + c_E & \\ b_{V3F1}b_{V1F1} + c_{EE} & b_{V3F1}b_{V2F1} & b_{V3F1}^2 + c_E \end{bmatrix},$$

and the implied covariance matrix $\hat{\Sigma}_{M2}$ for M2 is

$$\begin{bmatrix} b_{V1F1}^2 + c_E & & \\ b_{V2F1}b_{V1F1} & b_{V2F1}^2 + c_E & \\ b_{V3F1}b_{V1F1} + c_{EE} & b_{V3F1}b_{V2F1} + c_{EE} & b_{V3F1}^2 + c_E \end{bmatrix}.$$

The models are fit to

$$
\mathbf{S} = \begin{bmatrix} s^2_{V1} & & \\ s_{V2V1} & s^2_{V2} & \\ s_{V3V1} & s_{V3V2} & s^2_{V3} \end{bmatrix},
$$

where s^2_{V1}, s^2_{V2}, and s^2_{V3} are the variances and s_{V2V1}, s_{V3V1}, and s_{V3V2} are the covariances of the three observed variables. If a function exists that maps the parameters of M1 onto the parameters of M2, thereby yielding the same model-implied covariance matrix (i.e., $\hat{\Sigma}_1 = \hat{\Sigma}_2$) for all possible observed covariance matrices $\mathbf{S}$, then they are globally equivalent. In contrast, if a function cannot be found that transforms every parameter, then, at best, the two models are locally equivalent.

For example, in M1 the parameter b_{V1F1} can be shown to have an implied value of

$$
\frac{(s^2_{V1} - s_{V3V1})\sqrt{s^2_{V1} - s_{V2V1} - s_{V3V2}}}{s_{V3V2}},
$$

while in M2 the same parameter has an implied value of

$$
-\frac{s_{V2V1}}{\sqrt{s^2_{V2} - s_{V3V2}}}.
$$

The global equivalence of the two models is confirmed by noting that implied values of b_{V1F1} in M1 and M2 can be shown algebraically to be equal for all $\mathbf{S}$; that is $b_{V1F1}(M1) = b_{V1F1}(M2)$. Similarly, the implied values of the other four parameters in M1 (i.e., b_{V2F1}, b_{V3F1}, c_E, and c_{EE}) are algebraically equivalent to their counterparts in M2. Further, at one *df*, both M1 and M2 are over-identified, with each of their parameters having more than one implied value. These implied alternatives are also algebraically identical in M1 and M2.

The strategies described later in this chapter identify globally equivalent models; fewer strategies have been proposed for identifying locally equivalent models. The methods of identifying local equivalence depend on a model's so-called *Jacobian matrix*. What a Jacobian matrix is, and how it can be used to identify local equivalence, are both complex topics and well outside the purview of this chapter. We refer the reader to Bekker, Merckens, and Wansbeek (1994) as an excellent source for additional information on the topic of local equivalence.

STRATEGIES FOR IDENTIFYING EQUIVALENT MODELS

Two general categories of strategies for approaching the problem of equivalent models that have been considered in the literature are those that occur either *before* data collection or *after* data collection. It is our opinion that researchers should be strongly urged to at least try and detect models that are equivalent to one's theoretical model *before* hefty resources have been spent on a particular project. Thus, an ideal time for such an activity would be before data collection begins, when the study is still in the planning stage. Forewarned is forearmed: Becoming aware of the presence of equivalent models should at the very least motivate a researcher to (a) acknowledge the potentially severe inferential limitations regarding a model achieving satisfactory data-model fit; (b) consider revising the theoretical model as an attempt to decrease the number of equivalent models; (c) replace structural equation modeling as the method of statistical analysis with a technique that is not compromised by the presence of equivalent models; or (d) even cancel the study. Among these choices, alternative (a), although important, is rather minimal alone, whereas (b) could be considered highly desirable. Before deciding on a course of action, however, it is preferable that a method be first used to find or detect the presence of equivalent models without recourse to data. The next section describes such methods and illustrates strategies that have been proposed in the literature for determining the number and nature of equivalent models. In a subsequent section, other strategies for determining equivalent models before and after data collection are discussed.

The Replacing Rule and Structural Models

The four rules developed by Stelzl (1986) for completely recursive structural models may be simplified (and extended to models with nonrecursive relationships) by use of the *replacing rule* for locating covariance equivalent models. This more general rule was shown by Lee and Hershberger (1990) to subsume Stelzl's four rules. Before the replacing rule is explained, however, it is necessary that a number of important terms be defined. A structural model can be divided into three key blocks: a *preceding block*, a *focal block*, and a *succeeding block*. Figure 1.2 shows the location of these three blocks in the hypothesized "initial model." The focal block includes the relations we are interested in altering to produce an equivalent model; the preceding block consists of all the variables in the model that causally precede the variables in the focal block; and the succeeding block consists of all the variables in the model that causally succeed the variables in the focal block. Within the econometrics literature, the division of a structural model

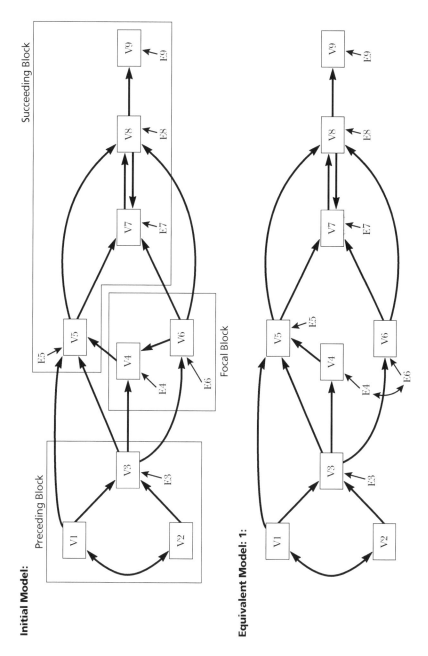

Figure 1.2 An example of the generation of equivalent structural models using the replacing rule.

into blocks, where recursiveness exists between and within blocks, is termed a *block recursive system*. For the application of the replacing rule, *limited block recursiveness* is required. In the limited block recursive model, the relations between the blocks and within the focal block are recursive; however, relations within the preceding and succeeding blocks may be nonrecursive (under certain conditions discussed below, relations within the focal block may be nonrecursive; for now, we will assume a fully recursive focal block).

The replacing rule is defined as follows. Let Vi and Vj represent any two variables in a focal block, where Vi and Vj within the focal block stand in the relation $Vi \rightarrow Vj$; Vi is a *source variable* and Vj is an *effect variable* in the focal block. Both Vi and Vj receive paths from other variables in the preceding and focal blocks, and they may also send paths to other variables in the succeeding block. Let Ei and Ej represent the residuals of the variables Vi and Vj. According to the replacing rule, the directed path between Vi and Vj (i.e., $Vi \rightarrow Vj$) may be replaced by their residual covariance (i.e., $Ei \leftrightarrow Ej$) if the predictors of the effect variable Vj are the same as or include those of the source variable Vi. The reverse application of the replacing rule will also produce an equivalent model; that is, a residual covariance, $Ei \leftrightarrow Ej$, may be replaced by a directed path, $Vi \rightarrow Vj$ or $Vi \leftarrow Vj$, when the predictors of the effect variable are the same as or include those of the source variable following the replacement.

An example of an equivalent model to which the replacing rule can be applied is shown in the second panel of Figure 1.2. In the initial model of Figure 1.2, V1, V2, and V3 are defined as a preceding block, V4 and V6 as a focal block, and V5, V7, V8, and V9 as a succeeding block. Within the focal block, the directed path V6 → V4 may be replaced by the residual covariance E6 ↔ E4; the predictor V3 influences both the source (V6) and effect (V4) variables in the focal block. This application of the replacing rule is represented as Model 1, which is a model that is equivalent to the initial model. This example also illustrates an important special case of the replacing rule, wherein both the source and effect variables have the same predictors. In this situation, the variables in the focal block are said to be *symmetrically determined*, and it is a matter of statistical indifference if the directed path between the source and effect variables is reversed or replaced by a residual covariance. The source and effect variables are said to be in a *symmetric focal block*.

Frequently, applications of the replacing rule can create equivalent models, that is, models that were not actually equivalent prior to application of the replacing rule. For example, let us consider a new focal block to be defined as consisting of V4 and V5, with V4 as the source variable and V5 as the effect variable. Within the initial model, application of the replacing rule to this focal block is not possible because both V4 and V5 have one predictor not shared with each other. After the alteration to the relation between V4 and V6 shown in Model 1, however, the effect variable V5 has as one of its predictors the only predictor (V3) of the source variable V4. In equivalent Model 2 of Figure 1.3, the directed path V4 → V5 has been replaced by the residual covariance E4 ↔ E5. The replacing rule will not only generate a model equivalent to an hypothesized model, but on occasion, a model that has other models equivalent to it—but that are *not* equivalent to the initially hypothesized model. Such was the case with Model 1, generated from the initial model by replacing V4 → V6 with E4 ↔ E6. Model 2 can be generated by replacing V4 → V5 of Model 1 with E4 ↔ E5. However, Model 2 is not equivalent to the initial model, suggesting that model equivalence is not transitive. That Model 2 is not equivalent to the initial model can be shown by fitting the two models to the same covariance matrix: the fit of the two models will not be identical. Because the initial model is not equivalent to Model 2, they may differ in the number of models equivalent to them.

This example also illustrates that focal blocks are not invariantly defined within a structural model: Any two variables within the same model may be designated as part of a focal block if the focal block meets the requirements of the replacing rule. For instance, neither V3 → V4, V5 → V7, nor V6 → V7 may be defined as a focal block for the application of the replacing rule; the source variables in all three cases have predictors not shared with the effect variable, and in addition, defining V5 → V7 or V6 → V7 as the focal block would require defining a succeeding block (V8 and V9) nonrecursively connected to a focal block.

Another special case of the replacing rule occurs when a preceding block is completely saturated, and all variables are connected by either directed paths or residual covariances. In this special case, the preceding block is defined as a focal block, and any residual covariance/directed path replacement yields an equivalent model. The replacing rule may be applied whenever the necessary conditions are met; the existence of a preceding or a succeeding block is not absolutely necessary. In the initial model (Figure 1.2), V1, V2, and V3 can be designated as a focal block. The results of applying the replacing rule to this saturated focal block are shown in Model 3 of Figure 1.3, where V1 ↔ V2 has been replaced by V1 → V2, V1 → V3 has been replaced by V1 ← V3, and V2 → V3 has been replaced by V2 ← V3. Further, application of the replacing rule to this focal block has provided yet another heretofore unavailable opportunity to generate an equivalent model. Variables V1 and V5 now define a symmetric focal block, thus permitting the substitution of V1 → V5 by E1 ↔ E5 or V1 ← V5, of which the latter substitution is shown in Model 4 of Figure 1.4.

Symmetric focal blocks also provide a basis for generating equivalent models when two variables are nonrecursively connected. Whenever two nonrecursively connected variables define a symmetric focal block (i.e., have all their predictors in common), the nonrecursive path between

Equivalent Model 2:

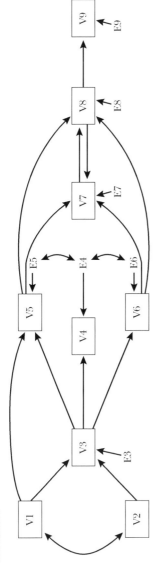

Equivalent Model 3:

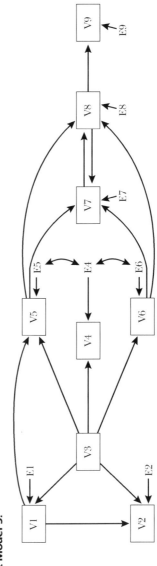

Figure 1.3 An example of equivalent models generated by the replacing rule, not equivalent to the initial model.

them may be replaced by a single directed path or a residual covariance. Importantly, the nonrecursive paths must be specified as equal due to the impossibility of uniquely identifying each path when both variables have the same predictors. The converse of this rule holds as well: Directed paths or residual covariances within a symmetric focal block may be replaced by equated nonrecursive paths (symbolized herein by ⇔). In Model 4, let V7 ⇔ V8 be a focal block where V7 → V8 = V7 ← V8; then V7 ⇔ V8 may be replaced by either V7 → V8, V7 ← V8, or E7 ↔ E8. Model 5 presents such an equivalent model where now V7 → V8. Thus, under the special circumstance of a symmetric focal block, the replacing rule can be applied to nonrecursive paths.

It is interesting to note that if, in Model 4, V8 ← V9 was specified instead of V8 → V9, then V7 and V8 would not have defined a symmetric focal block, thus preventing application of the replacing rule. In the alternative circumstance where V9 is an additional predictor of V8, V9 would have held the position of an *instrumental variable* allowing the unconstrained estimation of the nonrecursive paths between V7 and V8. Instrumental variables are an example of a more general strategy within SEM for the reduction of equivalent models. Instrumental variables are predictor variables that can be added to a model in order to help identify the model. To the degree that variables in a model do not share predictors, the number of equivalent models will essentially be reduced. Symmetric focal blocks exemplify the model equivalence difficulties that arise when focal block variables share all their predictors, and just-identified preceding blocks illustrate the problem even more so. In general, constraining the number of free parameters within a model will by definition reduce the number of possible equivalent models.

The Reversed Indicator Rule and Measurement Models

Measurement models are factor analysis models but with an important difference from traditional (exploratory) factor analysis models: The traditional factor analysis model is under-identified, with the resulting factor pattern potentially subject to an infinite number of rotational solutions. Therefore, an infinite number of equivalent solutions exist. On the other hand, the specific (confirmatory) pattern of factor loadings obtained from structural equation modeling is at least just-identified (or should be) and is unique up to and including rotation. For a given set of variables, and an hypothesized factor structure, equivalent measurement models would appear not to be problematic. However, model equivalence is still very much an issue for two reasons.

First, the pattern of correlations among a set of indicators may in fact indicate the existence of two or more distinct factors nested within the

Equivalent Model 4:

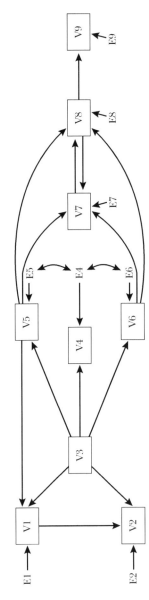

Equivalent Model 5:

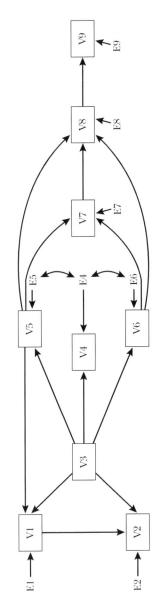

Figure 1.4 An example of the generation of equivalent structural models applying the replacing rule to a symmetric focal block.

Basic Model:

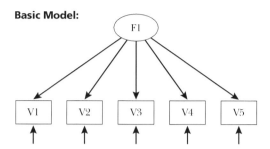

Figure 1.5 A proposed one common factor measurement model.

general common factor. As an example, consider the basic measurement model in Figure 1.5 that depicts a single common factor defined by five observed indicators. Assume that this model was tested on a set of data and rejected, the rejection apparently stemming from the presence of a relatively high correlation between V1 and V2 (i.e., higher than their correlation with the other three variables). In order to modify the model so as to accommodate the relatively high correlation between V1 and V2, any one of four mutually equivalent measurement models could be readily specified. These four equivalent models are shown in Figure 1.6. When tested against a set of data, identical fit statistics would be obtained for each presented model. Whenever a model is modified to accommodate a sizable but unmodeled correlation between two variables, the potential for model equivalence increases due to the many ways in which the source of the correlation may be represented. A number of possibilities might occur. For example, another factor may be specified (Model 1, Model 2, and Model 4), or the correlation between the two variables may be left as a residual covariance (Model 3). If the model with a second factor is selected, the choice then is whether the association between the two factors will be represented by an unanalyzed correlation (Model 1), a higher order factor (Model 2, with, e.g., $b_{F1F3} = b_{F2F3}$), or shared indicators (Model 4)—with appropriate constraints specified in each considered model.

The second reason why model equivalence will be a problem for measurement models is described by the *reversed indicator rule* (Hershberger, 1994). According to the reversed indicator rule, the direction of a path between a latent variable and an indicator is arbitrary for one and only one latent-indicator path in the measurement model of an exogenous factor, and therefore, this path may be reversed. A single recursive path may also be replaced by equated nonrecursive paths. Reversing the direction of the path, or replacing the recursive path with equated nonrecursive paths, results in an equivalent model. Two requirements exist, in addition to the reversal of only one indicator, for application of the reversed indicator rule: (1) the latent variable should be completely exogenous or affected by only

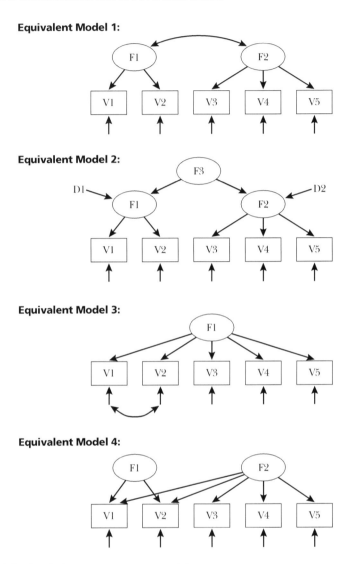

Figure 1.6 Equivalent measurement models.

one indicator before and after the rule's application, and (2) the exogenous latent variable must be uncorrelated with other exogenous latent variables before and after the rule's application.

In Figure 1.7, three equivalent measurement models are presented. These three models, although equivalent by the reversed indicator rule, imply very different conceptualizations. In Figure 1.7a, a latent variable exerts a common effect on a set of indicators. In contrast, in Figure 1.7b, a latent

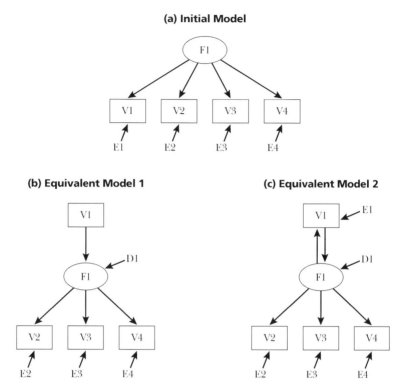

Figure 1.7 An example of the generation of equivalent measurement models using the reversed indicator rule.

variable is influenced by an indicator. Finally, Figure 1.7c shows another equivalent model, in which the latent variable influences, and is influenced by, an indicator. For more detail on such models, see the chapter by Kline in this volume.

Summary of Rules for Specifying Equivalent Structural Models

Replacing Rule—General Case. Given a directed path between two variables, V1 → V2 in a focal block, the directed path may be replaced by the residual covariance, E1 ↔ E2, if (1) only recursive paths exist between blocks; (2) only recursive paths exist within a focal block (this condition may be relaxed when the block is a symmetric focal block); and (3) the predictors of the effect variable are the same as or include those for the source variable.

Replacing Rule—Symmetric Focal Blocks. Given a directed path between two variables, V1 → V2 in a focal block, the directed path may be replaced by the residual covariance, E1 ↔ E2, or V1 ← V2, or V1 ⇔ V2, if (1) only recursive paths exist between blocks; (2) when the original specification is replaced by a nonrecursive relation, the two paths are equated; and (3) the source and effect variables have the same predictors.

Replacing Rule—Just-Identified Preceding Block. Given a directed path between two variables, V1 → V2, in a focal block that is a just-identified preceding block,[2] the directed path may be replaced by the residual covariance E1 ↔ E2, by V1 ← V2, or by V1 ⇔ V2, if (1) only recursive paths exist between the preceding (focal) block and succeeding block; and (2) when the original specification is replaced by a nonrecursive relation, the two paths are equated.

Summary of Rules for Specifying Equivalent Measurement Models

Reversed Indicator Rule. Given a directed path between a latent variable and an indicator, F1 → V1, the directed path may be replaced by F1 ← V1, or by F1 ⇔ V1, if (1) only the directed path for one indicator for each measurement model is altered; (2) the latent variable is affected by a single indicator, or is completely exogenous, before and after application of the rule; (3) the latent variable is uncorrelated with other exogenous latent variables; and (4) when the original specification is replaced by a nonrecursive relation, the two paths are equated.

We note that for models consisting of both a measurement and a structural part, the model equivalence rules may be applied to both portions, as long as the requirements of the rules for both are met.

Other Strategies Before and After Data Collection

The replacing rule and the reversed indicator rule are not the only possible strategies available for identifying equivalent models based on their structure alone, without resort to observed data. Other viable strategies include graph theory, matrix rank, and automated heuristic searches.

Graph Theory. Recent advances in graph theory have noticeably helped researchers expand their ability to detect equivalent models. These advances have resulted from efforts to provide heuristic searches of causal inferences with a more objective, mathematical foundation than had apparently been available in the past. We note that these heuristic search strategies do not specifically require numerical minimization or maximization of a data matrix—some of these methods are based upon the evaluation of tetrads (see, e.g., Glymour, Scheines, Spirtes, & Kelly, 1987; Scheines, Spirtes, Glymour, Meek, & Richardson, 1998; Spirtes, Scheines, & Glymour, 1990). These search approaches that make use of tetrads basically represent the relations among sets of four covariance elements (h, i, j, k) within any covariance matrix to be defined as simply $\tau_{hijk} = \sigma_{hi} \sigma_{jk} - \sigma_{hj} \sigma_{ik}$. When a tetrad is equal to zero, it is called a *vanishing tetrad*. Glymour et al. (1987) also developed the computer program TETRAD to explore the existence of all possible models (now in its version IV; for a recent review of the program, see Landsheer, 2010). Unfortunately, to date the TETRAD program is limited with regard to the number of variables in a model it can handle, thereby making its practical value rather restricted. In the next section we introduce some heuristic search strategies that can be used to detect equivalent models that, although they require numerical minimization or maximization of a data matrix, can handle models with large numbers of variables (for a complete overview of several heuristic search strategies and the conceptual and methodological details related to their application in a variety of settings and situations, see Marcoulides and Ing, 2012).

Thus, the strategy providing the foundation for this graph theory approach involves translating the causal relations implied by graph models (which are very similar to path models utilized in defining structural equation models) into statistical relations. The logic is that one can expect certain statistical relations only if a proposed causal relation is true. Zero-order partial correlations are important values here because a causal relation implies a particular pattern of partial correlations in the model. Further, for recursive models at least, partial correlations are sufficient: a set of zero-order partial correlations can fully characterize the statistical content of a structural model the causal relations embody in the model. This notion should come as no surprise, given that we found that the replacing rule worked because the only changes in a model implied by the rule did not alter the model's partial correlation structure. By implementing these algorithms, models that imply the same partial correlations but differ in causal structure can be readily found—and as it so happens, some of these models turn out to be equivalent.[3] The generality of the algorithms vary; some are restricted to completely recursive latent variable models (Spirtes, Glymour, & Scheines, 1993), others are suitable for nonrecursive latent variable models, complete with feedback loops and correlated errors (Richardson, 1997). Because the graph theory approach to identifying equivalent models is far too complex for further discussion in this chapter, we refer the reader to two excellent introductory texts on the topic by Glymour et al. (1987) and Shipley (2000); a somewhat more difficult but clear and comprehensive account is given by Pearl (2009).

Rank Matrix Approach. A second approach to identifying equivalent models makes use of the rank of the matrix of correlations among the parameters of the proposed model (Bekker et al., 1994).[4] In order to use this method one must combine the matrices from two potential models *suspected* of equivalence. If the rank of the combined matrix is less than the sum of the ranks of the separate matrices, the two models are considered minimally *locally equivalent.* Although with a correct software implementation the rank method is straightforward in application, it does have some disadvantages: (1) it is conceptually very difficult to understand without advanced knowledge of linear algebra; (2) only the local equivalence of two models is ultimately tested; and (3) models suspected of equivalence must be known beforehand.

Automated Heuristic Searches. Another approach to identifying equivalent models makes extensive use of computer intensive search algorithms. Some researchers refer to such automated approaches as *discovering structure in data* or *learning from data* while others simply call them *data mining* (Marcoulides, 2010). There is also a generally accepted taxonomy of the different kinds of automated approaches referred to as *supervised, semi-supervised,* and *unsupervised* methods (Larose, 2005; Marcoulides, 2005). In multivariate statistics the dividing line between supervised learning and unsupervised learning is the same as that distinguishing the techniques of discriminant analysis from cluster analysis. Supervised learning requires that the target variable is well defined and that a sufficient number of its values are given. In the case of a discriminant analysis, this would imply specifying a priori the number of groups to be considered and focusing on which variables can best be used for the discrimination of the groups. For unsupervised learning typically the target variable is unknown, which in the case of cluster analysis, corresponds to the number of groupings of observations based upon the selectively considered variables. Semi-supervised is a combination of both—the algorithms are provided with limited information concerning the must and/or cannot-link constraints on the model. Within the SEM literature, automated search approaches have received much attention in the literature (e.g., Marcoulides & Drezner, 2001, 2003; Marcoulides, Drezner, & Schumacker, 1998; Marcoulides & Ing, 2012; Scheines et al., 1998). This process of model exploration is also commonly referred to as a *specification search* (Long, 1983; Marcoulides & Drezner, 2001).

Generating and examining alternative models after fitting an initially hypothesized model to empirical data can be difficult, particularly when the number of variables is large. For many considered models, there might just be a small number of possibilities to consider. For example, with only two observed variables, there are only four possible models. Such a case is displayed in Figure 1.8. For 3 variables, there are 64 possible models. However, what happens when the number of possible combinations becomes prohibitively large? For example, even with just 6 observed variables there

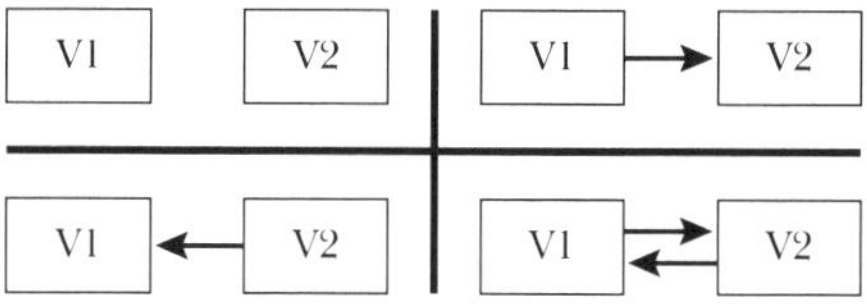

Figure 1.8 All possible combinations of models with 2 observed variables.

are 1,073,741,824 possible model combinations. In general, the total number of models among p variables is the number of possible ways each pair can be connected, to the power of the number of pairs of variables and is determined by $4^{|p(p-1)/2|}$ (Glymour et al., 1987).

In cases where enumeration procedures for examining all possible models are impractical, various heuristic optimization or automated search algorithms have been developed (Marcoulides et al., 1998) to examine the parameter space. Heuristic search algorithms are specifically designed to determine the best possible solution but do not guarantee that the optimal solution is found—though their performance using empirical testing or worst cases analysis indicates that in many situations they seem to be the only way forward to produce concrete results (Salhi, 1998). To date, heuristic procedures have been successfully applied to tackle difficult problems in business, economics, and industrial engineering. They have also recently made their way into the general SEM literature. Examples of such numerical heuristic procedures in SEM include: ant colony optimization (Leite, Huang, & Marcoulides, 2008; Leite & Marcoulides, 2009; Marcoulides & Drezner, 2003), genetic algorithms (Marcoulides & Drezner, 2001), ruin-and-recreate (Marcoulides, 2009), simulated annealing (Marcoulides & Drezner, 1999), and Tabu search (Marcoulides et al., 1998; Marcoulides & Drezner, 2004)—and over the years a great variety of modifications have been proposed to these heuristic search procedures (e.g., Drezner & Marcoulides, 2003; Marcoulides, 2010). Because the literature on these automated algorithms is vast and far too technically involved to discuss further in this chapter, we refer the reader to a recent overview of the available algorithms provided by Marcoulides and Ing (2012). In summary, all of these methods focus on the evaluation of an objective function, which is usually based upon some aspect of model fit and can provide users with a candidate list of, say, the top 10 or 20 feasible models.

CONCEPTUALIZATIONS OF MODEL EQUIVALENCE AND IDENTIFICATION

Rigorous mathematical treatment of the identification problem began in the early 1940s and 1950s with the work of Albert (1944a, b), Koopmans and

Reiersøl (1950), and Ledermann (1937). To summarize, identification basically consists of two specific aspects: existence and uniqueness (e.g., Bekker et al., 1994; Sato, 1992, 2004). Although identification issues can have major implications with respect to model fitting, the technical level of such developments can be quite challenging (Hayashi & Marcoulides, 2006). To simplify matters, some researchers often make a specific assumption about existence and focus mainly on uniqueness aspects. For example, existence within the context of factor analysis implies that a covariance factor decomposition exists for a given number of factors, whereas uniqueness assumes it is the only decomposition possible—in other words, it is commonly assumed that a factor decomposition does exist in the population. The topic of uniqueness in factor analysis and related structural equation models has essentially followed two early lines of research: one originating in Albert (1944a, b) and Anderson and Rubin (1956) and the other based on the work of Ledermann (1937)—for an overview see Hayashi and Marcoulides (2006) and the references therein.

Econometricians have also contributed much to the development of the identification literature; probably the best-known contribution was the introduction of criteria for the identification of simultaneous equation systems (e.g., the full rank condition). Fisher (1966) provided a thorough treatment of identification for linear equation systems and Bowden (1973) extended identification theory to embrace nonlinear models. Hannan and Deistler (1988) generalized the theory of identification for linear structural equations with serially correlated residual errors, establishing conditions for identifiability for the important case of auto-regressive moving average error terms. Some more recent, important contributions to the identification of structural equation models have been made by Kano (1997), Rigdon (1995), Sato (2004), ten Berge (2004), and Walter and Pronzato (1997).

As indicated above, *identification* refers to the uniqueness of the parameterization that generated the data. Hendry (1995) further specified three other aspects of identification: (1) uniqueness, (2) correspondence to a "desired entity," and (3) a satisfying interpretation of a theoretical model. As an analogy, the beta weight determined by regressing the quantity of a product sold on the price charged is uniquely determined by the data, but need not correspond to any underlying economic behavior, and may be incorrectly interpreted as a supply schedule due to rising prices.

We generally categorize a model's identifiability in one of two ways: (1) *global identification*, in which *all* of a model's parameters are identified, or in contrast, (2) *local identification*, in which at least one—but not all—of a model's parameters is identified. Globally identified models are locally identified, but locally identified models might or might not be globally identified. Global identification is a prerequisite for drawing inferences about an entire model. When a model is not globally identified, local identification of some of its parameters permits inferential testing in only that section of the model. Some researchers have referred to this as *partial identification* (Hayashi & Marcoulides, 2006). Kano (1997) has suggested that we focus more attention on examining what he referred to as *empirical identification*, rather than the more commonly considered notion of *mathematical identification*. At least in theory—but debatable in practice—nonidentified parameters do not influence the values of ones that are identified. Thus, the greater the local identifiability of a model, the stronger the inferences that can be drawn from the subset of identified equations.

Besides categorizing models as either globally or locally identified, models can also be classified as (1) *under-identified*, (2) *just-identified*, or (3) *over-identified*. Consider for example the model involving three ($p = 3$) observed variables presented in an earlier section. An observed covariance matrix for these three variables has altogether $p(p + 1)/2 = 3(4)/2 = 6$ nonredundant elements. Now let us consider the case of an under-identified model. Under-identification occurs when not enough relevant data are available to obtain unique parameter estimates. Because the degrees of freedom of any hypothesized model are the difference between the number of non-redundant elements in the covariance matrix and the number of parameters in the model, an under-identified model will have negative degrees of freedom. We note that when the degrees of freedom of a model are negative, at least one of its parameters is under-identified. Having positive degrees of freedom with any proposed model is a necessary but not a sufficient condition for identification. That is because having positive degrees of freedom does not guarantee that every parameter is identified. There can in fact be situations in which the degrees of freedom for a model are quite high and yet some of its parameters remain under-identified (Raykov & Marcoulides, 2006). Conversely, having negative degrees of freedom is a sufficient but not a necessary criterion for showing that a model is globally under-identified.

Saturated models have as many parameters as there are nonredundant data elements and are commonly referred to as *just-identified*. Just-identified models are always identified in a trivial way. This is because just-identification occurs when the number of data elements equals the number of parameters to be estimated. If the model is just-identified, a solution can always be found for the parameter estimates that will result in perfect fit—a discrepancy function equal to zero. This implies that both the chi-square goodness-of-fit test and its associated degrees of freedom will be zero. As such, there is no way one can really test the veracity of a just-identified model. In fact, all just-identified models are equivalent, but not all equivalent models are just-identified.

If theory testing is one's objective, the most desirable identification status of a model is *over-identification*, where the number of available data elements is more than what we need to obtain a unique solution. In an over-identi-fied model, the degrees of freedom are positive so that we can explicitly test model fit. An over-identified model implies that, for at least one parameter, there is more than one equation the estimate of a parameter must satisfy; only under these circumstances—the presence of multiple solutions—are models provided with the opportunity to be rejected by the data.

Though a family of equivalent models might be statistically identified, equivalent models are under-identified in their causal assumptions, sug-gesting that an intimate association is present among them. To begin, both under-identification and model equivalence are sources of indeterminacy: Under-identification requires the selection of a parameter's best estimate from an infinite number of estimates, while model equivalence requires the selection of the best model from among potential many (if not infinite; Raykov & Marcoulides, 2001) models. In one case, we cannot determine the correct population value of a parameter, and in the other we cannot de-termine the correct population model. A comparison of the mapping func-tions of identified parameter and equivalent model sets illustrates the simi-larity, and subtle difference, between identification and model equivalence. Recall that for models to be globally equivalent, a function must result in a one-to-one mapping associating (reparameterizing) each parameter of one model with every parameter of a second model, *without a remainder* (see Fig-ure 1.9). One-to-one mappings that exhaust every element of two sets are *bijective*. On the other hand, for a model to be identified, a one-to-one map-ping of each parameter onto a unique estimate of that parameter must be *injective*. We illustrate an injective one-to-one mapping in Figure 1.10a. An injective mapping associates two sets so that all members of set A are paired with a unique member of set B, *but with a remainder*. The set A elements do not exhaust the elements of set B; there are some elements within set B that are not paired with a member of set A. For a model to be identified, all of a model's parameters within set A must be paired with a unique parameter es-timate in set B. Figure 1.10b shows the mapping that would occur if a model

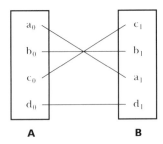

Figure 1.9   One-to-one mapping function of two models' parameters required for global equivalence.

(a) Identified

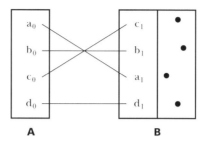

A   B

(b) Under-Identified

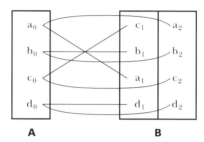

A   B

Figure 1.10 Mapping functions relating parameters (A) to estimates (B) for identified and under-identified models.

were under-identified: Each parameter within set *A* is potentially paired with each member of set *B*, which, though not shown, now has an infinite number of members that are potential parameter estimates.

Another association between under-identification and observational equivalence concerns the *uniqueness* of the one-to-one mapping function linking the parameters of two observationally equivalent models. If only one function can be found that reparameterizes Model A into Model B, then the parameters from both models are identified; that is, there is only one set of parameter estimates for each model. On the other hand, if more than one function exists, both models are under-identified, each having an infinite number of parameter estimates for each parameter.[5]

This discussion concerning similarities between model equivalence and model identification does not imply the ideas are simply "two sides of the same coin." Model equivalence and model identification are, in many respects, rather different. For example, a globally under-identified model is forever uninterpretable in its current formulation, worthless for theory testing.[6] In contrast, a model with many equivalent models can potentially provide some support and corroborating evidence for a theory, at least to the extent that the alternative hypotheses posed by the equivalent models can be ruled out on other (e.g., theoretical) grounds.

SELECTING FROM AMONG EQUIVALENT MODELS

Before Data Collection

As we indicated above, it is important and highly recommended for a researcher to get a sense of how many models are equivalent to an hypothesized model and understand the available preventative options before data collection actually begins. Even when only one equivalent model is possible, causal inferences from the hypothesized model are compromised. Unfortunately, unless the hypothesized model is significantly over-identified due to imposing restrictions on the model, we should expect at least some equivalent models to exist. However, the causal models typically specified in the behavioral and social sciences can hardly be characterized as significantly over-identified; many are only few degrees of freedom away from just-identified status. In the case of just-identification, the number of models equivalent to an hypothesized model is at a maximum. In situations that the number of equivalent models is "large," one should consider revising the hypothesized model to potentially reduce the number of equivalent models. Revising the hypothesized model may also in turn require revising the original data collection design by, for example, requiring the collection of data on an additional set of variables.

Collecting information on additional variables does not necessarily mean changing the originally hypothesized theory: the role of the additional variables can be comparable to the role of covariates introduced when random assignment has not been fully implemented. For example, in traditional regression modeling, to control for confounding by a variable one must include it in the structural part of the statistical model. The confounding variable is not itself of theoretical interest. We could also introduce the new variables into the model as instruments serving a purpose similar to that of instrumental variables when a model is just-identified or under-identified. If introducing the new variables results in a positive gain in the degrees of freedom of the model, and if each of the new variables is selectively associated with only a few of the original variables, the number of potentially equivalent models will decrease substantially. Of course, the revised model should retain the most important causal relations specified in the original model, or minimally, should not have specified relations theoretically inconsistent with the relations of the originally proposed model. In practice, reducing the number of equivalent models while at the same time retaining the integrity of the original model can be quite difficult. Consider also that although there might be many equivalent models initially, many will have nonsensical, illogical, or factually inconsistent causal relations. The presence of time going backward, for example, is a common *indiscretion* of equivalent models.

After Data Collection

If the data fit the hypothesized model (and its equivalent models) well, one can (a) retain the hypothesized model, (b) decide to select a particular equivalent model as the best model, or (c) assume a position of agnosticism with respect to which model should be selected. Most researchers would prefer (a) which, after all, was the theory testing purpose of the study. At this point, they might try "somehow" to justify, using quantitative criteria, retaining the hypothesized model. The strategies described below have been proposed in the literature as ways to help researchers rationalize retaining their hypothesized model when model equivalence is an issue.

Strategy 1. An Information Complexity Criterion (ICOMP). An information complexity criterion (*ICOMP*) has been proposed for selecting among equivalent models (Williams et al., 1996). The idea behind *ICOMP*, and complexity criteria in general, is that the value of a model should be based on the fit of the model to the data and the *complexity* of the parameter estimates: *ICOMP = model misfit + complexity*. Model misfit is defined in the usual way as a function of the difference between the observed and model-implied covariance matrices. *Complexity* is a measure of the correlations among the parameter estimates—higher correlations mean more complexity. These correlations are obtained from a model's information matrix, which is computed during the process of parameter estimation. While small discrepancies between the observed and model-implied covariance matrices suggest a good fit of the model to the data, large correlations among the parameter estimates do not. Correlations among the parameter estimates are not expected because, ideally, the estimation of one parameter should be independent of the estimation of the others. Problems begin when these correlations become high, a sign indicative of insufficient information in the data to estimate each parameter uniquely. Under these conditions, the same information (data) is used to estimate more than one parameter, thus inducing correlations among them. At a minimum, the model is locally under-identified. *ICOMP* is scaled so the model with the lowest value is selected, since a low *ICOMP* arises with a small discrepancy between $\mathbf{S}$ and $\hat{\mathbf{\Sigma}}$ (i.e., small misfit) and low correlations among the parameter estimates (i.e., small complexity). Among a set of equivalent models, which have identical misfit, the one selected with the lowest *ICOMP* will be the least complex of those models.

Strategy 2. Tortured iteration histories, negative variances, and constrained boundary values. Although equivalent models are all destined to arrive at the same model-implied covariance matrix, they travel over different roads. And some of these roads can be computationally bumpier than others. Indications of a difficult trip include an unusual number of iterations, characterized by haphazard improvement, negative parameter estimates for variances, and parameter estimates constrained by a modeling program (e.g., to 0 or 1) to avoid improper parameter estimates (e.g., constraining a variance to zero when, if freely estimated, would assign a negative parameter estimate to a variance). Both empirical research and anecdotal observation agree that these estimation difficulties frequently result from highly misspecified models (e.g., Green, Thompson, & Poirier, 1999).

It has been suggested that the computational problems associated with misspecification can actually be used to distinguish among equivalent models. Equivalent models are equivalent in fit, but almost necessarily not equivalent in structure. In fact, a good chance exists that many equivalent models will likely have nonsensical causal relations. Thus, the task of selecting the best model from a family of equivalent models becomes less ambiguous and arbitrary when models of dubious structure also show poor estimation histories. A model with a theoretically unlikely causal structure and with unlikely parameter estimates is, justifiably, unlikely to be selected as the "true" model.

Strategy 3. Comparing R^2 values among models. Jöreskog and Sörbom (1996) have suggested that R^2 values for specific structural equations might be used to select the optimal model from a set of equivalent alternatives. Because the R^2 values of structural equations are independent, they can be aggregated to reflect an overall R^2 for a model. While sensible, differences in aggregate R^2 as a criterion for selecting a model among a set of equivalent models has not been systematically investigated. One well-known disadvantage of using an R^2 selection criterion is the tendency for its value to fluctuate dramatically across samples. Such volatility may lead to the selection of different models for different samples, which is not at all a desirable situation. In contrast to the sample dependency of R^2 values, the values of information criteria such as *ICOMP* are very stable, suggesting that a numerical index to be used for the identification of an optimal model among equivalent alternatives should be based on characteristics like model parsimony and complexity instead of the amount of explained variance.

Strategy 4. Extended individual case residuals. Because the model-implied covariance matrix is identical among equivalent models, the residual covariance matrix denoting the discrepancy between the observed and model-implied covariance matrices is identical as well. Although these residuals are not helpful in discriminating among equivalent models, residuals computed at the individual level may be helpful in doing so. Raykov and Penev

(2001) proposed the use of *extended individual case residuals (EICR)* to help in this discrimination. An *EICR* for an individual, h_i, is the difference between an observed data point and its model-predicted value:

$$h_i = \Theta^{-1/2}[\mathbf{I} - \Lambda(\Lambda'\Theta^{-1}\Lambda)^{-1}\Lambda'\Theta^{-1}]y_i,$$

where Θ is an error covariance matrix, Λ is a matrix of factor loadings, $\mathbf{I}$ is an identity matrix, and y_i is the ith subject's score on an observed variable.[7]

The *EICR* for an individual will differ across equivalent models if (a) Θ is different for each equivalent model, (b) Λ is full rank in each model, and (c) the equivalent models have the same number of factors specified. An *EICR* is computed for each individual. While none of the individual *EICR* values will be identical across the equivalent models, the sum of squares of the *EICR* values will be. If the *EICR* values are available, we must decide how they can be used to select among equivalent models. Raykov and Penev (2001) have suggested that the model with the smallest average standardized *EICR* relative to its equivalent model alternatives should be preferred.

CONCLUSION

The problem of equivalent models should make researchers aware of the various issues and limitations that should be considered when confirming structural equation models. Arguably, model equivalence is one of the most compelling demonstrations available for showing why confirming a structural equation model, and ultimately the causal relations the model implies, can be so difficult: No matter how much better an hypothesized model fits the data in comparison with other competing models, no possibility exists of unequivocally confirming the hypothesized model if there are models equivalent to it. This is not to suggest that it is impossible to rule out equivalent models based on plausibility alone, but there will always be some degree of doubt to such drawn conclusions. The best method of reducing the number of potential equivalent models is to be as parsimonious as possible during model specification. Parsimonious specification acknowledges the inverse relation between the number of model constraints and the number of equivalent models. We caution readers that even a parsimonious specification approach to modeling is not completely without its pitfalls as a mechanism for differentiating among examined models. Rigid and routine applications of the parsimony principle can, in some cases, lead to conclusions favoring an incorrect model and implications that are incompatible with those of the correct model (for further details see Raykov & Marcoulides, 1999).

Unfortunately, it seems somewhat unlikely that consideration of equivalent models will become a standard part of the process of defining, testing, and modifying models. Despite nearly decades of reports arguing convincingly for the importance of model equivalence to the model-fitting enterprise, efforts to identify equivalent models are still infrequent. Not helping the situation are the limited capabilities of readily available commercial software packages that are capable of identifying equivalent models. To date it appears that only two commercially available programs can do so, even if they are somewhat limited in their capabilities and the size of the models they can handle; AMOS (Arbuckle, 2003), and TETRAD (Scheines, Spirtes, Glymour, & Meek, 1994). A freeware type program that promises to have extensive capabilities for identifying equivalent models is the open source program *Mx* (Boker, Neale, Maes, Wilde, Spiegel, Brick, Spies, Estabrook, Kenny, Bates, Mehta, & Fox, 2011; Neale, Boker, Xie, & Maes, 1999), which essentially works as an integral part of the R statistical software system. None of the other available popular SEM programs do. There really is no excuse for the omission of model equivalence in these commercial programs. Rules and algorithms available for finding equivalent models are readily available and programmable; these rules and algorithms typically identify no less than 99% of the possible equivalent models. Researchers should take the initiative, and routinely report the presence of equivalent models. If reporting equivalent models becomes common practice, software companies will soon respond by including automated methods for detecting equivalent models in their SEM programs. Until then, researchers will be forced to rely on specialized programs that are available only directly from methodologists.

NOTES

1. We note that although the principles of equivalent models apply to models with covariance structures and potential mean structures, for simplicity and without loss of generality mean structures are not explicitly treated in this chapter (for additional details see Raykov & Marcoulides, 2001; Raykov & Penev, 1999; Williams et al., 1996).

2. By definition, the variables of a just-identified block are symmetrically determined; a just-identified preceding (focal) block is a symmetric focal block.

3. All equivalent models have identical partial correlation structures but not all models with identical partial correlation structure are equivalent. The partial correlation structure of a model refers to the partial correlations that are equal to zero as implied by the model. For example, if one model implies the partial correlation $r_{xy.z} = 0$, then all of the models equivalent to it must also imply $r_{xy.z} = 0$.

4. It is not strictly true that this matrix, the Jacobian matrix referred to earlier, is a matrix of correlations among the model's parameters, but it is highly related to the matrix that is (the inverted information matrix).

5. Econometricians view the model equivalence-model identification relationship as a deterministic: *A model is identified if there is no observationally equivalent model.* In other words, whenever a model has models equivalent to it, the model must be under-identified. This view, which is not shared by most psychometricians, is based on the following reasoning. For a parameter to be identified, complete knowledge of the joint distribution of the random variables must provide enough information to calculate parameters uniquely. Further, every independent distribution of the same variables must also provide unique estimates. Yet, observationally equivalent models, by definition, always fit the same distribution equally well, *but do so with different parameter sets.* Although the distribution of the random variables is the same, the parameters of the equivalent models are different. Recall that observationally equivalent models are linked by a function that translates one model's configuration of parameters into another model's configuration exactly; the two models' parameter sets are simply transformations of each other and are therefore the same parameters, only configured (organized) differently. A model cannot be identified and have parameters with two solutions. Thus, an observationally equivalent model is always under-identified.

6. Of course, some parameters may be identified within an under-identified model, thereby rendering specific theories regarding those parameters still somewhat testable.

7. Raykov and Penev (2001) provided this expression for all models covered by the so-called Submodel 3B of the general structural equation model underlying the program LISREL (Jöreskog & Sörbom, 1996).

REFERENCES

Albert, A. A. (1944a). The matrices of factor analysis. *Proceedings of the National Academy of Science, USA, 30,* 90–95.

Albert, A. A. (1944b). The minimum rank of a correlation matrix. *Proceedings of the National Academy of Science, USA, 30,* 144–146.

Anderson, T. W., & Rubin, H. (1956). Statistical inferences in factor analysis. In J. Neyman (Ed.), *Proceedings of the Third Berkeley Symposium on Mathematical Statistics and Probability* (Vol. 5, pp. 111–150). Berkeley: University of California.

Arbuckle, J. L. (2003). *AMOS 5 user's guide.* Chicago, IL: SPSS.

Bekker, P. A., Merckens, A., & Wansbeek, T. J. (1994). *Identification, equivalent models, and computer algebra.* Boston: Academic Press.

Boker, S., Neale, M., Maes, H., Wilde, M., Spiegel, M., Brick, T., Spies, J., Estabrook, R., Kenny, S., Bates, T., Mehta, P., and Fox, J. (2011). OpenMx: An open source extended structural equation modeling framework. *Psychometrika, 76,* 306–317.

Bowden, R. (1973). The theory of parametric identification. *Econometrica, 41,* 1069–1074.

Breckler, S. J. (1990). Applications of covariance structure modeling in psychology: Cause for concern? *Psychological Bulletin, 107,* 260–273.

Chernov, M., Gallant, A. R., Ghysels, E., & Tauchen, G. (2003). Alternative models for stock price dynamics. *Journal of Econometrics, 116*, 225–257.

Drezner, Z., & Marcoulides, G. A. (2003). A distance-based selection of parents in genetic algorithms. In M. Resenda & J. P. Sousa (Eds.). *Metaheuristics: Computer decision-making* (pp. 257–278). Boston, MA: Kluwer Academic Publishers.

Fisher, F. M. (1966). *The identification problem in econometrics.* New York, NY: McGraw-Hill

Glymour, C., Scheines, R., Spirtes, R., & Kelly, K. (1987). *Discovering causal structure: Artificial intelligence, philosophy of science, and statistical modeling.* Orlando, FL: Academic Press.

Green, S. B., Thompson, M. S., & Poirier, J. (1999). Exploratory analyses to improve model fit: Errors due to misspecification and a strategy to reduce their occurrence. *Structural Equation Modeling, 6*, 113–126.

Hannan, E. J., & Deistler, M. (1988). *The statistical theory of linear systems.* New York, NY: Wiley.

Hayashi, K., & Marcoulides, G. A. (2006). Examining identification issues in factor analysis. *Structural Equation Modeling, 13*, 631–645.

Hendry, D. F. (1995). *Dynamic econometrics.* New York, NY: Oxford University Press.

Hershberger, S. L. (1994). The specification of equivalent models before the collection of data. In A. von Eye & C. Clogg (Eds.), *The analysis of latent variables in developmental research* (pp. 68–108). Beverly Hills, CA: Sage.

Hsiao, C. (1983). Identification. In Z. Griliches & M. D. Intriligator (Eds.), *Handbook of econometrics, Vol. 1* (pp. 224–283). Amsterdam: Elsevier Science.

Hu, L., & Bentler, P. M. (1999). Cutoff criteria for fit indexes in covariance structure analysis: Conventional criteria versus alternative. *Structural Equation Modeling, 6*, 1–55.

Jöreskog, K., & Sörbom, D. (1996). *LISREL 8: User's reference guide.* Chicago: Scientific Software.

Kano, Y. (1997). Exploratory factor analysis with a common factor with two indicators. *Behaviormetrika, 24*, 129–145.

Koopmans, T. C., & Reiersøl, O. (1950). The identification of structural characteristics. *Annals of Mathematical Statistics, 21*, 165–181.

Koopmans, T. C., Rubin, H., & Leipnik, R. B. (1950). Measuring the equation system of dynamic economics. In T. C. Koopmans (Ed.), *Statistical inference in dynamic economic models.* (pp. 53–237). New York, NY: Wiley.

Landsheer, J. A. (2010). The specification of causal models with TETRAD IV: A review. *Structural Equation Modeling, 17*, 630–640.

Larose, D. T. (2005). *Discovering knowledge in data: An introduction to data mining.* Hoboken, NJ: Wiley.

Ledermann, W. (1937). On the rank of reduced correlation matrices in multiple factor analysis. *Psychometrika, 2*, 85–93.

Lee., S., & Hershberger, S. L. (1990). A simple rule for generating equivalent models in covariance structure modeling. *Multivariate Behavioral Research, 25*, 313–334.

Leite, W. L., Huang, I. C., & Marcoulides, G. A. (2008). Item selection for the development of short-form of scaling using an Ant Colony Optimization algorithm. *Multivariate Behavioral Research, 43*, 411–431.

Leite, W. L., & Marcoulides, G. A. (2009, April). *Using the ant colony optimization algorithm for specification searches: A comparison of criteria.* Paper presented at the Annual Meeting of the American Education Research Association, San Diego, CA.

Levy, R., & Hancock, G. R. (2007). A framework of statistical tests for comparing mean and covariance structure models. *Multivariate Behavioral Research, 42,* 33–66.

Long, J. S. (1983). *Covariance structure models: An introduction to LISREL.* Beverly Hills, CA: Sage.

Luijben, T. C. W. (1991). Equivalent models in covariance structure analysis. *Psychometrika, 56,* 653–665.

MacCallum, R. C., Wegener, D. T., Uchino, B. N., & Fabrigar, L. R. (1993). The problem of equivalent models in applications of covariance structure analysis. *Psychological Bulletin, 114,* 185–199.

Marcoulides, G. A. (2005). Review of *Discovering knowledge in data: An introduction to data mining. Journal of the American Statistical Association, 100,* 1465–1465.

Marcoulides, G. A. (2009, May). *Conducting specification searches in SEM using a ruin and recreate principle.* Paper presented at the Annual Meeting of the American Psychological Society, San Francisco, CA.

Marcoulides, G. A. (2010, July). *Using heuristic algorithms for specification searches and optimization.* Paper presented at the Albert and Elaine Borchard Foundation International Colloquium, Missillac, France.

Marcoulides, G. A., & Drezner, Z. (1999). Using simulated annealing for model selection in multiple regression analysis. *Multiple Linear Regression Viewpoints, 25,* 1–4.

Marcoulides, G. A., & Drezner, Z. (2001). Specification searches in structural equation modeling with a genetic algorithm. In G. A. Marcoulides & R. E. Schumacker (Eds.), *New developments and techniques in structural equation modeling* (pp. 247–268). Mahwah, NJ: Lawrence Erlbaum Associates, Inc.

Marcoulides, G. A., & Drezner, Z. (2003). Model specification searchers using ant colony optimization algorithms. *Structural Equation Modeling, 10,* 154–164.

Marcoulides, G. A., & Drezner, Z. (2004). Tabu Search Variable Selection with Resource Constraints. *Communications in Statistics: Simulation & Computation, 33,* 355–362.

Marcoulides, G. A., Drezner, Z., & Schumacker, R. E. (1998). Model specification searches in structural equation modeling using Tabu search. *Structural Equation Modeling, 5,* 365–376.

Marcoulides, G. A., & Ing, M. (2012). Automated structural equation modeling strategies. R. Hoyle (Ed.), *Handbook of structural equation modeling* (pp. 690–704). New York, NY: Guilford Press.

Marcus, K. A. (2002). Statistical equivalence, semantic equivalence, eliminative induction and the Raykov-Marcoulides proof of infinite equivalence. *Structural Equation Modeling, 9,* 503–522.

Marsh, H. W., Hau, K. T., & Wen, Z. (2004). In search of golden rules: Comments on hypothesis-testing approaches to setting cutoff values for fit indexes and dangers of overgeneralizing Hu and Bentler's (1999) findings. *Structural Equation Modeling, 11,* 320–341.

McDonald, R. P. (2002). What can we learn from the path equations?: Identifiability, constraints, equivalence. *Psychometrika, 67,* 225–249.

Neale, M. C., Boker, S. M., Xie, G., & Maes, H. H. (1999). *Mx: Statistical modeling* (5th ed.). Richmond, VA: Virginia Commonwealth University.

Pearl, J. (2009). *Causality: Models, reasoning, and inference* (2nd ed.). New York, NY: Cambridge University Press.

Raykov, T. (1997). Equivalent structural equation models and group equality constraints. *Multivariate Behavioral Research, 32,* 95–104.

Raykov, T., & Marcoulides, G. A. (1999). On desirability of parsimony in structural equation model selection. *Structural Equation Modeling, 6,* 292–301.

Raykov, T., & Marcoulides, G. A. (2001). Can there be infinitely many models equivalent to a given covariance structure model? *Structural Equation Modeling, 8,* 142–149.

Raykov, T., & Marcoulides, G. A. (2006). *A first course in structural equation modeling (2nd ed.).* Mahwah, NJ: Erlbaum.

Raykov, T., & Marcoulides, G. A. (2007). Equivalent structural equation models: A challenge and responsibility. *Structural Equation Modeling, 14,* 527–532.

Raykov, T., & Penev, S. (1999). On structural equation model equivalence. *Multivariate Behavioral Research, 34,* 199–244.

Raykov, T., & Penev, S. (2001). The problem of equivalent structural equation models: An individual residual perspective. In G. A. Marcoulides & R. E. Schumacker (Eds.). *New developments and techniques in structural equation modeling* (pp. 297–321). Mahwah, NJ: Erlbaum.

Richardson, T. (1997). A characterization of Markov equivalence for directed cyclic graphs. *International Journal of Approximate Reasoning, 17,* 107–162.

Rigdon, E. E. (1995). A necessary and sufficient identification rule for structural models estimated in practice. *Multivariate Behavioral Research, 30,* 359–383.

Roesch, S. C. (1999). Modeling stress: A methodological review. *Journal of Behavioral Medicine, 22,* 249–269.

Salhi, S. (1998). Heuristic search methods. In G. A. Marcoulides (Ed.). *Modern methods for business research* (pp. 147–175). Mahwah, NJ: Erlbaum.

Sato, M. (1992). A study of identification problem and substitute use in principal component analysis in factor analysis. *Hiroshima Mathematical Journal, 22,* 479–524.

Sato, M. (2004). On the identification problem of the factor analysis model: A review and an application for an estimation of air pollution source profiles and amounts. In *Factor analysis symposium at Osaka* (pp. 179–184). Osaka: University of Osaka.

Scheines, R., Spirtes, P., Glymour, C., & Meek, C. (1994). *TETRAD II: Tools for Discovery.* Hillsdale, NJ: Lawrence Erlbaum Associates.

Scheines, R., Spirtes, P., Glymour, C., Meek, C., & Richardson, T. (1998). The TETRAD Project: Constraint based aids to causal model specification. *Multivariate Behavioral Research, 33,* 65–117.

Shipley, B. (2000). *Cause and correlation in biology.* New York, NY: Cambridge University Press.

Spirtes, P., Glymour, C., & Scheines, R. (1993). *Causation, prediction, and search.* New York, NY: Springer-Verlag.

Spirtes, P., Scheines, R., & Glymour, C. (1990). Simulation studies of the reliability of computer-aided model specification using the TETRAD II, EQS, and LISREL programs. *Sociological Methods and Research, 19,* 3–66.

Stelzl, I. (1986). Changing a causal hypothesis without changing the fit: Some rules for generating equivalent path models. *Multivariate Behavioral Research, 21,* 309–331.

Ten Berge, J. M. F., (2004). Factor analysis, reliability and unidimensionality. In *Factor analysis centennial symposium at Osaka* (pp. 55–69). Osaka: University of Osaka.

Walter, E., & Pronzato, L. (1997). *Identification of parametric models from experimental data.* London: Springer-Verlag.

Williams, L. J., Bozdogan, H., & Aiman-Smith, L. (1996). Inference problems with equivalent models. In G. A. Marcoulides & R. E. Schumacker (Eds.), *Advanced structural equation modeling: Issues and techniques* (pp. 279–314). Mahwah, NJ: Erlbaum.

CHAPTER 2

REVERSE ARROW DYNAMICS

Feedback Loops and Formative Measurement

Rex B. Kline

Life can only be understood backwards; but it must be lived forwards.
—Soren Kierkegaard (1843; quoted in Watkin, 2004, para. 31)

This chapter is about two types of special covariance structure models where some arrows (paths) point backwards compared with more standard models. The first kind is nonrecursive structural models with feedback loops where sets of variables are specified as causes and effects of each other in a cross-sectional design. An example of a feedback loop is the specification V1 ⇄ V2 where V1 is presumed to affect V2 and vice versa; that is, there is feedback. It is relatively easy to think of several real world causal processes, especially dynamic ones that may be based on cycles of mutual influence, including the relation between parental behavior and child behavior, rates of incarceration and crime rates, and violence on the part of protestors and police. In standard, or recursive, structural models

Structural Equation Modeling: A Second Course (2nd ed.), pages 41–79
Copyright © 2013 by Information Age Publishing

estimated in cross-sectional designs, all presumed causal effects are specified as unidirectional only so that no two variables are causes of each other.
Thus, it is impossible to estimate feedback effects when analyzing recursive structural models in cross-sectional designs. For example, a recursive
model for such a design could include either the path V1 → V2 or the path
V2 → V1, but not both.

The second type of special model considered in this chapter are formative measurement models—also known as emergent variable systems—
where observed variables (indicators) are specified as causes of underlying
factors (e.g., V1 → F1). The latter are referred to as latent composites because they are represented as the consequence of their indicators plus error
variance. This directionality specification is reversed compared with standard, or reflective, measurement models where factors are conceptualized
as latent variables (constructs) that affect observed scores (e.g., F1 → V1)
plus measurement error. It is reflective measurement that is based on classical measurement theory, but there are certain research contexts where the
assumption that causality "flows" from factors to indicators is untenable.
The specification that measurement is formative instead of reflective may
be an option in such cases.

The specification of either nonrecursive structural models with feedback
loops or formative measurement models potentially extends the range of
hypotheses that can be tested in structural equation modeling (SEM).
There are special considerations in the analysis of both kinds of models,
however, and the failure to pay heed to these requirements may render the
results meaningless. Thus, the main goal of this presentation is to help readers make informed decisions about the estimation of reciprocal causation
in cross-sectional designs or the specification of formative measurement in
SEM. Specifically, assumptions of both nonrecursive structural models with
feedback loops and formative measurement models are explained, and an
example of the analysis of each type of model just mentioned is described.
Readers can download all syntax, data, and output files for these examples
for three different SEM computer tools, EQS, LISREL, and Mplus, from a
freely-accessible Web page.[1] The computer syntax and output files are all
plain-text (ASCII) files that require nothing more than a basic text editor,
such as Notepad in Microsoft Windows, to view their contents. For readers
who use programs other than EQS, LISREL, or Mplus, it is still worthwhile
to view these files because (1) there are common principles about programming that apply across different SEM computer tools and (2) it can
be helpful to view the same analysis from somewhat different perspectives.
Contents of EQS, LISREL, and Mplus syntax files only for both examples
are listed in chapter appendices.

NONRECURSIVE MODELS WITH FEEDBACK LOOPS

There are two different kinds of feedback loops that can be specified in a nonrecursive structural model, direct and indirect. The most common are direct feedback loops where just two variables are specified as reciprocally affecting one another, such as V1 $\rightleftarrows$ V2. Indirect feedback loops involve three or more variables that make up a presumed unidirectional cycle of influence. In a path diagram, an indirect feedback loop with V1, V2, and V3 would be represented as a "triangle" with paths that connect them in the order specified by the researcher. Shown without error terms or other variables in the model, an example of an indirect feedback loop with three variables is:

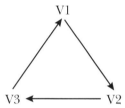

Because each variable in the feedback loop just illustrated serves as a mediator for the other two variables (e.g., V2 mediates the effect of V1 on V3, and so on), feedback is thus indirect. A structural model with an indirect feedback loop is automatically nonrecursive. Both direct and indirect feedback effects are estimated among variables measured concurrently instead of at different times; that is, the design is cross-sectional, not longitudinal. Although there are theoretical models that would seem to map well onto indirect feedback loops—for instance, Carson (1982) described maladaptive self-fulfilling prophesies in psychopathology as an "unbroken causal loop between social perception, behavioral enactment, and environmental reaction" (p. 576)—there are few reports of the estimation of models with indirect loops in the behavioral science literature. This is probably due to technical challenges in the analysis of such models. These same obstacles (elaborated later) also apply to models with direct feedback loops, but they are somewhat less vexing when there are only two variables involved in cycles of presumed mutual causation. Accordingly, only nonrecursive models with direct feedback loops are considered next.

Estimating reciprocal causality in a cross-sectional design by analyzing a nonrecursive model with a feedback loop can be understood as a proxy for estimating such effects in a longitudinal design by analyzing a cross-lagged panel model (Wong & Law, 1999). A feedback loop between V1 and V2 is presented in Figure 2.1a without disturbances or other variables. Note that (1) Figure 2.1a represents a fragment within a larger nonrecursive model and (2) it is not generally possible to estimate direct feedback without

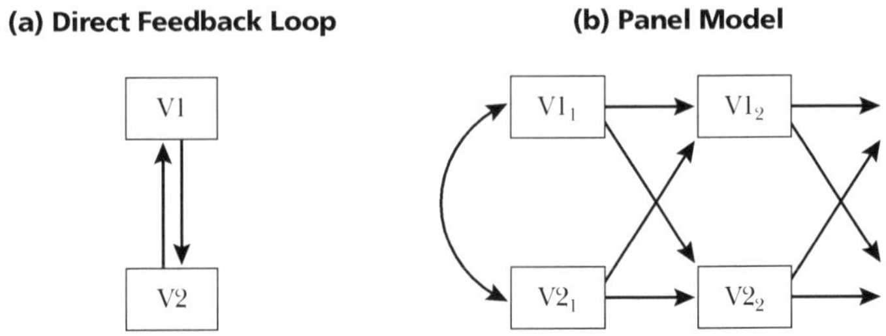

(a) Direct Feedback Loop　　　　　　**(b) Panel Model**

Figure 2.1 Reciprocal causal effects between V1 and V2 represented with (a) a direct feedback loop based on a cross-sectional design and (b) cross-lagged effects based on a longitudinal design (panel model), both models shown without disturbances or other variables.

other variables in the model due to identification. Variables V1 and V2 in Figure 2.1a are measured concurrently in a cross-sectional design, which implies the absence of temporal precedence, or the measurement of a presumed cause before that of the presumed effect. Evidence for reciprocal causation is indicated when estimates of both direct effects in a feedback loop, or V1 → V2 and V2 → V1 in Figure 2.1a, are of appreciable magnitude, given the research problem.

Presented in Figure 2.1b is a fragment from a cross-lagged panel model shown without disturbances or other variables and where V1 and V2 are each measured at different times in a longitudinal design. For example, V1 is measured at time 1 and again at time 2 designated by, respectively, $V1_1$ and $V1_2$ in Figure 2.1b; variable V2 is likewise measured twice ($V2_1$ and $V2_2$). The possibility to collect additional measurements of V1 and V2 at later times is also represented in Figure 2.1b. Presumed reciprocal causation is represented in the panel model of Figure 2.1b by the cross-lagged direct effects between V1 and V2 measured at different times, specifically, $V1_1$ → $V2_2$ and $V2_1$ → $V1_2$. Evidence for reciprocal causation is indicated when estimates of both cross-lagged direct effects in a panel model are of appreciable magnitude, again depending on the research area.[2]

Although it is theoretically and mathematically possible to estimate reciprocal causation in cross-sectional designs when analyzing nonrecursive models with direct feedback loops, there is controversy about the adequacy of those estimates (Wong & Law, 1999). The main reason is the absence of temporal precedence in cross-sectional designs. Causal effects are typically understood as taking place within a finite period, that is, there is a latency or lag between changes in causal variables and subsequent changes in outcome variables. With concurrent measurement there are no lags whatsoever, so in this sense the measurement occasions in cross-sectional designs are always wrong. This implies that the two-way paths that make up a direct feedback loop, such as V1 $\rightleftarrows$ V2 in Figure 2.1a, represent an instantaneous cycling process, but in reality there is no such causal mechanism (Hunter & Gerbing, 1982).[3] This may be true, but a feedback can still be viewed as a proxy or statistical model for a longitudinal process where causal effects occur within some definite latency. We do not expect statistical models to exactly mirror the inner workings of a complex reality. Instead, a statistical model is an approximation tool that helps researchers to structure their thinking (i.e., generate good ideas) in order to make sense of a phenomenon of interest (Humphreys, 2003). If the approximation is too coarse, then the model will be rejected.

In contrast to models with feedback for cross-sectional data, definite latencies (lags) for presumed causal effects are explicitly represented in panel models by the measurement periods in a longitudinal design. For example, variables $V1_1$ and $V2_2$ in Figure 2.1b represent the measurement of V1 at time 1 and V2 at some later time 2. This temporal precedence in measurement is consistent with the interpretation that the path $V1_1 \rightarrow V2_2$ corresponds to a causal effect. But it is critical to correctly specify the lag interval when estimating cross-lagged direct effects in longitudinal designs. This is because even if V1 actually causes V2, the observed magnitude of the direct effect $V1_1 \rightarrow V2_2$ may be too low if the measurement interval is either too short (causal effects take time to materialize) or too long (temporary effects have dissipated). The same requirement for a correct measurement interval applies to the other cross-lagged direct effect in Figure 2.1b, $V2_1 \rightarrow V1_2$.

The absence of time precedence in cross-sectional designs may not always be a liability in the estimation of reciprocal causation. Finkel (1995) argued that the lag for some causal effects is so short that it would be impractical to measure them over time. An example is reciprocal effects of the moods of spouses on each other. Although the causal lags in this example are not zero, they may be so short as to be virtually synchronous. If so, then the assumption of instantaneous cycling for feedback loops in nonrecursive designs would not be indefensible. Indeed, it may even be more appropriate to estimate reciprocal effects with very short lags in a cross-sectional design even when longitudinal panel data are available (Wong & Law, 1999). The true length of causal lags is not always known. In this case, longitudinal data collected according to some particular measurement schedule are not automatically superior to cross-sectional data. There is also the reality that longitudinal designs require more resources than cross-sectional designs. For many researchers, the estimation of reciprocal causation between variables measured simultaneously is the only viable alternative to a longitudinal design.

Identification Requirements of Models with Feedback Loops

Presented in Figure 2.2a is the most basic type of nonrecursive structural model with a direct feedback loop that is identified. This model includes observed variables only (i.e., it is path model), but the same basic configuration is required when variables in the structural model are latent each measured with multiple indicators (i.e., standard reflective measurement) and the whole model is a latent variable path model, not a measured variable path model (Kline, 2010, chap. 6). The two-headed curved arrows that exit and re-enter the same variable (⌒) in Figure 2.2a represent the variances of exogenous variables, which are generally free model parameters in SEM. The disturbances of the two variables that make up the feedback loop in Figure 2.2a, V3 and V4, are assumed to covary (D3 ⌣ D4). This specification makes sense for two reasons: (1) if variables are presumed to mutually cause one another, then it is plausible that there are common omitted causes of both; (2) some of the errors in predicting one variable in a direct feedback loop, such as V3 in Figure 2.2a, are due to the other vari-

(a) All possible disturbance correlations

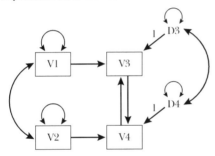

(b) All possible disturbance correlations within recursively related blocks

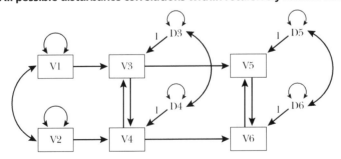

Figure 2.2 Two examples of nonrecursive path models with feedback loops.

able in that loop, or V4 in the figure, and vice versa (Schaubroeck, 1990). Although there is no requirement for correlated disturbances for variables involved in feedback loops, the presence of disturbance correlations in particular patterns in nonrecursive models helps to determine their identification status, a point elaborated next.

The model in Figure 2.2a satisfies the two necessary identification requirements for any type of structural equation model: (1) all latent variables are assigned a metric (scale), and (2) the number of free model parameters is less than or equal to the number of observations (i.e., the model degrees of freedom, df_M, are at least zero). In path models, disturbances can be viewed as latent (unmeasured) exogenous variables each of which requires a scale. The numerals (1) that appear in Figure 2.2a next to paths that point from disturbances to corresponding endogenous variables, such as D3 → V3, are scaling constants that assign to the disturbances a scale related that of the unexplained variance of its endogenous variable. With four (4) observed variables in Figure 2.2a, there are a total of $4(5)/2 = 10$ observations, which equals the number of variances and unique covariances among the four variables. The number of free parameters for the model in Figure 2.2a is 10, which includes the variances of four exogenous variables (of V1, V2, D3, and D4), two covariances between pairs of exogenous variables (V1 ⌣ V2, D3 ⌣ D4), and four direct effects, or

$$V1 \to V3, V2 \to V4, V3 \to V4, \text{ and } V4 \to V3.$$

Because $df_M = 0$ (i.e., 10 observations, 10 free parameters), the model in Figure 2.2a would, if actually identified, perfectly fit the data (i.e., the observed and predicted covariance matrices are identical), which means that no particular hypothesis would be tested in the analysis. It is usually only more complex nonrecursive models with positive degrees of freedom ($df_M > 0$)—which allow for the possibility of model-data discrepancies—that are analyzed in actual studies. But the basic pattern of direct effects in Figure 2.2a from external variables (those not in the feedback loop, or V1 and V2) to variables involved in the feedback loop (V3, V4) respect both of the requirements for identifying certain kinds of nonrecursive models outlined next.

If all latent variables are scaled and $df_M \geq 0$, then any recursive structural model is identified (e.g., Bollen, 1989, pp. 95–98). This characteristic of recursive structural models simplifies their analysis. Cross-lagged panel models are typically recursive, but they can be nonrecursive depending on their pattern of disturbance covariances (if any). Unfortunately, the case for nonrecursive models is more complicated. There are algebraic means to determine whether a nonrecursive model is identified (Berry, 1984), but these methods are practical only for very simple models. But there are two heuristics (rules) that involve determining whether a nonrecursive model meets certain requirements for identification that are straightforward to apply (Kline, 2010). These rules assume that $df_M \geq 0$ and all latent variables are properly scaled in a nonrecursive model. The first rule is for the order condition, which is a necessary requirement for identification. This means that satisfying the order condition does not guarantee identification, but failing to meet this condition says that the model is not identified. The second rule is for the rank condition, which is a sufficient requirement for identification, so a model that meets this requirement is in fact identified.

The heuristics for the order condition and the rank condition apply to either (1) nonrecursive models with feedback loops and all possible disturbance covariances or (2) nonrecursive models with ≥ 2 pairs of feedback loops with unidirectional paths between the loops but all possible disturbance covariances within each loop. The latter models are referred to as block recursive by some authors even though the whole model is nonrecursive. Consider the two nonrecursive path models in Figure 2.2. For both models, $df_M \geq 0$ and every latent variable is scaled, but these facts are not sufficient to identify either model. The model of Figure 2.2a has a direct feedback loop that involves V3–V4 and all possible disturbance correlations (1), or D3 ↶ D4. The model of Figure 2.2b is a block recursive model with two direct feedback loops, one that involves V3 and V4 and another loop made up of V5 and V6. Each block of these variable pairs in Figure 2.2b contains all possible disturbance correlations—D3 ↶ D4 for the first block, D5 ↶ D6 for the second—but the disturbances across the blocks are independent (e.g., D3 is uncorrelated with D5). The pattern of direct effects within each block in Figure 2.2b are nonrecursive (e.g., V3 ⇄ V4), but effects between the blocks are unidirectional (recursive) (e.g., V3 → V5). Thus, the two blocks of endogenous variables in the model of Figure 2.2b are recursively related to each other even though the whole model is nonrecursive.

Order condition. The order condition is a counting rule applied to each variable in a feedback loop. If the order condition is not satisfied, the equation for that variable is under-identified, which implies that the whole model is not identified. One evaluates the order condition by tallying the number of variables in the structural model (except disturbances) that have direct effects on each variable in a feedback loop versus the number that do not; the latter are excluded variables. The order condition requires for models with all possible disturbance correlations that the number of variables excluded from the equation of each variable in a feedback loop exceeds the total number of endogenous variables in the whole model minus 1. For example, the model in Figure 2.2a with all possible disturbance correlations has two variables in a feedback loop, V3 and V4. These two variables are the only endogenous variables in the model, so the total number of endogenous variables is 2. Therefore, variables V3 and V4 in Figure 2.2a must each have a minimum of $2 - 1 = 1$ other variable excluded from each of their equations, which is here true: V1 is excluded from the equation for V4, and V2 is excluded from the equation for V3. Because there is 1 variable excluded from the equation of each endogenous variable in Figure 2.2a, the order condition is satisfied.

The order condition is evaluated separately for each feedback loop in a block recursive model. For example, variables V3 and V4 are involved in the first feedback loop in the block recursive model of Figure 2.2b. For the moment we ignore the second feedback loop comprised of V5 and V6. The total number of endogenous variables for the first feedback loop is 2, or V3 and V4. This means that at least $2 - 1 = 1$ variables must be excluded for each of V3 and V4, which is true here: V1 is omitted from the equation for V4, and V2 is the excluded from the equation for V3. But the total number of endogenous variables for the second feedback loop comprised of V5 and V6 in Figure 2.2b is 4, or V3 through V6. This is because the first feedback loop with V3 and V4 in Figure 2.2b is recursively related to the second feedback loop with V5 and V6. Therefore, the order condition requires a minimum of $4 - 1 = 3$ excluded variables for each of V5 and V6, which is true here: V1, V2, and V4 are omitted from the equation for V5, and V1, V2, and V3 are excluded from the equation for V6. Thus, the block recursive model of Figure 2.2b satisfies the order condition.

Rank condition. Because the order condition is only necessary, it is still uncertain whether the models in Figure 2.2 are actually identified. Evaluation of the sufficient rank condition, however, will provide the answer. The rank condition is usually described in the SEM literature in matrix terms (e.g., Bollen, 1989, pp. 98–103). Berry (1984) devised an algorithm for checking the rank condition that does not require extensive knowledge of matrix operations. A non-technical description follows. The rank condition can be viewed as a requirement that each variable in a feedback loop has a unique pattern of direct effects on it from variables outside the loop. Such a pattern of direct effects provides a "statistical anchor" so that the free parameters of variables involved in feedback loops, including

$$V3 \rightarrow V4, V4 \rightarrow V3, Var(D3), Var(D4), \text{ and } D3 \smile D4$$

for the model in Figure 2.2a where "Var" means "variance," can be estimated distinctly from one another. Each of the two endogenous variables in Figure 2.2a has a unique pattern of direct effects on it from variables external to the feedback loop; that is, $V1 \rightarrow V3$ and $V2 \rightarrow V4$. If all direct effects on both endogenous variables in a direct feedback loop are from the same set of external variables, then the whole nonrecursive model would fail the rank condition.

Because the analogy just described does not hold for nonrecursive structural models that do not have feedback loops (e.g., Kline, 2010, pp. 106–

108), a more formal means of evaluating the rank condition is needed. Such a method that does not require knowledge of matrix algebra is described in Kline (2010). The application of this method to the model in Figure 2.2a is summarized in Appendix A of this chapter. The outcome of checking the rank condition is the conclusion that the model in Figure 2.2a is identified; specifically, it is just-identified because $df_M = 0$. See Kline (2010, p. 153) for a demonstration that the block recursive model in Figure 2.2b also meets the rank condition. Thus, the model is identified, specifically, it is also just-identified because $df_M = 0$. See Rigdon (1995), who described a graphical technique for evaluating whether nonrecursive models with direct feedback loops are identified and also Eusebi (2008), who described a graphical counterpart of the rank condition that requires knowledge of undirected, directed, and directed acyclic graphs from graphical models theory.

There may be no sufficient conditions that are straightforward to apply in order to check the identification status of nonrecursive structural models with disturbance covariances (including none) that do not match the two patterns described earlier. There are some empirical checks, however, that can be conducted to evaluate the uniqueness of a converged solution for such models (see Bollen, 1989, pp. 246–251; Kline, 2010, p. 233). These tests are only necessary conditions for identification. That is, a solution that passes them is not guaranteed to be unique.

Technical problems in the analysis are common even for nonrecursive models proven to be identified. For example, iterative estimation of models with feedback loops may fail to converge unless start values are quite accurate. This is especially true for the direct effects and disturbance variances and covariances of endogenous variables involved in feedback loops; see Kline (2010, pp. 172–175, 185) for an example of how to specify start values for a feedback loop. Even if estimation converges, the solution may be inadmissible, that is, it contains Heywood cases such as a negative variance estimate, an estimated correlation with an absolute value > 1.00, or other kinds of illogical results such as an estimated standard error that is so large that no interpretation is reasonable. Possible causes of Heywood cases include specification errors, nonidentification of the model, bad start values, the presence of outliers that distort the solution, or a combination of a small sample size and only two indicators when analyzing latent variable models (Chen, Bollen, Paxton, Curran, & Kirby, 2001).

Special Assumptions of Models with Feedback Loops

Estimation of reciprocal effects by analyzing a nonrecursive model with a feedback loop is based on two special assumptions. These assumptions are required because data from a cross-sectional design give only a "snapshot" of an ongoing dynamical process. One is that of *equilibrium*, which means that any changes in the system underlying a feedback relation have already manifested their effects and that the system is in a steady state. That is, the values of the estimates of the direct effects that make up the feedback loop do not depend upon the particular time point of data collection. Heise (1975) described equilibrium this way: it means that a dynamical system has completed its cycles of response to a set of inputs and that the inputs do not vary over time. This implies that the causal process has basically dampened out and is not just beginning (Kenny, 1979). The other assumption is that of *stationarity*, or the requirement that the underlying causal structure does not change over time.

It is important to know that there is generally no statistical way in a cross-sectional design to verify whether the assumptions of equilibrium or stationarity are tenable. Instead, these assumptions must be argued substantively based on the researcher's knowledge of theory and empirical results in a particular area. However, these special assumptions are rarely acknowledged in the literature where feedback effects are estimated with cross-sectional data in SEM (Kaplan, Harik, & Hotchkiss, 2001). This is unfortunate because results of two computer simulation studies summarized next indicate that violation of the special assumptions can lead to severely biased estimates of the direct effects in feedback loops:

1. Kaplan et al. (2001) specified a true dynamical system that was perturbed at some earlier time and was headed toward but had not attained yet equilibrium. Computer-generated data sets simulated the results of cross-sectional studies conducted at different numbers of cycles before equilibrium was reached. Estimates of direct effects in feedback loops varied widely depending upon when the simulated data were collected. This variation was greatest for dynamical systems in oscillation where the signs of direct effects swung from positive to negative and back again as the system moved back toward stability.

2. Wong and Law (1999) specified a cross-lagged panel model with reciprocal effects between two variables as the population (true) model and generated simulated cross-sectional samples in which nonrecursive models with a single feedback loop were analyzed under various conditions. They found that the adequacy of estimates from cross-sectional samples was better when (a) there was greater temporal stability in the cross-lagged effects (i.e., there is stationarity and equilibrium); (b) the nonrecursive model had a disturbance covariance as opposed to uncorrelated disturbances; (c) the sample size was larger (e.g., $N = 500$) rather than smaller (e.g., $N = 200$); and (d) the relative magnitudes of direct effects from prior variables

on variables in the feedback loop (e.g., V1 → V3 and V2 → V4 in Figure 2.2a) were approximately equal instead of disproportionate.

Example Analysis of a Nonrecursive Model with a Feedback Loop

Within a sample of 177 nurses out of 30 similar hospitals in Taiwan, H.-T. Chang, Chi, and Miao (2007) administered measures of occupational commitment (i.e., to the nursing profession) and organizational commitment (i.e., to the hospital that employs the nurse). Each commitment measure consisted of three scales, affective (degree of emotional attachment), continuance (perceived cost of leaving), and normative (feeling of obligation to stay). The affective, continuance, and normative aspects are part of a three-component theoretical and empirical model of commitment by Myer, Allen, and Smith (1993). Results of some studies reviewed by H.-T. Chang et al. (2007) indicate that commitment predicts turnover intentions concerning workers' careers (occupational turnover intentions) and place of employment (organizational turnover intentions). That is, workers who report low levels of organizational commitment are more likely to seek jobs in different organizations but in the same field, and workers with low occupational commitment are more likely to change careers altogether. H.-T. Chang et. al (2007) also predicted that organizational turnover intention and occupational turnover intention are reciprocally related: plans to change one's career may prompt leaving a particular organization, and vice-versa. Accordingly, H.-T. Chang et al. (2007) also administered measures of occupational turnover intention and of organizational turnover intention to the nurses in their sample. Presented in Table 2.1 are the correlations, standard deviations, and internal consistency (Cronbach's alpha) score reliability coefficients for all eight measures analyzed by H.-T. Chang et al. (2007).

H.T. Chang et al. (2007) hypothesized that (1) organizational turnover intention and occupational turnover intention are reciprocally related, that is, the intention to leave a place of employment affects the intention to remain in a profession, and vice versa; (2) the three components of organizational commitment (affective, continuance, normative) directly affect organizational turnover intention; and (3) the three components of occupational commitment directly affect occupational turnover intention. A nonrecursive path model that represents these hypotheses would consist of a direct feedback loop between organizational turnover intention and occupational turnover intention, direct effects of the three organizational commitment variables on organizational turnover intention, and direct effects of the three occupational commitment variables on occupational turnover intention. However, path analysis of a structural model of observed

TABLE 2.1. Input Data (Correlations, Standard Deviations, Score Reliabilities) for Analysis of a Nonrecursive Model of Organizational and Occupational Commitment and Turnover

Variable	1	2	3	4	5	6	7	8
Organizational commitment								
1. Affective	.82							
2. Continuance	−.10	.70						
3. Normative	.66	.10	.74					
Occupational commitment								
4. Affective	.48	.06	.42	.86				
5. Continuance	.08	.58	.15	.22	.71			
6. Normative	.48	.12	.44	.69	.34	.84		
Turnover intention								
7. Organizational	−.53	−.04	−.58	−.34	−.13	−.34	.86	
8. Occupational	−.50	−.02	−.40	−.63	−.28	−.58	.56	.88
SD	1.04	.98	.97	1.07	.78	1.09	1.40	1.50

Note: These data are from H.-T. Chang et al. (2007); $N = 177$. Values in the diagonal are internal consistency (Cronbach's alpha) score reliability coefficients.

variables does not permit the explicit estimation of measurement error in either exogenous variables (commitment) or endogenous variables (turnover intention). Fortunately, the availability of score reliability coefficients for these data (Table 2.1) allows specification of the latent variable model in Figure 2.3 with the features described next.

Each observed variable in Figure 2.3 is represented as the single indicator of an underlying factor. The unstandardized loading of each indicator is fixed to 1 in order to scale the corresponding factor. Measurement error variance for each indicator is fixed to equal the product of the sample variance of that indicator, or s^2, and one minus the score reliability for that indicator, or $1 - r_{XX}$. The latter quantity estimates the proportion of observed variance due to measurement error. For instance, the score reliability coefficient for the affective organizational commitment measure is .82 and the sample variance is $1.04^2 = 1.0816$ (see Table 2.1). The product $(1 - .82)$ 1.0816, or .1947, is the amount of total observed variance due to measurement error. Accordingly, the error variance for the affective organizational commitment indicator is fixed to .1947 for the model in Figure 2.3.[4] Error variances for the remaining 7 indicators in the figure were calculated in a similar way. So specified, estimation of the direct effects and disturbance variances and covariance for the model in Figure 2.3 explicitly controls for measurement error in all observed variables (see Kline, 2010, pp. 276–280, for additional examples). To save space, all possible covariances between

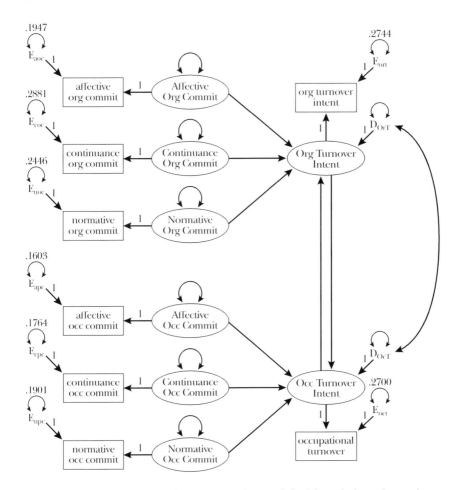

Figure 2.3 Initial latent variable nonrecursive model of the relation of organizational/occupational commitment to organizational/occupational turnover intention. Covariances among exogenous factors are omitted to save space. Unstandardized error variances for single indicators are fixed to the values indicated.

the 6 exogenous factors (15 altogether) are not shown in Figure 2.3, but they are specified in computer syntax for this analysis.

With eight observed variables, there are a total 8(9)/2, or 36 observations available to estimate the model's 32 free parameters, which include 8 variances (of 6 exogenous factors and 2 disturbances), 16 covariances (1 disturbance covariance and 15 exogenous factor covariances), and 8 direct effects, so df_M = 4. Each latent variable in the model of Figure 2.3 is assigned a scale, but we still do not know if the model is identified. Now looking only at the structural part of the model (i.e., the pattern of direct effects

among the 8 factors), there are two endogenous factors, so a minimum of
$2 - 1 = 1$ exogenous factor must be excluded from each equation for Occupational Turnover Intention and for Organizational Turnover Intention,
which is true (see Figure 2.3). For example, the affective, continuance, and
normative organizational commitment factors are all excluded from the
equation for the Occupational Turnover Intention factor. Thus, the model
in Figure 2.3 meets the necessary order condition. It can also be shown by
applying the method described in Appendix A to the structural model in
Figure 2.3 that the sufficient order condition is also satisfied. This observation plus other features of the model (e.g., each observed variable loads on
a single factor, measurement error variances are fixed to equal the values
shown in Figure 2.3) indicate that the whole model is identified, specifically, it is over-identified because $df_M > 0$.

The model in Figure 2.3 was fitted to the covariance matrix based on
the correlations and standard deviations in Table 2.1 using maximum likelihood (ML) estimation in EQS, LISREL, and Mplus. The EQS program is-
sued warnings about singular (nonpositive definite) parameter matrices in
early iterations, but then "recovered" to generate a converged and admissible solution. Both LISREL and Mplus generated converged and admissible
solutions without incident. Values of fit statistics and parameter estimates
are very similar across the three programs, so results from EQS only are
summarized here. Values of the model chi-square, Bentler Comparative Fit
Index (CFI), Jöreskog-Sörbom Goodness of Fit Index (GFI), standardized
root mean square residual (SRMR), and Steiger-Lind root mean square error of approximation (RMSEA) with its 90% confidence interval calculated
are listed next:

$$\chi^2_M(4) = 9.177, \, p = .057$$

$$CFI = .999; \, GFI = .987; \, SRMR = .018$$

$$RMSEA = .086 \, (0 - .160)$$

The model just "passes" the chi-square test at the .05 level, and values of
the CFI, GFI, and SRMR are generally favorable. In contrast, the RMSEA results are poor, specifically, the upper bound of its confidence interval (.160)
suggests poor fit within the limits of sampling error. Although none of the
absolute correlation residuals (calculated in EQS) exceeded .10, there were
several statistically significant standardized residuals (calculated by LISREL
and Mplus), including the residual for the association of the continuance
organizational commitment and occupational turnover intention variables.
A respecification consistent with this result is to add to the initial model in
Figure 2.3 a direct effect from continuance organizational commitment to
occupational turnover intention.

The model in Figure 2.3 was respecified by adding the direct effect just mentioned. All analyses in EQS, LISREL, and Mplus of the respecified model converged to admissible solutions that were all very similar. Listed in Appendices B–D is, respectively, EQS, LISREL, and Mplus syntax for this analysis. Values of selected fit statistics calculated by EQS for the respecified model are reported next:

$$\chi^2_M(3) = .814, \; p = .848$$

$$\text{CFI} = 1.000; \; \text{GFI} = .999; \; \text{SRMR} = .005$$

$$\text{RMSEA} = 0 \, (0 - .071)$$

The results just listed are all favorable. Also, the largest absolute correlation is .02, and there are no statistically significant standardized residuals for the revised model. The improvement in fit of the revised model relative to that of the initial model is also statistically significant, or

$$\chi^2_D(1) = 9.177 - .814 = 8.363, \; p = .004.$$

Based on all these results concerning model fit (Kline, 2010, chap. 8), the respecified nonrecursive model with a direct effect from continuance organizational commitment to occupational turnover intention is retained.

Reported in Table 2.2 are the ML estimates of the direct effects and the disturbance variances and covariance for the revised model. The best predictor of organizational turnover intention is normative organizational commitment. The standardized path coefficient is –.662, so a stronger sense of obligation to stay within the organization predicts lower levels of the intention to leave that organization. In contrast, the best predictor of occupational turnover intention is continuance occupational commitment, for which the standardized path coefficient is –.650. That is, higher estimation of the costs associated with leaving a discipline predicts a lower level of intention to leave that discipline. The second strongest predictor of occupational turnover intention is continuance organizational commitment and, surprisingly, the standardized path coefficient for this predictor is positive, or .564. That is, higher perceived costs of leaving an organization predicts a higher level of intention to leave the discipline. This result is an example of a suppression effect because although the Pearson correlation between continuance organizational commitment and occupational turnover intention is about zero (–.02; see Table 2.1), the standardized weight for the former is positive (.564) once other predictors are held constant. See Maasen and Bakker (2001) and Kline (2010, pp. 26–27, 160–172) for more information about suppression effects in structural models.

TABLE 2.2. Maximum Likelihood Estimates for a Nonrecursive Model of Organizational and Occupational Commitment and Turnover Intention

Parameter	Unstandardized	SE	Standardized
Direct effects			
AOC → OrgTO	−.065	.408	−.047
COC → OrgTO	.029	.173	.018
NOC → OrgTO	−1.028**	.403	−.662
OccTO → OrgTO	.037	.146	.040
APC → OccTO	−.727**	.216	−.513
CPC → OccTO	−1.395*	.626	−.650
NPC → OccTO	.051	.288	.036
OrgTO → OccTO	.281*	.143	.259
COC → OccTO	.969*	.440	.564
Disturbance variances and covariance			
OrgTO	.774**	.192	.459
OccTO	.472	.156	.236
OrgTO ⌣ OccTO	.231	.178	.382

Note: AOC, affective organizational commitment; COC, continuance organizational commitment; NOC, normative organizational commitment; APC, affective occupational commitment; CPC, continuance occupational commitment; NPC, normative occupational commitment; OrgTO, organizational turnover intention; OccTO, occupational turnover intention. Standardized estimates for disturbance variances are calculated as 1 minus the Bentler-Raykov corrected R^2.
* $p < .05$; ** $p < .01$.

As expected, the direct effects of organizational turnover intention and occupational turnover intention on each other are positive. Of the two, the standardized magnitude of the effect of organizational turnover on occupational turnover (.259) is stronger than the effect in the other direction (.040). If there is reciprocal causation between these two variables, then it is stronger in one direction than in the other. The EQS program prints values of the Bentler-Raykov corrected R^2, which controls for model-implied correlations between predictors and disturbances for variables in feedback loops (Kline, 2010, pp. 187–188). These corrected R^2 values are listed next: Organizational Turnover Intention factor, .541; Occupational Turnover Intention factor, .764. The prediction of occupational turnover intention is thus somewhat better than the prediction of organizational turnover intention by all prior variables of each endogenous factor in Figure 2.3.

FORMATIVE MEASUREMENT MODELS

A reflective measurement model for a single latent variable F1 with three indicators V1–V3 is presented in Figure 2.4a. This single-factor model is

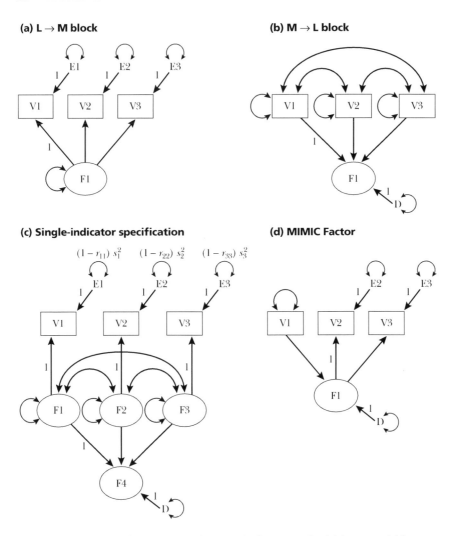

Figure 2.4   Directionalities of effects between indicators and a (a) latent variable and (b) latent composite. (c) Single-indicator specification of cause indicators. (d) A MIMIC factor. L, latent; M, manifest; MIMIC, multiple indicators and multiple causes.

identified but has no degrees of freedom ($df_M = 0$), but with ≥ 4 effect indicators it would be possible to test the fit of a single-factor reflective measurement model to the data in confirmatory factor analysis (CFA) because $df_M \geq 0$ in this case. The general goal of CFA is to explain (model) the observed covariances among the indicators. Specifically, CFA minimizes the differences between the observed covariances among the indicators and those predicted by a reflective measurement model. Because the direct effects in Figure 2.4a point from F1 to V1–V3, the latter are referred as reflective indicators or effect indicators. Grace and Bollen (2008) used the term L → M block (latent to manifest) to describe the association between factors and their effect indicators in reflective measurement models. Measurement error in reflective measurement models is represented at the indicator level as represented by the terms E1–E3 in Figure 2.4a. Reflective measurement is based on the domain sampling model (Nunnally & Bernstein, 1994, chap. 6). From this perspective, a set of effect indicators of the same factor should be internally consistent, which means that their intercorrelations should be positive and at least moderately high in magnitude; otherwise, estimates of the reliability of factor measurement (Hancock & Mueller, 2001) may be low. Sometimes intercorrelations are both positive and negative among indicators with a mix of positively- versus negatively-word items. For example, the item "My health is good" is positively worded, but the item "I am often worry about my health" is negatively worded. Participants who respond "true" to the first item may respond "false" to the second, and vice versa. In this case, the technique of reverse coding could be used to change the scoring of negatively-worded items to match that of positively worded items (or vice versa) so that intercorrelations among recoded items are all positive. The domain sampling model also assumes that effect indicators of the same construct with equal score reliabilities are interchangeable. This implies that indicators can be substituted for one another without appreciably affecting construct measurement.

The assumption that indicators are caused by factors is not always appropriate. Some indicators are viewed as cause indicators or formative indicators that affect a factor instead of the reverse. Consider this example by Bollen and Lennox (1991): The variables income, education, and occupation are used to measure socioeconomic status (SES). In a reflective measurement model, these observed variables would be specified as effect indicators that are caused by an underlying SES factor. But we usually think of SES as the outcome of these variables (and others), not vice versa. For example, a change in any one of these indicators, such as a salary increase, may affect SES. From the perspective of formative measurement, SES is a composite or index variable that is caused by its indicators. If it is assumed that SES is also caused by other, unmeasured variables besides income, education, and occupation, then SES would be conceptualized as a latent composite, which in a formative measurement model has a disturbance. Cause indicators are not generally interchangeable. This is because removal of a cause indicator is akin to removing a part of the underlying factor (Bollen & Lennox, 1991). Cause indicators may have any pattern of intercorrelations, including ones that are basically zero. This is because composites reflect the contribution of multiple dimensions through its cause indicators. That is, composites are not unidimensional, unlike latent variables in reflective measurement models. There are many examples of the analysis of composites in economics and business research (Diamantopoulos, Riefler, & Roth, 2005).

Presented in Figure 2.4b is a formative measurement model for a single latent composite F1 with three cause indicators V1–V3. It depicts an M → L block (manifest to latent) because F1 is assumed to be caused by its indicators. With no disturbance, F1 in Figure 2.4b would be just a linear combination of its cause indicators and thus not latent.[5] To scale the latent composite, the unstandardized direct effect of one of its cause indicators, V1, is fixed to 1.0. Cause indicators in formative measurement models are exogenous variables that are free to vary and covary in any pattern. Also, they have no measurement errors in the standard representation of formative measurement. Thus, it is assumed for the model in Figure 2.4b that cause indicators V1–V3 have perfect score reliabilities (i.e., $r_{XX} = 1.00$). The assumption of perfect score reliability is unrealistic for most observed variables. Consequently, measurement error in cause indicators is manifested in the disturbance term of the latent composite (D in Figure 2.4b). This means that measurement error in formative measurement is represented at the construct level, not at the indicator level as in reflective measurement. In other words, unreliability in formative indicators tends to increase the disturbance variance of the corresponding latent composite. Unlike a reflective measurement model, a formative measurement model does not explain the variances and covariances of the indicators. Instead, the formative indicators are specified as predictors of their corresponding latent composites, and the computer is asked to estimate regression coefficients for the indicators and the proportions of explained variance in the composites, not the indicators. Note that the formative measurement model in Figure 2.4b is not identified. In order to estimate its parameters, it would be necessary to embed it in a larger model. Identification requirements of formative measurement models are considered later.

There is an alternative to assuming that cause indicators of latent composites have perfect score reliabilities as in Figure 2.4b. This alternative is illustrated by the model in Figure 2.4c where

1. each cause indicator is specified as the single indicator of an underlying exogenous factor, such as F1 → V1;
2. the measurement error variance of each indicator is fixed to equal the product of one minus the score reliability and the observed variance, such as $(1 - r_{11})s_1^2$ where r_{11} is a reliability coefficient and s_1^2 is the sample variance for indicator V1;
3. factors F1–F3 are free to vary and covary; and
4. factors F1–F3 have direct effects on the latent composite F4.

The observed variables V1–V3 in Figure 2.4c are no longer technically cause indicators (they are effect indicators), but the corresponding factors F1–F3 are the cause indicators of the latent composite F4. Estimation of the disturbance term D in Figure 2.4c would control for measurement error in the cause indicators V1–V3, if this model were embedded in a larger structural equation model and corresponding identification requirements were met. Specifically, measurement error in the cause indicators V1–V3 of Figure 2.4c is estimated separately from the disturbance variance for the latent composite. This means that the disturbance in Figure 2.4c reflects only unexplained variance in the latent composite F4. In contrast, the disturbance of the latent composite F1 in Figure 2.4b reflects both unexplained variance and measurement error in V1–V3.

There is a kind of "compromise" between specifying that the indicators are either all effect indicators or all causal indicators. It is achieved by specifying a MIMIC (multiple indicators and multiple causes) factor with both effect indicators and cause indicators. A MIMIC factor with a single cause indicator V1 and two effect indicators V2–V3 is presented in Figure 2.4d. A MIMIC factor is endogenous, which explains why F1 has a disturbance. Note in the figure that effect indicators V2 and V3 have measurement errors (E2, E3), but cause indicator V1 does not. This means that measurement error in V2 and V3 is represented at the indicator level, but error in V1 is manifested at the construct level; specifically, in the disturbance of the MIMIC factor. The single-MIMIC-factor model in Figure 2.4d with 3 indicators is just-identified ($df_M = 0$). With more than 3 indicators, a measurement model with a single MIMIC factor and a single cause indicator is an equivalent version of a single-factor reflective measurement model where all the indicators are effect indicators (Kline, 2010, pp. 245–248). There are many examples in the SEM literature of the analysis MIMIC factors. For example, Hershberger (1994) described a MIMIC depression factor with indicators that represented various behaviors. Some of these indicators, such as "crying" and "feeling depressed," were specified as effect indicators because they are symptoms of depression. However, another indicator, "feeling lonely," was specified as a cause indicator. This is because "feeling lonely" may cause depression rather than vice versa.

Formative measurement and the analysis of composites is better known in areas such as economics, commerce, management, and biology than in psychology or education. For example, Grace (2006, chap. 6) and Grace and Bollen (2008) described the analysis of composites in the environmental sciences. There is a relatively recent special issue about formative measurement in the *Journal of Business Research* (Diamantopoulos, 2008). Jarvis, MacKenzie, and Podsakoff (2003) and others advised researchers in the consumer research area—and the rest of us, too—not to automatically specify factors with effect indicators only because doing so may result in specification error, perhaps due to lack of familiarity with formative measurement models. Familiarity with the idea of formative measurement makes researchers aware that there are options concerning the specification of directionalities of relations between latent variables and observed variables. This possibility should also prompt them to think long and hard about the nature of the constructs they are studying and appropriate means to measure them.

On the other hand, the specification of formative measurement is not a panacea. For example, because cause indicators are exogenous, their variances and covariances are not explained by a formative measurement model. This makes it more difficult to assess the validity of a set of cause indicators (Bollen, 1989), but the source just cited, Fayers and Hand (2002), and Diamantopoulos and Winklhofer (2001) offer some suggestions. The fact that error variance in formative measurement is represented at the construct level instead of at the indicator level as in reflective measurement is a related problem, but there is a way around this limitation (see Figure 2.4c). Howell, Breivik, and Wilcox (2007) noted that formative measurement models are more susceptible than reflective measurement models to interpretational confounding where values of direct effects of cause indicators on composites may change a great deal as the whole structural model is respecified. This means that the direct effects of cause indicators on their corresponding composites can change depending on which other variables are predicted by the composite. The absence of a nominal definition of a formative factor apart from the empirical values of loadings of its indicators exacerbates this potential problem. Having a reasoned operational definition of a concept, one based on extant theory and results of empirical studies, mitigates against interpretational confounding. That is, use your domain knowledge and good judgment when specifying measurement models. For these and other reasons, Howell et al. concluded that (1) formative measurement is not an equally attractive alternative to reflective measurement and (2) researchers should try to include reflective indicators whenever other indicators are specified as cause indicators of the same construct, but see Bagozzi (2007) and Bollen (2007) for other views.

Bollen and Bauldry (2011) described additional kinds of indicators beyond effect (reflective) indicators and cause (formative) indicators. Briefly, a composite indicator is a simple linear combination of observed variables. Such composites are not latent; instead, they are just total scores over a set of manifest variables. For the same reason, composite indicators do not have a disturbance. As their name suggests, composite indicators are specified in structural models as having direct effects on other variables, either observed or latent. Some authors, such as Grace and Bollen (2008), represent simple composites in model diagrams with hexagons. Although this is not a standard symbol, it does differentiate a composite from other observed variables. These same authors also distinguish between a fixed-weights composite where coefficients for the constituent observed variables are specified a priori (e.g., unit weighting) and an unknown weights composite where the weights are estimated with sample data. Bollen and Bauldry (2011) also distinguish covariates from cause indicators and composite indicators. Covariates are not measures of latent variables. Instead, they are variables thought to affect latent variables or their indicators the omission of which may bias the results. For example, women generally live longer than men, so gender may be included as a covariate in a model where survival time after a treatment is predicted. If gender covaries with other predictors of survival time, then the omission of this covariate would be a specification error. The best way for a researcher to decide whether an observed variable is an effect indicator, a cause indictor, or a covariate is to rely on his or her substantive knowledge (i.e., do your homework about measurement in the specification phase).

Identification Requirements of Models with Latent Composites

The main stumbling block to the analysis in SEM of measurement models where some factors have cause indicators only is identification. This is because it can be difficult to specify such a model that reflects the researcher's hypotheses and is identified. The need to scale a latent composite was mentioned, but meeting this requirement is not difficult (e.g., Figure 2.4b). In order for the disturbance variance of a latent composite to be identified, that composite must have direct effects on at least two other endogenous variables, such as endogenous latent variables with effect indicators only. This requirement is known as the 2+ emitted paths rule. If a factor measured with cause indicators only emits a single path, its disturbance variance will be underidentified. This means that it is impossible in SEM to analyze a model where a latent composite emits a single direct effect. Another requirement for models with two or more latent composites is that if factors measured with effect indicators only have indirect effects on other such factors that are mediated by latent composites, then some of the constituent direct effects may be underidentified (MacCallum & Browne, 1993).

One way to deal with the identification requirements just described is to fix the disturbance variance for the latent composite to zero, which drops the disturbance from the model and converts the latent composite to an observed composite, or just a total score across the cause indicators. However, this is not an ideal option. Recall that the disturbance of a latent composite reflects measurement error in its cause indicators. Dropping this disturbance basically forces the manifestation of measurement error in cause indicators to shift further "downstream" in the model, specifically, to the disturbances of endogenous variables on which the composite has direct effects. (This assumes that the cause indicators have no error terms; e.g., Figure 2.4b.) On the other hand, Grace (2006) argued that (1) error variance estimates for latent composites may have little theoretical significance in some contexts, and (2) the presence or absence of these error terms should not by itself drive decisions about the specification of composites in the analysis.

Another way to deal with identification problems is to add effect indicators for latent composites represented in the original model as measured by cause indicators only. That is, specify a MIMIC factor. For example, adding two effect indicators means that the formerly latent composite will emit at least two direct effects—see Diamantopoulos, Riefler, and K. Roth (2008) for examples. However, all such respecifications require a theoretical rationale. That is, a researcher should not switch the directionalities of effects between factors and indicators with the sole aim of obtaining an identified model.

Example Analysis of a Model with Risk as a Latent Composite

Worland, Weeks, Janes, and Strock (1984) administered measures of cognitive status (verbal reasoning) and scholastic achievement (reading, arithmetic, and spelling) within a sample of 158 adolescents. They also collected teacher reports about the classroom adjustment of these adolescents (motivation, harmony, and emotional stability) and measured family SES and the degree of parental psychiatric disturbance. For pedagogical reasons, I generated the hypothetical correlations and standard deviations reported in Table 2.3 to match the basic pattern of associations reported by Worland et al. (1984) among these nine variables.

Suppose that the construct of risk is conceptualized for this example as a latent composite with cause indicators family SES, parental psychopathology, and adolescent verbal IQ; that is, risk is determined by any combination of low family SES, parental psychiatric impairment, or low adolescent verbal IQ (plus error variance). Intercorrelations among these three variables are not all positive (see Table 2.3), but this is irrelevant for cause indicators. Presented in Figure 2.5 is a structural equation model where a latent risk composite has cause indicators only. Note in the figure that the risk composite emits two direct effects onto reflective factors (achievement, classroom adjustment) each measured with effect indicators only. This specification satisfies the 2+ emitted paths rule and identifies the disturbance variance for the risk composite. It also reflects the hypothesis that the association between adolescent achievement and classroom adjustment is spurious due to a common cause, risk. This assumption may not be plausible. For example,

TABLE 2.3. Input Data (Hypothetical Correlations and Standard Deviations) for Analysis of a Model with Risk as a Latent Composite

Variable	1	2	3	4	5	6	7	8	9
Risk									
1. Parental Psychiatric	1.00								
2. Low Family SES	.22	1.00							
3. Verbal IQ	−.43	−.49	1.00						
Achievement									
4. Reading	−.39	−.43	.58	1.00					
5. Arithmetic	−.24	−.37	.50	.73	1.00				
6. Spelling	−.31	−.33	.43	.78	.72	1.00			
Classroom adjustment									
7. Motivation	−.25	−.25	.39	.52	.53	.54	1.00		
8. Harmony	−.25	−.26	.41	.54	.43	.47	.77	1.00	
9. Stability	−.16	−.18	.31	.50	.46	.47	.60	.62	1.00
SD	13.00	13.50	13.10	12.50	13.50	14.20	9.50	11.10	8.70

Note: $N = 158$

achievement may directly affect classroom adjustment; specifically, students with better scholastic skills may be better adjusted at school. But including the direct effect just mentioned—or, alternatively, a disturbance covariance between the two reflective factors ($D_{Ac} \smile D_{CA}$)—would render the model in Figure 2.5 with a latent composite not identified.

There are a total of $9(10)/2$, or 45 observations available to estimate the 23 free parameters of the model in Figure 2.5. The latter include 12 variances (of 3 cause indicators, 3 disturbances, and 6 measurement errors), 3 covariances (among the cause indicators), 2 direct effects on the latent risk composite from its cause indicators, 2 direct effects emitted by the latent composite, and 4 factor loadings across the two reflective factors. Thus, $df_M = 45 - 23 = 22$ for this analysis. The model in Figure 2.5 was fitted to the covariance matrix based on the data in Table 2.3 using ML estimation in EQS, LISREL, and Mplus. The EQS syntax for this analysis is straightforward (Appendix B). However, it is trickier to specify a latent composite in LISREL SIMPLIS syntax or Mplus syntax. In LISREL, it is necessary to specify the "SO" option of the "LISREL Output" command (Appendix C). This option suppresses LISREL's automatic checking of the scale for each latent variable. If this option is omitted, LISREL will standardize the latent composite regardless of user specifications. A latent composite is specified in Mplus using its syntax for regression (keyword "on") instead of its syntax for factor loadings (keyword "by"), which is for reflective factors. However, use of the "by" keyword in Mplus is required to specify the direct effects emitted by a latent composite (Appendix D).

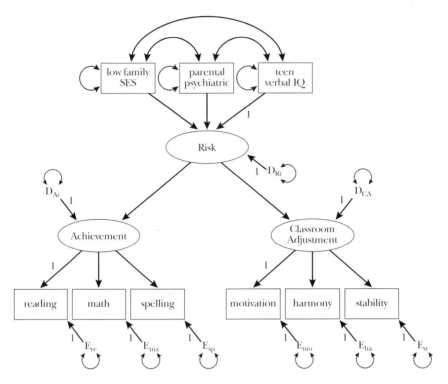

Figure 2.5 Initial identified model of risk as a latent composite with cause indicators only and two reflective constructs.

The analysis of models with latent composites may encounter technical difficulties. For example, the initial analysis of the model in Figure 2.5 failed in EQS because the default number of iterations (30) was reached before the solution converged. Increasing the number of iterations in EQS to 200 (see Appendix B) resulted in a converged and admissible solution. The first analysis in LISREL failed because the solution was found to be not admissible after 50 iterations. The error message suggested that setting "AD > 50" may solve the problem. This option of the LISREL Output command specifies the iteration number at which the admissibility of the solution is checked. Increasing this step in LISREL to number 100 (i.e., AD = 100; see Appendix C) resulted in a converged and admissible solution. The analysis in Mplus was uneventful and generated a converged and admissible solution. Values of fit statistics and parameter estimates were very similar across all three programs.

Values of selected fit statistics about the fit of the model in Figure 2.5 to the data in Table 2.3 calculated by LISREL are summarized next:

$$\chi_M^2(22) = 35.084, \; p = .038$$

$$CFI = .989; \; GFI = .953; \; SRMR = .031$$

$$RMSEA = .065 \, (.012 - .097)$$

The model "fails" the chi-square test, which indicates statistically significant discrepancies between observed and model-implied covariances. Values of the approximate fit indexes CFI, GFI, and SRMR are not bad, but the RMSEA results are marginal. No correlation residuals exceeded .10 in absolute value, but among larger positive residuals were ones for pairwise associations between indicators of achievements and indicators of classroom adjustment. This means that the model tends to under-predict these cross-factor correlations. Although only two standardized residuals were statistically significant, several others were almost so at the .05 level. Based on all these results, the model with the latent risk composite in Figure 2.5 is rejected.

As mentioned, the absence of a direct effect from the achievement factor to the classroom adjustment factor or a disturbance covariance between these two factors in Figure 2.5 may be a specification error, but adding either parameter to the model would make it not identified. Listed next are options to respecify the model in order to estimate either parameter just mentioned:

1. Respecify the latent risk composite as a MIMIC factor with at least one effect indicator, such as adolescent verbal IQ. This respecification makes sense because it assumes that family or parent variables (cause indicators) affect a characteristic of teenagers in those families (effect indicator) through a latent risk variable. After respecification of risk as a MIMIC factor, it can be demonstrated using the Lee-Hershberger replacing rules (Hershberger, 1994; see also Kline, 2010, pp. 225–288) that Model 1 with the direct effect

   Achievement → Classroom Adjustment

   but no disturbance covariance and Model 2 with the disturbance covariance $D_{Ac} \smile D_{CA}$ but no direct effect from achievement to classroom adjustment are equivalent. That is, Model 1 and Model 2 would have exactly the same fit to the data, so in practice they are not empirically distinguishable.
2. Drop the disturbance D_{Ri} from the model in Figure 2.5, which would convert the latent risk composite into a weighted combination of its cause indicators.

3. Drop the risk composite altogether from the model and replace it with direct effects from the three cause indicators in the original model to each of the two endogenous factors with effect indicators.

Analyzed next is a model with risk as a MIMIC factor with two cause indicators (low family SES, parental psychopathology) and one effect indicator (adolescent verbal IQ). Model 1 with a direct effect from the achievement factor to the classroom adjustment factor but no disturbance covariance was fitted to the data in Table 2.3 using ML estimation in EQS, LISREL, and Mplus. No special syntax is needed for this respecified model, and no problems were encountered in any analysis. Reported next are values of selected fit statistics calculated by LISREL:

$$\chi_M^2(23) = 35.084, \, p = .051$$

$$\text{CFI} = .990; \text{GFI} = .953; \text{SRMR} = .031$$

$$\text{RMSEA} = .057 \, (0 - .094)$$

The respecified Model 1 marginally "passes" the chi-square test, and values of approximate fit indexes are slightly better than those for the original model in Figure 2.4. There may be ways to further improve the fit of Model 1 that are not pursued here, but the estimated standardized direct effect of achievement on classroom adjustment calculated by LISREL is .627, which is of appreciable magnitude. Analysis of respecified Model 2 with a covariance between the disturbances of the achievement and classroom adjustment factors but with no direct effect between these factors generates exactly the same values of all fit statistics as Model 1. In the LISREL output for Model 2, the estimated disturbance correlation (i.e., the standardized disturbance covariance) is .293. Syntax, output, and data files for the analysis of Model 2 for EQS, LISREL, and Mplus can be freely downloaded over the Internet from the Web address listed in Note 1.

PLS Path Modeling as an Alternative to SEM in the Analysis of Composites

An alternative to SEM for analyzing models with composites is partial least squares (PLS) path modeling, also known as latent variable PLS. In this approach, constructs are estimated as linear combinations of observed variables, or composites. Although SEM is better for testing strong hypotheses about measurement, the PLS approach is well suited for situations where (1) prediction is emphasized over theory testing and (2) it is difficult to meet the requirements for large samples or identification in SEM.

For example, the partial-information estimators in PLS do not generally assume a particular distributional form and can be applied in smaller samples compared with ML and other full-information methods in SEM. There are generally no problems in PLS concerning inadmissible solutions. It is also possible to analyze composites in PLS path modeling without the strict identification requirements in SEM for doing so. One drawback of the PLS approach is that are generally no model fit statistics of the kind available in SEM. One could argue that PLS path modeling, which generally analyzes composites that are just linear combination of observed variables, does not really estimate substantive latent variables at all. See Rigdon's chapter on the PLS method in this volume for more information.

SUMMARY, PRACTICAL SUGGESTIONS, AND EXEMPLARS

The analysis of feedback loops in structural models or latent composites in formative measurement models are not without potential difficulties. However, there are times when the researcher's hypotheses about reciprocal causal effects or the directionality of effects between factors and indicators are better expressed in models where some arrows are backwards compared with more standard models in SEM. A problem to look out for in the analysis of models with feedback loops is nonconvergence of iterative estimation. This trouble may be prevented by providing reasonably accurate start values for direct effects in the feedback loop and associated disturbance variances and covariances. That is, do not rely exclusively on default start values of SEM computer programs. It also helps to build the model up over a series of steps, first without all direct effects in the feedback loop and then gradually adding paths until the whole model is "assembled." Also, carefully check a converged solution for admissibility. This is because SEM computer programs do not always issue warning messages when some estimates are mathematically impossible, such as negative error variances. For additional examples of the analysis of nonrecursive structural models with feedback loops, see Light, Grube, Madden, and Gover (2003), who estimated reciprocal effects between alcohol use and suicidal ideation in a sample of adolescents, and H. H. Chang and S. Chen (2008), who estimated reciprocal effects between trust and perceived risk in a model of online shopping (e-commerce).

The representation of formative measurement through the specification of MIMIC factors with both cause and effect indicators avoids some of the identification problems of models where some factors are measured by cause indicators only, that is, there is a latent composite. If the latter is required in a measurement model, then that factor should emit at least two direct effects onto factors each measured with effect indicators only.

Default limits on the maximum number of iterations in SEM computer programs may be too low when analyzing models with latent composites, but it is usually possible for users to specify a greater limit, such as at least 200. See Franke, Preacher, and Rigdon (2008) for more information about the estimation of direct effects of latent composites. Based on a series of computer simulation studies about possible interpretational confounding when analyzing model with latent composites, Kim, Shin, and Grover (2008) offer comprehensive guidelines for the specification and analysis of formative measurement models.

APPENDIX A
Method for Checking Rank Condition

The starting point for checking the rank condition for a nonrecursive model with all possible disturbance covariances is to construct a system matrix, in which the endogenous variables of the structural model are listed on the left side of the matrix (rows) and all variables in the structural model (excluding disturbances) along the top (columns). In each row, a "0" or "1" appears in the columns that correspond to that row. A "1" indicates that the variable represented by that column has a direct effect on the endogenous variable represented by that row. A "1" also appears in the column that corresponds to the endogenous variable represented by that row. The remaining entries are "0's," and they indicate excluded variables. The system matrix for the model in Figure 2.2a is presented here (I):

$$
\begin{array}{c}
 & \begin{array}{cc} V3 & V4 \end{array} \\
\begin{array}{c} V1 \\ V2 \end{array} & \begin{bmatrix} 1 & 0 \\ 0 & 1 \end{bmatrix}
\end{array}
\qquad (I)
$$

The rank condition is evaluated using the system matrix. Like the order condition, the rank condition must be evaluated for the equation of each endogenous variable. The steps to do so for a model with all possible disturbance correlations are summarized next:

1. Begin with the first row of the system matrix (the first endogenous variable). Cross out all entries of that row. Also cross out any column in the system matrix with a "1" in this row. Use the entries that remain to form a new, reduced matrix. Row and column labels are not needed in the reduced matrix.
2. Simplify the reduced matrix further by deleting any row with entries that are all zeros. Also delete any row that is an exact duplicate of another or that can be reproduced by adding other rows together. The number of remaining rows is the rank. The rank condition is met for the equation of this endogenous variable if the rank of the reduced matrix is greater than or equal to the total number of endogenous variables minus 1.
3. Repeat steps 1 and 2 for every endogenous variable. If the rank condition is satisfied for every endogenous variable, then the model is identified.

Steps 1 and 2 applied to the system matrix for the model of Figure 2.2a are outlined here (II). Note the we are beginning with V1:

$$
\rightarrow V1 \begin{array}{c} V3 \quad V4 \\ \begin{bmatrix} \cancel{1} & \cancel{0} \\ \cancel{0} & 1 \end{bmatrix} \end{array} \rightarrow \begin{bmatrix} 1 \end{bmatrix} \rightarrow \text{Rank} = 1 \tag{II}
$$

The rank of the equation for endogenous variable V1 in Figure 2.2a is thus 1. Steps 1 and 2 applied to the system matrix for endogenous variable V2 are outlined here (III):

$$
\rightarrow V2 \begin{array}{c} V3 \quad V4 \\ \begin{bmatrix} 1 & \cancel{0} \\ \cancel{0} & \cancel{1} \end{bmatrix} \end{array} \rightarrow \begin{bmatrix} 1 \end{bmatrix} \rightarrow \text{Rank} = 1 \tag{III}
$$

The rank of the equation for endogenous variable V2 in Figure 2.2a is also 1. Because the rank of the equations for each of V1 and V2 is 1, or exactly the minimum required by the rank condition, we conclude that the model in Figure 2.2a is identified. See Kline (2010, chap. 6) for additional examples.

APPENDIX B
EQS Syntax

Feedback Loop

```
/Title
 Kline, Reverse Arrow Dynamics
 SEM: A 2nd Course (2nd ed.)
 Feedback Loop (Figure 2.3, Final Model)
/Specifications
 variables = 8; cases = 177;
 datafile = 'chang-eqs.ess';
/Labels
 V1 = aoc; V2 = coc; V3 = noc;
 V4 = apc; V5 = cpc; V6 = npc;
 V7 = orgto; V8 = occto;
 F1 = AOC_L; F2 = COC_L; F3 = NOC_L;
 F4 = APC_L; F5 = CPC_L; F6 = NPC_L;
 F7 = ORGTO_L; F8 = OCCTO_L;
/Equations
 ! associates each indicator with its latent:
   V1 = F1 + E1; V2 = F2 + E2; V3 = F3 + E3;
   V4 = F4 + E4; V5 = F5 + E5; V6 = F6 + E6;
   V7 = F7 + E7; V8 = F8 + E8;
```

```
! structural model for latent variables:
   F7 = *F1 + *F2 + *F3 + *F8 + D7;
   F8 = *F4 + *F5 + *F6 + *F7 + *F2 + D8;
/Variances
   F1-F6 = *; D7-D8 = *;
 ! fixes error variances of single indicators to
 ! constants that equal (1 - r_xx) s**2:
   E1 = .1947; E2 = .2881; E3 = .2446;
   E4 = .1603; E5 = .1764; E6 = .1901;
   E7 = .2744; E8 = .2700;
/Covariances
  F1-F6 = *; D7,D8 = *;
/Print
  fit = all; effect = yes;
/End
```

Formative Measurement

```
/Title
 Kline, Reverse Arrow Dynamics
 SEM: A 2nd Course (2nd ed.)
 Formative Measurement (Figure 2.5)
/Specification
 variables = 9; cases = 158;
 datafile = 'worland-eqs.ess';
/Labels
 V1 = parpsy; V2 = lowses; V3 = verbaliq;
 V4 = read; V5 = math; V6 = spell; V7 = motiv;
 V8 = harmony; V9 = stable;
 F1 = Risk; F2 = Achieve; F3 = ClassAdj;
/Equations
 ! cause indicators of latent composite:
   F1 = *V1 + *V2 + V3 + D1;
 ! reflective indicators of latent variables:
   V4 = F2 + E4; V5 = *F2 + E5;
   V6 = *F2 + E6;
   V7 = F3 + E7; V8 = *F3 + E8;
   V9 = *F3 + E9;
 ! structural model:
   F2 = *F1 + D2; F3 = *F1 + D3;
/Variances
 V1-V3 = *; D1 = *; D2-D3 = *; E4-E9=*;
/Covariances
 V1-V3 = *;
/Print
 fit = all;
/Technical
 ! option itr increases iteration limit:
   itr = 200;
/End
```

APPENDIX C
LISREL (SIMPLIS) Syntax

Feedback Loop

```
Kline, Reverse Arrow Dynamics
SEM: A 2nd Course (2nd ed.)
Feedback Loop (Figure 2.3, Final Model)
Observed Variables
 aoc coc noc apc cpc npc orgto occto
Latent Variables
 AOC COC NOC APC CPC NPC ORGTO OCCTO
Correlation Matrix from file chang-lisrel.dat
Standard Deviations from file chang-lisrel.dat
Sample Size is 177
Paths
 ! associates each indicator with its latent:
   aoc = 1*AOC
   coc = 1*COC
   noc = 1*NOC
   apc = 1*APC
   cpc = 1*CPC
   npc = 1*NPC
   orgto = 1*ORGTO
   occto = 1*OCCTO
 ! structural model for latent variables:
   ORGTO = AOC COC NOC OCCTO
   OCCTO = APC CPC NPC ORGTO COC
Let the errors of ORGTO and OCCTO covary
 ! fixes error variances of single indicators to
 ! constants that equal (1 - r_xx) s**2:
   Set the error variance of aoc to .1947
   Set the error variance of coc to .2881
   Set the error variance of noc to .2446
   Set the error variance of apc to .1603
   Set the error variance of cpc to .1764
   Set the error variance of npc to .1901
   Set the error variance of orgto to .2744
   Set the error variance of occto to .2700
LISREL Output: ND = 3 SC RS EF
Path Diagram
End of Problem
```

Formative Measurement

```
Kline, Reverse Arrow Dynamics
SEM: A 2nd Course (2nd ed.)
Formative Measurement (Figure 2.5)
Observed Variables
 parpsy lowses verbaliq read math spell
```

```
 motiv harmony stable
Latent Variables
 Risk Achieve ClassAdj
Correlation Matrix from file worland-lisrel.dat
Standard Deviations from file worland-lisrel.dat
Sample Size is 158
Paths
 ! reflective indicators of latent variables:
   read = 1*Achieve
   math = Achieve
   spell = Achieve
   motiv = 1*ClassAdj
   harmony = ClassAdj
   stable = ClassAdj
 ! cause indicators of latent composite:
   Risk = 1*verbaliq
   Risk = parpsy lowses
 ! structural model:
   Achieve ClassAdj = Risk
 ! the SO option correctly scales composites;
 ! option AD allows more iterations:
LISREL Output: SO ND = 3 SC RS EF AD = 100
Path Diagram
End of Problem
```

APPENDIX D
Mplus Syntax

Feedback Loop

```
Title: Kline, Reverse Arrow Dynamics
       SEM: A 2nd Course (2nd ed.)
       Feedback Loop (Figure 2.3, Final Model)
Data:
  File is "chang-mplus.dat";
  Type is Stdeviations Correlation;
  Ngroups = 1; Nobservations = 177;
Variable:
  Names are aoc coc noc apc cpc npc orgto occto;
Analysis:
  Type is General;
Model:
 ! labels error variances of single indicators,
 ! associates each indicator with its latent:
   aoc (e1); AOC_L by aoc;
   coc (e2); COC_L by coc;
   noc (e3); NOC_L by noc;
   apc (e4); APC_L by apc;
   cpc (e5); CPC_L by cpc;
   npc (e6); NPC_L by npc;
```

```
   orgto (e7); ORGTO_L by orgto;
   occto (e8); OCCTO_L by occto;
 ! structural model for latent variables:
   ORGTO_L on AOC_L COC_L NOC_L OCCTO_L;
   OCCTO_L on APC_L CPC_L NPC_L ORGTO_L COC_L;
   ORGTO_L with OCCTO_L;
Model Constraint:
 ! fixes error variances of single indicators to
 ! constants that equal (1 - r_xx) s**2:
   e1 = .1947; e2 = .2881; e3 = .2446;
   e4 = .1603; e5 = .1764; e6 = .1901;
   e7 = .2744; e8 = .2700;
Model Indirect:
 ORGTO_L via AOC_L COC_L NOC_L APC_L CPC_L
  NPC_L OCCTO_L ORGTO_L;
 OCCTO_L via AOC_L COC_L NOC_L APC_L CPC_L
  NPC_L ORGTO_L OCCTO_L;
Output: Sampstat Residual Stdyx;
```

Formative Measurement

```
Title: Kline, Reverse Arrow Dynamics
       SEM: A 2nd Course (2nd ed.)
       Formative Measurement, Figure 2.5
Data:
  File is "worland-mplus.dat";
  Type is Stdeviations Correlation;
  Ngroups = 1; Nobservations = 158;
Variable:
  Names are parpsy lowses verbaliq read math
  spell motiv harmony stable;
Analysis:
  Type is General;
Model:
 ! reflective indicators of latent variables:
   Achieve by read math spell;
   ClassAdj by motiv harmony stable;
 ! cause indicators of latent composite:
   Risk on verbaliq@1 parpsy lowses;
   parpsy with lowses;
   verbaliq with lowses;
   parpsy with verbaliq;
 ! structural model;
 ! * required to free first path:
   Risk by Achieve* ClassAdj;
Output: Sampstat Residual Stdyx;
```

NOTES

1. http://tinyurl.com/reversearrow
2. Statistical significance is not an adequate criterion for comparing direct effects in a feedback loop or cross-lagged direct effects in a panel model because (1) large effects could fail to be significant in small samples but effects of trivial magnitude could be significant in large samples and (2) effects sizes considered to be nontrivial depend on the particular research area.
3. For a different view see Rosenberg's (1998) discussion of Immanuel Kant's arguments about the possibility of simultaneous causation.
4. It is necessary to fix both the factor loadings and measurement error variances in order to identify the model in Figure 2.2.
5. Grace and Bollen (2008) used the term L → C block (manifest to composite) to describe the association between cause indicators and a composite that is not latent.

REFERENCES

Bagozzi, R. P. (2007). On the meaning of formative measurement and how it differs from reflective measurement: Comment on Howell, Breivik, and Wilcox (2007). *Psychological Methods, 12,* 229–237. doi: 10.1037/1082-989X.12.2.229

Berry, W. D. (1984). *Nonrecursive causal models.* Beverly Hills, CA: Sage.

Bollen, K. A. (1989). *Structural equations with latent variables.* New York: Wiley.

Bollen, K. A. (2007). Interpretational confounding is due to misspecification, not to type of indicator: Comment on Howell, Breivik, and Wilcox (2007). *Psychological Methods, 12,* 219–228. doi: 10.1037/1082-989X.12.2.219

Bollen, K. A., & Bauldry, S. (2011). Three Cs in measurement models: Causal indicators, composite indicators, and covariates. *Psychological Methods, 16,* 265–284. doi: 10.1037/a0024448

Bollen, K., A., & Lennox, R. (1991). Conventional wisdom on measurement: A structural equation perspective. *Psychological Bulletin, 110,* 305–314. doi:10.1037/0033-2909.110.2.305

Carson, R. (1982). Self-fulfilling prophecy, maladaptive behavior, and psychotherapy. In J. Anchin & D. Kiesler (Eds.), *Handbook of interpersonal psychotherapy* (pp. 64–77). New York: Pergamon.

Chang, H. H., & Chen, S. W. (2008). The impact of online store environment cues on purchase intention: Trust and perceived risk as a mediator. *Online Information Review, 32,* 818–841. doi 10.1108/14684520810923953

Chang, H.-T., Chi, N.W., & Miao, M.C. (2007). Testing the relationship between three-component organizational/occupational commitment and organizational/occupational turnover intention using a non-recursive model. *Journal of Vocational Behavior, 70,* 352–368. doi:10.1016/j.jvb.2006.10.001

Chen, F., Bollen, K. A., Paxton, P., Curran, P. J., & Kirby, J. B. (2001). Improper solutions in structural equation models: Causes, consequences, and strategies. *Sociological Methods and Research, 29,* 468–508. doi:10.1177/0049124101029004003

Diamantopoulos, A. (Ed.). (2008). Formative indicators [Special issue]. *Journal of Business Research, 61*(12).

Diamantopoulos, A., Riefler, P., & Roth, K. P. (2005). The problem of measurement model misspecification in behavioral and organizational research and some recommended solutions. *Journal of Applied Psychology, 90,* 710–730. doi:10.1037/0021-9010.90.4.710

Diamantopoulos, A., Riefler, P., & Roth, K. P. (2008). Advancing formative measurement models. *Journal of Business Research, 61,* 1203–1218. doi:10.1016/j.jbusres.2008.01.009

Diamantopoulos, A., & Winklhofer, H. M. (2001). Index construction with formative indicators: An alternative to scale development. *Journal of Marketing Research, 38,* 269–277. doi:10.1509/jmkr.38.2.269.18845

Eusebi, P. (2008). A graphical method for assessing the identification of linear structural equation models. *Structural Equation Modeling, 15,* 403–412. doi:10.1080/10705510801922589

Fayers, P. M., & Hand, D. J. (2002). Causal variables, indicator variables and measurement scales: An example from quality of life. *Journal of the Royal Statistical Society: Series A. Statistics in Society, 165,* 233–261. doi:10.1111/1467-985X.02020

Finkel, S. E. (1995). *Causal analysis with panel data.* Thousand Oaks, CA: Sage.

Franke, G. R., Preacher, K. J., & Rigdon, E. E. (2008). Proportional structural effects of formative indicators. *Journal of Business Research, 61,* 1229–1237. doi:10.1016/j.jbusres.2008.01.011

Grace, J. B. (2006). *Structural equation modeling and natural systems.* New York: Cambridge University Press.

Grace, J. B., & Bollen, K. A. (2008). Representing general theoretical concepts in structural equation models: The role of composite variables. *Environmental and Ecological Statistics, 15,* 191–213. doi:10.1007/s10651-007-0047-7

Hancock, G. R., & Mueller, R. O. (2001). Rethinking construct reliability within latent variable systems. In R. Cudeck, S. du Toit, & D. Sörbom (Eds.), *Structural Equation Modeling: Present and future. A Festschrift in honor of Karl Jöreskog* (pp. 195–216). Lincolnwood, IL: Scientific Software International.

Heise, D. R. (1975). *Causal analysis.* New York: Wiley.

Hershberger, S. L. (1994). The specification of equivalent models before the collection of data. In A. von Eye & C. C. Clogg (Eds.), *Latent variables analysis* (pp. 68–105). Thousand Oaks, CA: Sage.

Howell, R. D., Breivik, E., & Wilcox, J. B. (2007). Reconsidering formative measurement. *Psychological Methods, 12,* 205–218. doi:10.1037/1082-989X.12.2.205

Humphreys, P. (2003). Mathematical modeling in the social sciences. In S. P. Turner & P. A. Roth (Eds.), *The Blackwell guide to the philosophy of the social sciences* (pp. 166–184). Malden, MA: Blackwell Publishing.

Hunter, J. E., & Gerbing, D. W. (1982). Unidimensional measurement, second order factor analysis, and causal models. *Research in Organizational Behavior, 4,* 267–320.

Jarvis, C. B., MacKenzie, S. B., & Podsakoff, P. M. (2003). A critical review of construct indicators and measurement model misspecification in marketing and consumer research. *Journal of Consumer Research, 30,* 199–218. doi:10.1086/376806

Kaplan, D., Harik, P., & Hotchkiss, L. (2001). Cross-sectional estimation of dynamic structural equation models in disequilibrium. In R. Cudeck, S. Du Toit, and D. Sörbom (Eds.), *Structural equation modeling: Present and future. A Festschrift in honor of Karl Jöreskog* (pp. 315–339). Lincolnwood, IL: Scientific Software International.

Kenny, D. A. (1979). *Correlation and causality.* New York: Wiley.

Kim, G., Shin, B., & Grover, V. (2008). Investigating two contradictory views of formative measurement in information systems research. *MIS Quarterly, 34,* 345–365.

Kline, R. B. (2010). *Principles and practice of structural equation modeling* (3rd ed.). New York: Guilford Press.

Light, J. M., Grube, J. W., Madden, P. A., & Gover, J. (2003). Adolescent alcohol use and suicidal ideation: A nonrecursive model. *Addictive Behaviors, 28,* 705–724. doi:10.1016/S0306-4603(01)00270-2

Maasen, G. H., & Bakker, A. B. (2001). Suppressor variables in path models: Definitions and interpretations. *Sociological Methods and Research, 30,* 241–270. doi:10.1177/0049124101030002004

MacCallum, R. C., & Browne, M. W. (1993). The use of causal indicators in covariance structure models: Some practical issues. *Psychological Bulletin, 114,* 533-541. doi:10.1037/0033-2909.114.3.533

Meyer, J. P., Allen, N. J., & Smith, C. A. (1993). Commitment to organizations and occupations: extension and test of a three-component conceptualization. *Journal of Applied Psychology, 78,* 538–551. doi:10.1037//0021-9010.78.4.538

Nunnally, J. C., & Bernstein, I. H. (1994). *Psychometric theory* (3rd ed.). New York: McGraw-Hill.

Rigdon, E. E. (1995). A necessary and sufficient identification rule for structural models estimated in practice. *Multivariate Behavioral Research, 30,* 359–383. doi:10.1207/s15327906mbr3003_4

Rosenberg J. F. (1998). Kant and the problem of simultaneous causation. *International Journal of Philosophical Studies, 6,* 167–188. doi 10.1080/096725598342091

Schaubroeck, J. (1990). Investigating reciprocal causation in organizational behavior research. *Journal of Organizational Behavior, 11,* 17–28. doi: 10.1002/job.4030110105

Watkin, J. (2004). Kierkegaard quotations and questions. Retrieved from http://www.utas.edu.au/docs/humsoc/kierkegaard/resources/Kierkquotes.html

Wong, C.S., & Law, K. S. (1999). Testing reciprocal relations by nonrecursive structural equation models using cross-sectional data. *Organizational Research Methods, 2,* 69–87. doi: 10.1177/109442819921005

Worland, J., Weeks, G. G., Janes, C. L., & Stock, B. D. (1984). Intelligence, classroom behavior, and academic achievement in children at high and low risk for psychopathology: A structural equation analysis. *Journal of Abnormal Child Psychology, 12,* 437–454. doi:10.1007/BF00910658

CHAPTER 3

PARTIAL LEAST SQUARES PATH MODELING

Edward E. Rigdon

At the conceptual level, structural models generally consist of relations among a number of idealized constructs (Bagozzi, 2011). Before such a model can be evaluated empirically, a researcher must find some way to represent those constructs in terms of data. In structural equation modeling (SEM), constructs are most often represented by the communality or shared variance among a set of observed variables—in other words, by a common factor.

But alternatives exist. One alternative approach is to craft representations of the theoretical variables as weighted composites of observed variables. Despite the dominance of factor-based approaches in SEM, decades of literature contrasts factor-based and composite-based approaches to data analysis, revealing strengths and weaknesses on both sides (e.g., Velicer & Jackson 1990). Factor-analytic method promise statistical advantages, but those benefits come at the cost of restrictive assumptions, which do not always hold. In some circumstances, researchers will have little hope of meeting the requisite conditions; in others, the researcher's focus may be on something other than statistical optima. Researchers in possession of promising data but lacking either a well-developed theoretical framework or the luxury of multiple rounds of measure development may find in composite-

Structural Equation Modeling: A Second Course (2nd ed.), pages 81–116
Copyright © 2013 by Information Age Publishing

81

based methods the opportunity for insight, using methods that are convenient, fast, and tolerant of less-than-ideal conditions. So, while researchers will continue to value the perfect world strengths of factor-based SEM, researchers should also appreciate the real world advantages of composite-based methods.

Among the wide range of possible composite-based approaches, Partial Least Squares (PLS) path modeling is one of the best-known. PLS path modeling has become a popular approach particularly within certain disciplines such as information systems (Gefen, Rigdon, & Straub, 2011), and to a lesser extent marketing (Hair, Sarstedt, Ringle, & Mena, 2012). Besides using composites instead of factors to represent theoretical constructs, PLS path modeling differs in employing a piecewise approach to parameter estimation, as opposed to the simultaneous estimation methods (such as maximum likelihood) most commonly used in SEM applications. PLS path modeling estimates standard errors empirically through bootstrapping, rather than by computing standard errors using theoretical formulas built upon distributional assumptions as is most typical in SEM practice. It is important to understand how these differences in construct representation and parameter and standard error estimation affect the behavior of this method and to examine its strengths and weaknesses as compared to the available alternatives.

Just as the SEM community has its own jargon, the PLS path modeling literature tends to use a distinctive language. Within the bulk of the PLS literature, PLS path modeling and common factor-based SEM are both counted as instances of SEM. Factor-based SEM is identified as *covariance-based structural equation modeling* (CBSEM) while PLS path modeling is identified as PLS-SEM (e.g., Hair, Ringle, & Sarstedt, 2011). Other differences in terminology will be mentioned when they are relevant to this discussion.

Origin of PLS Path Modeling

Many readers will be familiar with the story of how Karl Jöreskog set aside his plans to teach high school mathematics and instead became a graduate student in statistics under the supervision of econometrician Herman Wold at Uppsala University (Sörbom, 2001). Jöreskog's major contributions toward the development of SEM inspired Wold to develop a composite-based alternative (Wold, 1988). This effort blossomed during Wold's tenure as visiting scholar at the University of Pennsylvania (Dijkstra, 2010). Unfortunate circumstances stalled the public development of PLS path modeling for many years. Recently, however, this method has seen an explosion of innovation across many dimensions (Esposito Vinzi, Chin, Henseler, & Wang,

2010) which should certainly incline researchers to take another look at PLS path modeling.

As Wold (1985) stated, "PLS was initially designed for research contexts that are simultaneously data-rich and theory-primitive" (p. 589), envisioning a discovery-oriented process that is "a dialogue between the investigator and the computer" (p. 590). Rather than commit to a specific model a priori and frame the statistical analysis as a hypothesis test, Wold imagined a researcher estimating numerous models in the course of learning something about the data and about the phenomena underlying the data. Exploratory context aside, the development of PLS path modeling was also motivated by a desire to minimize restrictive distributional assumptions and to avoid reliance on asymptotic properties (Wold, 1985). While some authors have continued to assert PLS path modeling's exploratory character (e.g., Esposito Vinzi, Trinchera, & Amato, 2010), others argue that PLS path modeling is suitable for both exploratory and confirmatory analysis (e.g., Chin, 2010). Published applications (e.g., Fornell, Johnson, Anderson, Cha, & Bryant, 1996; Reinartz, Krafft, & Hoyer, 2004; Venkatesh & Agarwal, 2006) have generally argued their case in confirmatory terms. Then again, some of this might reflect an academic bias in favor of findings presented in confirmatory terms (see, e.g., Greenwald, Pratkanis, Leippe, & Baumgardner, 1986). Choices in statistical methods tend to involve tradeoffs, and Wold recognized both strengths and limitations in his technique. It is important for modern users to consider these same issues.

HOW DOES PLS PATH MODELING WORK?

Constructs

An SEM analysis generally begins with constructs (Rigdon, 1998). By contrast, as Wold's comment suggested, researchers might choose PLS path modeling because they are confronted by a dataset containing sets of conceptually-related observed variables, and because they are seeking to extract or uncover meaning within the data. Still, fundamentally, the constructs that figure in PLS path models are of the same character as those that figure in structural equation models. These constructs are abstract idealized variables like Customer Satisfaction or Perceived Ease of Use, which are not identical to any given observed variable to be found in a dataset. Users of both PLS path modeling and SEM might concur with Embretson's (1983) simple definition of a construct as, "a theoretical variable that may or may not be a source of individual differences" (p. 180).

Are the constructs employed in PLS path modeling *latent variables*? As a point of comparison, SEM authors define their constructs in terms of idealizations and then empirically represent them with common factors, which cannot be derived as exact functions of any set of observed variables. Thus, both conceptually and empirically, SEM features constructs that are clearly not present in the data and therefore are latent. Empirically, PLS path modeling replaces constructs with *proxies*, explicit weighted composites of the observed variables in the model. While these proxies are obviously not latent, given that they are formed from the data at hand, they are specifically designated as proxies. In other words, they are distinct from the constructs themselves, and remain so except at the limit (Hui & Wold, 1982). At the conceptual level, the constructs in PLS path modeling are indeed latent variables—idealizations that are not directly represented in the dataset. Wold (1975) commonly referred to both composites and factors as "latent variables," and the PLS community is essentially unanimous in characterizing PLS path modeling as a latent variable method.

Relations Among Constructs

Structural equation models most commonly employ linear relations among constructs, and the same is true in PLS path modeling. Methods exist for modeling quadratic and interaction effects among constructs in SEM, and such methods also exist in PLS path modeling (Chin, Marcolin, & Newsted, 2003). PLS path modeling applications are typically limited to *recursive models* (a term coined by Wold); these exclude reciprocal relations, feedback loops, and correlated structural errors (Rigdon, 1995), which cannot be specified using many popular PLS path modeling software packages. Procedures exist for estimating PLS path models with reciprocal relations or feedback loops (Boardman, Hui, & Wold, 1981), and some packages will include the option.

Figure 3.1 is designed to help illustrate aspects of the PLS path modeling approach. In this figure four constructs, labeled A, B, C, and D, are

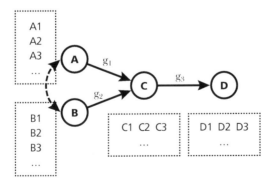

Figure 3.1 Components of a partial least squares path model.

related through a structural model that can be described in two regression equations:

$$C = g_1 A + g_2 B + e_C$$

$$D = g_3 C + e_D$$

Here, g_1, g_2, and g_3 are parameters to be estimated. The figure shows a dashed curved line representing the covariance between A and B. As in regression, covariation among predictors plays a role in determining results, but these covariances are not parameters to be manipulated as they tend to be in SEM. This figure shows a dotted box beside each construct. This box contains labels (e.g., A1, A2, A3) that represent observed variables associated with each construct. For now, relations between constructs and observed variables remain unspecified.

Observed Variables

PLS path modeling was designed for situations where the available data do not conform to the multivariate normality assumption underlying maximum likelihood, the estimation method most commonly used in SEM (Wold, 1985). Like ordinary least squares (OLS) regression, PLS path modeling requires no overall distributional assumptions about the data. Instead, distributional assumptions focus on the residuals, the differences between actual and model-implied values for the dependent variables. Wold (1985) argued that the use of nonparametric methods to estimate standard errors freed PLS path modeling even from the assumption of independence of observations (see Wold, 1988, for example). In fact, besides causal homogeneity, the only distributional requirement upon which Wold insisted was *predictor specification* (Wold, 1975; 1988) which asserts only that the conditional expectation for each dependent variable is defined by its predictors, as specified in the dependent variable's equation. This amounts to little more than assuming that the specified equations are correct. PLS path modeling's nickname, *soft modeling*, was meant to reflect these very limited distributional assumptions.

The limited theory base that Wold envisioned for PLS path modeling analyses naturally has implications for the quality of observed variables analyzed with this method. Strong prior theory aids not only model specification but also the design and evaluation of indicators (Bagozzi, 2011). A weak theory base is unlikely to produce sets of observed variable indicators that conform closely to the proportionality constraints imposed by factor-based SEM with multiple indicators (Franke, Preacher, & Rigdon, 2008; Jöreskog, 1979). To elaborate briefly, when multiple observed variables load on a single common factor with no cross-loadings or correlated error terms, covariances among those observed variables must adhere to certain proportionality constraints. If y_1 and y_2 are two such observed variables, then the ratio of their covariances to a third observed variable y_t, that is, $cov(y_1, y_t)/cov(y_2, y_t)$, must be a constant within sampling error across all other observed variables in the model (Jöreskog, 1979). Failure of a dataset to conform to the proportionality constraints implied by the factor model is a primary driver of poor data-model fit in SEM. Component-based methods such as PLS path modeling, on the other hand, do not impose similar constraints. As such, a researcher who lacks observed variables that closely conform to these proportionality constraints may well prefer PLS path modeling as an analytical method.

Given the potential conceptual uncertainty associated with a weak theory base, one might anticipate an approach where indicators are allowed to associate with multiple constructs, as Asparouhov and Muthén's (2009) exploratory SEM (see chapter by Morin, Marsh, & Nagengast, this volume). In PLS path modeling, however, each observed variable is assigned exclusively to a single construct. As a method, PLS does include the possibility of extracting multiple components from a single set of observed variables (Apel & Wold, 1982; Wold, 1985), but few modern PLS applications exploit this potential.

SEM's need for high-quality and specially developed observed variables means that SEM generally yields poor results when it is applied to archival or secondary data, which are typically collected without the benefit of a theoretical framework and are not designed especially for SEM analysis. In the computer age, researchers sometimes encounter datasets that hold the promise of yielding insights about research questions but that were collected by organizations in the course of doing business. This may include transactional data, data in financial statements, and data from consumer choice studies and surveys of customers, employees and other populations. PLS path modeling might be the better choice for after-the-fact structural modeling of such secondary data.

Model Estimation

The PLS estimation algorithm can appear complex, but the process aims to solve a fairly complex problem—estimating relations among multiple composites of observed variables—by using very simple methods. Again, both PLS path modeling and SEM face the challenge of deriving single entities from groups of observed variables to represent these constructs. One can imagine a wide range of approaches. A researcher familiar with regression might simply standardize and then sum the observed variables in each set. Under certain conditions (Rozeboom, 1979), such naïve weights do not perform much worse than optimal weights in terms of predicting some dependent variable, and this approach avoids the whole messy business of estimation (Armstrong, in press).

In PLS path modeling, observed variables are typically standardized. Standardization is crucially important for a method of composites. Forming proxies as weighted averages of the original observed variables, differences in scale could dramatically increase the influence of one observed variable at the expense of others. By contrast, in SEM, with scale-invariant models researchers can largely disregard scale differences across variables (although, in practice, substantial differences in scaling can induce anomalous behavior in the software used to estimate structural equation models). For the remainder of this chapter, measured variables are assumed to be standardized unless otherwise noted.

In the simple case of two sets of observed variables, canonical correlation provides a ready solution, choosing weights to optimize the correlation between composites. With more groups of variables, however, simple solutions are not so obvious. Consider again Figure 3.1, where construct C stands as dependent on constructs A and B and also as a predictor of construct D, and each construct is associated with a set of observed variables. The same set of weights for the observed variables in the C set probably will not optimize all relations (represented by parameters g_1, g_2, and g_3) in the model. So how is a researcher to choose a single best set of weights?

In solving this problem, estimation of weights in PLS path modeling proceeds through iterations, with each iteration consisting of a series of stages that successively estimate: (1) the parameters of the *inner model*, quantifying relations between the constructs; (2) the parameters of the *outer model*, quantifying relations between the constructs and their indicators; and (3) *case values* for the constructs, expressing each as an updated function of its indicators (see Figure 3.2). During each stage, each construct is replaced by a proxy, formed as an exact weighted composite of other variables. Proxies are updated stage by stage.

In PLS path modeling, each construct, F_k, $k = 1 \ldots K$, is uniquely associated with a set of one or more observed variables y_{jk}, $j = 1 \ldots J$. Unlike SEM, which can model higher-order factors with no directly associated observed indicators, PLS path modeling cannot proceed with a construct that is not directly connected to at least one observed variable because the method uses associated observed variables in the process of forming proxies.

To begin estimation, the method must establish a first proxy F_k^* for each construct. This first proxy, which acts like a start value, will typically be an equal-weight sum of the associated observed (standardized) variables,

$$F_k^* = \sum_{j=1}^{J} w_{jk} y_{jk},$$

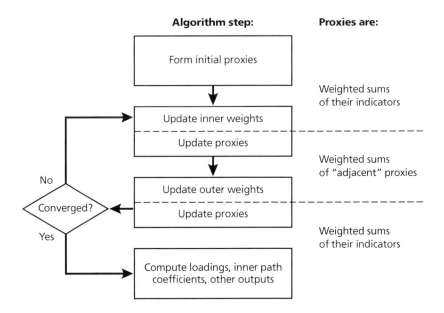

Figure 3.2 The partial least squares estimation algorithm.

with weights scaled for each of the K proxies so as to standardize the proxy (i.e., mean 0, standard deviation 1). So, in the example of Figure 3.1, each of the constructs A, B, C, and D would be replaced by a standardized simple average of its associated standardized indicators in order to start the estimation process.

In the first stage of each iteration, the method estimates the inner relations among the constructs using OLS regression. Applying this technique to the constructs' proxies for the example in Figure 3.1 would involve two OLS regressions, one for C* regressed on A* and B*, and one for D* regressed on C*. Because the model that is linking the construct proxies is recursive, the weights can be estimated one equation at a time.

The final task in this first stage is to update the construct proxies. For each construct, a new proxy is formed as a weighted sum of the proxies for those other constructs that are "adjacent" to it. Constructs are adjacent to a given construct when they are connected to it by direct paths, either as dependent variables or as predictors. (Covariances between predictors, indicated by the dashed curved arrow in Figure 3.1, do not count as paths for this purpose.) In Figure 3.1, constructs A, B, and D each has a direct connection to only one construct, C; thus, the new proxies for A, B, and D would each be based only on the proxy for construct C. Ultimately, as all proxies become weighted sums of the constructs' associated observed variables, at this stage the proxies for constructs A, B, and D would all be essentially the same weighted sum of the observed variables associated with construct C. By contrast, construct C is directly connected to all three other constructs, so its new proxy would be based on a weighted sum of the proxies for A, B, and D.

Standard PLS implementations offer three choices regarding precisely how these "inner weights" are specified. In the *factor weighting* scheme, weights are proportional to zero-order correlations for each pair of proxies. The *path weighting* scheme uses zero-order correlations for constructs that predict the given construct but uses standardized path coefficients for constructs that are dependent on the given construct. Wold's original *centroid weighting* scheme uses weights that are proportional to unit weights, with signs based on the sign of the correlation between each pair of proxies (see Rigdon, Ringle, & Sarstedt, 2010). Recall in the Figure 3.1 model, constructs A, B, and D are each directly connected to only one other construct, C. So for those three constructs, their new proxy will be simply the proxy for construct C from the preceding stage. For construct C, by contrast, the difference between the factor and path weighting schemes would lie in the path between C and D. The factor weighting scheme would build the new proxy for C using zero-order correlations for all three components, while the path weighting scheme would use the value of g_3 as the weight for the proxy of D. Under the centroid weighting scheme, the new proxy would be a weighted sum with equal weights for the proxies of A, B, and D, adjusted for sign. Understandably, the choice of weighting scheme for inner relations often makes little difference in practice (Rigdon et al., 2010).

The second stage in each iteration is to update the weights linking construct proxies to the observed variables assigned to each construct. Here again the researcher has a choice, but the PLS path modeling literature imbues this choice with greater significance. Standard PLS path modeling software gives the researcher the option, for each construct, between *Mode A* and *Mode B*, explanatory diagrams for which appear in Figure 3.3. It will be easier to explain the difference by starting with Mode B.

With Mode B, a construct's proxy is regressed on the construct's associated indicators:

$$F_k^* = \sum_{j=1}^{J} b_{jk} y_{jk} + \varepsilon_k.$$

As is typical in regression, covariances among the observed variable predictors play a role in the estimation, although, as with covariances among predictor constructs, these covariances are not manipulable parameters of the model. The PLS literature associates Mode B with the use of formative measurement models (see, e.g., Esposito Vinzi et al., 2010), as depicted in Figure 3.3(a). SEM can incorporate such formative relations only in limited terms (MacCallum & Browne, 1993; see also the chapter by Kline, this

(a) **Mode B**

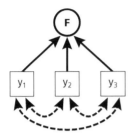

(b) **Mode A as usually depicted**

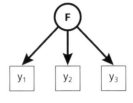

(c) **Mode A with explict**
error covariances

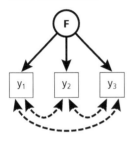

(d) **Mode A as Mode B with**
predictors uncorrelated

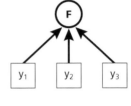

Figure 3.3 Graphical representations of Mode B and Mode A.

volume). For researchers wishing to model formative relations, PLS path modeling offers a straightforward approach.

With Mode A, each observed variable is regressed individually on the proxy of the associated construct:

$$y_{jk} = \lambda_{jk} F_k^* + \delta_{jk}, \text{ for } j = 1 \ldots J.$$

Notice that the parameters linking observed variables and proxies here are loadings, not weights per se (though the loadings are used to produce weights later in this stage). Mode A is generally depicted visually as shown in Figure 3.3(b). SEM users may see in this equation and diagram an example of a factor-based reflective measurement model. The PLS literature, while recognizing that the method is based on composites, still tends to support the factor-like characterization of Mode A—Esposito Vinzi et al. (2010) asserted that this mode "reproduces the factor analysis model" (p. 50).

That said, there are important distinctions between Mode A and a factor model. The equations of Mode A are not estimated simultaneously, and there are no constraints across equations/observed variables. The proportionality constraints implied by a factor analytic measurement model need not hold. Mode A operates as if error terms are free to covary in whatever way is convenient but without the loss of statistical identification that would occur if one attempted to estimate a factor model with all error covariances free. Wold (1985) allowed that error covariance constraints are not formally part of PLS path modeling. Thus, one may view Mode A as a factor measurement model with errors correlated as convenient, as illustrated in Figure 3.3(c).

But a more parsimonious view, illustrated in Figure 3.3(d), is that Mode A is a formative alternative to Mode B that ignores covariance among the observed predictor variables (Rigdon & Gefen, 2011). Recall that, in OLS regression, in a simple model with a single predictor and with standardized variables, the formula for the slope reduces to the correlation between the two variables—regardless of which variable is predictor and which is dependent. While a formative model has multiple observed variables predicting one proxy, treating those multiple predictors as being uncorrelated again means that the direction of relation is immaterial. In an SEM analysis, such disregard for elements of the observed variable covariance matrix would contribute to poor data-model fit, but unlike SEM, PLS path modeling does not aim to faithfully reproduce the covariance matrix of the observed variables (e.g., Esposito Vinzi et al., 2010).

This cycle of updating inner weights and outer weights, and swapping proxies, continues until the convergence criterion is satisfied. Convergence is indicated when the sum of the absolute differences in outer weights across successive iterations is less than some small quantity (Henseler, 2010). While this algorithm shares some of the reliability of least squares estimation, users cannot be completely confident that convergence will occur, except in special cases (Henseler, 2010). Small variations in the algorithm and different choices for start values and different inner or outer weighting schemes can alter convergence behavior (Henseler, 2010), so researchers who encounter convergence problems should try varying these choices. Still, convergence is highly likely when sample size is large.

Once the estimation process has converged, and final values for all construct proxies have been obtained, it is a simple matter to produce statistical outputs, such as R^2 values for dependent constructs. The final proxy values are, indeed, weighted sums of their associated indicators. Where Mode B was used, the weights are simply those produced from the regression of the prior-stage proxy on the observed variables. Where Mode A was used, the weights are chosen to be proportional to the loadings produced by Mode A—in other words, proportional to the zero-order correlation between each indicator and the proxy from the prior stage. Thus, Mode A weights are similar to what Dana and Dawes (2004) called *correlation weights*, which effectively ignore the covariance among the indicators, while Mode B weights are *regression weights* that automatically accommodate the indicators' degree of covariation.

The above approach does not exactly reproduce a factor model, because a factor model implies structured covariance among the predictors. Moreover, in a factor analysis, the factor represents only common variance among a set of observed variables, while unique systematic variance and random error variance are excluded. By contrast, proxies in PLS path modeling are weighted composites of observed variables, and thus may include a measure of both unique systematic variance and random error. Still, if data are consistent with a factor model, Mode A results may approximate those of a factor model, yielding the *deliberate approximation* that Wold (1982, pp. 27–29) originally described. Further, PLS path analysis might well yield results in situations where factor-based methods simply fail. This in itself might be another important reason to consider PLS path modeling as an analytical alternative.

Case values point to another distinction between SEM and PLS path modeling. In using common factors to represent constructs, SEM avoids some of the problems caused by the error that taints observed variables (Rigdon, 1994), but it also abstracts away from the level of observed variables (Bagozzi, 2011). This is a one-way trip. Common factors cannot be expressed as functions of observed variables, except approximately. By contrast, the proxies used in PLS path modeling always end as exact linear functions of the observed variables associated with a construct. For researchers who need to stay in touch with the observable plane, PLS path modeling is the more natural approach.

Unlike the choice among the factor method, the path method, and the centroid method for weighting inner relations, the choice between Mode A and Mode B has immediate consequences. The Mode B approach is equivalent to OLS regression, and so certain optimal properties apply. Researchers seeking to maximize R^2 for dependent constructs will tend to find that Mode B performs somewhat better than Mode A because Mode B discounts observed variables that are highly redundant, while Mode A ignores this collinearity (Dijkstra, 2010). On the other hand, the forecasting literature demonstrates that correlation weights can often outperform regression weights in true cross-validation prediction tasks (e.g., Dana & Dawes, 2004), so researchers should not be too quick to discount the value of Mode A estimation. Moreover, collinearity among the observed variables can yield Mode B weights with unexpected values. Often enough, researchers theorize in zero-order terms ("impact of X on Y") but estimate partialed coefficients ("impact of X on Y when controlling for W"). Mode A is less likely to produce unexpected weights precisely because it ignores collinearity.

Overall, PLS path modeling produces parameter estimates that are known to be biased in finite samples (Chin, 1998). If data are generated according to a correct factor-based model, using PLS path modeling to estimate the parameters of the same model will tend to produce estimates of outer relations (between constructs and associated observed variables) that are over-estimated, and estimates of inner relations (among constructs) that are under-estimated (Chin, 1998). This is predictable, given that PLS path modeling does not make the same distinctions between variance components in observed variables that a factor model makes, and because the PLS model is not equivalent to a factor model—in other words, because the model is misspecified.

Parameter estimation was not a primary goal for Wold in developing PLS path modeling. Wold rated prediction of dependent constructs and reproduction of the data as more important. Nevertheless, Wold argued that PLS path modeling, under its limited assumptions, produces parameter estimates that are *consistent at large* (Hui & Wold, 1982). This means that parameter estimates approach population values as both sample size and the number of observed variables per construct approach infinity. This is not, however, a property peculiar to PLS path modeling. Indeed, as the number of predictors in a set increases, the distinction between optimal regression weights and simple unit weights tends to disappear (Rozeboom, 1979). Thus, it is certainly plausible that large sample size and many indicators—which one might encounter in an archival dataset—will tend to blur the distinction between SEM and PLS path modeling results. McDonald (1996) illustrated the expected decrease in bias for PLS estimates with hypothetical increase in the number of indicators per construct. Granted, published applications of PLS path modeling often involve few observed variables per construct. But the improvement in results as the number of indicators increases could be an attractive feature for researchers facing large datasets with many variables.

Standard Errors

Eschewing distributional assumptions, PLS path modeling relies on bootstrapping to estimate standard errors. Repeated resampling with replacement from the dataset produces many sets of parameter estimates. Routines typically filter or adjust these results to weed out effects such as sign-switching (Temme, Kreis, & Hildebrandt, 2010) which occurs because standardized quadratic forms are indeterminate with respect to sign (e.g., the same positive x^2 is equally consistent with both positive x and negative x). In SEM, reference variables—observed variables with loadings fixed to positive values—prevent this indeterminacy, but this option is not available in PLS path modeling.

The empirical distribution of parameter estimates across these bootstrap replications provides a basis for estimating the variability of parameter estimates. The PLS algorithm offers the rapid convergence of OLS, and the speed of modern computers ensures that this operation is not excessively time-consuming. However, this reliance on bootstrapping re-emphasizes a contrast between PLS path modeling and SEM. Given an assumption of multivariate normality, SEM can proceed with only summary moments—a covariance matrix and perhaps a vector of means—yielding a full range of outputs. PLS path modeling, on the other hand, needs individual data in order to estimate both standard errors and case values for proxies. PLS offers a more narrow range of results when only summary data are available.

Evaluating Results

SEM users apply a standard set of criteria in evaluating results from structural models. The criteria used in PLS path modeling overlap those in SEM partially, but not entirely. Detailed and current reviews of evaluative criteria used in PLS path modeling are available from Chin (2010) and Gefen et al. (2011). In SEM, evaluation is likely to begin with some form of the χ^2 statistic which can provide a basis for overall inferential model testing. PLS path modeling lacks such an overall fit statistic, though there have been proposals (e.g., Tenenhaus, Esposito Vinzi, Chatelin, & Lauro, 2005). Still, PLS users examine a number of statistics and diagnostics when evaluating modeling results.

Given the emphasis on prediction in PLS path modeling, R^2 values for dependent constructs will be important in evaluating a solution. Chin (2010) recommended the f^2 measure of effect size from regression:

$$f^2 = \frac{R^2}{1-R^2}$$

using incremental analysis to evaluate the contribution of individual predictors. That is, as in regression, one can evaluate the contribution of each predictor construct to variance explained for a given dependent construct by adding predictor constructs one at a time and observing the change in R^2. These predictors freely correlate, so it is always important to add predictors in an order that makes theoretical sense (Cohen, Cohen, West, & Aiken, 2003). While parameter estimation is not the focus in PLS path modeling, researchers will still examine the sign and magnitude of parameter estimates as compared with expectations, as well as the statistical significance of parameter estimates based on bootstrap standard errors. Because the PLS literature views Mode A estimation as analogous to reflective measurement in SEM (cf. Hair, Anderson, Tatham, & Black, 1995, pp. 641–642), researchers are directed to report for each kth construct (where Mode A is used) the composite reliability (CR_k):

$$CR_k = \frac{\left(\sum_{j=1}^{J}\lambda_j\right)^2}{\left(\sum_{j=1}^{J}\lambda_j\right)^2 + \sum_{j=1}^{J}(1-\lambda_j^2)}$$

and the average variance extracted (AVE_k):

$$AVE_k = \frac{1}{J}\sum_{j=1}^{J}\lambda_j^2$$

assuming all proxies and parameter estimates are standardized. Users are advised to rely on rules of thumb that would be familiar to SEM users: CR above 0.6 (Götz, Liehr-Gobbers, & Krafft, 2010, p. 695) and AVE above 0.5 (Götz et al., 2010, p. 696). If, as is argued here, Mode A is not reflective measurement, then the specific logic behind the application of these procedures in SEM does not necessarily translate to the PLS path modeling context. Moreover, more recent SEM literature has proposed better alternatives to composite reliability, such as coefficient H (Hancock & Mueller, 2001), but these innovations have not taken hold within the PLS path modeling literature. Researchers using Mode A are also advised to examine cross-loadings, here defined as correlations between observed variables and non-associated proxies. Observed variables are supposed to correlate more highly with the proxy of their associated construct than with the proxies of other constructs, in order to demonstrate discriminant validity. Given that each proxy is ultimately defined as a weighted sum of its associated indicators, this seems like a mild condition. Even if all relations between observed variables and proxies are formative, it might give a researcher pause to see an observed variable demonstrating a stronger correlation with a different construct's proxy.

For each construct, PLS users might also compute a *communality* index (Tenenhaus et al., 2005), which is the average squared correlation between each observed indicator variable (y_{jk}) and the associated proxy (F_k^*). Averaging gives the index a range of 0 to 1, and makes the communality index computationally equivalent to average variance extracted, but the communality index may be computed for all blocks. Except for this averaging, communality in PLS path modeling is equivalent to communality in factor analysis (Mulaik, 2010, p. 134).

For proxies of dependent constructs, PLS users may also compute a *redundancy* index (Tenenhaus et al., 2005), which is equal to the product of the proxy's communality and its R^2. The redundancy index summarizes, for each dependent construct, the extent to which the observed variables are predicted by the model's predictor constructs. Given a rationale which emphasizes prediction of observed variables in dependent blocks, this must be a key performance metric in PLS path modeling, Obviously, communality is bounded by the scale of squared correlations. Being multiplied by R^2, redundancy must be even smaller.

Wold's emphasis on reproduction of the data also led to his adoption of the Q^2 measure of *predictive relevance* by Stone (1974) and Geisser (1975), as described below (see also Chin, 2010; Wold, 1982). Some SEM users might see a connection between this approach, which is also labeled *blindfolding*, and the posterior predictive checking that is a part of Bayesian SEM (Muthén & Asparouhov, in press). Through all of the calculation and swapping and updating of proxies, Wold wanted to ensure that the model was still faithful to the data, meaning that the model and its parameter estimates could reproduce the data. A model that failed such a standard would be like one of Anscombe's (1973) famous regression examples—fine enough in reproducing summary moments but not descriptive of the data, and not to be trusted in predicting new data points drawn from the same population as that of the original dataset. Indeed, the optimal estimation properties of maximum likelihood methods may only be available through sacrificing the ability to reproduce the actual data, a consequence of the reliance on essential supporting assumptions (Wold, 1975).

This predictive relevance technique begins by selecting a subset of the observed variables, which might be the observed variables associated with a particular construct. Starting with a complete data matrix for these variables, this procedure uses a systematic random sampling approach to delete or obscure individual data points, with the aim of testing the ability of the model to reproduce those data points. To establish a basis for evaluating the quality of this reproduction, this approach creates a worst case where the obscured data points are replaced by mean imputation (i.e., replacing each missing value with the variable's observed mean). Then the model, with the observations omitted, is re-estimated, and those estimates are used to reconstruct the missing data points. Once these operations have been performed for all data points in the set, the model and the worst case are compared in terms of sum of squared error:

$$Q^2 = 1 - \frac{\Sigma(d - e_M)^2}{\Sigma(d - e_0)^2}$$

where $(d - e_M)^2$ is the squared deviation between the actual data point and the predicted value based on the model, while $(d - e_0)^2$ is the squared deviation between the same data point and the sample mean for the variable. Summation is across every data point and all observed variables used in the analysis (e.g., those associated with a given construct). Obviously, one would expect the model to outperform the worst case, implying that Q^2 should be positive for every construct (Chin, 2010).

The PLS path modeling literature treats this analysis as a form of cross-validation. If the obscured data points are predicted by the proxy for the construct with which the observed variables are associated, this is called *cross-validated communality*. It is said to demonstrate that the observed variables indeed are represented in the proxy. If the construct is a dependent variable in the model, then the obscured data points may also be predicted by the proxies for the predictor constructs. This is labeled *cross-validated redundancy*, and is taken to show that the predictors effectively predict not only the dependent proxy but also the observed variables that make up the dependent proxy. As Chin (2010) recommended, "redundancy Q^2 above 0.5 is indicative of a predictive model" (p. 680).

Diagnostics

In SEM, with its emphasis on fit in the sense of accurately reproducing a covariance matrix, diagnostics center around the impact of model changes on fit in this sense. Thus, SEM features modification indices (and expected parameter change statistics) and residuals defined most commonly as the difference between the empirical covariance and the covariance implied by the model. In PLS path modeling, diagnostics focus on reproduction of the data and improving variance explained. SEM is primarily confirmatory; unsurprisingly, diagnostics in SEM highlight small modifications which improve "fit" without fundamentally reconstructing the model. By contrast, PLS path modeling was conceived as a method of exploration and discovery: Wold imagined the PLS user adding, discarding, and rearranging observed variables and constructs. Speedy, reliable least squares estimation would allow the researcher to repeatedly re-estimate the model, looking for improvement in key indices and elimination of unexpectedly signed parameter estimates. So while diagnostics in SEM involve tinkering with a model that already rests on strong theoretical support, diagnostics in theory-primitive PLS path modeling derive from comparing results across multiple possible models: "Soft modeling is an evolutionary process. At the outset, the arrow scheme of a soft model is more or less tentative. The investigator uses the first design to squeeze fresh knowledge from the data, thereby putting flesh on the bones of the model, getting indications for modifications and improvements, and gradually consolidating the design. For example, the case values of [a construct] may show high correlation with an observable that hasn't as yet been included among its indicators. Or the residuals of a causal-predictive relation may be highly correlated with an observable that could be included in an inner relation as [a construct] with a single indicator" (Wold 1982, p. 29). Moreover, with bootstrapping used routinely to estimate standard errors, it is computationally simple to derive diagnostics at the level of the individual respondent. While outlier detection methods have been introduced for SEM (Bollen & Arminger, 1991), respondent-level diagnostics have received little attention, outside of mixture modeling.

EMPIRICAL EXAMPLE: NATIONAL PARTNERSHIP FOR REINVENTING GOVERNMENT EMPLOYEE SURVEY

To demonstrate PLS path modeling and to compare results from this method to results from SEM, data were drawn from the 1998 National Partnership for Reinventing Government Employee Survey (United States Office of Personnel Management, 1998), study #3419 in the archives of the Inter-University Consortium for Political and Social Research (ICPSR) at the University of Michigan. The questionnaire included items tapping job satisfaction and assessments of the unit's procedures and performance, and was completed by more than 13,000 employees employed at 48 federal agencies. This demonstration uses responses from National Park Service employees. For this demonstration, incomplete cases were excluded, leaving 315 completed responses.

The general model links four constructs, labeled Respect, Teamwork, Performance, and Job Satisfaction, as seen in Figure 3.4. The conceptual model argues that treating employees with respect and encouraging effective teamwork should enhance performance. Moreover, being treated with respect and feeling good about the quality of work being performed should contrib-

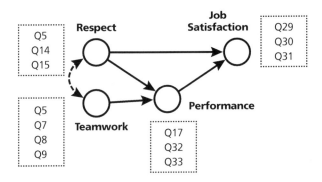

Figure 3.4 Empirical example model of job satisfaction.

TABLE 3.1   Questionnaire Items Used in Empirical Demonstration

Teamwork

Q6 "A spirit of cooperation and teamwork exists in my immediate work unit."

Q7 "Teams are used to accomplish organizational goals, when appropriate."

Q8 "Employees are rewarded for working together in teams (for example . . .)."

Q9 "Employees in different work units participate in cross-functional teams to accomplish work objectives."

Respect

Q5 "At the place I work, my opinions seem to count."

Q14 "Differences among individuals (for example . . .) are respected and valued."

Q15 "Supervisors/team leaders understand and support employees' family/personal life responsibilities."

Performance

Q17 "In the past two years, the productivity of my work unit has improved."

Q32 "Overall, how good a job do you feel is being done by your immediate supervisor/team leader?"

Q33 "How would you rate the overall quality of work being done in your work group?"

Job Satisfaction

Q29 "Considering everything, how satisfied are you with your job?"

Q30 "How satisfied are you with your involvement in decisions that affect your work?"

Q31 "How satisfied are you with the recognition that you receive?"

ute to job satisfaction. This demonstration uses responses to questionnaire items that relate to each of these general themes; these items are shown in Table 3.1. Responses were recorded using five category Likert-type response scales, and generally show some skew toward the positive end of the scale (as is typical with surveys of current employees). Thus, the observed variables might be most properly considered ordinal. While SEM includes special techniques for working with ordinal data, the declared distribution-free nature of PLS path modeling has been taken to include indifference to measurement scale. Then again, PLS path estimates are known to be biased except at the limit, and SEM solutions for dealing with ordinal data sometimes involve an assumption about an underlying statistical distribution—an assumption that would be anathema within the PLS path modeling literature. Therefore, in this demonstration, the observed variables are treated as being interval scaled in both PLS path modeling and in SEM. All PLS analyses were carried out using SmartPLS 2 (Ringle, Wende, & Will, 2005), while all SEM analyses were carried out with Mplus 6 (Muthén & Muthén, 2010).

The pattern of correlations among these observed variables suggests that a factor model, with observed indicator variables as shown, will not be entirely consistent with the data, as seen in Table 3.2. For example, while the

TABLE 3.2. Moments of the Observed Variables

	Q6	Q7	Q8	Q9	Q5	Q14	Q15	Q29	Q30	Q31	Q17	Q32	Q33
												Correlations	
Q7	0.578												
Q8	0.424	0.573											
Q9	0.422	0.606	0.53										
Q5	0.515	0.581	0.495	0.467									
Q14	0.281	0.363	0.253	0.335	0.428								
Q15	0.398	0.429	0.352	0.355	0.539	0.464							
Q29	0.493	0.455	0.448	0.295	0.587	0.365	0.503						
Q30	0.552	0.535	0.539	0.418	0.746	0.368	0.564	0.718					
Q31	0.494	0.485	0.635	0.387	0.592	0.313	0.449	0.634	0.676				
Q17	0.443	0.289	0.345	0.302	0.346	0.159	0.409	0.469	0.412	0.407			
Q32	0.573	0.483	0.418	0.340	0.621	0.301	0.503	0.525	0.639	0.533	0.335		
Q33	0.574	0.418	0.366	0.359	0.437	0.322	0.403	0.486	0.479	0.450	0.494	0.531	
Mean	3.502	3.375	2.717	3.044	3.305	3.559	3.559	3.590	3.127	2.819	3.032	3.371	3.949
SD	1.256	1.118	1.226	1.157	1.286	1.042	1.160	1.187	1.322	1.353	1.141	1.282	0.938

correlations of Q32 and Q33, two indicators of the Performance construct, with other variables tend to be similar in size, the correlation of Q32 with Q5 and Q30 is much larger than the correlation of Q33 with those variables. Similar violations of the proportionality constraints implied by the factor model can be found elsewhere in the correlation matrix as well. Given the sample size of 315, it is reasonable to expect that a constrained factor model, with each observed variable loading on a single construct, will show poor data-model fit.

Overall Results

For this comparison, two PLS path models were estimated, one using Mode A to link indicators to constructs and one using Mode B. These two sets of results are compared to an SEM analysis using a true factor analytic measurement model and maximum likelihood estimation. Judging by the χ^2 statistic and the data-model fit indices, overall fit is somewhat poor— $\chi^2 = 259.2$ ($df = 60$, $p < .0001$), CFI = .909, RMSEA = .103, SRMR = .055. In an actual application, this weak fit would raise doubts about interpretation of results, but these results still provide a basis for the current comparison.

A comparison of results across these analyses can start with the R^2 values for the dependent constructs, Performance and Satisfaction, as seen in Table 3.3. PLS path modeling users expect Mode B estimation to produce somewhat higher R^2 values for dependent constructs than Mode A estima-

TABLE 3.3. Selected Results from PLS Path Model and SEM Analysis

	Mode A	Mode B	SEM
Performance R^2	.522	.592	.799
Respect	.360 (.0657)	.402 (.0547)	.609 (.132)
Teamwork	.436 (.0567)	.450 (.0557)	.323 (.134)
Satisfaction R^2	.627	.669	.875
Respect	.478 (.0551)	.552 (.0561)	.676 (.144)
Performance	.395 (.0553)	.334 (.0564)	.284 (.149)
Communality			
Teamwork	.638	.559	.530
Respect	.646	.583	.475
Performance	.635	.606	.455
Satisfaction	.784	.756	.677
Redundancy			
Performance	.207	.229	.364
Satisfaction	.312	.281	.592

tion, because Mode B accounts for collinearity among the observed variable predictors while Mode A does not. Here the advantage for Mode B amounts to .070 for Performance and .042 for Satisfaction. The advantage for SEM is much larger—at least .200 higher than the Mode B R^2 for either dependent construct.

The communality and redundancy statistics for the two PLS models point to different trade-offs between Mode A and Mode B. Communality—the average squared correlation between a proxy and its indicators—is higher for Mode A for all constructs. However, the R^2 advantage for Mode B means that the redundancy statistics (the product of communality and R^2 for dependent constructs) are mixed for Mode A and Mode B. It is easy to compute communality and redundancy values for the SEM results. With each observed variable loading exclusively on one common factor, and with errors uncorrelated, the standardized loading is equal to the correlation between observed variable and factor. The SEM results are like the Mode B results, only more so: communalities are even lower than those in the Mode B results, while R^2 values are even higher, in the end producing higher redundancy values than those obtained from either Mode A or Mode B.

Parameter Estimates

A comparison of standardized parameters estimates yields mixed results. The standardized parameter estimate of the regression slope for the Respect construct predicting both Performance and Satisfaction is larger in the SEM results, but other SEM path estimates are smaller than the estimates from PLS path modeling. Correlations among the factors in SEM are universally larger than the correlations between proxies in either Mode A or Mode B (see Table 3.4), consistent with the tendency of PLS path modeling to underestimate the inner model while overestimating the outer model.

Comparisons of loadings from SEM with those from PLS Mode A are also consistent with this bias, as seen in Table 3.5. The Mode A loadings are universally larger than the SEM loadings. The weights produced by Mode A are roughly proportional to the Mode A loadings, and thus are somewhat homogeneous. By contrast, the Mode B weights cover a wide range, with some weights approximating zero. The difference between Mode A and Mode B appears to reflect the impact of collinearity on the Mode B weights.

Standard Errors

Estimated standard errors also suggest some fundamental differences across the analytical alternatives of Mode A, Mode B, and SEM. It is worth

TABLE 3.4   Factor/Proxy Correlations from SEM/PLS Path Model Analysis

	Teamwork	Respect	Performance
Mode A			
Respect	.646		
Performance	.668	.642	
Satisfaction	.684	.732	.702
Mode B			
Respect	.628		
Performance	.703	.685	
Satisfaction	.679	.781	.712
SEM			
Respect	.824		
Performance	.825	.875	
Satisfaction	.792	.925	.876

TABLE 3.5   Standardized Loadings / Weights (Standard Errors) from SEM and PLS Path Modeling Analysis

	SEM Loadings	Mode A Loadings	Mode A Weights	Mode B Weights
Teamwork				
Q6	.716 (.033)	.809 (.017)	.404 (.023)	.734 (.060)
Q7	.811 (.026)	.857 (.019)	.306 (.017)	.128 (.087)
Q8	.701 (.034)	.766 (.027)	.286 (.021)	.265 (.086)
Q9	.676 (.036)	.760 (.035)	.253 (.022)	.055 (.083)
Respect				
Q5	.848 (.024)	.865 (.015)	.512 (.025)	.745 (.050)
Q14	.503 (.036)	.699 (.044)	.280 (.027)	.000 (.049)
Q15	.674 (.025)	.838 (.021)	.431 (.019)	.377 (.063)
Performance				
Q17	.548 (.045)	.719 (.035)	.353 (.021)	.291 (.059)
Q32	.772 (.031)	.822 (.020)	.484 (.025)	.686 (.061)
Q33	.684 (.036)	.844 (.015)	.413 (.019)	.246 (.065)
Satisfaction				
Q29	.794 (.024)	.884 (.016)	.368 (.011)	.198 (.074)
Q30	.900 (.016)	.912 (.010)	.412 (.013)	.688 (.075)
Q31	.768 (.026)	.860 (.018)	.348 (.011)	.210 (.064)

remembering that the products of bootstrapping will vary from one instance to the next, even starting with the same dataset. Still, standard errors for the SEM loadings and for the Mode A loadings are comparable. Combined with the positive bias in Mode A loadings, this suggests that PLS Mode A is more likely to find statistically significant loadings. There is a large difference in standard errors for the Mode B weights and the Mode A weights, with the Mode A standard errors being substantially smaller. In regression, standard errors tend to be larger when R^2 is low and when collinearity is high (e.g., Cohen, Cohen, West, & Aiken, 2003). With the Mode A estimation method ignoring collinearity, it should not be surprising to see much smaller standard errors for the Mode A weights.

Stone-Geisser Q²

The Stone-Geisser approach can be applied to a given PLS path model in many ways. Here, the ultimate dependent construct is Satisfaction, so it makes sense to focus on that construct and its three indicators. As mentioned previously, Chin (2010) recommended Q^2 redundancy values of 0.5 or above. Using an exclusion distance of seven (that is, excluding every 7th value from the responses across the three indicators of Satisfaction, Q^2 communality values are all above 0.5, while Q^2 redundancy values are below 0.5, though they are substantially above 0 (see Table 3.6). This might suggest that the model needs more predictors in order to have high predictive relevance.

TABLE 3.6 Stone-Geisser Q² Applying Exclusion to the Indicators of Satisfaction

	Mode A	Mode B
Communality		
Teamwork	.638	.559
Respect	.646	.583
Performance	.635	.606
Satisfaction	**	**
Redundancy		
Performance	.326	.362
Satisfaction	.478	.492

**: Not calculated

WHY USE PLS PATH MODELING?

So, then, why should a researcher consider using PLS path modeling, when SEM is available? One might start with a broader question: Why choose a composite-based approach to modeling instead of a factor-based approach? PLS path modeling is only one of several composite-based approaches that researchers might consider, so the choice between a factor-based method like SEM and a composite-based method logically comes first.

There are a number of plausible arguments for favoring a composite-based method over a factor-based method. In fact, it has been argued, though not in the context of structural modeling, that composite-based methods ought to be routinely preferred (Velicer & Jackson, 1990). Factor-based methods offer unique benefits or rewards, but sometimes those rewards might not be valued. These rewards include: (a) unbiased parameter estimates when the model is correct and other assumptions hold; (b) flexibility in modeling error structures; (c) structural consistency between statistical model and motivating theory; and (d) statistical hypothesis testing of the model's constraints. Perhaps a researcher is not primarily concerned with parameter estimates, or perhaps the researcher views the parameters and the model itself as a convenient contrivance that facilitates insight but does not necessarily represent a precise reality. Similarly, perhaps the researcher does not value statistical hypothesis testing in the current research context. In an early stage of research, or in an exploratory mode, researchers might value "getting a feel for the data" much more than being able to conduct statistical hypothesis tests. Some researchers might reject the very notion that any statistical model bears an exact likeness to the reality being examined. A researcher who rejects the possibility of verisimilitude may be driven much more by convenience than by statistical benefits that come at a price.

Many within the SEM community do not expect constrained factor models to fit exactly in a given population. Steiger (1990) estimated the probability of achieving exact fit with a constrained model at "essentially zero" (p. 41). Concern about the assumption that the model fits exactly in the population provides the rationale for widely used alternative fit indices like RMSEA or CFI. Hypothesis testing loses some of its attraction when the tested hypothesis is believed a priori false. Misspecification may also induce bias in SEM parameter estimates, so that the choice between factor-based and composite-based parameter estimates is a choice between one set of biased estimates and another. In addition, researchers might simply reject factor analysis as a plausible measurement model (Rigdon, Ringle, Sarstedt, & Gudergan, 2011). Even if a particular composite-based approach does not exactly reflect the model believed to hold, it could in fact represent a better approximation than the factor model.

The statistical benefits of factor-based methods follow only when underlying assumptions hold. Correctness of the model in the population is one major assumption. Other key assumptions, varying by estimation method, relate to the distribution of the observations and the size of the sample. Violation of distributional assumptions figures prominently within the PLS literature as a reason to choose PLS path modeling over SEM, but of course those assumptions are very much a matter of estimation method and not of SEM itself. While maximum likelihood estimation leans on an assumption of multivariate normality, it has also been shown to be robust to mild or even moderate deviations from multivariate normality (e.g., see chapter by Finney & DiStefano, this volume), and other estimation methods require only milder assumptions. Granted, some of these alternatives make other difficult demands in terms of sample size, though not all do. On the other hand, independence of observations is a fundamental distributional assumption across SEM estimation methods, while the PLS path modeling literature insists that independence of observations is irrelevant for this method (Chin, 2010; Wold, 1982)—non-independence primarily undermines standard error estimates, but PLS's reliance on bootstrapping for standard errors renders this issue moot. SEM includes procedures for accounting for clustering of observations, but those methods do carry their own assumptions and limitations (see the chapter by Stapleton, this volume). Then again, the general insistence that PLS path modeling is largely indifferent to data issues has discouraged the development of special procedures for addressing data that are ordinal or clustered or missing. Some issues, such as the problem of ordinal data, certainly must have implications for PLS path modeling, just as they have implications for SEM (Finney & DiStefano, this volume). Thus, a researcher who chooses composite-based methods on distributional grounds might only be swapping one set of problems for another.

The choice between factor-based and composite-based methods might also be affected by sample size. Factor-based methods often rely on properties that hold only asymptotically, and these methods might be unstable when sample size is too low. By contrast, the simplest composite-based methods make very limited demands. A widely cited heuristic suggests that minimum sample size might be 10 observations per predictor within a single equation. Using this guideline, one might determine which of the many regressions within a PLS path model involves the largest number of predictors and multiply that number by 10 (Chin, 1998). It has also been argued, in the special context of modeling interactions, that PLS path modeling offer greater statistical power than regression (Chin et al., 2003), though this claim has been challenged (Goodhue, Lewis, & Thompson, 2007). More recently, Chin (2010) has recommended using power tables from regression to determine minimum sample size, again focusing on the individual regression with the fewest number of predictors. Built upon a foundation of OLS regression, PLS path modeling is likely to be stable at very low sample size. Moreover, other composite-based modeling approaches (some of which are described later in this chapter) are also available, and some of those alternatives may provide basic functionality when sample size is even lower. Thus, a researcher with a limited number of observations who finds SEM methods unworkable might certainly turn to composite-based methods to obtain some results.

More generally, there are research contexts where factor-based methods perform relatively poorly or fail altogether, where composite-based methods represent a useful alternative. Analysis of archival or secondary data and exploratory analysis are two examples. Factor-based measurement models, especially the highly constrained measurement models that seem to be favored, require multiple rounds of measure development and revision (Churchill, 1979). With archival or secondary data, the opportunity for revising and refining may not exist. Composite-based methods might provide results and spark insights where factor-based methods cannot.

Wold originally conceived PLS path modeling as an exploratory alternative to Jöreskog's confirmatory factor analysis. Yet, factor analysis itself originated as an exploratory tool. Exploratory factor analysis (EFA) allows observed variables to load on multiple factors, and offers more room for sparking insight than for model testing per se (Rozeboom, 1992). Recall that PLS does not permit a single observed variable to load on multiple constructs, but makes less stringent distributional assumptions. Moreover, methods exist for combining exploratory factor analysis with a structural model (Asparouhov & Muthén, 2009; Rozeboom, 1991). There is room to argue that exploratory methods are undervalued in many fields, and that many researchers waste a great deal of time and energy trying to make confirmatory methods work when exploratory methods are better suited to the job. As Velicer and Jackson (1990) stated:

> Exploratory analytic approaches...should be preferred except for those cases where a well-defined theory exists. Exploratory approaches avoid a confirmation bias, do not force a theory-oriented approach prematurely, and represent a conservative strategy. Confirmation approaches should be employed in a more limited role. (p. 21)

There is also reason to believe that the true relative utility of factor-based versus composite-based exploratory methods is at least an open question (Steiger, 1990; Velicer & Jackson, 1990), and that further research is needed to address this question in light of recent methodological innovations on both sides.

Finally, consider the issue of factor scores. Representing constructs as exact weighted functions of the observed variables is built into the PLS path modeling estimation algorithm. It has been argued that composite methods generally have a substantial advantage over factor-based methods in producing these factor scores (Steiger, 1990; Velicer & Jackson, 1990). Factor based methods suffer from *factor indeterminacy*, the intrinsic mathematical problem in trying to use p observed variables to render m common factors plus p error terms. The extent of this indeterminacy—the degree to which different sets of factor scores for the same data and model can disagree among themselves—can be substantial, especially when loadings are weak and there are few observed variables per factor (Velicer & Jackson, 1990). Any factor scores derived as functions of the observed variables are only approximation to the underlying factors. By contrast, the proxy scores from PLS path modeling have a deterministic relation to the observed variables and a straightforward derivation. With these scores neatly summarizing the observed variables within the context of the structural model, a researcher might proceed to further analyses of many sorts. This might seem, then, to be a capability unique to composite methods. When comparing scores derived from the same data using principal components versus factor analysis, for example, the scores tend to be similar to the point that substantive conclusions are essentially the same across both methods (Velicer & Jackson, 1990). Still, composite methods retain the general advantage of producing scores that are easy to explain, especially to a lay audience.

RECENT DEVELOPMENTS IN PLS PATH MODELING

PLS path modeling seems to have gone through a long period of stagnation after its initial development, but recent years have seen innovations along many dimensions. Wold insisted on homogeneity—all data conforming to the same statistical model—as one of the few distributional assumptions for PLS path modeling: "For model estimation and evaluation in general, including PLS modeling, it is a general requisite that the data are homogeneous and uniform" (Wold, 1985, p. 587). Today, multiple approaches are available for PLS modeling of heterogeneous data (Rigdon et al., 2010). A variety of procedures have been proposed for multiple group analysis (Sarstedt, Henseler, & Ringle, 2011) where respondents are assigned to populations a priori, but none of the current approaches permits the across-group constraints that are familiar in multiple group SEM (Hwang & Takane, 2004). Procedures have also been proposed for PLS path modeling with interactions (Chin et al., 2003; Goodhue et al., 2007) where the strength of a relation between two variables changes with the value of a third variable.

A number of approaches have been proposed for dealing with latent heterogeneity (Jedidi, Jagpal, & DeSarbo, 1997), where data are believed to represent a mixture of multiple populations and where affiliation of respondents with populations is unknown. The *finite mixture PLS* (FIMIX-PLS) method applies Jedidi et al.'s (2003) algorithm to the case values resulting from a standard PLS analysis (Rigdon et al., 2010; Rigdon et al., 2011). Note that while finite mixture modeling results in SEM are largely driven by differences in observed variable means (see Pastor & Gagné, this volume), the routine standardization in PLS path modeling means that FIMIX-PLS results are driven by differences in slope parameters. The *response-based unit segmentation* (REBUS-PLS) algorithm assigns respondents to latent classes in order to optimize communality and R^2 (Esposito Vinzi et al., 2010). In *PLS typological path modeling*, respondents are assigned to classes so as to minimize squared residuals in the prediction of dependent variables (Squillacciotti, 2010). All of these latent heterogeneity techniques benefit from the explicit nature of case values in PLS path modeling.

Combination of PLS case values and data mining techniques have also been proposed. A genetic algorithm approach to latent heterogeneity avoids some of the limitations of the methods cited above (Ringle, Sarstedt, & Schlittgen, 2010). In this approach, different allocations of respondents to latent classes are encoded in an analogy to the genetic material of living things. Then processes analogous to reproduction and mutation lead to eventual selection of a solution that minimizes unexplained error variance in dependent variables.

A neural network approach has been employed for discovering nonlinearity relations among construct proxies (Buckler, 2010; Buckler & Hennig-Thurau, 2008). Starting with a PLS framework and researcher-provided prior probabilities, this approach discovers quadratic, cubic, and interaction relations that significantly improve prediction. For the applied user, these developments in accommodating and discovering different forms of heterogeneity may provide another strong reason to consider PLS path modeling and related methods.

COMPOSITE-BASED ALTERNATIVES
TO PLS PATH MODELING

While much of the available literature on PLS path modeling portrays a dichotomy, with composite-based PLS on the one hand and factor-based SEM on the other, there are actually many composite-based approaches to structural modeling. The essential problem is how to derive a single score for each construct from a set of observed variable values. As Dijkstra (2009) stated: "Both within and outside the PLS family there are many, many ways to construct indices" (p. 3). PLS Mode A and Mode B represent only two possibilities. One elementary alternative is the unit-weighted sum. A unit-weighted sum or average simplifies estimation and conserves degrees of freedom, but this simple approach has been shown to be competitive with regression weights under known conditions (Dana & Dawes, 2004; Rozeboom, 1979).

There are other approaches to representing relations among groups of observed variables. Canonical correlation represents the relation between two sets of variables. Analogous to principal components analysis, canonical correlation defines maximal correlations between successively extracted orthogonal components of the two sets of variables. In a model with two constructs and two sets of observed variables, the construct proxies produced by PLS Mode B are equivalent to the first canonical variates, with a between-proxy correlation equal to the first canonical correlation, but Mode B estimation will understate the correlation between sets of variables if the second and succeeding canonical correlations are nonzero (Rigdon & Gefen, 2011). An alternative procedure, *set correlation*, captures the relation between sets of variables across all canonical variates (Cohen, 1982). Set correlation includes the means to partial out the impact of additional sets of observed variables, just as relations are partialed in regression analysis.

Just as there are many ways to construct indices, there are many ways to invoke an overall fit criterion within composite variable modeling (McDonald, 1996). Generalized structured component analysis (GSCA) is an alternative to PLS path modeling (Hwang & Takane, 2004). Like PLS, GSCA minimizes distributional assumptions and uses bootstrapping to estimate standard errors. Unlike PLS path modeling, GSCA parameter estimation is driven by a single overall least squares criterion, maximizing the sum of variance explained for all observed variables linked to dependent constructs. In its two-step estimation algorithm, GSCA computes weights for all observed variables, and also estimates loadings for designated constructs. Instead of using proxies, GSCA uses an alternating least squares estimation algorithm, alternately updating observed variable weights in one step and inner model weights and observed variable loadings in the other step. Also unlike PLS, GSCA incorporates multiple group analysis with user-specified across-group constraints.

SOFTWARE FOR PLS PATH
MODELING AND RELATED TECHNIQUES

From the beginning, computational simplicity has been one of the virtues of PLS path modeling (Wold, 1985), indeed for composite-based modeling generally (Velicer & Jackson, 1990). It is easy to create routines to perform this analysis. Yet, for many years, PLS path modeling software largely consisted of LVPLS (Lohmöller, 1987), a Fortran program with a punch card-like user interface. A second generation of software simply formatted command files and data for LVPLS execution (Temme et al., 2010). Third generation packages, such as SmartPLS (Ringle et al., 2005) and PLS-Graph (Chin, 2001), are self-contained and offer a graphical interface and updated analytical features as well as flexible blindfolding. Users draw model structures with simple graphical tools and access features through menus, as seen in Figure 3.5. SmartPLS, for example, includes the FIMIX-PLS procedure for modeling latent heterogeneity, as well as a convenient tool for specifying multiplicative interaction models. SmartPLS is currently available as a free download (http://www.smartpls.de/).

PLS path modeling tools are also available as part of larger statistical packages. Users of the open-source R package (see Venables, Smith, & R Core Development Team, 2011) have access to the semPLS package (Monecke, 2012) which offers basic PLS functionality as well as a measure of interoperability with other R packages and with SmartPLS. The package PLS-PM performs PLS path modeling analysis as part of XLSTAT, a statistical analysis add-in for the Microsoft Excel spreadsheet package (http://www.xlstat.com/en/products-solutions/plspm.html). Several PLS packages rely on Microsoft Excel to format data files, so incorporating PLS analysis within Microsoft Excel offers additional convenience. Users need to be careful to distinguish between routines that perform PLS path modeling and those that perform PLS regression, a distinct but related procedure.

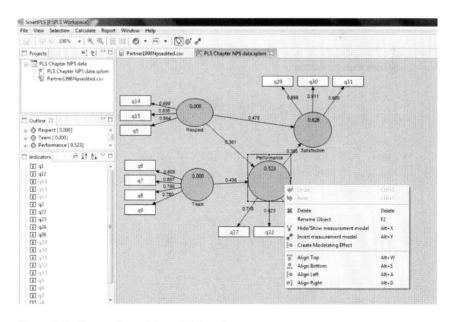

Figure 3.5 Screenshot of SmartPLS package.

Software is also available for the composite-based alternatives to PLS mentioned here. Systat13 (http://www.systat.com/) includes a routine for conducting Cohen's set correlation analysis. The NEUSREL package for exploring nonlinear relations, which uses Microsoft Excel to format both data and its command file, is available from http://www.neusrel.com/. The website for GeSCA, web-based software for conducting generalized structured component analysis, is http://www.sem-gesca.org/.

CONCLUSION

There seems to be a wide gulf separating the literatures on factor-based and component-based approaches to structural modeling, and that is a shame. It surely means that researchers are making inferior choices based on limited knowledge of alternatives. Composite-based methods like PLS path modeling have advantages in robustness and in the straightforward representation of constructs in terms of observed variables. SEM users who reject the conventional factor-based measurement model may find that PLS path modeling comes closer to representing the measurement model that they have in mind. Users who truly are in an exploratory mode, who are faced with the challenge of seeking insight from a trove of data but who are not ready to be constrained within a single model and attendant distributional assumptions, may find that PLS path modeling is the best tool for addressing their specific research needs. Indeed, it could be that the SEM world as a whole would benefit if it were tilted substantially away from its confirmatory, factor-based inclination toward a more exploratory and composite-based orientation.

ACKNOWLEDGEMENTS

The author would very much like to thank Marko Sarstedt, Theo Dijkstra, and Frank Buckler for their exceptionally helpful comments on this chapter. The chapter is better because of their contributions.

REFERENCES

Anscombe, F. J. (1973). Graphs in statistical analysis. *The American Statistician, 27,* 17–21.

Apel, H., & Wold, H. (1982). Soft modeling with latent variables in two or more dimensions: PLS estimation and testing for predictive relevance. In K. G. Jöreskog & H. Wold (Eds.), *Systems under indirect observation: Causality, structure, prediction* (Part II) (pp. 209–248). Amsterdam: North-Holland.

Armstrong, J. S. (in press). Illusions in regression analysis. *International Journal of Forecasting.*

Asparouhov, T., & Muthén, B. (2009). Exploratory structural equation modeling. *Structural Equation Modeling: A Multidisciplinary Journal, 16,* 397–438.

Bagozzi, R. P. (2011). Measurement and meaning in information systems and organizational research: Methodological and philosophical foundations. *MIS Quarterly, 35,* 261–292.

Boardman, A. E., Hui, B. S., & Wold, H. (1981). The partial least squares-fix point method of estimating interdependent systems with latent variables. *Communications in Statistics: Theory and Methods, 10,* 613–639.

Bollen, K. A., & Arminger, G. (1991). Observational residuals in factor analysis and structural equation modeling. *Sociological Methodology, 21,* 235–262.

Buckler, F. (2010). *NEUSREL 3.2 universal structural modeling software.* Cologne: Neusrel Causal Analytics.

Buckler, F., & Hennig-Thurau, T. (2008). Identifying hidden structures in marketing's structural models through universal structural modeling: An explorative Bayesian neural network complement to LISREL and PLS. *Marketing—Journal of Research and Management, 4,* 47–66.

Chin, W. W. (1998). The partial least squares approach for structural equation modeling. In G. Marcoulides (Ed.), *Modern methods for business research* (pp. 295–336). Mahwah, NJ: Lawrence Erlbaum Associates.

Chin, W. W. (2001). *PLS-Graph 3.0 user's guide.* Houston: Soft Modeling, Inc.

Chin, W. W. (2010). How to write up and report PLS analyses. In V. Esposito Vinzi, W. W. Chin, J. Henseler, & H. Wang (Eds.), *Handbook of partial least squares: Concepts, methods and applications* (pp. 655–690). Berlin: Springer-Verlag.

Chin, W. W., Marcolin, B. L., & Newsted, P. R. (2003). A partial least squares latent variable modeling approach for measuring interaction effects: Results from a Monte Carlo simulation study and an electronic mail / emotion-adoption study. *Information Systems Research, 14,* 189–217.

Churchill, G. A., Jr. (1979). A paradigm for developing better measures of marketing constructs. *Journal of Marketing Research, 16,* 64–73.

Cohen, J. (1982). Set correlation as a general data analytic method. *Multivariate Behavioral Research, 17,* 301–341.

Cohen, J., Cohen, P., West, S. G., & Aiken, L. S. (2003). *Applied multiple regression / correlation analysis for the behavioral sciences* (3rd ed.). Mahwah, NJ: Lawrence Erlbaum Associates.

Dana, J., & Dawes, R. M. (2004). The superiority of simple alternatives to regression for social science prediction. *Journal of Educational and Behavioral Statistics, 29,* 317–331.

Dijkstra, T. K. (2009, September). PLS for path diagrams revisited, and extended. In V. Esposito Vinzi, M. Tenenhaus, R. Guan, L. Guo, & B. Yi, 6th international conference on partial least squares and related methods, Beijing.

Dijkstra, T. K. (2010). Latent variables and indices: Herman Wold's basic design and partial least squares. In V. Esposito Vinzi, W. W. Chin, J. Henseler, & H. Wang (Eds.), *Handbook of partial least squares: Concepts, methods and applications* (pp. 23–46). Berlin: Springer-Verlag.

Embretson (Whitely), S. (1983). Construct validity: Construct representation versus nomothetic span. *Psychological Bulletin, 93*, 179–197.

Esposito Vinzi, V., Chin, W. W., Henseler, J., & Wang, H. (Eds.) (2010). *Handbook of partial least squares: Concepts, methods and applications.* Berlin: Springer-Verlag.

Fornell, C., Johnson, M., Anderson, E., Cha, J., & Bryant, B. (1996). The American Customer Satisfaction Index: Nature, purpose, and findings. *Journal of Marketing, 60*, 7–18.

Franke, G. R., Preacher, K. J., & Rigdon, E. E. (2008). Proportional structural effects of formative indicators. *Journal of Business Research, 61*, 1229–1237.

Gefen, D., Rigdon, E. E., & Straub, D. W. (2011). An update and extension to SEM guidelines for administrative and social science research. *MIS Quarterly, 35*, iii–xiv.

Geisser, S. (1975). The predictive sample reuse method with applications. *Journal of the American Statistical Association, 70*, 320–328.

Goodhue, D., Lewis, W., & Thompson, R. (2007). Statistical power in analyzing interaction effects: Questioning the advantage of PLS with product indicators. *Information Systems Research, 18,* 211–227.

Gotz, O., Liehr-Gobbers, K., & Krafft, M. (2010). Evaluation of structural equation models using the partial least squares approach. In V. Esposito Vinzi, W. W. Chin, J. Henseler, & H. Wang (Eds.), *Handbook of partial least squares: Concepts, methods and applications* (pp. 691–711). Berlin: Springer-Verlag.

Greenwald, A. G., Pratkanis, A. R., Leippe, M. R., & Baumgardner, M. H. (1986). Under what condition does theory obstruct research progress? *Psychological Review, 93*, 216–229.

Hancock, G.R., & Mueller, R.O. (2001). Rethinking construct reliability within latent variable systems. In R. Cudeck, S. du Toit & D. Sörbom (Eds.), *Structural equation modeling: Present and future* (pp. 195–216). Chicago: SSI.

Hair, J.F., Jr., Anderson, R.E., Tatham, R.L., & Black, W.C. (1995). *Multivariate data analysis* (4th ed.). Englewood Cliffs, NJ: Prentice-Hall.

Hair, J. F., Ringle, C. M., & Sarstedt, M. (2011). PLS-SEM: Indeed a silver bullet. *Journal of Marketing Theory and Practice, 19*, 139–152.

Hair, J. F., Sarstedt, M., Ringle, C. M., & Mena, J. A. (2012). An assessment of the use of partial least squares structural equation modeling in marketing research. *Journal of the Academy of Marketing Science, 40*, 414–433.

Henseler, J. (2010). On the convergence of the partial least squares path modeling algorithm. *Computational Statistics, 25*, 107–120.

Hui, B. S., & Wold, H. O. (1982). Consistency and consistency at large of partial least squares estimates. In K. G. Jöreskog & H. Wold (Eds.), *Systems under indirect observation: Causality, structure, prediction* (Part II) (pp. 119–130). Amsterdam: North-Holland.

Hwang, H., & Takane, Y. (2004). Generalized structured component analysis. *Psychometrika, 69*, 81–99.

Jedidi, K., Jagpal, H. S., & DeSarbo, W. S. (1997). Finite-mixture structural equation models for response-based segmentation and unobserved heterogeneity. *Marketing Science, 16*, 39–59.

Jöreskog, K. G. (1979). Basic ideas of factor and component analysis. In K. Jöreskog & D. Sörbom (Eds.), *Advances in factor analysis and structural equation modeling* (pp. 5–20). Cambridge, MA: Abt Books.

Lohmöller, J.-B. (1987). *LVPLS program manual* (version 1.8). Köln: Zentralarchiv für Empirische Sozialforschung.

MacCallum, R. C., & Browne, M.W. (1993). The use of causal indicators in covariance structure models: Some practical issues. *Psychological Bulletin, 114,* 533–541.

McDonald, R. P. (1996). Path analysis with composite variables. *Multivariate Behavioral Research, 31,* 239–270.

Monecke, A. (2012). Package 'semPLS. http://cran.r-project.org/web/packages/semPLS/semPLS.pdf

Mulaik, S.A. (2010). *Foundations of factor analysis* (2nd. Ed.). Boca Raton, FL: Taylor & Francis.

Muthén, B., & Asparouhov, T. (in press). Bayesian SEM: A more flexible representation of substantive theory. *Psychological Methods.*

Muthén, L. K., & Muthén, B. O. (2010). *Mplus 6 user's guide.* Los Angeles: Muthén & Muthén.

Reinartz, W. J., Krafft, M., & Hoyer, W.D. (2004). The customer relationship management process: Its measurement and impact on performance. *Journal of Marketing Research, 41,* 293–305.

Rigdon, E. E. (1994). Demonstrating the effects of unmodeled random measurement error. *Structural Equation Modeling: A Multidisciplinary Journal, 1,* 375–380.

Rigdon, E. E. (1995). A necessary and sufficient identification rule for structural models estimated in practice. *Multivariate Behavioral Research, 30,* 359–383.

Rigdon, E. E. (1998). Structural equation modeling. In G. A. Marcoulides (Ed.), *Modern methods for business research* (pp. 251–294). Mahwah, NJ: Lawrence Erlbaum Associates.

Rigdon, E. E., & Gefen, D. (2011). Questioning some claims associated with PLS path modeling. In S. M. Noble and C. H. Noble (Eds.), *Proceedings of the 2011 American Marketing Association Summer Educators Conference* (pp. 14–15). Chicago: American Marketing Association.

Rigdon, E. E., Ringle, C. M., & Sarstedt, M. (2010). Structural modeling of heterogeneous data with partial least squares. *Review of Marketing Research, 7,* 255–296.

Rigdon, E. E., Ringle, C. M., Sarstedt, M., & Gudergan, S. P. (2011). Assessing heterogeneity in customer satisfaction studies: Across industry similarities and within industry differences. *Advances in International Marketing, 22,* 169–194.

Ringle, C. M., Sarstedt, N., & Schlittgen, R. (2010). Finite mixture and genetic algorithm segmentation in partial least squares segmentation: Identification of multiple segments in complex path models. In A. Fink, B. Lausen, W. Seidel, & A. Ultsch (Eds.), *Advances in data analysis, data handling and business intelligence* (pp. 167–176). Berlin: Springer-Verlag.

Ringle, C. M., Wende, S., & Will, A. (2005), SmartPLS Version 2.0.M3. Hamburg. http://www.smartpls.de/.

Rozeboom, W. W. (1979). Sensitivity of a linear composite of predictor items to differential item weighting. *Psychometrika, 44,* 289–296.

Rozeboom, W. W. (1991). HYBALL: A method for subspace-constrained factor rotation. *Multivariate Behavioral Research, 26,* 163–177.

Rozeboom, W. W. (1992). The glory of suboptimal factor rotation: Why local minima in analytic optimization of simple structure are more blessing than curse. *Multivariate Behavioral Research, 27,* 585–599.

Sarstedt, M., Henseler, J., & Ringle, C. M. (2011). Multigroup analysis in partial least squares (PLS) path modeling: Alternative methods and empirical results. *Advances in International Modeling, 22,* 195–218.

Sörbom, D. (2001). Karl Jöreskog and LISREL: A personal story. In R. Cudeck, S. du Toit, & D. Sörbom (Eds.), *Structural equation modeling: Present and future: A festschrift in honor of Karl Jöreskog* (pp. 3–10) Lincolnwood, IL: Scientific Software.

Squillacciotti, S. (2010). Prediction-oriented classification in PLS path modeling. In V. Esposito Vinzi, W. W. Chin, J. Henseler, & H. Wang (Eds.), *Handbook of partial least squares: Concepts, methods and applications* (pp. 219–233). Berlin: Springer-Verlag.

Steiger, J.H. (1990). Some additional thoughts on components, factors and factor indeterminacy. *Multivariate Behavioral Research, 25,* 41–45.

Stone, M. (1974). Cross-validatory choice and assessment of statistical predictions. *Journal of the Royal Statistical Society, Series B (Methodological), 36,* 111–147.

Temme, D., Kreis, H., & Hildebrandt, L. (2010). A comparison of current PLS path modeling software: Features, ease-of-use, and performance. In V. Esposito Vinzi, W. W. Chin, J. Henseler, & H. Wang (Eds.), *Handbook of partial least squares: Concepts, methods and applications* (pp. 737–756). Berlin: Springer-Verlag.

Tenenhaus, M., Esposito Vinzi, V., Chatelin, Y.-M., & Lauro, C. (2005). PLS path modeling. *Computational Statistics & Data Analysis, 48,* 159–205.

United States Office of Personnel Management (1998). National partnership for reinventing government employee survey [Computer file]. ICPSR version. Washington, DC: United States Office of Personnel Management [producer]. http://www.icpsr.umich.edu/icpsrweb/ICPSR/studies/3419/detail

Velicer, W. F., & Jackson, D. N. (1990). Component analysis versus common factor analysis: Some issues in selecting an appropriate procedure. *Multivariate Behavioral Research, 25,* 1–28.

Venables, W. N., Smith, D. N., & R Core Development Team (2011). An introduction to R. R Core Development Team. http://www.r-project.org/.

Venkatesh, V., & Agarwal, R. (2006). Turning visitors into customers: A usability-centric perspective on purchase behavior in electronic channels. *Management Science, 52,* 367–382.

Wold, H. O. (1975), Path models with latent variables: The NIPALS approach. In H. M. Blalock, A. Agenbegian, F. M. Borodkin, R. Boudon, & V. Capecchi (Eds.), *Quantitative sociology: International perspectives on mathematical and statistical sociology* (pp. 307–357). New York, NY: Academic Press.

Wold, H. O. (1982). Soft modeling: The basic design and some extensions. In K.G. Jöreskog & H. Wold (Eds.), *Systems under indirect observation: Causality, structure, prediction* (Part II) (pp. 1–54). Amsterdam: North-Holland.

Wold, H. O. (1985). Partial least squares. In S. Kotz & N. L. Johnson (Eds.), *Encyclopedia of statistical sciences, vol. 6* (pp. 581–591).New York, NY: Wiley.

Wold, H. O. (1988). Predictor specification. In S. Kotz & N. L. Johnson (Eds.), *Encyclopedia of statistical sciences, vol. 8* (pp. 587–599).New York, NY: Wiley.

CHAPTER 4

POWER ANALYSIS IN STRUCTURAL EQUATION MODELING

Gregory R. Hancock and Brian F. French

POWER ANALYSIS IN STRUCTURAL EQUATION MODELING

Within the structural equation modeling (SEM) literature, methodological examinations of what constitutes adequate sample size, and the determinants thereof, are plentiful (see, e.g., Gagné & Hancock, 2006; Jackson, 2003; MacCallum, Widaman, Zhang, & Hong, 1999; Marsh, Hau, Balla, & Grayson, 1998). Predictably, practitioners gravitate toward those sources providing evidence and/or recommendations supporting the acceptability of smaller sample sizes. Such recommendations, however, tend to be based upon issues such as model convergence and parameter bias; they do not typically address statistical power. Thus, while there might be relatively large models for which, say, $n = 50$ yields reliable rates of model convergence with reasonable parameter estimates, such a sample size might be entirely inadequate in terms of power for statistical tests of interest.

Within SEM, two contexts for power analysis are relevant: tests of data-model fit for an entire model and tests of specific parameters within a given

Structural Equation Modeling: A Second Course (2nd ed.), pages 117–159
Copyright © 2013 by Information Age Publishing

model. For each context, a researcher might have a need for *post hoc power analysis* or *a priori power analysis*. For the former, data have been gathered and the statistical tests have already been conducted; the researcher now wishes to estimate the power associated with those particular statistical tests. For the latter, more directly termed *sample size planning*, the researcher's goal is to budget (e.g., financially, logistically) for the necessary sample size to achieve sufficient power for all statistical tests of interest. It is our belief for power analysis in SEM (and across other analytical contexts, for that matter) that a priori methods are *far* more important than post hoc methods. Conducting a post hoc power analysis after a null hypothesis has been rejected seems mainly to be establishing what is already apparent to the researcher, that there was sufficient power to detect what was believed to be there all along. Similarly, conducting a post hoc power analysis after a null hypothesis has been retained also is merely establishing what is already apparent to the researcher, in this case that there was insufficient power to detect what is likely still believed to be there. Thus, put rather cynically, post hoc power analysis seems either self-congratulatory or self-pitying, and seldom belief-changing. In addition, estimates of post hoc power are just that—estimates—and as such are subject to sampling variability that can yield misleading and even inaccurate assessments of power (e.g., Hoenig & Heisey, 2001; Yuan & Maxwell, 2005). For these reasons, philosophical and statistical, the current chapter will not address post hoc power estimation for SEM.

More positively, a priori power analysis can prove extremely useful in estimating the sample size necessary for conducting statistical tests of interest that, in SEM, can be for data-model fit as well as for specific model parameters. Indeed, conducting such analyses prior to entering into complex modeling endeavors makes sound practical sense, and is becoming increasingly recognized as indispensable for researchers seeking funding for their applied research. In the current chapter, a priori power analysis will be addressed as it relates to testing data-model fit for an entire model as well as to testing parameters within a model. Because of the increasingly wide array of models to which this framework applies, the focus of the chapter will be on the most common scenario of the single-sample covariance structure model analyzed using maximum likelihood (ML) estimation, which rests on the assumptions about the data (conditional multivariate normality) and the model (exact or reasonably close fit; see Curran, Bollen, Paxton, Kirby, & Chen, 2002; Steiger, Shapiro, & Browne, 1985; Yuan & Bentler, 2004). Some assumptions will be able to be relaxed a bit using Monte Carlo approaches to power analysis, which also will be addressed in this chapter. Finally, extensions of the core SEM power analysis principles presented to alternative scenarios, such as models involving multiple samples and models with mean structures, will be touched upon briefly later in the chapter.

To begin, a priori power analysis (hereafter simply *power analysis*) in SEM is built upon the same basic ideas as power analysis in other settings (e.g., analysis of variance). These are (a) null and alternative hypotheses, (b) test statistics to assess the null hypothesis, and (c) central and noncentral distributions. Each of these is described briefly below in the context of SEM, after which they are addressed in detail in the context of tests of data-model fit and tests of model parameters.

Null and Alternative Hypotheses

For any research question involving specific characteristics of a population (e.g., means, variances), we can conceive of appropriate null and alternative hypotheses. Within the context of SEM, these hypotheses may concern the fit of the entire model or specific parameters within that model. For the entire model, data-model fit might be summarized by a popular index, such as the root mean-square error of approximation (RMSEA; Steiger & Lind, 1980) or comparative fit index (CFI; Bentler, 1990), where the null hypothesis could be that the model fits perfectly (or "acceptably" by some criterion) in the population and the alternative hypothesis is that the model does not. For a parameter within a model, generically denoted as θ, the typical (and nondirectional) null hypothesis would be that H_0: $\theta = 0$, while the alternative hypothesis would be the converse, H_1: $\theta \neq 0$.

Test Statistics to Assess the Null Hypothesis

Sometimes the population characteristic of interest is in a metric that is familiar and easy to work with for power analysis, whereas other times it is not. The latter is especially true in the case of power analysis methods that rely on known asymptotic distributions rather than those derived empirically (e.g., as in bootstrapping). Thus, the actual test statistic for statistically assessing data model fit might be, say, the sample RMSEA or CFI, which are already in metrics with which the structural modeler is acquainted. In the case of a model parameter, however, estimates of θ tend to be less useful for power analysis purposes, and instead more familiar test statistics might be employed. Most commonly these include a parameter estimate's z-value or the likelihood ratio (χ^2 difference) test statistic assessing the model with and without that parameter (i.e., with and without that parameter constrained to its null hypothesized value, typically 0). For the purposes of this chapter, let us generically refer to the test statistic of interest that is used to test a null hypothesis, whether for data-model fit or a specific parameter, as *T*.

Central and Noncentral Distributions

When a null hypothesis is true, the test statistic T follows a sampling distribution described as a *central* distribution, often with familiar distributional characteristics when certain underlying (and rather ideal) assumptions hold. In Figure 4.1, the left side of the figure shows the case where the null hypothesis H_0 is true, in which observed test statistic values labeled as T_0 follow the central sampling distribution and have expected value $E(T_0)$ (which is 5.00 in the figure). Conversely, when H_0 is false and instead a specific alternative H_1 is true, the sampling distribution for the test statistic differs from that under the null condition; such a sampling distribution is referred to as a *noncentral* distribution. In Figure 4.1, a particular noncentral sampling distribution of observed test statistic values (labeled as T_1) is depicted to the right of the central distribution, with expected value $E(T_1)$ (which is 17.83 in the figure). Such distributions are often described in terms of the expected value's location with respect to that of the central distribution, expressed as a *noncentrality parameter* λ, where

$$\lambda = E(T_1) - E(T_0). \tag{4.1}$$

For the noncentral distribution depicted in Figure 4.1, the noncentrality parameter is $\lambda = 12.83$.

Power

The above basic principles regarding null and alternative hypotheses, test statistics, and distributions that are central and noncentral are fundamental to all of power analysis, whether in SEM or elsewhere. As seen in Figure 4.1, under conditions when a null hypothesis H_0 is true, a point of particular interest within the central distribution is the critical value at the $(1 - \alpha) \times 100$th percentile. This point, labeled as $_{(1-\alpha)}T_0$ in the figure, corresponds to 11.07 for the commonly used $\alpha = .05$ level. Relative to the central distribution, the noncentral distribution of T_1 test statistics under the alternative hypothesis H_1 is shifted to the right considerably. In fact, because of its relatively large degree of noncentrality ($\lambda = 12.83$), most of the noncentral distribution of T_1 values exceeds the point defined by the α-level critical value $_{(1-\alpha)}T_0$ in the central distribution. The proportion of the noncentral distribution exceeding the point $_{(1-\alpha)}T_0$ is, by definition, the power π of the statistical test of H_0 given the condition that this specific alternative hypothesis H_1 is true. Symbolically,

$$\pi = \text{pr}[T_1 > _{(1-\alpha)}T_0 | H_1]. \tag{4.2}$$

For the noncentral distribution in Figure 4.1, the power is $\pi = .80$, implying that if one repeatedly conducted the α-level statistical test of interest un-

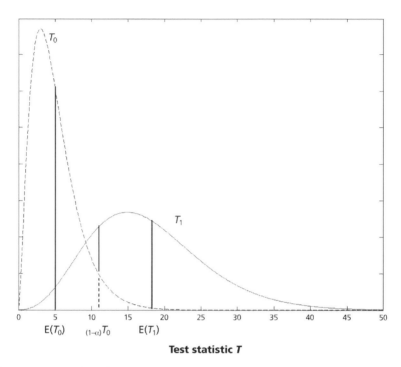

Test statistic _T_

Figure 4.1 Sampling distribution for test statistic _T_ under null (T_0) and alternative (T_1) conditions.

der the specific nonnull (and hence noncentral) condition shown, 80% of those tests would be expected to lead to a rejection of H_0 (given that all salient assumptions regarding, for example, distributions, are met). Thus, if we know the distributional form of the test statistic _T_ under both the null (central) and the specific alternative (noncentral) conditions, then knowledge of the noncentrality parameter λ will allow for the determination of power. This holds true whether dealing with z, t, F, or χ^2 distributions, with the latter holding particular interest for power analysis within SEM. These principles are drawn upon below to facilitate power analysis for testing data-model fit as well as parameters within a model. Because in the practice of SEM satisfactory data-model fit should be established before evaluating individual parameters, power analysis at the model level will be presented first.

POWER ANALYSIS FOR TESTING DATA-MODEL FIT

As presented above, there are three common elements to consider prior to addressing power analysis: null and alternative hypotheses, test statistics to assess the null hypothesis, and central and noncentral distributions. Each of these will be addressed specifically as they bear on testing data-model fit in SEM, setting the stage for methods of power analysis for testing data-model fit.

Null and Alternative Hypotheses

A possible, albeit rather optimistic, null hypothesis regarding one's structural equation model for p variables with parameters in vector θ, is that the model is correct in the population. If this is indeed the case, then the $p \times p$ covariance matrix implied by the model ($\hat{\Sigma}(\theta)$, or simply $\hat{\Sigma}$) and the corresponding information observed in the population data (Σ) are identical; that is, there is exact fit in the population such that $F_{ML} = 0$, where

$$F_{ML} = \ln\left|\hat{\Sigma}\right| + tr(\Sigma\hat{\Sigma}^{-1}) - \ln|\Sigma| - p. \tag{4.3}$$

Unfortunately, there is not a reciprocal relation between model truth and model fit. That is, as discussed by Hershberger and Marcoulides (this volume), many different models can give rise to the same data moments, and thus while a correct model implies equal model-implied and observed moments, the converse need not be true: equal model-implied and observed moments need not imply a correct model. In the end, then, while a researcher might wish to express null and alternative hypotheses in terms of population model-reality correspondence, the null and alternative models tested are only about population data-model fit. In this framework a rejection of the null hypothesis of exact data-model fit leads to the reasonable inference of an incorrect model (i.e., no model-reality correspondence); a failure to reject the null hypothesis leads merely to a tentative retention of that model as one in a set of models that could exactly represent the population processes at work.

Even though we can express null and alternative hypotheses in terms of exact data-model fit, as described above, in fact there are two primary drawbacks of doing so. The first is the backwards nature of the hypothesis testing to which it leads. Specifically, whereas rejecting a null hypothesis in favor of an alternative is typically the desired outcome in an analytical endeavor, in testing exact fit in SEM one would wish to retain the null hypothesis as it represents consonance between population data and the theoretical model of interest. Given that statistical power refers to the probability of rejecting a hypothesized model rather than retaining it, relative to more traditional analytical scenarios the power analysis for testing data-model fit as a whole in SEM needs to be reframed (as will be done below).

The second drawback of testing exact data-model fit, as alluded to above, is its unrealistic optimism. In general, researchers do not expect of their model a complete fidelity to the underlying population mechanisms, but rather, that the model is a reasonable approximation to reality. Theoretically trivial internal misspecifications (e.g., omitted error covariances or cross loadings), while indicative of a false theoretical model, might be deemed quite acceptable at a practical level. For this reason, rather than discussing models as true or false, practicality suggests discussing models as having *acceptable* or *unacceptable* data-model fit. In fact, as MacCallum, Browne, and Sugawara (1996) suggested, for testing purposes it is useful to define the null hypothesis for a given model as containing population data-model fit at the threshold between acceptable and unacceptable fit (termed "close fit" and "not close fit," respectively, by MacCallum et al.). The researcher's goal, then, becomes the rejection of this null hypothesis in favor of an alternative in which the model of theoretical interest has acceptable data-model fit, and in executing the study itself to amass sufficient power for such a favorable rejection.

So, in order to differentiate between acceptable and unacceptable fit, a metric for data-model fit needs to be established. One could characterize data-model fit using the fit function F_{ML} directly, where relatively smaller values are indicative of less discrepancy between observed and model-implied moments and hence more acceptable fit, and relatively larger values indicate a greater discrepancy in the fit between data and model and thus less acceptable fit. Unfortunately, the metric for fit functions is not particularly intuitive, in part because their values are tied to the complexity of the models themselves. As a result, Steiger and Lind (1980) recommended the root mean square error of approximation (RMSEA), ε, as an index to characterize the degree of discrepancy between model-implied and observed moments, and hence a metric for assessing degree of acceptability of data-model fit. The RMSEA represents a dispersal of data-model discrepancy (F_{ML}) across degrees of freedom:

$$\varepsilon = \sqrt{\frac{F_{ML}}{df}}. \tag{4.4}$$

Using the RMSEA as a metric for data-model fit, we may define the value of ε_0 as representing the transition from acceptable to unacceptable data-model fit, which in turn will help to articulate the null and alternative hypotheses. One could use, for example, the null hypothesis $H_0: \varepsilon = \varepsilon_0$, that the population data-model fit is at the boundary between acceptability and unacceptability. The alternative hypothesis in this case, $H_1: \varepsilon \neq \varepsilon_0$, contains two possibilities: $H_1: \varepsilon < \varepsilon_0$, that the data-model fit is acceptable, and $H_1: \varepsilon > \varepsilon_0$, that the data-model fit is unacceptable. For the purposes of power analysis, however, interest is in being able to reject data-model fit at the threshold in favor of acceptable data-model fit. As such, we will hereafter define the alternative hypothesis as $H_1: \varepsilon < \varepsilon_0$, and hence the corresponding null hypothesis as $H_0: \varepsilon \geq \varepsilon_0$.

As for the numerical choice of ε_0, some authors have suggested .05 as representing a reasonable threshold between acceptable ($< .05$) and unacceptable ($> .05$) data-model fit (Browne & Cudeck, 1993; Browne & Mels, 1990; Steiger, 1989). This value will be adopted for power analysis in the current chapter (although a way to choose other values will be made available later in the chapter as well). That is, for this chapter the null and alternative hypotheses regarding population data-model fit will be $H_0: \varepsilon \geq .05$ and $H_1: \varepsilon < .05$, respectively (although options will exist for choosing other thresholds). Thus, sample information must be gathered in a general attempt to reject the null hypothesis containing unacceptability in favor of the alternative of acceptability. Test statistics for such hypothesis testing will be discussed next.

Test Statistics to Assess the Null Hypothesis

If one held as the null hypothesis of choice that there was exact data-model fit in the population, this could be represented in a variety of ways—$F_{ML} = 0$ and $\varepsilon = 0$, among others. To test these one could conceive of using sample estimates $\hat{F}_{ML}$ or $\hat{\varepsilon}$, respectively. Regarding the former, $\hat{F}_{ML}$ for a covariance structure model with $p \times p$ sample covariance matrix $\mathbf{S}$ is the familiar

$$\hat{F}_{ML} = \ln\left|\hat{\Sigma}\right| + tr(\mathbf{S}\hat{\Sigma}^{-1}) - \ln\left|\mathbf{S}\right| - p, \tag{4.5}$$

whose expected value may be derived (see Browne & Cudeck, 1993) to be

$$E[\hat{F}_{ML}] = F_{ML} + \frac{df}{n-1}. \tag{4.6}$$

Rearranging Equation 4.6, this means that the best estimator of the population data-model fit F_{ML} is

$$\hat{F}_{ML} - \frac{df}{n-1}, \tag{4.7}$$

and hence the RMSEA estimate $\hat{\varepsilon}$ follows accordingly from Equations 4.4 and 4.7 as

$$\hat{\varepsilon} = \sqrt{\frac{\hat{F}_{ML} - \dfrac{df}{(n-1)}}{df}} = \sqrt{\frac{(n-1)\hat{F}_{ML} - df}{df(n-1)}}. \tag{4.8}$$

(where, if the numerator is negative, it is customarily set to 0). This test statistic, whose sampling distribution is discussed below, may in turn be used for hypothesis testing for data-model fit.

As mentioned previously, though, because a retention of the null hypothesis is actually the desired outcome in the exact data-model fit testing paradigm, a shift is required—quite literally a distributional shift to a tolerable and rejectable level of data-model misfit. As this was framed using the RMSEA, specifically $\varepsilon_0 \geq .05$, $\hat{\varepsilon}$ is the appropriate test statistic of choice. That is, if $\hat{\varepsilon}$ is statistically significantly less than .05 at some acceptable α level, then we will reject the null hypothesis containing unacceptable population data-model fit in favor of the alternative of acceptable population data-model fit, $\varepsilon_0 < .05$. Otherwise, the null hypothesis containing unacceptable data-model fit will be retained. The last components necessary for power analysis, then, are the central and noncentral distributions associated with $\hat{\varepsilon}$.

Central Distributions (for H_0) and Noncentral Distributions (for H_1)

H_0 and central $\hat{\varepsilon}$ distributions. Under a null condition that $\varepsilon_0 = .05$ (ignoring, for the moment, $\varepsilon_0 > .05$), we expect $\hat{\varepsilon}$ to fluctuate randomly around .05. This sampling distribution is thus a central distribution with respect to this null condition, but with its own distributional characteristics. Fortunately, under distributional and model conditions mentioned at the beginning of the chapter, $\hat{\varepsilon}$ may be expressed as a translations of the model χ^2 statistic, and hence its sampling behavior is a function of the corresponding χ^2 distribution. Specifically, because

$$\chi^2 = (n-1)\hat{F}_{\mathrm{ML}} \qquad (4.9)$$

with distributional noncentrality parameter

$$\lambda = (n-1)F_{\mathrm{ML}}, \qquad (4.10)$$

by substitution of Equation 4.9 into Equation 4.8 we get the RMSEA estimate expressible as

$$\hat{\varepsilon} = \sqrt{\frac{\chi^2 - df}{df(n-1)}}. \qquad (4.11)$$

Further, the central distribution describing the variation of $\hat{\varepsilon}$ around ε_0 can be translated to a more familiar noncentral χ^2 distribution with noncentrality parameter λ_0. Specifically, following Equations 4.4 and 4.10:

$$df(n-1)\varepsilon_0^2 = \lambda_0. \tag{4.12}$$

So, for example, if one has designated the customary threshold $\varepsilon_0 = .05$ for a model with $df = 5$ that is to be evaluated using data from $n = 1,027$ subjects, the corresponding noncentral χ^2 distribution with 5 df has noncentrality parameter $\lambda_0 = (5)(1,027 - 1)(.05^2) = 12.83$. This is, in fact, the noncentral distribution pictured in Figure 4.1. Thus, in this example the central fluctuation of $\hat{\varepsilon}$ statistics around .05 can be represented by the noncentral fluctuation of model χ^2 statistics around 17.83, their expected value in a distribution with $df = 5$ and $\lambda = 12.83$.

H_1 and noncentral $\hat{\varepsilon}$ distributions. Although the alternative hypothesis of H_1: $\varepsilon < .05$ contains many possibilities for ε, power analysis requires the designation of a specific alternative numerical value. That is, one must state how much better (smaller) than $\varepsilon = .05$ one is willing to consider the data-model fit in the population. Most optimistic would be the designation of an alternative value $\varepsilon_1 = 0$, implying the belief (or hope) that the model has exact data-model fit in the population. Somewhat less optimistic, but more realistic, are values between 0 and .05, the choice of which can be based on one's literature-informed expectations and/or general conservatism. Sampling distributions of $\hat{\varepsilon}$ under conditions with ε_1 closer to 0 are more distinct from those under the null $\varepsilon_0 = .05$ condition, while sampling distributions of $\hat{\varepsilon}$ under conditions with ε_1 closer to .05 are considerably less distinct from those under the null $\varepsilon_0 = .05$ condition.

Just as in the null case where the central sampling distribution of $\hat{\varepsilon}$ values translates to a χ^2 distribution, so too does the noncentral distribution of $\hat{\varepsilon}$ values in the alternative case. Following from Equation 4.12, the corresponding χ^2 distribution has noncentrality parameter λ_1, where

$$\lambda_1 = df(n-1)\varepsilon_1^2. \tag{4.13}$$

Thus, for a study with data from $n = 1,027$ subjects, the optimistic value of $\varepsilon_1 = 0$ for a model with $df = 5$ would correspond to the designation of $\lambda_1 = 0$, meaning that the noncentral sampling distribution of $\hat{\varepsilon}$ statistics when $\varepsilon_1 = 0$ transforms to a central χ^2 distribution (with $df = 5$, in this case). On the other hand, the same study with an alternative value of, say, $\varepsilon_1 = .02$, would have its noncentral sampling distribution of $\hat{\varepsilon}$ statistics transform to a 5-df noncentral χ^2 distribution with $\lambda_1 = 5(1,027 - 1)(.02)^2 = 2.05$. Thus, the central fluctuation of $\hat{\varepsilon}$ statistics when $\varepsilon_1 = .02$ under these study conditions can be represented by the noncentral fluctuation of model χ^2 statistics around 7.05, their expected value in a distribution with $df = 5$ and $\lambda = 2.05$. This distribution sits somewhere between those pictured in Figure 4.1, and thus has a greater degree of overlap with the $\varepsilon = .05$ distribution to the right than does the $\varepsilon = 0$ distribution to the left. The designation of ε_1, therefore, will have obvious implications for power, as described next.

Power and Examples

A priori power analysis for testing data-model fit addresses the question of how many participants one needs in order to achieve a desired level of power to reject the null hypothesis containing unacceptable data-model fit. The current chapter will present tables useful for conducting sample size planning for tests of data-model fit as a whole. Following Hancock and Freeman's (2001) adaptation of the SAS programs offered by MacCallum et al. (1996), which draw from noncentral χ^2 distributions, a sample size table for an $\alpha = .05$ level one-tailed test was created for $\varepsilon_1 = .00$, .02, and .04., where ε_0 was again set to the recommended value of .05. For each level, model degrees of freedom were varied in increments of 5 from $df = 5$ to 250, while tabled power levels included $\pi = .70$, .80, and .90. Sample sizes required for these variations of ε_1, df, and π appear in Table 4.1.

As an example, imagine that a researcher was planning to evaluate data-model fit of a model with $df = 15$, and the desired power is .80 for an $\alpha = .05$-level test associated with the RMSEA. As presented previously, the sample size required would depend upon the degree to which the model's level of data-model fit, and hence the reference distribution's degree of noncentrality, deviate from the null $\varepsilon_0 = .05$. For a perfectly specified model ($\varepsilon_1 = .00$), Table 4.1 shows that a minimum of $n = 530$ is required. As perfect fit is rather optimistic, one might more reasonably consider $\varepsilon_1 = .02$, for which a minimum sample size of $n = 741$ is necessary. On the most conservative side tabled, true data-model fit of $\varepsilon_1 = .04$ would require a considerably larger $n = 4,349$ in order to achieve power of .80. For researchers interested in planning to have sufficient power to reject $\varepsilon_0 \geq .05$ in favor of acceptable data-model fit, selecting $\varepsilon_1 = .02$ seems like a desirable balance between the generally unrealistic optimism of $\varepsilon_1 = .00$ and the frequent impracticality associated with the recommended sample size for $\varepsilon_1 = .04$.

For conducting a power analysis, the levels of ε_1 covered by Table 4.1 should allow for an adequate picture of sample size necessary in planning a typical study. However, in situations where more precision is needed for df values not directly tabled, two options exist. First, one may interpolate sample sizes nonlinearly using exponential functions derived empirically and presented by Hancock and Freeman (2001). Second, and much more simply, Preacher and Coffman (2006) offered an on-line utility for RMSEA-based sample size planning at http://www.quantpsy.org/rmsea/rmsea.htm (which may be used in lieu of Table 4.1 altogether). Consider, for example, a researcher interested in a model with $df = 8$, who chooses to plan a study

TABLE 4.1 Sample Size for Models with *df* = 5 to 250 at Power Levels π = .70, .80, .90

df	$\varepsilon_1 = .00$			$\varepsilon_1 = .02$			$\varepsilon_1 = .04$		
	.70	.80	.90	.70	.80	.90	.70	.80	.90
5	929	1,099	1,354	1,379	1,707	2,234	9,616	12,568	17,328
10	586	690	842	821	997	1,273	4,930	6,406	8,786
15	452	530	644	616	741	934	3,364	4,349	5,936
20	377	441	535	505	604	756	2,579	3,319	4,510
25	329	384	464	435	518	644	2,106	2,699	3,653
30	294	343	414	386	458	567	1,789	2,285	3,081
35	268	312	376	349	414	510	1,562	1,988	2,672
40	247	288	347	321	379	466	1,391	1,765	2,364
45	231	268	323	297	351	431	1,257	1,591	2,125
50	217	252	303	278	328	402	1,150	1,451	1,933
55	205	238	286	262	309	377	1,061	1,336	1,775
60	195	226	271	249	292	357	987	1,240	1,643
65	186	216	259	237	278	339	924	1,158	1,532
70	178	207	248	226	266	323	870	1,088	1,236
75	171	199	238	217	254	309	823	1,027	1,165
80	165	191	229	209	245	297	781	974	1,103
85	159	185	221	201	236	286	744	926	1,049
90	154	179	214	194	228	276	711	884	1,000
95	150	173	207	188	220	267	682	846	956
100	145	168	201	183	213	258	655	811	917
110	138	159	190	173	202	244	608	752	848
120	131	152	181	164	191	231	569	702	790
130	125	145	173	157	183	220	535	659	741
140	120	139	166	150	175	211	506	622	699
150	116	134	160	144	168	202	481	590	662
160	112	129	154	139	162	194	459	561	630
170	108	125	149	134	156	188	439	536	601
180	105	121	144	130	151	181	421	514	575
190	102	117	140	126	146	176	405	493	552
200	99	114	136	122	142	170	390	475	531
225	93	107	127	114	133	159	359	435	486
250	88	101	120	108	125	150	333	403	450

using ε_1 = .02. Further, a level of power of π = .80 is desired for the α = .05 level test associated with the RMSEA. Using Hancock and Freeman's (2001) interpolation methods or using Preacher and Coffman's (2006) on-line utility (setting Alpha to .05, Degrees of Freedom to 8, Desired Power to .80, Null RMSEA to .05, and Alt. RMSEA to .02) yields a minimum sample size of n = 1,182. That is, to have a .80 probability of rejecting $\varepsilon_0 \geq .05$ in favor of acceptable data-model fit (using an $\alpha = .05$ level test), given that the true level of data model fit in the population is $\varepsilon_1 = .02$, the study would require a minimum of 1,182 subjects. As the reader may also verify, had the researcher been interested in assuming exact data-model fit in the population, $\varepsilon_1 = .00$, the minimum required sample size would have been $n = 800$.

Issues and Extensions

Although the above section presented a fairly common approach to power analysis for the model as a whole, other perspectives and strategies exist as well. Briefly, the RMSEA was employed as a tool to assess data-model fit and to plan sample sizes accordingly. Other fit indices exist, of course, and these could be utilized in a similar manner; the interested reader is referred to MacCallum and Hong (1997) and Kim (2005). Also, fit indices such as the RMSEA could be used to plan not just for the test of data-model fit, as presented here, but also for the precision of the estimate of the population's data model fit; work by Kelley and Lai (2011) discussed such extensions. Further, the RMSEA was used above specifically for assessing the fit of a single-sample covariance structure model. Additional uses of the RMSEA can include (a) models with mean structures and/or multisample models, using the fit functions associated with those more complex models in an extension of the methods presented above, and (b) competing model comparisons and planning adequate sample size for those statistical comparisons. Details of these comparison approaches can be found in, for example, MacCallum, Browne, and Cai (2006) and, more recently, in Li and Bentler (2011). Alternatively, for strategies that avoid this traditional testing framework altogether, seeking instead to use information criteria for model assessment and sample size planning, the reader is directed to interesting work by Preacher, Cai, and MacCallum (2007). Finally, for techniques that are less reliant upon distributional assumptions such as those associated with the ML estimation paradigm, Monte Carlo simulation procedures are addressed later in the chapter.

POWER ANALYSIS FOR TESTING
PARAMETERS WITHIN A MODEL

Power analysis for testing parameters within a model, just as with power analysis for data-model fit as a whole, may be addressed in terms of the three foundational ideas: null and alternative hypotheses, test statistics to assess the null hypothesis, and central and noncentral distributions. Each of these is described below in the specific context of parameter testing in

SEM, after which power analysis is developed and illustrated for testing parameters within a model.

Null and Alternative Hypotheses

Consider the example of a two-tailed test of an unstandardized path in a latent variable path model, where that path is from factor F1 to factor F2. In this case the null hypothesis would commonly be H_0: $b_{F2F1} = 0$, while the corresponding alternative would be H_1: $b_{F2F1} \neq 0$. More generally, for any single parameter θ one can imagine a null hypothesis H_0: $\theta = \theta_0$ and its corresponding alternative as H_1: $\theta \neq \theta_0$, where the most frequent choice of θ_0 is 0. That is, researchers most commonly care about detecting any degree of nonzero contribution of a given parameter to the model; the current chapter will focus on power analysis associated with this purpose.

The above expression regarding a single parameter extends to any number of parameters as contained in a parameter vector θ. Consider a model with a "full" set of parameters in vector θ_F, and a model with a "reduced" set of parameters in vector θ_R, where elements in θ_R constitute a subset of elements in θ_F; the models therefore have a hierarchical (nested) relation. The reduced model contains what we might term the *peripheral* parameters, which define a context for the parameters of key interest; the full model contains the peripheral parameters as well as the *focal* parameters, the latter being those for which power is of interest. As an example, θ_F might contain all the parameters contained in an oblique confirmatory factor analysis (CFA) model (i.e., factor covariances, factor variances, loadings, and error variances), while θ_R could contain the parameters in a competing orthogonal CFA model (i.e., factor variances, loadings, and error variances); the focal parameters are thus the factor covariances. The null hypothesis in this case could be expressed directly as pertaining to the factor covariances contained in θ_F, specifically that they are all zero; correspondingly, the alternative hypothesis would be that one or more are nonzero. More generally, and more useful for the purposes of this chapter, the null and alternative hypotheses may be framed in terms of the overall badness of fit of the population data to the model with θ_F and to the model with θ_R. Expressed generically, we may write the null hypothesis for the population as

H_0: *Badness of data-model fit with* θ_R = *Badness of data-model fit with* θ_F.

This would occur in the above CFA case, for example, when the factor covariances are all zero (i.e., orthogonality exists). The corresponding alternative hypothesis for the population is thus

H_1: *Badness of data-model fit with θ_R > Badness of data-model fit with θ_F,*

where a false null hypothesis implies that the badness of fit associated with the reduced model (with relatively fewer parameters in θ_R) is necessarily greater than that associated with the full model (with relatively more parameters in θ_F). This would occur in the above CFA case, for example, when at least one of the factor covariances is nonzero (i.e., some obliqueness exists). Thus, the null hypothesis is that the reduced and full models have equivalent data-model fit in the population, while the alternative hypothesis is that the reduced model has worse data-model fit in the population.

To be more concrete, when assessing the population-level data-model fit for models with full and reduced sets of parameters, one starts with the fit function associated with the desired method of estimation. For ML estimation, the relevant degree of data-model fit at the population level for a covariance structure model involving π measured variables is the familiar F_{ML} presented in Equation 4.3, and repeated here for convenience:

$$F_{ML} = \ln\left|\hat{\Sigma}\right| + tr(\Sigma\hat{\Sigma}^{-1}) - \ln\left|\Sigma\right| - p. \tag{4.14}$$

Fitting population data Σ to both reduced and full models would yield population fit function values $F_{ML}(\theta_R)$ and $F_{ML}(\theta_F)$, respectively, where $F_{ML}(\theta_R) - F_{ML}(\theta_F) \geq 0$ as the full model has additional parameters and hence cannot have more badness of data-model fit in the population. That is, for testing one or more parameters in a covariance structure model, the null hypothesis may be expressed as H_0: $F_{ML}(\theta_R) = F_{ML}(\theta_F)$, or equivalently

$$H_0: F_{ML}(\theta_R) - F_{ML}(\theta_F) = 0, \tag{4.15}$$

indicating that the constraint of parameters in the reduced model has no effect on the data-model fit in the population. In the orthogonal and oblique CFA example, the null hypothesis states that the factors do not covary in the population, and therefore restricting them from doing so has no effect on the population-level data-model fit. The corresponding alternative hypothesis may be expressed as H_1: $F_{ML}(\theta_R) > F_{ML}(\theta_F)$, or equivalently

$$H_1: F_{ML}(\theta_R) - F_{ML}(\theta_F) > 0, \tag{4.16}$$

indicating that the additional parameters characterizing the difference between the full and reduced models do have at least some nonzero degree of impact on population-level data-model fit. For the CFA example, the alternative hypothesis states that there is at least some degree of covariation among the factors, and therefore restricting them from covarying entirely would have a deleterious effect on the population-level data-model fit. This framing of the null and alternative hypotheses, in terms of relative data-model fit of full and reduced models in the population, will prove particularly useful for power analysis.

Test Statistics to Assess Null Hypotheses

In the example of a two-tailed test of H_0: $b_{F2F1} = 0$, the population parameter b_{F2F1} is estimated by that value derived from sample data, $\hat{b}_{F2F1}$; this is in turn typically transformed to an observed z-statistic, where

$$z = \frac{\hat{b}_{F2F1}}{\text{SE}(\hat{b}_{F2F1})}$$

(assuming ML estimation). In fact, for any parameter θ, to test H_0: $\theta = 0$ the parameter estimate derived from sample data, $\hat{\theta}$, is customarily transformed to

$$z = \frac{\hat{\theta}}{\text{SE}(\hat{\theta})}$$

automatically within the SEM software.

To be more general, we return to parameter assessment framed in terms of comparative data-model fit for full and reduced models in the population, specifically focusing on the statistical comparison of models with reduced and full sets of parameters (θ_R and θ_F, respectively). For testing the null hypothesis in Equation 4.15, the test statistic $\hat{F}_{ML}(\hat{\theta}_R) - \hat{F}_{ML}(\hat{\theta}_F)$ may be determined from sample data that have been fit to both reduced and full models. This estimated difference in data-model fit, which has $df_{diff} = df_R - df_F$, will fluctuate from sample to sample with its own sampling distribution; fortunately, under the data and model conditions mentioned at the beginning of the chapter, this statistic converts to a χ^2 distribution (with test statistic χ^2_{diff}) as

$$(n-1)[\hat{F}_{ML}(\hat{\theta}_R) - \hat{F}_{ML}(\hat{\theta}_F)] = \chi^2_{diff}. \qquad (4.17)$$

The χ^2_{diff} statistic allows one to evaluate hypotheses regarding differences in data-model fit for models with and without focal parameters (or, more generally speaking, with and without constraints on those parameters). Note that one might also compare models using data-model fit indices such as information criteria (e.g., AIC, BIC); however, for the purpose of providing a *statistical* comparison upon which power analysis will be grounded, this χ^2_{diff} test statistic is essential.

Central Distributions (for H$_0$) and Noncentral Distributions (for H$_1$)

H$_0$ and central distributions. As discussed previously, when a null hypothesis is true, the statistic used to test H$_0$ follows a central sampling distribution, often with familiar distributional characteristics when certain underlying (and rather ideal) assumptions hold. Within the context of SEM, the test statistic (z) for evaluating H$_0$: $b_{F2F1} = 0$, for example, will asymptotically approach a standard normal distribution under assumed conditions. Correspondingly, the square of this test statistic, z^2, will asymptotically approach a χ^2 distribution (with 1 df) under the same distributional conditions, thereby approximating the difference in data-model fit with and without that particular parameter.

More generally, the test statistic facilitating the comparison of data-model fit for reduced and full models (as per Equation 4.17) will asymptotically approach a χ^2 distribution under data and model conditions mentioned previously. When H$_0$ is true, this is a central χ^2 distribution, with sample χ^2 values fluctuating randomly around their expected value which is equal to the number of degrees of freedom, $E[\chi^2] = df$. For comparing reduced and full models, the null condition is that the set of additional parameters contained in θ_F is unnecessary; that is, $F_{ML}(\theta_R) = F_{ML}(\theta_F)$. In such cases, one possibility is that the reduced and full models both have exact fit in the population (i.e., $F_{ML}(\theta_R) = F_{ML}(\theta_F) = 0$), and as such their corresponding χ^2 sampling distributions are centrally located at df_R and df_F, respectively. When fitting sample data to both models, the χ^2 difference statistic $\chi^2_{diff} = \chi^2_R - \chi^2_F$ follows a central χ^2 sampling distribution located at df_{diff}, where $df_{diff} = df_R - df_F$. A second possibility is that, although $F_{ML}(\theta_R) = F_{ML}(\theta_F)$, neither the reduced model nor the full model is correct in the population (i.e., $F_{ML}(\theta_R) = F_{ML}(\theta_F) \neq 0$). In such cases the reduced and full model test statistics do not follow central χ^2 distributions; however, it is their difference statistic whose behavior is important. As simulation evidence by Yuan and Bentler (2004) has indicated, in this case the behavior of χ^2_{diff} can be approximated by a central χ^2 distribution, but only as long as the full (and thus reduced) model is reasonably close to correct. In such cases the inference relying upon this test statistic ought to be acceptably valid.

H$_1$ and noncentral distributions. As discussed previously, when H$_0$ is false and instead a specific alternative H$_1$ is true, the sampling distribution for the test statistic differs from that under the null condition and is instead a noncentral distribution. Within the context of SEM, when H$_0$: $b_{F2F1} = 0$ is false, for example, the test statistic (z) will ideally follow a noncentral distribution, which also is normal in shape but is centered at a point other than zero. More generally, when a null hypothesis H$_0$ is false and a particular alternative H$_1$ is true, the χ^2 test statistic facilitating the comparison of data-model fit for reduced and full models will follow a noncentral distribution under assumed conditions, with sample χ^2 values fluctuating randomly around their expected value which equals the number of degrees of freedom plus some degree of noncentrality, $E[\chi^2] = df + \lambda$.

With respect to comparing models with and without specific parameters (i.e., with and without parameters constrained to zero), the alternative condition is when one or more of the focal parameters contained in θ_F are not zero, and hence $F_{ML}(\theta_R) > F_{ML}(\theta_F)$. In such cases one possibility is that the full model has exact fit in the population ($F_{ML}(\theta_F) = 0$) while the reduced model does not ($F_{ML}(\theta_R) > 0$). Here the model χ^2 sampling distribution for the full model is centrally located at df_F, while that for the reduced model is noncentral with expected value $df_R + \lambda$. The noncentrality parameter λ results from the improperly omitted (or, more generally, improperly constrained) parameters in the reduced model, and may be determined as

$$\lambda = \lambda_R - \lambda_F = (n-1)F_{ML}(\theta_R) - (n-1)F_{ML}(\theta_F) = (n-1)[F_{ML}(\theta_R) - F_{ML}(\theta_F)], \quad (4.18)$$

which equals

$$\lambda = (n-1)[F_{ML}(\theta_R)] \quad (4.19)$$

when the full model is correct. When fitting sample data to both models, the χ^2 difference statistic $\chi^2_{diff} = \chi^2_R - \chi^2_F$ asymptotically approaches a noncentral χ^2 sampling distribution located at $df_{diff} + \lambda = (df_R - df_F) + \lambda$. When comparing models under a false null hypothesis, where the full model has better data-model fit in the population than the reduced model, the second possibility is that neither model is actually correct (i.e., $F_{ML}(\theta_F) > 0$ and $F_{ML}(\theta_R) > 0$). For this nonnull scenario, simulation evidence by Yuan and Bentler (2004) implied that χ^2 difference tests would only be satisfactorily approximated by noncentral χ^2 distributions for full models that have fairly close fit. Power analysis proceeds under the key assumed condition that the full model is reasonably well specified.

Power

The above basic principles regarding null and alternative hypotheses, test statistics, and distributions that are central and noncentral, are fundamental to determining the necessary sample size to conduct the α-level parameter test(s) of interest with a desired level of power π. For a given number of degrees of freedom, corresponding to the number of focal parameters to be assessed in a single test, an α-level χ^2 test has power as per Equation 4.2 that is governed by the degree of noncentrality associated with the specific alternative hypothesis H_1. To achieve a target level of power, customarily $\pi = .80$, there must be a specific degree of noncentrality for the alternative distribution; such necessary noncentrality parameters appear in Table 4.2

TABLE 4.2. Noncentrality parameters for .05-level χ^2 tests

π								*df*							
	1	2	3	4	5	6	7	8	9	10	20	30	40	50
.10	.43	.62	.78	.91	1.03	1.13	1.23	1.32	1.40	1.49	2.14	2.65	3.08	3.46
.15	.84	1.19	1.46	1.69	1.89	2.07	2.23	2.39	2.53	2.67	3.77	4.63	5.35	5.98
.20	1.24	1.73	2.10	2.40	2.67	2.91	3.13	3.33	3.53	3.71	5.18	6.31	7.26	8.10
.25	1.65	2.26	2.71	3.08	3.40	3.70	3.96	4.21	4.45	4.67	6.45	7.82	8.97	9.99
.30	2.06	2.78	3.30	3.74	4.12	4.46	4.77	5.06	5.33	5.59	7.65	9.24	10.57	11.75
.35	2.48	3.30	3.90	4.39	4.82	5.21	5.56	5.89	6.19	6.48	8.81	10.60	12.10	13.42
.40	2.91	3.83	4.50	5.05	5.53	5.96	6.35	6.71	7.05	7.37	9.96	11.93	13.59	15.06
.45	3.36	4.38	5.12	5.72	6.25	6.72	7.15	7.55	7.92	8.27	11.10	13.26	15.08	16.68
.50	3.84	4.96	5.76	6.42	6.99	7.50	7.97	8.40	8.81	9.19	12.26	14.60	16.58	18.31
.55	4.35	5.56	6.43	7.15	7.77	8.32	8.83	9.30	9.73	10.15	13.46	15.99	18.11	19.98
.60	4.90	6.21	7.15	7.92	8.59	9.19	9.73	10.24	10.71	11.15	14.71	17.43	19.71	21.72
.65	5.50	6.92	7.93	8.76	9.48	10.12	10.70	11.25	11.75	12.23	16.05	18.96	21.40	23.55
.70	6.17	7.70	8.79	9.68	10.45	11.14	11.77	12.35	12.89	13.40	17.50	20.61	23.23	25.53
.75	6.94	8.59	9.76	10.72	11.55	12.29	12.96	13.59	14.17	14.72	19.11	22.44	25.25	27.71
.80	7.85	9.63	10.90	11.94	12.83	13.62	14.35	15.02	15.65	16.24	20.96	24.55	27.56	30.20
.85	8.98	10.92	12.30	13.42	14.39	15.25	16.04	16.77	17.45	18.09	23.20	27.08	30.33	33.19
.90	10.51	12.65	14.17	15.41	16.47	17.42	18.28	19.08	19.83	20.53	26.13	30.38	33.94	37.07
.95	12.99	15.44	17.17	18.57	19.78	20.86	21.84	22.74	23.59	24.39	30.72	35.52	39.54	43.07

for many different numbers of degrees of freedom for $\alpha = .05$ level tests. For the common case of $df = 1$, for example, the alternative distribution must have noncentrality of at least $\lambda = 7.85$ to yield a minimum of $\pi = .80$ power for a .05-level test (assuming standard distributional conditions).

The key in sample size planning, then, is to determine what conditions lead to the required noncentrality. Following from the previous presentation, and from the reader's prior experience with power analysis, one expects the degree of falseness of the null hypothesis H_0 (i.e., $F_{ML}(\theta_R) - F_{ML}(\theta_F)$) and the sample size n to play key roles in noncentrality, and hence power. With regard to H_0, the difference between the focal parameter values in the full model's parameter vector θ_F and those in the reduced model's parameter vector θ_R (usually values of zero) yields a difference in fit functions, $F_{ML}(\theta_R) - F_{ML}(\theta_F)$, whose magnitude directly relates to noncentrality and power. To elaborate briefly, the parameters of the full model in θ_F yield model-implied population moment information in $\hat{\Sigma}(\theta_F)$. Using this population moment information as input for the full and reduced models yields $F_{ML}(\theta_R)$ and $F_{ML}(\theta_F)$, respectively. Given that the moment information was based on θ_F, we expect $F_{ML}(\theta_F) = 0$; however, the data-model fit associated with the reduced model, $F_{ML}(\theta_R)$, should be nonzero under a false null hypothesis. The magnitude of $F_{ML}(\theta_R)$, then, becomes critical. As per Equation 4.19, multiplying this fit function difference by the quantity $n - 1$ yields the noncentrality parameter λ; this relation will prove key in a priori power analysis.

We now rearrange the information in the above paragraph as needed. A researcher knows how many focal parameters are being tested using an α-level test, and thus df_{diff} is known; from this information the necessary noncentrality parameter λ may be determined (from, for example, Table 4.2) in order to yield the desired level of power. Next, the numerical values of the parameters in θ_R and θ_F must be specified; this includes the focal parameters being tested and the peripheral parameters not being tested. For the focal parameters, typically (although not necessarily) the null values are 0; that is, the researcher is interested in testing whether the focal parameters are nonzero. As for the alternative values of those focal parameters, the researcher must posit specific numerical values. Just as in sample size planning in other areas of statistical data analysis, researchers must make educated guesses as to the nature of the effect(s) they wish to have sufficient power to detect. These guesses might be made based on theory or prior research, or might be selected to be conservative so as to ensure sufficient power to detect effects of at least that magnitude or of practical importance. Some helpful strategies in making these guesses and/ or simplifying the process are offered by Lai and Kelley (2011), including one discussed later in this chapter.

As stated above, numerical values must be specified not just for the focal parameters, but for *all* remaining peripheral parameters as well. These peripheral parameters might include factor loadings, error variances, inter-factor covariances, and so forth, and they define the context in which the tests of the focal parameters are being conducted. The numerical values of these parameters can have a bearing on the noncentrality associated with the focal parameter tests, and hence must be chosen with care as well (see below). Once designated, these values are part of both the reduced and full model parameter vectors, θ_R and θ_F, respectively, thus being held constant while determining $\hat{\Sigma}(\theta_F)$ and hence the resulting fit function value $F_{ML}(\theta_R)$.

Finally, following from Equation 4.19, the necessary sample size may be determined as

$$n = 1 + \left[\frac{\lambda}{F_{ML}(\theta_R)} \right]. \tag{4.20}$$

That is, the sample size is determined that magnifies the badness of fit of the reduced model so as to provide sufficient noncentrality to achieve the desired level of power for the α-level test of the focal parameter(s). The practical steps involved in this process are detailed more explicitly below, followed by an example using a latent variable path model. Note that the researcher must initially decide whether α-level testing of the whole set of focal parameters simultaneously is of interest, or if testing each focal parameter individually is the objective for the sample size planning. In this latter and more common case, there will exist pairs of full and reduced models each differing by a single focal parameter and where the remaining focal parameters are temporarily treated as peripheral. This process for planning sample size for tests of all focal parameters is represented in Figure 4.2, and summarized below in a series of steps.

Step 1. This step involves critical preliminary decisions. First, decide upon the desired level of power π and the α level for the focal parameter test(s), which are customarily .80 and .05, respectively (we do remind the reader, however, that these levels are somewhat arbitrary and can be modified within reason given substantive research concerns; see, e.g., Cohen, 1988). Second, determine the corresponding target noncentrality parameter(s); for .05-level tests, the values in Table 4.2 are useful. For tests involving a single parameter, setting power to $\pi = .80$ leads to $\lambda = 7.85$.

Third, and most critical, researchers must select numerical values for all parameters of the full model. This is probably the most difficult part of any a priori power analysis, and even more so in SEM as values for both focal and peripheral parameters must be selected. To help, for many models researchers might find it easier to select parameter values in a standardized rather than unstandardized metric (e.g., standardized path coefficients); exceptions to this will certainly exist, however, such as for latent growth models where the units are an essential part of the model. Even if research-

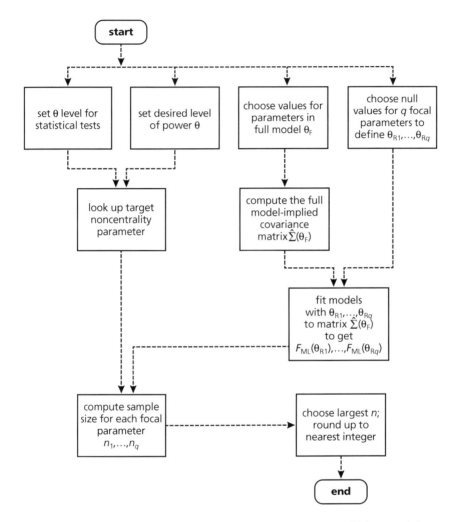

Figure 4.2 Flowchart for power analysis for testing parameters within a model.

ers do find it more convenient to think in terms of standardized units, they might find it challenging to think within the context of the model under examination. For example, while one might have a theory-based sense of the magnitude of the relation between V1 and V4, estimating that magnitude after controlling for the roles of V2 and V3 might be more challenging. Prior research might not inform this choice, as V2 and V3 might not have been included in prior analyses. A further challenge is that the values researchers posit for focal and/or peripheral parameters might wind up being logically impossible, yielding local problems such as endogenous variable variances in excess of 1 (when attempting to choose parameter values on a standardized metric), or more global problems such as a non-positive definite model-implied population covariance matrix $\hat{\Sigma}(\theta_F)$. Again, we suggest relying on past research, theory, and/or practical importance in selecting these values. Given that the challenges of parameter designation for the full model have been surmounted, the values of the focal parameters for the reduced model must be assigned. This task is easier, given that values of zero are most commonly chosen. In the end, if there are q focal parameters and sample size planning is to be conducted for the test of each individual parameter, then there will be a reduced parameter vector associated with each: $\theta_{R1}, \ldots, \theta_{Rq}$.

Step 2. From the values assigned in θ_F, the relevant model-implied moment information must be determined for the population. For covariance structure models this is $\hat{\Sigma}(\theta_F)$, which, if standardized path values were chosen, will resemble a correlation matrix with values of 1 along the diagonal. These moment values can be determined by path tracing, the algebra of expectations using the structural equations, or in some cases from the modeling software itself. With regard to the latter, one might fix all parameters in the full model to the values in θ_F and request the model-implied covariance matrix; or, one might use the values in θ_F as start values, set the number of iterations to zero, and again request the model-implied covariance matrix. In either of the software strategies, input information must be supplied (i.e., a covariance matrix for covariance structure models), but this input is essentially irrelevant as the program will not iterate to fit that information. By whatever means, then, the researcher has determined the necessary moment information in $\hat{\Sigma}(\theta_F)$.

Step 3. Having obtained the relevant model-implied population moment information from the full model, that information must now be used as input data to estimate the population data-model fit for each reduced model (with parameters in $\theta_{R1}, \ldots, \theta_{Rq}$). Note, because the necessary sample size is not yet known, an arbitrary sample size (e.g., $n = 1,000$) will generally have to be assigned in order for the modeling software to analyze the supplied population moment information. Now, the only source of badness of data-model fit associated with this reduced model comes directly from the constraint of the focal parameter to zero; as such, each reduced model will iterate to yield a model fit function value $F_{ML}(\theta_R)$. From these fitted values $F_{ML}(\theta_{R1}), \ldots, F_{ML}(\theta_{Rq})$, along with the target noncentrality parameter λ, the necessary sample size for each reduced model may now be estimated using Equation 4.20, yielding $n_1, \ldots, n_q$. Choosing the largest of these sample sizes, and rounding up to the nearest integer, will ensure sufficient power for tests of all focal parameters (under the assumed distributional and model conditions).

Example of Power Analysis for Testing Parameters Within a Model

Imagine that a researcher is planning to assess the impact of latent Reading Self-Concept and latent Math Self-Concept on ninth graders' latent Math Proficiency. In the past, three questionnaire items have been used as measured indicators of each of the self-concept constructs (V1 through V3 for reading, and V4 through V6 for math), although additional questionnaire items are available. As for Math Proficiency, two well-established standardized tests are available for use with the students of interest. The full model the researcher plans to use is depicted in Figure 4.3. The focal parameters will be those relating the latent variables: c_{F1F2}, b_{F3F1}, and b_{F3F2} (seen in Figure 4.3 in rounded dashed boxes). Each of these will be of individual interest; that is, necessary sample size should be determined for the test of each parameter separately.

Step 1. The level of power desired for these single-parameter (1 df) tests will be set at the customary $\pi = .80$ for .05-level tests. The corresponding target noncentrality parameter to be used for all parameters, as shown in Table 4.2, is $\lambda = 7.85$. For the peripheral parameters in the measurement model, imagine that past research has shown standardized loadings for Reading Self-Concept indicators to be approximately $b_{V1F1} = .50$, $b_{V2F1} = -.70$, and $b_{V3F1} = .60$; corresponding error variances for a standardized model would be $c_{E1} = .75$, $c_{E2} = .51$, and $c_{E3} = .64$. For the Math Self-Concept indicators, prior evidence has shown standardized loadings to be approximately $b_{V4F2} = .80$, $b_{V5F2} = .90$, and $b_{V6F2} = -.60$, thus having corresponding error variances for a standardized model of $c_{E4} = .36$, $c_{E5} = .19$, and $c_{E6} = .64$. The two Math Proficiency tests have reported reliability estimates of .88 and .93, thus suggesting standardized loading values of $b_{V7F3} = .94$ and $b_{V8F3} = .96$ (the square roots of the reliabilities), respectively; corresponding error variances would in turn be $c_{E7} = .12$ and $c_{E8} = .07$.

For the focal parameters, the researcher expects a negative relation between the Reading and Math Self-Concept factors of standardized magnitude $c_{F1F2} = -.30$. The expected direct effects on Math Proficiency, in a standardized metric, are set at $b_{F3F1} = .20$ for Reading Self-Concept and .50 for $b_{F3F2} = $ Math Self-Concept. Although the researcher suspects the true relations might be a bit higher, slightly conservative values are chosen to ensure sufficient power. To preserve the standardized framework, the remaining peripheral parameters in the latent portion of the model are $c_{F1} = 1$, $c_{F2} = 1$, and $c_{D3} = .77$ (which the reader may verify using, for example, path tracing).

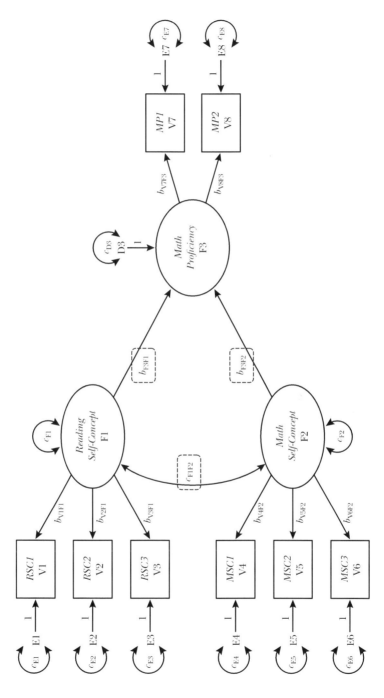

Figure 4.3   Latent variable path model for example of a priori power analysis for testing parameters within a model.

Step 2. Having articulated theoretical values for all parameters in the full model, the model-implied covariance matrix $\hat{\Sigma}(\theta_F)$ may be determined. In this example, as the reader may verify, the matrix (shown to two decimals) is

$$\hat{\Sigma}(\theta_F) = \begin{bmatrix} 1.00 & & & & & & & \\ -.35 & 1.00 & & & & & & \\ .30 & -.40 & 1.00 & & & & & \\ -.12 & .17 & -.14 & 1.00 & & & & \\ -.14 & .19 & -.16 & .72 & 1.00 & & & \\ .09 & -.13 & .11 & -.48 & -.54 & 1.00 & & \\ .02 & -.03 & .03 & .33 & .37 & -.25 & 1.00 & \\ .02 & -.03 & .03 & .34 & .38 & -.25 & .90 & 1.00 \end{bmatrix}.$$

Step 3. Now, using the implied matrix as input (with more decimals than shown above), three reduced models are fitted in turn: one with c_{F1F2} constrained to zero, one with b_{F3F1} constrained to zero, and one with b_{F3F2} constrained to zero. In doing so, the resulting maximum likelihood fit function values reflecting the badness of data-model fit associated with each single constraint (to three decimal places) are $F_{F1F2} = .052$, $F_{F3F1} = .026$, and $F_{F3F2} = .196$ (using EQS 6.2).

Finally, following from Equation 4.20, the necessary sample sizes (rounded up to the next integer) are determined as $n_{F1F2} = 152$, $n_{F3F1} = 306$, and $n_{F3F2} = 42$. Thus, the sample size needed to ensure at least .80 power for tests of all focal parameters, is $n = 306$. This makes the standard distributional assumption of conditional multivariate normality, assumes that the model at hand is correct, and assumes that the parameter values used for the peripheral and focal parameters also are correct (if, in fact, any focal parameters are actually larger than planned, power will exceed .80 at this sample size).

A Simplification for Models with Latent Variables

As evident from the above presentation, a considerable amount of information must be assumed in order to determine sample size in a power analysis. For models with latent variables, where the relations among latent variables will almost always contain the focal parameters, assumptions must be made about the nature of the supporting measurement model both in terms of each factor's number of indicator variables and the magnitude of those indicators' loadings. In practice, however, such information might not be well known. In fact, a researcher might actually wonder what quality of measurement model is necessary to achieve the power desired for tests within the structural portion of the model. While the quality of a factor's measurement model has been operationalized in a variety of ways (e.g., Fornell & Larcker, 1981), Hancock and Kroopnick (2005) demonstrated that one particular index is directly tied to noncentrality and power analysis in covariance structure models (see also Penev & Raykov, 2006). This measure is referred to as *maximal reliability* in the context of scale construction (e.g., Raykov, 2004), and as the measure of construct reliability *coefficient H* in the context of latent variable models (see, e.g., Hancock & Mueller, 2001). Specifically, Hancock and Kroopnick illustrated that the precise nature of the measurement model is actually irrelevant; instead, the researcher may simply specify a single number as a place holder for the measurement model of each factor in the latent variable path model, where that number reflects an anticipated (or minimum achievable) construct reliability associated with each factor's measurement model.

For a given factor with k indicators, the value of coefficient H may be computed using standardized loadings ℓ_i as follows:

$$H = \frac{\sum_{i=1}^{k}\left(\dfrac{\ell_i^2}{1-\ell_i^2}\right)}{1+\sum_{i=1}^{k}\left(\dfrac{\ell_i^2}{1-\ell_i^2}\right)}. \tag{4.21}$$

There are a number of interpretations for this index, detailed by Hancock and Mueller (2001). Most simply, it represents the proportion of variance in the construct that is theoretically explainable by its indicator variables. As such it is bounded by 0 and 1, is unaffected by loadings' sign, can never decrease with additional indicators, and is always greater than (or equal to) reliability of strongest indicator (i.e., $\max(\ell_i^2)$). Thus, the assessment of construct reliability draws information from the measured indicators only inasmuch as that information is germane to the construct itself. For Reading Self-Concept (F1), Math Self-Concept (F2), and Math Proficiency (F3) from the previous example, these values may be computed (to two decimals) to be $H_{F1} = .65$, $H_{F2} = .87$, and $H_{F3} = .95$, respectively.

Following from Hancock and Kroopnick (2005), as far as power is concerned, how one achieves the particular level of construct reliability for a given factor is irrelevant. A construct reliability of $H = .65$ could be achieved with three indicators of standardized loadings of .50, −.70, and .60, as in the previous example. It could also be achieved, for instance, with 10 indicators each having a standardized loading of .40, or with a single indicator of standardized loading .81 $(= \sqrt{.65})$. Assessments of sample size necessary to achieve a desired level of power are the same. For simplicity, then, the model in the previous example could have been depicted as shown in Figure 4.4, where factors have single indicators each with a standardized loading equal to the square root of the factor's construct reliability. As these single indicators are placeholders for the respective measurement models, they are designated as "V" terms.

To be explicit, then, in this case the peripheral measurement model parameters are set to $b_{V1F1} = .81$ ($= \sqrt{.65}$) for F1 (with $c_{E1} = .35$), $b_{V2F2} = .93$ ($= \sqrt{.87}$) for F2 (with $c_{E2} = .13$), and $b_{V3F3} = .98$ ($= \sqrt{.95}$) for F3 (with $c_{E3} = .05$). These values, along with the focal parameters and peripheral structural parameters of the full model from the previous example, yield the following considerably simpler model-implied covariance matrix (shown here to two decimal places):

$$\hat{\Sigma}(\theta_F) = \begin{bmatrix} 1.00 & & \\ -.23 & 1.00 & \\ .04 & .40 & 1.00 \end{bmatrix}.$$

As the reader may now verify, in going through Step 3 in the outlined process for power analysis for testing parameters within a model, the same values for F_{ML} will arise for the focal parameters, and hence the same required sample sizes will be derived.

While this simplification is somewhat intriguing in and of itself, its practical value is the reason it is presented in this chapter. Specifically, for models with latent variables where the focal parameters are contained in the structural portion of the model, the researcher does not need to specify a measurement model a priori in order to conduct sample size planning. The researcher merely needs to estimate the level of construct reliability achievable for each factor, or perhaps articulate a commitment to achieving a minimum quality of measurement model for each individual factor.

This simplification also will allow a preliminary sensitivity analysis, of sorts, whereby the researcher can determine the degree of sensitivity of the sample size estimate to the quality of the measurement models. In the current example, we can easily determine what would happen if, for example, the construct reliability associated with Reading Self-Concept (F1) were to be increased to, say, .80. As the reader may verify, the new fit function values for the focal parameters (to three decimal places) would become $F_{F1F2} = .065$, $F_{F3F1} = .032$, and $F_{F3F2} = .201$, leading to required sample sizes of $n_{F1F2} = 123$, $n_{F3F1} = 245$, and $n_{F3F2} = 41$. The sample size now needed to ensure at least .80 power for tests of all focal parameters is $n = 245$. That means that by improving the quality of the measurement model for Reading Self-Concept from $H_{F1} = .65$ to $H_{F1} = .80$, the desired level of power may be ensured with 61 fewer subjects. This increase in construct reliability

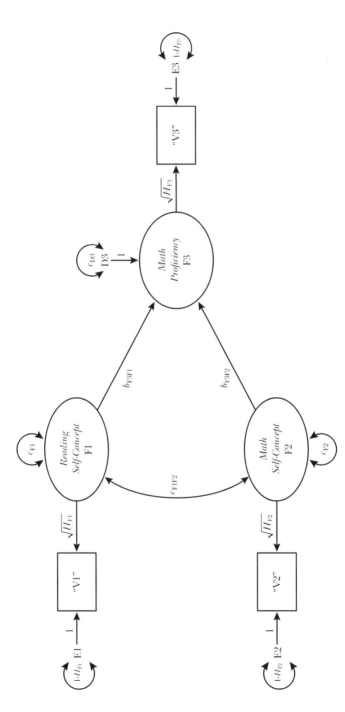

Figure 4.4 Simplified latent variable path model for example of a priori power analysis for testing parameters within a model

might be accomplished in practice, for example, by replacing RSC1 in the original data with an indicator of standardized loading .85, or by adding to the original set a fourth indicator of standardized loading .83, or by adding four indicators of standardized loadings .60. In fact, for purposes of power analysis, how it is done is irrelevant.

In the end, then, taking advantage of this simplification requires some familiarity with the coefficient H measure, but gaining such familiarity may indeed be a worthwhile investment in this case (see Hancock & Mueller, 2001). Doing so will afford the researcher less specificity in terms of number and quality of measured indicator variables, making a priori power analysis for testing parameters within a model a simpler process.

Issues and Extensions

The above approaches, which are applicable both to sets of parameters as a whole as well as to individual parameters, can also extend to multisample models (e.g., for invariance testing) as well as to latent variable mean structure models (see, e.g., Hancock, 2001). Most commonly individual parameter tests will be of interest, and for that reason the above presentation illustrated sample size planning in that context. The reader should be aware that, although the presentation above was framed in terms of the change in χ^2 associated with the constraint of each individual parameter, other ways of course exist to evaluate individual parameters, the most obvious being its estimate's z-value. As mentioned previously, the square of this test statistic, z^2, will asymptotically approach a χ^2 distribution (with 1 df), and practically speaking would require that only the full model be analyzed rather than a reduced model for each focal parameter. Indeed this would be easier, but there is room for concern with applying such an approach to models with latent variables, and this concern is two-fold. First, as has been discussed by de Pijper and Saris (1982) and Gonzalez and Griffin (2001), the z-values can vary considerably depending upon which scale referent is chosen for a latent variable. Second, as Yuan and Hayashi (2006) discussed, the standard errors upon which the z-values are based can themselves be rather unstable. Thus, except perhaps in the case where a model would become under-identified when imposing the restrictions associated with a reduced model, using z-values for sample size planning is generally not favored over the more tedious but more stable process presented above.

Further, as indicated in the context of tests of the model as a whole, sample size planning may be completed not just for the purposes of the statistical tests but also for the precision of the estimates of interest. In the case of parameters within a model, one could plan the overall sample size to obtain interval estimates for focal parameters that are sufficiently narrow for the researcher's inferential purposes. Details about these strategies are presented in the context of SEM by Lai and Kelley (2011), and more generally by Maxwell, Kelley, and Rausch (2008).

Finally, the power analysis methods presented for testing parameters rest, as for testing the model as a whole, on assumptions regarding the models as well as regarding the underlying distributions of the data. For techniques that are less dependent upon distributional assumptions such as those associated with the ML estimation paradigm, Monte Carlo simulation procedures for sample size planning are presented next.

MONTE CARLO APPROACHES TO POWER ANALYSIS IN SEM

The power analysis approaches addressed above can be quite useful in planning for the appropriate sample size for testing models as a whole as well as the parameters within those models. A potential drawback of these approaches, however, as the astute reader might have realized, is that the veracity of the planned sample size relies on how well one has met the approaches' assumed conditions. To start, such methods are built within an ML framework that makes assumptions about the normality of the population data. Deviations from these ideal distributional conditions have implications for both tests of data-model fit as well as tests of parameters within those models, as thoroughly detailed elsewhere (e.g., Finney & DiStefano, this volume). More importantly for the purposes of this chapter, distributional deviations affect the sample size needed for model and parameter testing, rendering suggested sample sizes potentially inaccurate. Further, for parameter testing specifically, one must make a host of assumptions about both the focal and peripheral parameters. Unfortunately, the more distant the actual parameter values are from the assumed values, the more misguided the samples size estimates will be, either in a positive direction (higher power) or negative direction (lower power). Finally, and as mentioned previously, the approaches discussed earlier do not easily facilitate planning for a desired level of precision of data-model fit indices or of the model parameter estimates themselves.

Fortunately, reasonably straightforward and accessible methods for extending the previous power analysis procedures are available that can take into account various data and parameter conditions. Specifically, one can utilize Monte Carlo (MC) methods to address statistical power in a manner similar to how researchers have used MC studies to determine the behavior of statistical tests or numerical indices in methodological studies across the social sciences. For a thorough presentation of MC methods in SEM the reader is referred to the chapter by Bandalos and Leite in this volume, and for examples of applications of these methods in the context of invariance testing under many conditions (e.g., model misspecification, non-normal data, sample sizes, degree of invariance), the interested reader is referred to Cheung and Rensvold (2002), Fan and Sivo (2009), and French and Finch (2008, 2011). A main benefit of sample size planning both for data-model fit and individual parameters through an MC framework is that several criteria, including precision of parameter estimates and power, can be simultaneously accommodated for a given structural equation model (Muthén & Muthén, 2002). And while other related approaches have been proposed to study the stability of estimated parameters, the empirical distribution of certain statistics, and related power (e.g., bootstrapping, Mooijaart, 2003; Yuan & Hayashi, 2003), the MC approach presented in this chapter allows for the accuracy of fit indices and parameter estimates to be examined under many conditions in order to estimate the necessary sample size.

The utilization of the MC approach follows the same basic steps as was outlined for estimating sample size for Figure 4.3 and 4.4. Step 1 is to specify the model, the associated population values for all model parameters, and a starting sample size; the population covariance matrix does not have to be calculated as the computer software selected to conduct the MC procedure will complete this component of the process. Once the necessary values are selected, Step 2 requires this information to be specified in a modeling program such as Mplus or LISREL, from which the MC routines of that program can generate numerous random samples based on the model and population values specified. In Step 3, which is automated within the software, the model is estimated in each of the randomly generated samples and the results of the data-model fit indices and parameter estimates are aggregated across the analyses and saved to a file. Step 4 requires examining the results file to determine the power of the tests of individual parameters as well as how certain fit indices (e.g., RMSEA, χ^2) behave under the assumed model values and specified sample size. During this step, some hand calculation might be required depending on the amount of information desired, as illustrated below. Step 2 through Step 4 can then be repeated specifying various sample sizes of interest, or other model parameter and assumption changes, to determine where data-model fit and power are optimized given model and data assumptions. The examination of the "what-if" scenarios can be quite informative in terms of how various assumptions influence the behavior (e.g., bias, variability) of parameter estimates and statistical power.

An example of the MC approach follows, which is based on the model in Figure 4.3 and the parameter values presented in the associated text. These parameter values were submitted to Mplus version 7 as seen in the input code in the chapter Appendix. Normality was assumed for this example. The number of datasets (replications) was set to 10,000 with a starting sample size of 306, which is size that was determined above for testing the b_{F3F1} parameter. While the number of replications might seem large, computer processing time is a minor concern and the wait will be at most a few minutes for simple models. Additionally, as is common in simulation studies, models are likely to fail to converge across some replications for a variety of reasons; hence, a relatively large number of replications assists in ensuring adequate data for characterizing the expected behavior of fit indices and parameter estimates. Note also that several criteria are typically employed to determine the necessary sample size through this process, beyond the targeted level of power. These criteria, as discussed by Muthén and Muthén (2002), include (a) parameter or standard error bias does not exceed 10%, (b) the standard error bias for a focal parameter for which power is assessed does not exceed 5%, and (c) the proportion of 95% confidence intervals capturing the true parameter value (i.e., the *coverage*) remains between 0.91 and 0.98. If these criteria are met then the sample size is selected where power is at the desired level ($\pi = .80$, in this example).

Evaluation of Data-Model Fit

With respect to data-model fit, the Mplus output file contains results pertaining to the behavior of fit indices (e.g., χ^2, AIC, BIC, SRMR, RMSEA); a highly edited version of this output file appears in the chapter Appendix. The focus here will be on the model χ^2, recognizing that the results can be interpreted similarly for the other indices. For all indices, the mean, standard deviation, number of successful computations, and expected and observed proportions and percentiles are provided. For the model χ^2 specifically, the expected proportion for tests at the $\alpha = .05$ level is .05, that is, for a model with 17 degrees of freedom there should be a .05 expected probability of obtaining a χ^2 value greater than the critical value of 27.59. As the computer has successfully completed 10,000 calculations based on the population model, the results can be examined to determine the proportion of replications for which the critical value was surpassed. In this case, the observed proportion is (to three decimal places) .050, with the 95th percentile point in the empirical distribution of model χ^2 values across replications falling at 27.61. Given that the expected and observed proportions are virtually identical, there is strong support that the χ^2 distribution with 17 degrees of freedom closely approximates the behavior of the model χ^2 test statistic when the sample size is 306. Percent bias also can be assessed by subtracting the expected value (27.59) from the observed value (27.61), dividing this difference by the population value and then multiplying by 100. Percent bias in this example is .07%, well below the recommended tolerance of 10% (Muthén & Muthén, 2002) and supporting again that the

χ^2 distribution was well approximated. Note that the average model χ^2 value across replications also is reported as 17.13 ($SD = 5.80$) and demonstrates that the data-model fit would be deemed acceptable (i.e., $17.13 < 27.59$). Again, the other indices (e.g., average RMSEA = .013; $SD = .017$) can be reviewed to determine data-model fit at a given sample size under the specified parameter and distributional conditions.

For illustrative purposes, Step 2 through Step 4 were repeated assuming a necessary sample size of 673 with power set at $\pi = .80$ for $\varepsilon_1 = .02$ and $df = 17$ as would be obtained in Table 4.1. In this case, the observed proportion is (to three decimal places) .052 with an empirical 95th percentile value of $\chi^2 = 27.60$. The percent bias, which may be calculated to be .03%, again supports the use of the χ^2 distribution. Sample sizes can be adjusted quickly to determine various scenarios (e.g., $n = 50$, $bias_{\chi^2} = 7.32\%$; $n = 2,000$, $bias_{\chi^2} = .02\%$). Thus, fit indices can be evaluated at different sample sizes to determine where data-model fit indices meet criteria. However, this should be examined in the context of the power associated with parameter estimates as it would be rare that a researcher is interested in data-model fit without examining parameters in the model.

Evaluation of Power for Testing Parameters Within a Model

The other component of the results in an MC approach that should be considered in combination with the data-model fit is the power for testing the specific (focal) parameters. Unlike the above procedures, the MC methods allow for all parameter (focal and peripheral) values to be evaluated simultaneously over the replications (10,000 in this case). This takes advantage of the simulation environment where models are evaluated in a manner that most closely reflects how the models and parameters will be evaluated in practice.

Given that models were estimated for evaluating data-model fit with all parameters present (i.e., free), the results are readily available in the results file. In Mplus, for example, and as seen in the highly edited output in the chapter Appendix, after data-model fit summary information, tables are produced with several columns of information that contain each population parameter value (i.e., the value specified), average parameter estimates, standard deviations of estimates, average standard errors (SE), and mean-square error (MSE) averaged across replications. Also reported for each parameter is the 95% coverage rate, indicating the proportion of estimates for which the 95% confidence interval contains the population parameter, as well as the percentage of parameter estimates that were statistically significantly different from zero ($\pi < .05$)—that is, the power. Percent bias for parameters, as well as for their standard errors, is determined just as described previously for fit indices.

For this example the focal parameters that were used previously will be used here as well: c_{F1F2}, b_{F3F1}, and b_{F3F2}. Recall that a sample size was derived using the model-based approach for each of the three focal parameters, with 306 (for testing b_{F3F1}) being the largest. Examining the results for c_{F1F2}, whose population value was −.30, reveals that the average estimate across replications was −0.297 with a standard deviation of .073, an average SE of .071, and a MSE of .005. This indicates that bias is .70% and SE bias is 2.74%. Additionally, coverage for c_{F1F2} is .94 and the proportion of parameter estimates statistically significantly different from zero (i.e., the power) is .98. For another focal parameter, b_{F3F1}, the percent bias for the parameter is 1.35% and for the SE is 3.79%. Additionally, coverage is .94 and the power is .80, this latter value being expected for this example based on the previous model-based results. The added benefit of the MC approach is information about the precision of the estimate. For the focal parameter of b_{F3F2}, the percent bias for the parameter is 1.04% and for the SE is 1.87%, while the coverage is .94 and power is effectively 1.00. This level of power is consistent with the previous model-based results, which suggested that a sample size of only 42 was adequate for testing this focal parameter. As seen before, Step 2 through Step 4 can be repeated at various sample sizes and data conditions to evaluate data-model fit and parameter precision in relation to power. The model presented in Figure 4.4 using the H statistic also can be specified in the same manner in the MC framework to take advantage of the elegant way of assessing the influence of reliability on power and data-model fit.

In this example, a selected sample size to begin the MC process was available from the prior model-based power analysis. The manner in which to proceed without such an estimate would be to begin with a certain sample size, perhaps based on anticipated resources, and then continue adjusting that sample size until data-model fit, precision of estimates, and power are at an appropriate balance and level, recognizing that fit indices, parameter values, and standard errors can have different thresholds of stability. Recall that the sample size of 50 resulted in bias below 10%, which is desirable; however, power for the focal parameter of b_{F3F1} was only .19. This point emphasizes why it is important to consider the criteria outlined above when determining sample size. The benefit of employing an MC approach to determining sample size is that many conditions can be simulated to reflect what assumptions the data may actually meet. Specifically, non-normal data, different estimation procedures, missing data, and non-continuous variables can be accounted for in sample size planning simply by specifying different conditions in the simulation request. The literature should guide the range of conditions that are assessed just as it guides the choice of the effect size and parameter estimates specified in the analysis. Once these values and conditions are determined, the process is relatively straightforward and publications (e.g., Muthén & Muthén, 2002) and software manuals provide valuable information and examples to employ the MC approach.

FINAL THOUGHTS REGARDING POWER ANALYSIS IN STRUCTURAL EQUATION MODELING

The purpose of the current chapter was to acquaint the reader with power analysis in SEM, both at the model level and the parameter level, in order to plan for adequate sample sizes. As is the case with any power analysis there are *many* assumptions involved, and as the current chapter should make clear—the more complex the models the more complex the array of assumptions upon which power analysis for those models rests. This might make sample size planning for SEM seem like a house of cards, increasing in tenuousness with each new layer of supposition until you get to a final, but rather wobbly, *n*. In response to this perspective, however, we offer three points. First, whatever stability can be achieved in sample size planning comes from viewing it as more about the process than the product. Yes, the end goal is to get a final sample size, but one should not view that sample size as some ultimate truth any more than one views any of the distribution or parameter choices as truths. Rather, the process of power analysis is part crystal ball gazer and part mechanic, translating a range of hazy potentialities into suggestions for sample size through trial-and-error sensitivity analysis computation. In the end the *n* selected for planning purposes is merely one among a range of values, ultimately chosen not to represent what is typical but instead what is more conservative in that range while still being feasible financially and logistically.

Second, the SEM power analysis framework is extremely versatile, having applications not just within SEM but also to more traditional measured variable analyses. Within SEM one can accommodate multi-sample models and mean structure models, as mentioned previously, as well as designs with planned missing data (e.g., Davey & Savla, 2010). These methods are also very useful for planning purposes for more traditional analyses, as discussed in thoughtful work by, for example, Duncan, Duncan, and Li (2003), Duncan, Duncan, Strycker, and Li (2002), and Miles (2003). Although these traditional analyses often involve small-sample test statistics (e.g., *t* and *F*), the versatility of the SEM power analysis framework can often accommodate more complex traditional designs, thus providing useful sample size approximations where none existed through more traditional power analysis methods.

Third, for whatever uncertainty exists in the power analysis process, the cost of failing to plan can be far higher. For those studies that get an unnecessarily large sample size, any benefits through minute improvements in power might be far outweighed by the expense of the extra participants. On the other hand, the cost of getting too small a sample size can be both financial and scientific. The financial expense comes from the execution of a study that was largely a waste of resources, pre-destined to find little or nothing for all its potentially considerable efforts. And the scientific costs, of course, include at best an inconsistent literature around key variables' relations, and at worst a literature appearing to solidify consistently around what are incorrectly believed to be null relations. As Maxwell (2004) stated in his wonderful article on the continued prevalence of underpowered studies, if we are "to continue to develop a coherent and accurate body of scientific literature, it is imperative that further attention be given to the role of power in designing studies and interpreting results" (p. 161). By examining a thoughtful range of possibilities for models of interest, researchers have the ability to ensure that they enter their studies hoping for the best and prepared for the worst. We hope that the current chapter will help researchers in these endeavors.

ACKNOWLEDGEMENTS

We wish to thank Ken Kelley and Nick Meyers for their helpful comments regarding the second edition of this chapter.

APPENDIX:
Mplus Input and Output for MC Power Analysis Example

Input

```
TITLE: Figure4.3.inp
MONTECARLO: NAMES ARE y1-y8; NOBSERVATIONS = 306;
NREPS = 10000; SEED = 53487;
SAVE = MCExample.sav; ANALYSIS: TYPE = general; ESTIMATOR = ML;
MODEL MONTECARLO:
f1 BY y1*.5;
f1 BY y2*-.7;
f1 BY y3*.6;
f2 BY y4*.8;
f2 BY y5*.9;
f2 BY y6*-.6;
y1*.75;
y2*.51;
y3*.64;
y4*.36;
y5*.19;
y6*.64;
y7*.12;
y8*.07;
f3 By y7*.938;
f3 By y8*.964;
f1@1 f2@1;
f3*.77;
f1 WITH f2*-.30;
f3 on f1*.20;
f3 on f2*.50;
MODEL:
f1 BY y1*.5;
f1 BY y2*-.7;
f1 BY y3*.6;
f2 BY y4*.8;
f2 BY y5*.9;
f2 BY y6*-.6;
y1*.75;
y2*.51;
y3*.64;
y4*.36;
y5*.19;
y6*.64;
y7*.12;
y8*.07;
f3 By y7*.938;
f3 By y8*.964;
f1@1 f2@1;
f3@.77;
f1 WITH f2*-.30;
f3 on f1*.20;
f3 on f2*.50;
```

Output (highly edited)

```
TESTS OF MODEL FIT
Number of Free Parameters  19
Chi-Square Test of Model Fit
    Degrees of freedom  17
    Mean  17.129
    Std Dev  5.802
    Number of successful computations  10000
```

	Proportions		Percentiles	
Expected	Observed	Expected	Observed
0.990	0.991	6.408	6.410
0.980	0.981	7.255	7.295
0.950	0.952	8.672	8.735
0.900	0.906	10.085	10.223
0.800	0.806	12.002	12.091
0.700	0.710	13.531	13.664
0.500	0.510	16.338	16.473
0.300	0.309	19.511	19.649
0.200	0.205	21.615	21.736
0.100	0.103	24.769	24.905
0.050	0.050	27.587	27.609
0.020	0.020	30.995	30.930
0.010	0.010	33.409	33.412

```
RMSEA (Root Mean Square Error Of Approximation)
    Mean  0.013
    Std Dev  0.017
    Number of successful computations  10000
```

	Proportions		Percentiles	
Expected	Observed	Expected	Observed
0.990	1.000	-0.026	0.000
0.980	1.000	-0.021	0.000
0.950	1.000	-0.014	0.000
0.900	1.000	-0.008	0.000
0.800	1.000	-0.001	0.000
0.700	0.457	0.004	0.000
0.500	0.407	0.013	0.000
0.300	0.309	0.022	0.020
0.200	0.239	0.027	0.027
0.100	0.145	0.035	0.035
0.050	0.083	0.041	0.041
0.020	0.037	0.048	0.048
0.010	0.019	0.052	0.053

```
MODEL RESULTS
              ESTIMATES
                            Std.     S.E.                 95%     % Sig
          Population  Average  Dev.    Average  M.S.E.    Cover   Coeff
F3  ON
F1          0.200     0.2027  0.0764   0.0735   0.0058    0.945   0.798
F2          0.500     0.5052  0.0746   0.0732   0.0056    0.949   1.000


F1  WITH
F2         -0.300    -0.2979  0.0727   0.0708   0.0053    0.942   0.981
```

REFERENCES

Bentler, P. M. (1990). Comparative fit indexes in structural models. *Psychological Bulletin, 107,* 238–246.

Browne, M. W., & Cudeck, R. (1993). Alternative ways of assessing model fit. In K. A. Bollen & J. S. Long (Eds.), *Testing structural equation models* (pp. 136–162). Newbury Park, CA: Sage.

Browne, M. W., & Mels, G. (1990). *RAMONA user's guide.* Unpublished report, Department of Psychology, The Ohio State University.

Cheung, G. W., & Rensvold, R. B. (2002). Evaluating goodness-of-fit indexes for testing measurement invariance. *Structural Equation Modeling: A Multidisciplinary Journal, 9,* 233–255.

Cohen, J. (1988). *Statistical power analysis for the behavioral sciences.* Hillsdale, NJ: Erlbaum.

Curran, P. J., Bollen, K. A., Paxton, P., Kirby, J., & Chen, F. (2002). The noncentral chi-square distribution in misspecified structural equation models: Finite sample results from a Monte Carlo simulation. *Multivariate Behavioral Research, 37,* 1–36.

Davey, A., & Savla, J. (2010). *Statistical power analysis with missing data: A structural equation modeling approach.* New York: Routledge.

de Pijper, W. M., & Saris, W. E. (1982). The effect of identification restrictions on the test statistics in latent variable models. In K. G. Jöreskog & H. Wold (Eds.), *Systems under indirect observation,* (part 1, pp. 175–182). Amsterdam: North Holland.

Duncan, T. E., Duncan, S. C., & Li, F. (2003). Power analysis models and methods: A latent variable framework for power estimation and analysis. In Z. Sloboda & W. J. Bukosk (Eds.), *Handbook of drug abuse prevention (Handbooks of sociology and social research)* (pp. 609–626). New York: Kluwer Academic/Plenum Publishers.

Duncan, T. E., Duncan, S. C., Strycker, L. A., & Li, F. (2002). A latent variable framework for power estimation within intervention contexts. *Journal of Psychopathology and Behavioral Assessment, 24,* 1–12.

Fan, X., & Sivo, S. A., (2009). Using goodness-of-fit indexes in assessing mean structure invariance. *Structural Equation Modeling: A Multidisciplinary Journal, 16,* 54–69.

Fornell, C., & Larcker, D. F. (1981). Evaluating structural equation models with unobservable variables and measurement error. *Journal of Marketing Research, 18,* 39–50.

French, B. F., & Finch, W. H. (2008). Multi-group confirmatory factor analysis: Locating the invariant referent. *Structural Equation Modeling: A Multidisciplinary Journal, 15,* 96–113.

French, B. F., & Finch, W. H. (2011). Model misspecification and invariance testing via confirmatory factor analytic procedures. *Journal of Experimental Education, 79,* 404–428.

Gagné, P., & Hancock, G. R. (2006). Measurement model quality, sample size, and solution propriety in confirmatory factor models. *Multivariate Behavioral Research, 41,* 65–83.

Gonzalez, R., & Griffin, D. (2001). Testing parameters in structural equation modeling: Every "one" matters. *Psychological Methods, 6,* 258–269.

Hancock, G. R. (2001). Effect size, power, and sample size determination for structured means modeling and MIMIC approaches to between-groups hypothesis testing of means on a single latent construct. *Psychometrika, 66,* 373–388.

Hancock, G. R., & Freeman, M. J. (2001). Power and sample size for the RMSEA test of not close fit in structural equation modeling. *Educational and Psychological Measurement, 61,* 741–758.

Hancock, G. R., & Kroopnick, M. H. (2005, April). *A simplified sample size determination for tests of latent covariance structure parameters.* Paper presented at the annual meeting of the American Educational Research Association, Montréal, Canada.

Hancock, G. R., & Mueller, R. O. (2001). Rethinking construct reliability within latent variable systems. In R. Cudeck, S. H. C. du Toit, & D. Sörbom (Eds.), *Structural equation modeling: Past and present. A Festschrift in honor of Karl G. Jöreskog* (pp. 195–261). Chicago, IL: Scientific Software International.

Hoenig, J. M., & Heisey, D. M. (2001). The abuse of power: The pervasive fallacy of power calculations for data analysis. *American Statistician, 55,* 19–24.

Jackson, D. L. (2003). Revisiting sample size and number of parameter estimates: Some support for the *N:q* hypothesis. *Structural Equation Modeling: A Multidisciplinary Journal, 10,* 28–141.

Kelley, K., & Lai, K. (2011). Accuracy in parameter estimation for the root mean square of approximation: Sample size planning for narrow confidence intervals. *Multivariate Behavioral Research, 46,* 1–32.

Kim, K. H. (2005). The relation among fit indexes, power, and sample size in structural equation modeling. *Structural Equation Modeling: A Multidisciplinary Journal, 12,* 368–390.

Lai, K., & Kelley, K. (2011). Accuracy in parameter estimation for targeted effects in structural equation modeling: Sample size planning for narrow confidence intervals. *Psychological Methods, 16,* 127–148.

Li, L., & Bentler, P. M. (2011). Quantified choice of root-mean-square errors of approximation for evaluation and power analysis of small differences between structural equation models. *Psychological Methods, 16,* 116–126.

MacCallum, R. C., Browne, M. W., & Cai, L. (2006). Testing differences between nested covariance structure models: Power analysis and null hypotheses. *Psychological Methods, 11,* 19–35.

MacCallum, R. C., Browne, M. W., & Sugawara, H. M. (1996). Power analysis and determination of sample size for covariance structure modeling. *Psychological Methods, 1,* 130–149.

MacCallum, R. C., & Hong, S. (1997). Power analysis in covariance structure modeling using GFI and AGFI. *Multivariate Behavioral Research, 32,* 193–210.

MacCallum, R. C., Widaman, K. F., Zhang, S., & Hong, S. (1999). Sample size in factor analysis. *Psychological Methods, 4,* 84–99.

Marsh, H. W., Hau, K-T., Balla, J. R., & Grayson, D. (1998). Is more ever too much? The number of indicators per factor in confirmatory factor analysis. *Multivariate Behavioral Research, 33,* 181–220.

Maxwell, S. E. (2004). The persistence of underpowered studies in psychological research: Causes, consequences, and remedies. *Psychological Methods, 9,* 147–163.

Maxwell, S. E., Kelley, K., & Rausch, J. R. (2008). Sample size planning for statistical power and accuracy in parameter estimation. *Annual Review of Psychology, 59,* 537–563.

Miles, J. (2003). A framework for power analysis using a structural equation modeling procedure. *Medical Research Methodology, 3,* 27.

Mooijaart, A. (2003). Estimating the statistical power in small samples by empirical distributions. In H. Yanai, A. Okada, K. Shigemasu, Y. Kano, & J. J. Meulman (Eds.), *New developments in psychometrics* (pp. 149–156). Tokyo: Springer-Verlag.

Muthén, L., & Muthén, B. (2002). How to use a Monte Carlo study to decide on sample size and determine power. *Structural Equation Modeling: A Multidisciplinary Journal, 9,* 599–620.

Penev, S., & Raykov, T. (2006). Maximal reliability and power in covariance structure models. *British Journal of Mathematical and Statistical Psychology, 59,* 75–87.

Preacher, K. J., Cai, L., & MacCallum, R. C. (2007). Alternatives to traditional model comparison strategies for covariance structure models. In T. D. Little, J. A. Bovaird, & N. A. Card (Eds.), *Modeling contextual effects in longitudinal studies* (pp. 33–62). Mahwah, NJ: Lawrence Erlbaum Associates.

Preacher, K. J., & Coffman, D. L. (2006, May). Computing power and minimum sample size for RMSEA [Computer software]. Available from http://quantpsy.org/.

Raykov, T. (2004). Estimation of maximal reliability: A note on a covariance structure modelling approach. *British Journal of Mathematical and Statistical Psychology, 57,* 21–27.

Steiger, J. H. (1989). *EzPATH: A supplemental module for SYSTAT and SYSGRAPH.* Evanston, IL: SYSTAT.

Steiger, J. H., & Lind, J. M. (1980, June). *Statistically based tests for the number of common factors.* Paper presented at the annual meeting of the Psychometric Society, Iowa City, IA.

Steiger, J. H., Shapiro, A., & Browne, M. W. (1985). On the multivariate asymptotic distribution of sequential chi-square statistics. *Psychometrika, 50,* 253–263.

Yuan, K., & Bentler, P. M. (2004). On chi-square difference and z tests in mean and covariance structure analysis when the base model is misspecified. *Educational and Psychological Measurement, 64,* 737–757.

Yuan, K.-H., & Hayashi, K. (2003). Bootstrap approach to inference and power analysis based on three statistics for covariance structure models. *British Journal of Mathematical and Statistical Psychology, 56,* 93–110.

Yuan, K.-H., & Hayashi, K. (2006). Standard errors in covariance structure models: Asymptotics versus bootstrap. *British Journal of Mathematical and Statistical Psychology, 59,* 397–417.

Yuan, K., & Maxwell, S. E. (2005). On the post hoc power in testing mean differences. *Journal of Educational and Behavioral Statistics, 30,* 141–167.

PART II

EXTENSIONS

CHAPTER 5

EVALUATING BETWEEN-GROUP DIFFERENCES IN LATENT VARIABLE MEANS

Marilyn S. Thompson and Samuel B. Green

Researchers often seek to compare groups hypothesized to differ on one or more outcome variables. In the simplest form, groups may be compared on a single, directly observable variable that is believed to be measured reliably. On the other hand, research questions frequently involve group differences on multiple outcomes that are thought to represent one or more constructs. Although multivariate analysis of variance (MANOVA) persists as a commonly applied method for analyzing mean differences on a set of outcomes, methodologists have argued that researchers should use structural equation modeling (SEM) methods for comparing groups on latent variable means if multiple dependent variables represent each of the underlying constructs of interest (e.g., Aiken, Stein, & Bentler, 1994; Cole, Maxwell, Arvey, & Salas, 1993; Green & Thompson, 2003a; Hancock, 1997). In this chapter, we discuss the utility of SEM for evaluating between-group differences on latent variable means and illustrate SEM approaches for doing so. Specifically, we address the following points:

Structural Equation Modeling: A Second Course (2nd ed.), pages 163–218
Copyright © 2013 by Information Age Publishing

- SEM provides a flexible approach for comparing latent variable means that accounts for unreliability of measures and allows for coherent conclusions to be reached at the construct level;
- Between-group differences in latent variable means can be evaluated using a structured means modeling (SMM) approach or a multiple-indicator multiple-cause (MIMIC) modeling approach;
- The SMM approach is more flexible than the MIMIC approach in that it allows for partial measurement invariance across groups;
- MIMIC and SMM approaches can be extended to include covariates and multifactorial designs.

We illustrate the MIMIC and SMM approaches for the two-group case, but note that these methods extend to any number of populations and can be applied in experimental, quasi-experimental, and nonexperimental designs.

Both MIMIC and SMM approaches can be applied broadly to assess mean differences between groups. These methods can be used to evaluate group differences on a single dependent variable and, in this context, offer the advantage of circumventing assumptions required by standard analysis of variance (ANOVA) approaches (Fan & Hancock, 2012; Green & Thompson, 2012). We focus not on these univariate applications of MIMIC and SMM approaches but rather on their use with multiple dependent variables that are hypothesized to be effect indicators of constructs. Although models may also be specified to assess differences in latent means for repeated measures problems, this chapter focuses on methods for between-subjects designs. To avoid confusion with the differing meaning of the term "factor" across the SEM and ANOVA/MANOVA literatures, we use the term "grouping variable" in this chapter to refer to a variable that differentiates individuals into groups and generally reserve the term "factor" to represent a construct or dimension underlying the dependent variables.

GETTING IN THE MOOD FOR MODELING: SOME ADVANTAGES OF SEM FOR ASSESSING DIFFERENCES IN LATENT MEANS

We suspect that because SEM offers a general framework for specifying models, working within SEM encourages researchers to devote greater thought to specification of relationships between the constructs and measured variables, and, additionally, to alternative models that may explain the data nearly as well, equally as well, or better. Careful thought and decisions are required at every stage of the SEM process, including model specification, testing, and interpretation. In this section, we offer a few thoughts on the benefits of taking an active modeling perspective to assessing differences in latent means.

First, it is essential for researchers to choose a model and analytic approach that are conceptually consistent with the directionality of hypothesized relationships between the constructs and measured variables. In most applications of SEM, including the evaluation of differences in latent means discussed in this chapter, constructs are thought to be causal agents of the dependent measures. In these *latent variable systems*, the measured variables are *effect indicators* that are presumed to be correlated because they share common causes, that is, the same underlying factors. In these latent variable systems, the measured variables are hypothesized to be linear combinations of factors plus error, so arrows are directed from factors and errors to the measured variables. Researchers commonly conceptualize human behaviors to be effect indicators of underlying factors, such that the observed indicators are affected by the latent trait. For example, social self-concept is a latent trait that might be hypothesized to affect measures of shyness, comfort with adults, and comfort with peers. The measurement model relating factors to their effect indicators can be validated by collecting data on the measured variables and conducting confirmatory factor analysis (CFA) to assess whether the intended factor structure accounts for the covariances among the effect indicators.

In contrast, the direction of the relationship between the measures and the concept may be reversed such that measured variables are contributory agents of construct. This implies an *emergent variable system* with indicators that combine additively to form either: (a) an emergent latent variable system with causal indicators and a disturbance on the latent variable, or (b) a composite variable that is a linear combination of the composite indicators, with no disturbance on the composite variable (Bollen & Bauldry, 2011). Model specifications in which effects flow from the measures to emergent latent variables or composite variables have also been referred to as *formative systems* (e.g., Bollen & Lennox, 1991; Cohen, Cohen, Teresi, Marchi, & Velez, 1990; Cole et al., 1993; Howell, Breivik, & Wilcox, 2007; MacKenzie, Podsakoff, & Jarvis, 2005). Evaluating emergent variable systems with causal indicators within an SEM framework presents challenges with respect to model identification and is not within the scope of this chapter (see the chapter by Kline, this volume), but methods are being developed to improve the ease with which these models may be analyzed using SEM (e.g., Bollen & Bauldry, 2011; Treiblmaier, Bentler, & Mair, 2011).

Perhaps most commonly, researchers employ MANOVA for evaluating differences in means in emergent variable systems. The emergent variables in MANOVA are the mathematically derived linear combinations of the measured variables on which groups are maximally separated (i.e., the discriminant functions). However, because these linear combinations are not necessarily the combinations of interest to the researcher (Hancock, 2003), the results of MANOVA may not be interpretable with respect to the latent constructs of interest. Additionally, MANOVA yields substantive conclusions similar to SEM only for a single factor model with no measurement bias between groups (Cole et al., 1993; Green & Thompson, 2003b). Conclusions may differ dramatically across the methods to the extent that some of the measures also share variance from a source not accounted for in the model (such as method variance that can be represented in SEM by correlated errors) or that measurement bias between groups is present in one or more measured variables (which implies unequal intercepts, as will be discussed with the SMM approach).

Second, researchers should consider the advantage SEM offers for modeling measurement error. Each measured variable, regardless of how it is related to the construct, presumably has measurement error associated with it. Linear composites in emergent variable systems, such as those assessed in MANOVA, may be heavily influenced by the propagation of measurement errors of the indicators. Tests for population differences in construct means based on these linear composites may fall short of finding true differences because unreliability of the composites can obscure true differences between groups. In contrast, latent variables in SEM are theoretically error-free representations of constructs. Accordingly, SEM in many cases may have greater power than MANOVA approaches for assessing differences in latent variable means (Hancock, 2003; Hancock, Lawrence, & Nevitt, 2000; Kano, 2001).

Third, data might be fit to alternative models. After fitting the data to the initially hypothesized model, the analyst can empirically explore which parts of the model contribute to lack of fit and may choose to relax certain constraints imposed on the initial model, allowing for nested model comparisons. Alternative, non-nested model configurations may also be examined.

Fourth, as we discuss later in this chapter, covariates can be included in SEM and the covariates may be measures or factors underlying multiple measures. The user has flexibility in specifying relations between the covariates and the other variables included in the model.

Finally, SEM encourages researchers to be more cognizant of the model they are specifying in that, regardless of which software package is used for conducting SEM, the researcher must indicate the model fully through program syntax or by employing dialogue boxes or path diagrams to assist with syntax generation. In summary, taking an SEM approach to assessing differences in latent means puts you in the mood for modeling.

INTRODUCING MODELS FOR EVALUATING
DIFFERENCES IN FACTOR MEANS

Differences in factor means can be evaluated using two types of models: a multiple indicators and multiple causes (MIMIC) model or the structured means model (SMM). The MIMIC model is the simpler of the two. The data for MIMIC models are variances and covariances among measured variables, and the model parameters include variances, covariances, and path coefficients. In contrast, SMM is a multiple-group model that is conducted on means in addition to variances and covariances among measured variables, and the model includes intercept parameters as well as variances, covariances, and path coefficients.

The choice between MIMIC and SMM approaches for assessing factor mean differences is partially dictated by the extent to which model parameters are equal, or *invariant*, between groups. Meredith (1993) described three levels of measurement invariance: weak, strong, and strict. With *weak measurement invariance*, only factor loadings are required to be the same across groups. With *strong measurement invariance*, intercepts of measures as well as factor loadings are equivalent across group. Finally, for *strict measurement invariance*, in addition to factor loadings and intercepts of measures, variances and covariances of errors are required to be equivalent between groups. It should be noted that measurement invariance does not pertain to the variances and covariances among the factors or the factor means for a confirmatory factor analysis model. We refer to *complete invariance* as invariance of all model parameters (i.e., both the measurement and structural parts of the model).

We cannot reach valid conclusions about between-group differences in factor means without requiring some degree of measurement invariance. As we will illustrate, the MIMIC model is very restrictive. As typically applied, it assumes not only strict measurement invariance, but also invariance of the structural part of the model. In contrast, the SMM model allows for much greater flexibility. The variances and the covariances among errors and among factors (or disturbances of factors) may differ between groups. For reasons we describe later, it is ideal if strong measurement invariance holds; that is, factor loadings and intercepts for measures are equivalent between-groups. Nonetheless, as illustrated by Byrne, Shavelson, and Muthén (1989), valid statistical inferences can be made under partial invariance of factor loadings and intercepts. A minimum requirement for partial measurement invariance is (a) the loadings between each factor and one of its measures are equivalent between groups and (b) the intercepts of one measure for each factor are equivalent between groups. Although MIMIC models can accommodate differences in intercepts (i.e., inclusion of direct effects from the dummy-coded variable to measures), SMM models have much greater flexibility in their ability to handle between-group variability in model parameters.

In this section, we introduce the structured means and the MIMIC models for evaluating differences among factor means. When introducing these models, we assume strong invariance with the SMM approach and complete invariance with the MIMIC model approach. Once we have a better understanding of these models, we will discuss in the following section a method for assessing measurement invariance with the SMM method and discuss the implications associated with testing factor mean differences with less restrictive structured means models.

We created a fictitious study and data to illustrate the MIMIC and SMM approaches. The study involves evaluating the relationship between day-care use and positive outcome variables for preschool children raised in two-parent families. The preschool children are divided into two groups within this nonexperimental study: (a) those with two working parents who consistently use day-care facilities (day-care group) and (b) those with one stay-at-home parent and minimal use of day-care facilities (home-care group). Six outcome variables are included to assess academic and social school readiness. Indicators of academic readiness include scales measuring vocabulary level (V1), knowledge of letters and numbers (V2), and classification skills (V3), whereas indicators of social readiness include scales assessing self-control (V4), positive interaction with adults (V5), and positive interaction with peers (V6). Our developmental theory suggests that the readiness factors are latent and affect their respective measures. For example, the covariances among the academic measures can be viewed as a function of academic readiness. Data for these indicators are presented in Table 5.1 for the day-care group ($N_1 = 200$) and home-care group ($N_2 = 200$), as well as for the combined sample ($N = 400$). Later in this chapter, we introduce different data for this scenario to illustrate additional analyses.

Structured Means Modeling for Assessment of Mean Differences

We take a building blocks approach to explain structured means modeling. We begin with a confirmatory factor analysis model for a single group and then expand the model to assess differences between factor means for independent populations using multiple-group analyses.

One-group CFA. We begin by conducting a CFA with a mean structure on measures for a single group of individuals. The data—means, variances, and covariances among the academic and social school readiness measures—are presented in the top section of Table 5.1 for the day-care group. The CFA model for these data is presented as a path diagram in Figure 5.1.

TABLE 5.1 Covariance Matrices and Means for Measured Variables for the Day-Care Group, Home-Care Group, and Total Sample—Dataset 1

Measured variables	Covariance Matrix						Means
	V1	**V2**	**V3**	**V4**	**V5**	**V6**	
Group 1—Day-Care Group (N_1 = 200)							
V1	138.00						50.40
V2	45.58	80.49					79.60
V3	35.19	23.56	56.34				98.88
V4	45.13	32.00	16.64	232.17			74.06
V5	35.33	10.49	10.56	79.74	149.16		49.12
V6	73.34	28.73	33.21	117.30	79.90	324.36	120.10
Group 2—Home-Care Group (N_2 = 200)							
V1	127.61						53.70
V2	58.49	76.81					81.32
V3	29.09	19.72	54.29				101.82
V4	45.84	28.94	31.82	223.09			77.69
V5	20.33	13.27	5.81	62.25	135.72		51.62
V6	50.64	55.45	30.15	126.74	62.16	337.37	123.37

Total—Combined for Day-Care and Home-Care Groups (N = 400)

	Covariance Matrix						
	V1	**V2**	**V3**	**V4**	**V5**	**V6**	**V7**
V1	135.19						
V2	53.33	79.21					
V3	34.48	22.86	57.33				
V4	48.37	31.96	26.83	230.35			
V5	29.82	12.93	9.99	73.08	143.64		
V6	64.53	43.40	34.00	124.69	72.90	332.72	
V7	.83	.43	.73	.91	.62	.82	.25

Note: V1: Vocabulary; V2: Letters/Numbers; V3: Classification; V4: Self-Control; V5: Adult interaction; V6: Peer interaction; V7: Day (0) versus home (1) care.

In your first SEM course, you might not have been exposed to path diagrams that include arrows emanating from a triangle with a 1 inside, as shown in Figure 5.1. The triangle represents a unit predictor, which is equal to 1 for all individuals. The unit predictor has no variance and therefore is sometimes called a *pseudo-variable*. Arrows from the unit predictor can be directed toward factors or measures, and the model parameters associated with these arrows are intercepts (i.e., additive constants) in a prediction equation. The symbol for an intercept is an *a;* the subscript of *a* indicates the factor or measure with which it is associated.

Intercepts allow us to reproduce the means of the measured variables in models. As we will soon illustrate, these means must be perfectly reproduced

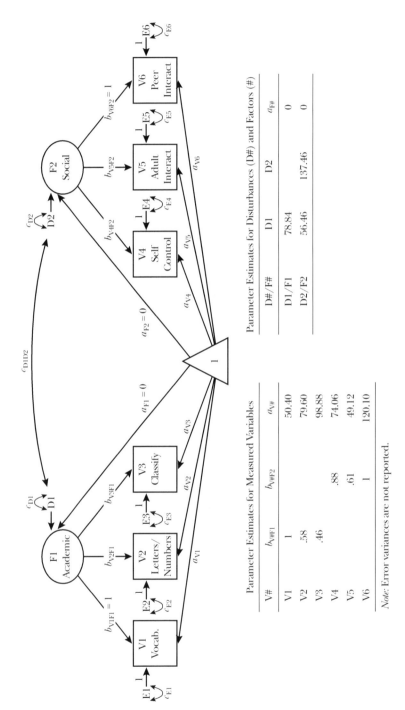

Parameter Estimates for Measured Variables

V#	$b_{V\#F1}$	$b_{V\#F2}$	$a_{V\#}$
V1	1		50.40
V2	.58		79.60
V3	.46		98.88
V4		.88	74.06
V5		.61	49.12
V6		1	120.10

Note: Error variances are not reported.

Parameter Estimates for Disturbances (D#) and Factors (#)

D#/F#	D1	D2	$a_{F\#}$
D1/F1	78.84		0
D2/F2	56.46	137.46	0

Figure 5.1 Results for one-group CFA with means for Dataset 1.

for our single group model and, thus, the part of the model involving intercepts is not open to empirical scrutiny (i.e., not falsifiable). For the one-sample CFA, we could have excluded intercepts from our model and analyzed only the variances and covariances of the measures, that is, excluding the means as data, but chose to include intercepts in our model for pedagogical purposes as a building block for the multiple-group CFA model.

Means, variances, and covariances of factors. Following the model specification approach described by Bentler (1995), the factors in our model are treated as dependent variables rather than independent variables because they are a function of the unit predictor as well as a disturbance term. Accordingly, the means, variances, and covariances of factors are not model parameters, but are equal to model parameters that are the intercepts, disturbance variances, and disturbance covariances of the factors, respectively.

To better understand the factor intercepts, let's examine the equation for F1 in our example:

$$F1 = a_{F1}\ 1 + D1 \tag{5.1}$$

where 1 is the unit predictor, a_{F1} is the intercept for F1, and D1 is the disturbance term for F1. Given Equation 5.1 and assuming the mean of the disturbance is zero, the mean of the factor is a_{F1}:

$$\mu_{F1} = \mu_{a_{F1}1+D1} = \mu_{a_{F1}1} + \mu_{D1} = a_{F1} + 0 = a_{F1}. \tag{5.2}$$

Similarly, the intercept for the second factor is the mean of that factor $(a_{F2} = \mu_{F2})$. The mean for a factor is arbitrary for a single-group analysis and must be defined by the researcher; otherwise, the model is underidentified. A mean for a factor may be defined by setting it to zero. As implied by Equation 5.2, a factor mean is set to zero by fixing the intercept for that factor to be zero.

The variances and covariances among the disturbances for the factors are equivalent to the variances and covariances among the factors (i.e., $c_{D1} = \sigma_{F1}^2$, $c_{D2} = \sigma_{F2}^2$, and $c_{D1D2} = \sigma_{F1F2}$). To avoid having an underidentifed model, we must specify the metric of the factors. As shown in Figure 5.1, we specified the metric in our model by fixing a loading for each factor to 1, but could have done so by setting the disturbance variance for each factor to 1.

For clarity and simplicity, we frequently refer to factor means, variances, and covariances in the text without necessarily indicating the equivalent model parameters. However, we do present the factor intercepts, disturbance variances, and disturbance covariances when specifying the parameters of models to maximize mathematical transparency. It should be noted that an alternative approach to model specification is to treat factors as independent variables, with the means, variances, and covariances of factors being model parameters (see, e.g., Byrne, 2012, Chapter 7; Muthén & Muthén, 2010).

Factor loadings, intercepts, and error variances of measures. Each of the first three measures (V1 to V3) is a function of the academic readiness factor (F1), the unit predictor, and measurement error (E1 to E3), and each of the last three measures (V4 to V6) is a function of the social readiness factor (F2), the unit predictor, and measurement error (E4 to E6). The factor loadings for the first and sixth measures are fixed to 1 to define the metric of F1 and F2, respectively. More specifically, the equations for the six measures are as follows:

$$
\begin{aligned}
V1 &= a_{V1}1 + 1 \quad\ \ F1 + E1 \\
V2 &= a_{V2}1 + b_{V2F1}\ F1 + E2 \\
V3 &= a_{V3}1 + b_{V3F1}\ F1 + E3 \\
V4 &= a_{V4}1 + b_{V4F2}\ F2 + E4 \\
V5 &= a_{V5}1 + b_{V5F2}\ F2 + E5 \\
V6 &= a_{V6}1 + 1 \quad\ \ F2 + E6
\end{aligned}
\tag{5.3}
$$

Again we focus on intercepts in that these parameters might not have been stressed in your first course. The intercepts allow for the reproduction of the means of the measured variables. For illustration purposes, let's examine the intercept for the second measure. The mean of V2 is equal to its intercept in that (a) the F1 intercept was fixed to zero for identification purposes and thus its mean is equal to zero and (b) the mean of E2 is assumed to be equal to zero:

$$
\begin{aligned}
\mu_{V2} &= \mu_{a_{V2}1 + b_{V2F1}F1\, +\, E2} \\
&= \mu_{a_{V2}1} + b_{V2F1}\mu_{F1} + \mu_{E2} \\
&= a_{V2} + b_{V2F1}(a_{F1}) + \mu_{E2} \\
&= a_{V2} + b_{V2F1}(0) + 0 \\
&= a_{V2}
\end{aligned}
\tag{5.4}
$$

Equation 5.4 illustrates that means of measures must be perfectly reproduced due to the inclusion of freely estimated intercepts of measures.

Fitting the model and testing factor mean differences. The parameters of the model are estimated so that the reproduced means, variances, and covariances among the measured variables based on the model parameters are as similar as possible to the sample means, variances, and covariances among the measured variables. The estimated parameters (except for error variances) are reported in the tables below the path diagram in Figure 5.1. Lack of fit between these matrices is a function of the constraints imposed on the model (e.g., no factor loading between F1 and V4). We assessed global lack of fit using four indices: chi-square statistic (χ^2), comparative fit index (CFI), standardized root mean square residual (SRMR), and root mean square error of approximation (RMSEA). For our data, the model fits very well, $\chi^2(8) = 10.78$, $p = .21$, CFI = .99, SRMR = .03, RMSEA = .04 with 90% CI [.00, .10].

It should be noted that the estimated intercepts for measures are equal to the means for these measures. As discussed earlier, these intercepts permit the model to reproduce perfectly the means of the measures, and consequently the means part of the model cannot produce lack of fit. An indication that the means part of the model is just-identified is that it contributes no degrees of freedom. For our example, the means part of the model contributes an additional six pieces of data—the means of V1 through V6—but requires an equal number of additional model parameters—the intercepts for V1 through V6. Because the means part of the model is just-identified, it adds no new fit information over and above a CFA excluding means of measures. Accordingly, researchers frequently exclude the means of these measures and analyze only the variances and covariances among the measures, making intercepts unnecessary in the CFA model.

Multiple-group CFA. We now consider how to evaluate differences in factor means using a multiple-group CFA approach. The data for a two-group CFA consists of means, variances, and covariances among measures for the two groups, as shown in Table 5.1 for our example. In this section, we specify the model to have the same form between groups and the values of the parameters demonstrate strong invariance. A model has the same form (i.e., configural invariance) if the specification of the parameters within groups is the same for all groups (Bollen, 1989). In the context of our example, the model, as specified in Figure 5.2, should fit for both the day-care and home-care groups. Strong invariance requires not only the form of the model to be equivalent between groups, but also that the paths between factors and measures (i.e., factor loadings) and the intercepts for measures to be equivalent between groups. All other parameters are allowed to differ between groups. If strong invariance holds, we can assess between-group differences in the factor means. As mentioned earlier (and illustrated later), the requirement of strong invariance does not have to hold; that is, some factor loadings and measure intercepts can be allowed to differ between groups to assess differences in factor means. Syntax for the SMM analyses is in Appendix A for EQS 6.1, Mplus 6, and the SIMPLIS language of LISREL 8.

The model for the SMM approach. We could present the same path diagram showing the relations between the factors and the measured variables

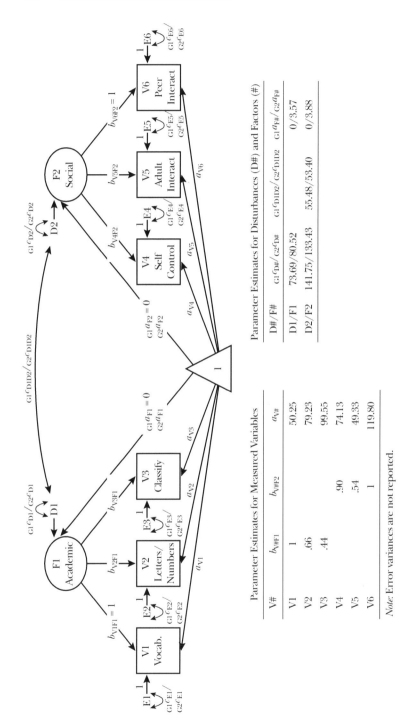

Parameter Estimates for Measured Variables

V#	$b_{V\#F1}$	$b_{V\#F2}$	$a_{V\#}$
V1	1		50.25
V2	.66		79.23
V3	.44		99.55
V4		.90	74.13
V5		.54	49.33
V6		1	119.80

Note: Error variances are not reported.

Parameter Estimates for Disturbances (D#) and Factors (#)

D#/F#	$G1^cD\#/G2^cD\#$	$G1^cD1D2/G2^cD1D2$	$G1^aF\#/G2^aF\#$
D1/F1	73.69/80.52		0/3.57
D2/F2	141.75/133.43	55.48/53.40	0/3.88

Figure 5.2 Results for structured means model for Dataset 1.

for each group. However, because the form of the model is the same for both groups, we present a single diagram for our example in Figure 5.2 and, where appropriate, indicate notational differences in the parameters across groups. Under strong invariance, we constrain the factor loadings and the intercepts of measured variables to be equal across groups, but permit the remaining parameters to differ between groups. We denote distinctions between parameters for groups by the pre-subscripts G1 and G2 for groups 1 and 2, respectively, but exclude these pre-subscripts for parameters that are constrained to be equal between groups. For example, the value of the intercept for V1 is constrained to be the same for both groups and accordingly is denoted as a_{V1}, regardless of group. In contrast, the factor means for groups 1 and 2 are allowed to differ across groups; therefore, the factor means (i.e., specified by the factor intercepts) for groups 1 and 2 are $_{G1}a_{F1}$ and $_{G2}a_{F1}$ for F1 and $_{G1}a_{F2}$ and $_{G2}a_{F2}$ for F2.

We again concentrate on the intercepts of models in that readers are likely to have had less exposure to these parameters. Different from the one-group model, lack of fit can occur with a multiple-group model due to fitting means of measures and specifically due to the between-group equality constraints on the intercepts for measures. Similar to the one-group CFA with means, factor intercepts for a multiple-group model are factor means, and the factor intercepts in one group—arbitrarily defined as the first group—are constrained to zero so that the model is identified, regardless of what other constraints are imposed. However, the factor intercepts for the second group can be freely estimated. These factor intercepts are not only the factor means in the second group, but also differences in factor means between the first and second groups because the factor means in the first group are equal to zero.

Although intercepts for measures are constrained to be equal across groups, it is useful to examine them separately in the different groups. The intercepts for measures in the first group are interpreted as the means for these measures in this group based on the model. For example, V2 in group 1 is a function of its intercept, the first factor, and error: $V2 = a_{V2}1 + b_{V2F1}F1 + E2$. Because the factor means are equal to zero in the first group (due to restricting the factor intercepts to zero in group 1) and the mean of errors is equal to zero, the mean of V2 in the first group is equal to the intercept for V2.

$$_{G1}\mu_{V2} = \mu_{a_{V2}1 + b_{V2F1}F1 + E2} \tag{5.5}$$

$$= \mu_{a_{V2}1} + b_{V2F1}(_{G1}\mu_{F1}) + \mu_{E2}$$

$$= a_{V2} + b_{V2F1}(_{G1}a_{F1}) + \mu_{E2}$$

$$= a_{V2} + b_{V2F1}(0) + 0$$

$$= a_{V2}$$

As a cautionary note, this interpretation is appropriate only if the model is correct. For example, if intercepts for measures are unequal, then the intercept for a measure should not be interpreted as its mean in the first group. In practice, this interpretation should not be made if the data-model fit is deemed inadequate.

Because the intercepts for V1 through V6 are equated across groups, the intercepts for the measures in group 2 also must represent the means for the measures in group 1. Under these circumstances, the only way to reproduce means for measures in the second group that differ from those in the first group is as a function of factor means (i.e., factor intercepts). For V2 in the second group,

$$\begin{aligned}
_{G2}\mu_{V2} &= \mu_{a_{V2}1+b_{V2F1}F1+E2} & (5.6)\\
&= \mu_{a_{V2}1} + b_{V2F1}(_{G2}\mu_{F1}) + \mu_{E2}\\
&= a_{V2} + b_{V2F1}(_{G2}a_{F1}) + 0\\
&= _{G1}\mu_{V2} + b_{V2F1}(_{G2}a_{F1})
\end{aligned}$$

Given the model is correct, any between-group difference in means on V2 (i.e., $_{G2}\mu_{V2} - _{G1}\mu_{V2}$) is a function of $_{G2}a_{F1}$, the difference in means between groups for F1.

Assessing data-model fit and testing factor mean differences. Before evaluating the model in Figure 5.2 that requires strong measurement invariance, we begin by evaluating a less restrictive model, one that requires configural invariance. It is good practice to first assess the data-model fit in each group, and then to estimate a multiple-group model. We have already assessed the form of the model in the first group in our illustration of a single-group CFA and found that it fit reasonably well, $\chi^2(8) = 10.78$, $p = .21$, CFI = .99, SRMR = .03, RMSEA = .04 with 90% CI [.00, .10]. The model also fits well in group 2, $\chi^2(8) = 15.27$, $p = .054$, CFI = .97, SRMR = .04, RMSEA = .07 with 90% CI [.00, .12]. We next use a multiple-group approach to assess fit for a model encompassing both groups without applying between-group constraints; the results for this model are a function of the two single-group analyses. Accordingly, the configurally-invariant model based on the multiple-group analysis also demonstrates good fit, $\chi^2(16) = 26.05$, $p = .053$, CFI = .98, SRMR = .04, RMSEA = .06 with 90% CI [.00, .19].

It is important to note that when using an incremental fit index, such as the CFI, to evaluate fit of a multiple-group model with mean structure, proper specification of the restricted null model is critical to ensure that it is nested within the researcher's most restricted model (Widaman & Thompson, 2003). Given the necessity of evaluating strong measurement invariance in comparing latent means, we have computed CFIs in SMM analyses such that they evaluate the improvement in fit for the researcher's model in comparison with a null model in which the factor loadings are equal to zero and intercepts are constrained to be equal between groups. The null model used by SEM software in the computation of the CFI may not be appropriate for a given analysis, so users are advised to carefully identify the proper null model for use in computing the CFI. We discuss the issues concerning the choice of null models for incremental fit indices in greater detail in Appendix B.

Next we assess the more restrictive model requiring strong measurement invariance, as shown in Figure 5.2. For our data, the model fits quite well, $\chi^2(24) = 37.10$, $p = .04$, CFI $= .97$, SRMR $= .05$, RMSEA $= .05$ with 90% CI [.01, .08]. The parameter estimates for this model (except for the error variances) are shown in the bottom part of Figure 5.2. Because the model requiring strong measurement invariance is nested within the model requiring configural invariance, a chi-square difference test can be conducted to evaluate the between-group equality constraints of the more restrictive model. The change in the fit indices was minimal, and the chi-square difference test was nonsignificant, $\chi^2(24 - 16 = 8) = 37.10 - 26.05 = 11.05$, $p = .20$. In practice and as described later in the chapter, researchers frequently evaluate separately the equality of factor loadings and the equality of intercepts for measures rather than combining the two types of parameters together in a single difference test.

If the data-model fit had not been adequate, we would have needed to consider releasing some of the between-group constraints; we describe an approach for handling partial invariance in a later section. In this example, our fit was adequate and we can proceed to evaluate the factor mean differences. Because the mean for any factor is fixed to zero in the first group, the difference between group means for this factor is equal to the factor mean in the second group. For our example, the means are greater in the second group on both factors: 3.57 for F1 and 3.88 for F2. We can test whether the difference in population means for a factor is equal to zero by conducting a chi-square difference test for that factor. First, a second model must be fitted in which the mean for the factor of interest is constrained to be equal to zero in the second group, that is, the same value as the factor mean in the first group. Then the chi-square for the model without this constraint is subtracted from the chi-square for the model with this constraint. The difference is approximately distributed in large samples as a chi-square with degrees of freedom equal to the difference in the degrees of freedom for the two respective models. For our example, $\chi^2(25) = 48.68$, $p < .01$ for the model with equal means for F1, and $\chi^2(25) = 44.42$, $p < .01$ for the model with equal means for F2. The chi-square difference tests for the factor means are both significant at the .05 level: $\chi^2(25 - 24 = 1) = 48.68 - 37.10 = 11.58$, $p < .01$ for the first factor mean and $\chi^2(25 - 24 = 1) = 44.42 - 37.10 = 7.32$, $p < .01$ for the second factor mean. Alternatively, Wald z tests, which are typically reported by SEM packages, can be used to test the factor means in the second group when they are freely estimated; the squared z statistics are estimates of the chi-square difference values. Although the Wald test is generally less accurate (Chou & Bentler, 1990), it is asymptotically equivalent to the chi-square difference test and is easier to obtain because it is not necessary to respecify the model. In this case, the Wald test yields very similar results: $z = 3.40$, $p < .01$ for the first factor mean and $z = 2.67$, $p < .01$ for the second factor mean (z^2 of 11.56 and 7.13 for F1 and F2, respectively).

Based on these results, we know the null hypothesis that factor means are equal across groups in the population can be rejected for both factors. However, we do not know the strength of the effects of day-care versus home-care on school and academic readiness factors. To compute a standardized effect size for a factor, the difference in factor means for the two groups is divided by the square root of the variance (i.e., standard deviation) of the disturbance for that same factor (Hancock, 2001). In terms of the model parameters, the standardized effect size is the difference in intercepts for a factor divided by the square root of the disturbance variance for that factor. In computing effect size statistics for our example, we have a choice among denominators: the disturbance variance of the home-care group, the disturbance variance of the day-care group, or the disturbance variance pooled across the two groups. We had allowed the disturbance variances to differ between the day-care and the home-care groups in our analyses: 73.69 and 80.52 for F1 and 141.75 and 133.43 for F2, respectively. These disturbance variances for either factor do not differ dramatically and therefore might be pooled by constraining them to be equal in a multiple-group analysis. Nevertheless, we decided to allow them to differ in our assessment of factor mean differences because conceptually they should differ. Consistent with this decision, we computed the effect size statistics based on the disturbance variance for a particular group rather than based on the disturbance variance pooled across groups. Although the decision was somewhat arbitrary, we chose the factor disturbance variances for the day-care pre-school children in that this group was thought of as the reference or comparison group. Thus, the standardized effect sizes for our example are

$$\text{ES}_{\text{F1}} = \frac{{}_{G2}a_{\text{F1}} - {}_{G1}a_{\text{F1}}}{\sqrt{{}_{G1}c_{\text{D1}}}} = \frac{3.57 - 0}{\sqrt{73.69}} = .42$$

and $$\hspace{4cm} (5.7)$$

$$\text{ES}_{\text{F2}} = \frac{{}_{G2}a_{\text{F2}} - {}_{G1}a_{\text{F2}}}{\sqrt{{}_{G1}c_{\text{D2}}}} = \frac{3.88 - 0}{\sqrt{141.75}} = .33$$

See Keselman, Algina, Lix, Wilcox, and Deering (2008) for a more general discussion of choice of denominators for effect sizes with heterogeneous variances.

Based on these analyses, the standardized effect size is slightly greater on the academic readiness factor. The means for the home-care group were .42 day-care standard deviation units higher than the day-care group on academic readiness and .33 day-care standard deviation units higher on social readiness. If we can assume the factors are normally distributed, we could estimate that the average home-care child would be at the 66th and 63rd percentiles among day-care children on academic readiness and social readiness, respectively. Presumably, because factors are error free, the size of an effect should be larger than those found with measured variables. Accordingly, although these standardized effect sizes might be judged as small to moderate based on Cohen's (1988) guidelines, they might be considered small for a structured means analysis.

The MIMIC Model for Assessment of Mean Differences

Analyses can be conducted to assess differences between factor means with the much simpler MIMIC model, which does not require a multiple-group approach. With the MIMIC model, the measured variables include not only indicators of the factors, but also coded variables that differentiate among groups. The number of coded variables included in an analysis is equal to the number of groups minus one. With two groups, a single coded variable is included in the model that has direct effects on the factors, and those direct effects allow for differences in factor means. In our example, the coded variable is a 1-0 dummy-coded variable, although other types of coding could also be used. We arbitrarily coded 0 for all individuals in the day-care group and 1 for all individuals in the home-care group. As shown in Figure 5.3, the dummy-coded variable has direct effects on the two readiness factors.

Before discussing the model further, we need to consider the format of the data. To conduct the analyses, scores on all variables from the multiple groups (N_1 = 200, N_2 = 200) are combined together in a single data set (N = 400). For our example, the combined data set includes scores on V1 through V6 (indicators of school readiness) as well as V7 (dummy-coded, child-care variable) for both the day-care and the home-care groups. A covariance matrix among the measured variables (i.e., V1 through V7) may then be computed, and, for our example, is shown at the bottom of Table 5.1.

It might seem confusing initially to be able to reach conclusions about factor mean differences by analyzing a covariance matrix. The covariance matrix contains two parts: (a) the variances and the covariances among the indicators of the factors and (b) the covariances between the dummy-coded

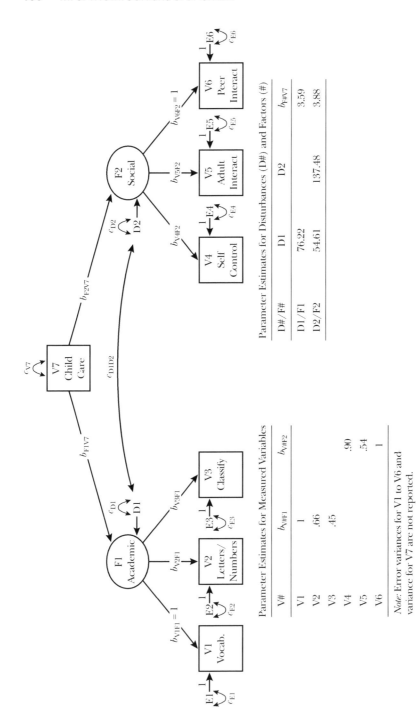

Parameter Estimates for Disturbances (D#) and Factors (#)

D#/F#	D1	D2	$b_{F#V7}$
D1/F1	76.22		3.59
D2/F2	54.61	137.48	3.88

Parameter Estimates for Measured Variables

V#	$b_{V#F1}$	$b_{V#F2}$
V1	1	
V2	.66	
V3	.45	
V4		.90
V5		.54
V6		1

Note: Error variances for V1 to V6 and variance for V7 are not reported.

Figure 5.3 Results for MIMIC model for Dataset 1.

variable and the indicators as well as the variance of dummy-coded variable. The covariance matrix among indicators is a total covariance matrix; that is, it is a combination of the between-group covariance matrix among the means on the indicator variables and the within-group covariance matrix among indicator scores deviated about their means. In contrast to the SMM, no distinction is necessary between the covariance matrices for the two groups (i.e., the within-day-care and within-home-care covariance matrices) because the MIMIC model assumes that the population covariance matrices are homogeneous and, therefore, can be pooled. Inclusion of the variance of the dummy-coded variable and the covariances between the dummy-coded variable and indicators allows the MIMIC analysis to differentiate between the between-group and within-group covariance matrices and, therefore, reach conclusions about differences in means.

The MIMIC model as shown in Figure 5.3 requires complete invariance (except for differences in factor means). In contrast, the SMM approach is more flexible. For example, the SMM approach in the previous section required only strong measurement invariance. We would expect the structured means model and the MIMIC model to yield more similar results to the extent that the data are consistent with complete invariance. For these data, the MIMIC model fit similarly to the structured means model in that the data are relatively consistent with complete invariance. In particular, the MIMIC model fit quite well, $\chi^2(12) = 15.28$, $p = .23$, CFI = .99, SRMR = .03, RMSEA = .03 with 90% CI [.00, .06] and, as shown in Figure 5.3, the parameter estimates are similar to those found with the SMM approach. Given the reasonable data-model fit, parameter estimates may be interpreted.

To answer our question about differences in factor means, we interpret the unstandardized parameter estimates between V7 (dummy-coded variable) and F1 of 3.59 and between V7 and F2 of 3.88. Individuals in the home-care group (coded as 1) score, on average, 3.59 units higher on the academic readiness factor and 3.88 units higher on the social readiness factor in comparison with individuals in the day-care group (coded as 0). The Wald z tests are significant for both the mean difference on Factor 1, $z = 3.53$, $p < .01$ and on Factor 2, $z = 2.75$, $p < .01$. The asymptotically equivalent chi-square difference tests, with the path from the dummy-coded variable set to zero for each factor separately, yield similar results. The programs for the MIMIC analysis are in Appendix A.

Based on the significance tests, the population factor means differ between groups, but standardized effect size statistics are necessary to assess how strong the effects are. To compute a standardized effect size, we divide the unstandardized weight between the dummy-coded variables and the factor by the square root of the disturbance variance for a factor (Hancock, 2001). For our example, the standardized effect sizes are

$$\text{ES}_{\text{F1}} = \frac{b_{\text{F1V7}}}{\sqrt{c_{\text{D1}}}} = \frac{3.59}{\sqrt{76.22}} = .41 \quad \text{and} \quad \text{ES}_{\text{F2}} = \frac{b_{\text{F2V7}}}{\sqrt{c_{\text{D2}}}} = \frac{3.88}{\sqrt{137.29}} = .33 \qquad (5.8)$$

MIMIC models do not permit disturbance variances for factors to differ between groups, and thus the denominators represent pooled within-group factor standard deviations. In contrast, the denominators for effect sizes with the SMM analyses were based on the estimated factor standard deviation for the day-care group.

Assessment of Differences in Factor Means Under Partial Invariance

As previously discussed, we cannot reach valid conclusions about between-group differences in factor means without requiring some model parameters to be equal between groups (Meredith, 1993). A minimum requirement is that the loadings and intercepts for one measure for each factor are equivalent between groups. In this section, we consider SMM methods for analyzing data when the factor loadings and intercepts for measures are partially invariant, that is, when loadings and/or intercepts for at least one measure are allowed to differ between groups.

We again use the day-care versus home-care example, but analyze a different data set, as presented in Table 5.2. For these data, the numbers of observations for the day-care and the home-care groups are 250 and 150, respectively. Models requiring strong measurement invariance fail to fit these data and, thus, require respecification to allow for partial invariance.

Stepwise Method for Testing Differences in Factor Means under Partial Invariance

In practice, researchers who use the SMM approach must make decisions about between-group constraints on the factor loadings and intercepts for measures. In Table 5.3, we present a four-step approach for assessing differences in factor means that incorporates decisions about these between-group constraints. We illustrate this stepwise approach in Table 5.4 using the second data set. Multiple structured means models are specified in the various steps and are labeled to indicate in what step they are used. For example, models explored in Step 2 are labeled Model 2, 2a, 2b, and so on.

Some overall comments about Tables 5.3 and 5.4 might be useful before we discuss individual steps. First, Table 5.3 describes a general procedure for analyzing factor mean differences between two independent groups, but is relatively easily extended for analyzing mean differences among three

TABLE 5.2   Covariance Matrices and Means for Measured Variables for the Day-Care Group, Home-Care Group, and Total Sample—Dataset 2

Measured variables	Covariance Matrix						Means
	V1	V2	V3	V4	V5	V6	
Group 1—Day-Care Group (N_1 = 250)							
V1	154.54						49.14
V2	44.75	90.23					82.60
V3	40.98	22.77	78.76				104.95
V4	41.35	2.83	7.92	220.12			78.58
V5	23.28	9.12	.75	61.46	159.88		54.95
V6	47.08	22.88	16.05	125.08	84.31	332.26	119.91
Group 2—Home-Care Group (N_2 = 150)							
V1	124.93						55.01
V2	52.19	80.67					81.22
V3	64.45	42.85	83.56				97.01
V4	59.95	33.10	38.34	290.65			72.67
V5	32.32	16.09	18.29	124.71	169.17		46.57
V6	87.15	39.70	51.82	174.20	108.39	355.22	124.28

Total—Combined for Day-Care and Home-Care Groups (N = 400)

	Covariance Matrix						
	V1	V2	V3	V4	V5	V6	V7
V1	151.18						
V2	45.52	86.88					
V3	38.69	32.78	95.18				
V4	40.06	16.03	30.27	254.09			
V5	15.04	14.41	22.94	96.55	179.46		
V6	67.94	27.69	21.22	137.05	84.49	344.48	
V7	1.83	−.32	−1.87	−1.39	−1.97	1.03	.235

Note: V1: Vocabulary; V2: Letters/Numbers; V3: Classification; V4: Self-Control; V5: Adult interaction; V6: Peer interaction; V7: Day (0) versus home (1) care.

or more groups and higher-way designs that exclude or include covariates. We briefly describe these extensions later in the chapter. Except for Step 1, these decisions are based on the relative fit of nested models. Just like the chi-square test for assessing fit of a model, the chi-square difference test for assessing differential fit of nested models is strongly influenced by sample size. Accordingly, differential fit should be assessed using not only the chi-square difference test, but also differences in other fit indices, such as CFI, SRMR, and RMSEA (e.g., see Chen, 2007; Cheung & Rensvold, 2002). Third, decisions about whether to constrain parameters based on fit indices are subjective, and different researchers can reach different decisions based on the same model results. Fourth, care should be taken in interpreting the

TABLE 5.3 Steps for Assessing Between-Group Differences in Factor Means

Steps	Model Specification[a]	Decision Rule
	Model 1	**Fit of Model 1**
Step 1. Equivalence of forms	Variances and covariances of factors and errors • Freely estimate all variances and covariances in all groups. Factor loadings • Fix loadings of one measure for each factor to 1.0 in all groups. • Freely estimate all remaining loadings. Intercepts for measures and factors ⇒ Freely estimate intercepts of measures in all groups. ⇒ Fix intercepts of factors to be 0 in all groups.	• Adequate fit in all groups: Proceed to Step 2 in that conceptually meaningful model was consistent with data in all groups. • Inadequate fit in at least one group: Cannot proceed to Step 2 unless model fits adequately in all groups and is conceptually meaningful.
	Model 2	**Difference in Fit between Model 1 and Model 2**
Step 2. Equivalence of loadings	Variances and covariances of factors and errors • Freely estimate all variances and covariances in all groups. Factor loadings • Fix loadings of one measure for each factor to 1.0 in all groups. ⇒ Constrain remaining loadings to be equal between groups. Intercepts for measures and factors ⇒ Freely estimate intercepts of measures in all groups. ⇒ Fix intercepts of factors to be 0 in all groups.	• Minimal difference in fit: Conclude essential equivalence of factor loadings between groups and proceed to Step 3. • Non-minimal difference in fit: Conclude at least one set of factor loadings between a particular measure and factor is not essentially equivalent. Based on MI tests for Model 2, choose the between-group constraint on loadings between a measure and factor that would maximize an increase in fit if relaxed in Model 2a. Specify Model 2a with this focal constraint relaxed.

(continued)

TABLE 5.3   Steps for Assessing Between-Group Differences in Factor Means (continued)

Steps	Model Specification[a]	Decision Rule
Step 2. Equivalence of loadings (*continued*)	**Model 2a, Model 2b, and Beyond** Variances and covariances of factors and errors • Freely estimate all variances and covariances in all groups. Factor loadings • Maintain same constraints on factor loadings as specified in the previous model with one exception as described next. ⇒ Relax focal constraint on factor loading according to MI tests based on previously specified model. Intercepts for measures and factors ⇒ Freely estimate intercepts of measures in all groups. • Fix intercept of factors to be 0 in all groups.	**Difference in Fit between Previous and Current Models** • Minimal difference in fit: Conclude that the relaxation of equality constraints on loadings in the current model is unnecessary based on the data, and that these loadings are essentially equivalent between groups. Proceed to Step 3. • Non-minimal difference in fit. Conclude that the relaxation of equality constraints on loadings in the current model are consistent with the data, and that these loadings are not essentially equivalent. If only one between-group constraint on loadings remains for each factor, assume the remaining constraints are appropriate (with caution) and proceed to Step 3. Otherwise, based on MI tests for the current model, choose the between-group equality constraint on loadings between a measure and factor that would maximize an increase in fit if relaxed in the next specified model. Specify the next model with this focal constraint relaxed.
Step 3. Equivalence of intercepts for measures	**Model 3** Variances and covariances of factors and errors • Freely estimate all variances and covariances in all groups. Factor loadings • Allow loadings to vary between groups if they were found to be not essentially equivalent between groups in Step 2. • Constrain all other factor loadings to be equal between groups.	**Difference in Fit between Final Model from Step 2 and Model 3** • Minimal difference in fit: Conclude essential equivalence of intercepts of measures that were constrained to be equal between groups for Model 3 and proceed to Step 4. • Non-minimal difference in fit: Conclude at least one set of intercepts of measures that were constrained to be equal for Model 3 is not essentially equivalent. Based on MI tests for Model 3, choose the between-group constraint on intercepts

(continued)

TABLE 5.3 Steps for Assessing Between-Group Differences in Factor Means (continued)

Steps	Model Specification[a]	Decision Rule
Step 3. Equivalence of intercepts for measures (*continued*)	Intercepts for measures and factors ⇒ Freely estimate intercepts of measures if the loadings for these measures were allowed to differ between groups. ⇒ Constrain remaining intercept to be equal between groups. • Fix intercepts of factors to be 0 in one group. ⇒ Freely estimate intercepts of factor in remaining groups. **Model 3a, Model 3b, and Beyond** Variances and covariances of factors and errors • Freely estimate all variances and covariances in all groups. Factor loadings • Allow loadings to vary between groups if they were found to be not essentially equivalent between groups in Step 2. • Constrain all other factor loadings to be equal between groups. Intercepts for measures and factors • Maintain same constraints on intercepts for measures as specified in the previous model with one exception as described next. ⇒ Relax focal constraint on intercepts for measures according to MI tests based on previously specified model. • Fix intercepts of factors to be 0 in one group. • Freely estimate intercepts of factor in remaining groups.	for measure that would maximize an increase in fit if relaxed in Model 3a. Specify Model 3a with this focal constraint relaxed. **Difference in Fit between Previous and Current Models** • Minimal difference in fit: Conclude that the relaxation of equality constraints on intercepts on measures in the current model is unnecessary based on the data, and that these intercepts are essentially equivalent between groups. Proceed to Step 4. • Non-minimal difference in fit. Conclude that the relaxation of equality constraints on intercepts of measures in the current model are consistent with the data, and that these intercepts are not essentially equivalent. If only one between-group constraint on intercepts remains for each factor, assume the remaining intercepts are appropriate (with caution) and proceed to Step 4. Otherwise, based on MI tests for the current model, choose the between-group equality constraint on intercepts for measures that would maximize an increase in fit if relaxed in the next specified model. Specify the next model with this focal constraint relaxed.

(continued)

TABLE 5.3 Steps for Assessing Between-Group Differences in Factor Means (continued)

Steps	Model Specificationᵃ	Difference in Fit between Final Model from Step 3 and Current Model in Step 4
Step 4. Equivalence of factor means/factor interceptsᵇ	**Model 4a, 4b, and Beyond** Variances and covariances of factors and errors • Freely estimate all variances and covariances in all groups. Factor loadings • Same as models in Step 3 Intercepts for measures and factors • Allow intercepts for measures to vary between groups if they were presumed or found to be not essentially equivalent between groups in Step 3. • Constrain all other intercepts for measures to be equal between groups. ⇒ Restrict intercepts for the focal factor to zero in all groups. ⇒ Restrict intercepts for all other factors to be equal to zero in first group, but freely estimate intercepts for these factors in all other groups.	• Minimal difference in fit: Conclude that the relaxation of the between-group equality constraints on intercepts of focal factor in the current model is unnecessary based on the data, and thus the means on this factor are essentially equivalent between groups. Proceed to next model in Step 4. • Non-minimal difference in fit. Conclude that the relaxation of equality constraints of intercepts on focal factor in the current model are relatively consistent with the data, and that the means on this factor are not essentially equivalent. Proceed to next model.

Note: The described procedure is for analyses with two independent groups, but can be extended for more complex designs, as discussed at the end of the chapter. In the description, it is presumed that the factor model has no within-group constraints imposed on the error variances, factor variances, and factor covariances—other than those necessary to identify the metric of the factor—and includes no error covariances. The steps can be easily modified if these model specifications do not hold.
ᵃ The symbol ⇒ indicates a primary change in the model for the current model in comparison with the previous model.
ᵇ The factor intercepts are the factor means. In some SEM programs, the factor means are directly specified.

TABLE 5.4   Results for Assessing Between-Group (B-G) Differences in Factor Means with Dataset 2

Steps	Statistical Results[a]	Discussion
Step 1. Equivalence of forms	**Model 1.** Same model form for both groups *Total:* $\chi^2(16) = 17.89$, $p = .33$, CFI = .997, SRMR = .033, RMSEA = .024 Group 1: $\chi^2(8) = 8.06$, $p = .43$, CFI = 1.00, SRMR = .027, RMSEA = .005 Group 2: $\chi^2(8) = 9.84$, $p = .28$, CFI = .993, SRMR = .040, RMSEA = .039	• Model 1 specifies same model for each group. b_{V1F1} and b_{V6F2} are fixed to 1 to define metric of factors; no b-g equality constraints are imposed on remaining loadings. Intercepts for measures are freely estimated. a_{F1} and a_{F2} are fixed to 0 in both groups. All variances and covariances are freely estimated. • Model fit was better for group 1, but very good in both groups. • Concluded model form was appropriate for both groups, and therefore proceeded to Step 2.
Step 2. Equivalence of loadings	**Model 2.** B-G equivalence of loadings $\chi^2(20) = 26.21$, $p = .16$; CFI = .990, SRMR = .042, RMSEA = .039 Difference between Models 1 and 2: $\Delta\chi^2(4) = 8.32$, $p = .08$; ΔCFI = −.007; ΔSRMR = +.009, ΔRMSEA = +.015 MI based on Model 2: Largest v^2 for loading on V3, $\chi^2(1) = 4.14$	• Specifications for Models 1 and 2 are identical except for factor loadings. Model 2 specifies b-g equality constraints on b_{V2F1}, b_{V3F1}, b_{V4F2}, b_{V5F2}, with $b_{V1F1}=1$ and $b_{V6F2}=1$ in both groups. • Results indicate difference in fit between Models 1 and 2 was not large; however, we decided the decrease in fit was sufficient to examine b-g differences in loadings for individual measures. • Based on the MI tests with Model 2, we allowed the loadings for V3 to differ between groups for Model 2a. The MI chi square was significant at the .05 level without controlling for Type I error across the multiple tests (critical $\chi^2(1) = 3.84$).

(continued)

TABLE 5.4   Results for Assessing Between-Group (B-G) Differences in Factor Means with Dataset 2 (continued)

Steps	Statistical Results[a]	Discussion
Step 2. Equivalence of loadings (*continued*)	**Model 2a.** B-G equivalence of loadings except for λ_{V3F1}: $\chi^2(19) = 22.16$, $p = .28$; CFI = .995, SRMR = .037, RMSEA = .029 Difference between Models 2 and 2a: $\Delta\chi^2(1) = 4.05$, $p = .04$; ΔCFI = +.005, ΔSRMR = −.005, ΔRMSEA = −.010 MI based on Model 2a: Largest χ^2 for loading on V6, $\chi^2(1) = 2.60$	• Model 2a is specified similarly to Model 2 except no b-g equality constraint is imposed on λ_{V3F1}. • Results indicate marginal improvement in fit of Model 2a over Model 2. The loadings for V3 were .48 and .73 for the groups. We allowed loadings for V3 to differ in subsequent models. • Based on the MI tests with Model 2a, we permitted the loadings for V6 (in addition to V3) to differ between groups for Model 2b, although the MI χ^2 value was nonsignificant at the .05 level.
	Model 2b. B-G equivalence of loadings except for λ_{V3F1} and λ_{V6F2}: $\chi^2(18) = 19.26$, $p = .36$; CFI = .998, SRMR = .035, RMSEA = .019 Difference between Models for 2a and 2b: $\Delta\chi^2(1) = 2.90$, $p = .09$; ΔCFI = +.003, ΔSRMR = −.002, ΔRMSEA = −.010 MI based on Model 2b: Largest χ^2 for loading on V1 or V2, $\chi^2(1) = 1.14$	• Model 2b is specified similarly to Model 2a except λ_{V6F2} is freely estimated in both groups and $\lambda_{V6F2} = 1$ in both groups. • Results indicate non-substantial increase in fit with Model 2b versus Model 2a. The loadings for V6 differed for the two groups (i.e., 1.37 and .96). To avoid misspecification, we permitted the loadings for V6 to differ in subsequent models. • Based on the MI tests with Model 2b, the expected increase in fit by allowing b-g differences in loadings for V1 or for V2 would be greatest and identical, but the MI χ^2 values for these loadings were minimal. We arbitrarily chose the loadings for V1 (in addition to V3 and V6) to be freely estimated in both groups for Model 2c, but expected minimal change in the model χ^2 values.

(continued)

TABLE 5.4   Results for Assessing Between-Group (B-G) Differences in Factor Means with Dataset 2 (continued)

Steps	Statistical Results[a]	Discussion
Step 2. Equivalence of loadings (*continued*)	**Model 2c.** B-G equivalence of loadings except for b_{V1F1}, b_{V3F1}, and b_{V6F2}: $\chi^2(17) = 17.96$, $p = .39$; CFI = .998; SRMR = .033, RMSEA = .017 Difference between Models 2b and 2c: $\Delta\chi^2(1) = 1.30$, $p = .25$; ΔCFI $< +.001$; ΔSRMR = −.002, ΔRMSEA = −.002 MI based on Model 2c: Largest χ^2 for loadings on V4 or V5, $\chi^2(1) = .07$	• Model 2c is specified similarly to Model 2b except b_{V1F1} is freely estimated in both groups and $b_{V3F1} = 1$ in both groups. • The loadings for V1 were .91 and 1.21 for the groups. Results indicate improvement in fit was negligible for Model 2c versus Model 2b. Given these results, we imposed b-g equality constraints on b_{V1} in all subsequent models. • The MI test with Model 2c yielded extremely small χ^2 values for the remaining b-g equality constraints. We proceeded to Step 3 allowing differences in b_{V3} and b_{V6}.
Step 3. Equivalence of intercepts for measures	**Model 3.** B-G equivalence of intercepts except a_{V3} and a_{V6}: $\chi^2(20) = 54.91$, $p < .01$, CFI = .942; SRMR = .060, RMSEA = .093 Difference between Models 2b and 3: $\Delta\chi^2(2) = 35.65$, $p < .01$; ΔCFI = −.056; ΔSRMR = +.025; ΔRMSEA = +.097 MI based on Model 3: Largest χ^2 for intercepts for V1 or V2, $\chi^2(1) = 23.65$	• Model 3 is specified similarly to Model 2b with the following exceptions: b-g equality constraints a_{V1}, a_{V2}, a_{V4}, and a_{V5}. Also the intercepts on factors are restricted to 0 in group 1 only. • Difference in fit between Models 2b and 3 is substantial and indicates that one or more of the b-g equality constraints on a_{V1}, a_{V2}, a_{V4}, and a_{V5} are problematic. • MI tests based on Model 3 indicated that releasing the b-g constraints on a_{V1} or a_{V2} would improve fit most. We arbitrarily permitted a_{V1} (in addition to a_{V3} and a_{V6}) to differ between groups for Model 3a.

(continued)

TABLE 5.4 Results for Assessing Between-Group (B-G) Differences in Factor Means with Dataset 2 (continued)

Steps	Statistical Results[a]	Discussion
Step 3. Equivalence of intercepts for measures *(continued)*	**Model 3a.** Constraints on all intercepts except a_{V1}, a_{V3}, and a_{V6}: $\chi^2(19) = 29.64$, $p = .06$, CFI = .982; SRMR = .044, RMSEA = .053 Difference between Models 3 and 3a: $\Delta\chi^2(1) = 25.27$, $p < .01$; ΔCFI = +.040; ΔSRMR = −.016; ΔRMSEA = −.040 MI based on Model 3a: Largest χ^2 for intercepts for V4 or V5, $\chi^2(1) = 10.30$ **Model 3b.** Constraints on all intercepts except a_{V1}, a_{V3}, a_{V4}, and a_{V6}: $\chi^2(18) = 19.26$, $p = .38$, CFI = .998; SRMR = .035, RMSEA = .019 Difference between Models 3a and 3b: $\Delta\chi^2(1) = 10.38$, $p < .01$; ΔCFI = +.016; ΔSRMR = −.009; ΔRMSEA = −.013	• Model 3a is specified comparably to Model 3 except a_{V1} is freely estimated in both groups. • Improvement in fit was substantial for Model 3a versus Model 3. Accordingly, we allowed a_{V1} to be freely estimated in both groups in subsequent models. • MIs based on Model 3a were greatest and identical for intercepts for V4 and V5. We arbitrarily permitted the intercept for V4 (as well as V1, V3, and V6) to differ between groups for Model 3b. • Model 3b is specified comparably to Model 3a except a_{V4} is freely estimated in both groups. • The improvement in fit was substantial for Model 3b versus Model 3a. Accordingly, we allowed the intercepts for V1 to differ in subsequent models. • Only one between-group intercept constraint remained per factor. We assumed (with caution) that these constraints were appropriate and proceeded to Step 4.
Step 4. Equivalence of means for factors	**Model 4a.** Factor intercepts restricted to 0 except a_{F2} in group 2: $\chi^2(19) = 21.41$, $p = .31$, CFI = .996; SRMR = .039, RMSEA = .025 Difference between Models 3b and 4a: $\Delta\chi^2(1) = 2.15$, $p = .14$; ΔCFI = −.002; ΔSRMR = +.004; ΔRMSEA = +.006	• Model 4a is specified comparably to Model 3b except a_{F1} is fixed to 0 in group 2. • Difference in fit between Models 3b and 4a was minimal. We cannot conclude that the means on F1 are different between groups in the population. The best estimate of the difference in means on F1 (i.e., factor intercept for group 2) is −1.38 based on Model 3b.

(continued)

TABLE 5.4 Results for Assessing Between-Group (B-G) Differences in Factor Means with Dataset 2 (continued)

Steps	Statistical Results[a]	Discussion
Step 4. Equivalence of means for factors (*continued*)	**Model 4b.** Factor intercepts restricted to 0 except a_{F1} in group 2: $\chi^2(19) = 56.70$, $p < .01$, CFI = .937; SRMR = .090, RMSEA = .100 Difference between Models 3b and 4b: $\Delta\chi^2(1) = 37.44$, $p < .01$; ΔCFI = −.059; ΔSRMR = +.055; ΔRMSEA = +.081	• Model 4b is specified comparable to Model 3b except a_{F2} is fixed to 0 in group 2. • Difference in fit between Models 3b and 4b was minimal. We concluded that the means on F1 are different between groups in the population. The best estimate of the difference in means on F2 (i.e., intercept of F1 for group 2) is −12.95 based on Model 3b.

[a] All analyses were conducted with Mplus. Differences in the fit indices of CFI, SRMR, and RMSEA were computed by subtracting the fit index value for the earlier model from the fit index value of the later model.

results in the table in that sometimes models become more complex as analyses proceed and sometimes less complex. Fifth, modification indices (MI), also referred to as Lagrange multiplier (LM) tests, are used to assess which between-group equality constraints on factor loadings or intercepts of measures should be removed to maximize improvement in fit (Chou & Bentler, 1990). It is advantageous, when possible, to be less dependent on these empirical search methods and rely more on theory and past empirical findings to make these decisions. Next, we briefly describe the stepwise approach for assessing differences in factor means.

In Steps 1 and 2, the focus is on modeling the variances and the covariances among the measures and not their means. The data in these steps could simply be the variances and covariances among measures (and not means of the measures) for each group. Intercepts are unnecessary to reproduce these variances and covariances and are excluded from models in these steps if means are not part of the data. To facilitate a more seamless transition to subsequent steps, we chose to analyze means of measures as well as their variances and covariances. We specify all intercepts of measures to be freely estimated and restrict factor means (i.e., factor intercepts) to be equal to 0 in all groups. With this specification, the intercepts of measures are the means of these measures and must perfectly reproduce them. Alternatively stated, the mean structure of the model is saturated, with the number of estimated parameters equal to the number of means. It is important to recognize that the model does not assess in any fashion whether the factor means are equal even though the means for a factor (i.e., factor intercepts) are restricted to the same value (i.e., 0) across all groups. The choice of values for factor means is arbitrary and has no effect on the fit of the model; that is, the factor means for Groups 1 and 2 could be restricted to 0 and 9, respectively, and the fit would be identical.

In Step 1, we assess whether the same form of the factor model fits adequately in all groups (i.e., configural invariance). If it does not, we cannot proceed. In addition, we assess whether the factor model fits adequately over all groups by conducting a multiple-group analysis with no between-group constraints on the factor model. The fit from the multiple-group analysis for this step is necessary to compare it with the model fit at Step 2.

In Step 2, we begin by assessing whether factor loadings are equivalent between groups by conducting a chi-square difference test. The fit of Model 1 with no between-group equality constraints of loadings is compared with the first model in Step 2 (i.e., Model 2) with between-group equality constraints on all loadings. If the hypothesis of equality of between-groups factor loadings is rejected, a search is conducted to evaluate which loadings between factors and measures can be freely estimated across groups. The search for which loadings to freely estimate and which loadings to constrain to be equal across groups is a process that potentially involves multiple models (i.e., Model 2a, 2b, and so on). For any one factor, a loading for one measure must be constrained to be equivalent between groups in order to have an identified model. It is advantageous, however, to have more than two loadings that are equivalent for each factor. As shown in the results for Model 2b in our example (i.e., Table 5.4), when only two between-group constraints are imposed on loadings per factor, the modification indices (MIs) indicate the same chi-square value for both loadings and are not informative.

Based on the results of Step 2, factor loadings are freely estimated or constrained between groups in models in Step 3 (and Step 4). Step 3 is similar to Step 2 in that it assesses what parameters should be freely estimated across groups, except the focal parameters are intercepts of measures rather than factor loadings. If factor loadings for measures are found to differ in Step 2, the intercepts for these measures are not evaluated for between-group equivalence in Step 3, but assumed to differ. It is very unlikely that intercepts of measures would be equivalent between groups if their loadings (i.e., the slopes) differ across groups.

The first model in Step 3 (i.e., Model 3) constrains intercepts of all measures to be equal between groups except the intercepts of measures that have loadings that differ between groups. The fit for Model 3 is compared to the fit for the model selected in Step 2 to assess the between-group equality constraints of intercepts of measures. Additional models may be investigated to evaluate what other constraints on intercepts of measures should be allowed to differ across groups. However, for any one factor, an intercept for one measure must be constrained to be equivalent between groups in order to have an identified model. Also as shown in Step 3 in Table 5.4, when two between-group constraints are imposed on intercepts within a factor, the MIs are identical for both intercepts and are not informative.

Some confusion might occur in understanding Model 3 and its nesting relationship with the model selected based on Step 2. To better understand this relation, it is important to recognize that the models in Step 2 can be reparameterized with respect to their intercepts and yield equivalent fit. Rather than freely estimating all intercepts of measures (as with models in Step 2), we could impose between-group equality constraints on intercepts for one measure per factor and freely estimate all remaining measure intercepts in the reparameterization. In addition, in place of restricting all factor means to 0, we could restrict the factor means to 0 in only one group and allow the factor means in the other groups to be freely estimated. For both the original and reparameterized models in Step 2, the intercepts part of the model is saturated, with the number of freely estimated parameters equal to the number of means of measures. The implication is that the reparameterization yields an equivalent model and, accordingly, has no effect on model fit. Now we can impose additional between-group constraints on the intercepts for measures of the reparameterized model selected in

Step 2 to obtain Model 3. Thus, Model 3 is nested within the model selected based on Step 2, and a chi-square difference test can be computed to assess whether freeing up additional between-group constraints improves fit in the population, as illustrated in Table 5.4.

We also note that intercepts for a measure should not generally be constrained if the corresponding factor loadings are allowed to vary between groups. If the factor loadings (i.e., slopes of the regression of the measured variable on the factor) differ between groups for a measure, then the expected measured variable score could be the same for only one value of the factor, that is, where the regression lines intersect. It is unlikely that this intersection would occur exactly at the origin where the factor score equals zero—the value that defines the intercepts. Accordingly, the intercepts for a measure are typically unconstrained between groups if the factor loadings for the measure were found to be noninvariant in the previous step. In Step 4, we evaluate the differences in factor means between groups by comparing models with the means (i.e., intercepts) of the focal factor restricted to 0 in all groups versus restricted to 0 in only one group. In conducting these tests, we specify these models so that constraints imposed on the factor loadings and intercepts for the measures are consistent with the results found in Steps 2 and 3.

To help interpret the magnitude of the differences in factor means, standardized effects sizes are computed (Hancock, 2001). The standardized effect sizes were calculated using the estimated factor disturbance variance for the day-care group:

$$\text{ES}_{F1} = \frac{_{G2}a_{F1} - _{G1}a_{F1}}{\sqrt{_{G1}c_{D1}}} = \frac{-2.40 - 0}{\sqrt{81.70}} = -.26$$

$$\text{and} \tag{5.9}$$

$$\text{ES}_{F2} = \frac{_{G2}a_{F2} - _{G1}a_{F2}}{\sqrt{_{G1}c_{D2}}} = \frac{-12.95 - 0}{\sqrt{92.23}} = -1.35$$

The mean for academic readiness is .26 day-care standard deviations smaller for the home-care group in comparison with the day-care group, whereas the mean for social readiness is 1.35 day-care standard deviation units smaller for the home-care group.

Problems with Interpreting Mean Differences

Problems with interpreting mean differences of observed measures. We briefly consider next under what conditions differences in means of measures are accurate or misleading representations of differences in factor means. In discussing these conditions, we assume correct decisions were made concerning parameters that were constrained to be equal between groups in our example. First, if the factor loadings and intercepts for a measure are equivalent between groups, the difference in means for this measure reflects on average the difference in factor means. In our example, V2 meets this condition in that $_{G2}\mu_{V2} - _{G1}\mu_{V2} = b_{V2F1}(_{G2}\mu_{F1} - _{G1}\mu_{F1})$. Even in this condition, the standardized effect size for the observed measure generally is smaller than the standardized effect size for its underlying factor due to the contribution of the measure's error variance to the denominator (in addition to the factor variance). Second, if factor loadings for a measure are equivalent between groups, but intercepts for the measure differ, the difference in means on that measure is on average a function of not only the difference in factor means, but also the difference in intercepts for the measure. In our example, V1 meets this condition, and $_{G2}\mu_{V1} - _{G1}\mu_{V1} = b_{V1F1}(_{G2}\mu_{F1} - _{G1}\mu_{F1}) + (_{G2}a_{V1} - _{G1}a_{V1})$. In this condition, some source other than the factor is causing differences in means for a measure, and, in this sense, the measure is biased. Third, if a measure has different factor loadings in different groups, the intercepts for that measure almost certainly would differ too, yielding differences in means for a measure that are complexly determined and not reflective of the differences in factor means. In our example, V3 meets this condition: $_{G2}\mu_{V3} - _{G1}\mu_{V3} = [_{G2}b_{V3F1}(_{G2}\mu_{F1}) - _{G1}b_{V3F1}(_{G1}\mu_{F1})] + (_{G2}a_{V3} - _{G1}a_{V3})$. In summary, the mean differences of observed measures are inaccurate representations of mean differences of their underlying factor in the latter two conditions. Even in the first condition in which the mean difference in a measure is proportional to the mean difference in its underlying factor, the standardized effect size is attenuated by the measure's error variance.

Problems with interpreting mean differences on factors. If factors underlie a set of measures, we have argued in favor of the evaluation of the mean differences in these factors rather than mean differences of the observed measures. However, researchers can encounter hurdles reaching accurate conclusions based on differences in factor means. Estimates of differences in factor means can be badly biased if models used to estimate these differences are misspecified and, more specifically, if parameters are incorrectly constrained to be equal between groups. From this perspective, it might seem reasonable to use a strategy in Steps 2 and 3 (as presented in Table 5.3) that allows for between-group differences in factor loadings and intercepts of measures if researchers have any uncertainty concerning the equality constraints on these parameters.

The strategy to minimize the number of equality constraints is ill advised for a number of reasons. First, if we take the strategy to the extreme, we eventually must make decisions about what measure should be the referent indicator for each factor. The referent indicator is the variable for a factor that is assumed for identification purposes to have equal loadings and intercepts across groups. Although strategies have been suggested to address the problem of defining the referent indicator (e.g., Cheung & Rensvold, 1999; Rensvold & Cheung, 2001; Yoon & Millsap, 2007), these methods can be labor intensive and are fallible (for further discussion, see Millsap, 2011). Second, decisions concerning equality constraints can be incorrect and may lead to erroneous conclusions. Monte Carlo research indicates it is more difficult to make these decisions to the extent that factor loadings, differences in parameter values, sample size, and the proportion of invariant parameters are small (e.g., Johnson, Meade, & DuVernet, 2009; Lo & Thompson, 2009; Yoon & Millsap, 2007). Third, the interpretability of factor mean differences is limited to the extent that we rely on a minimal number of indicators with invariant parameters. We next illustrate these problems with our second dataset.

We generally cannot decide based on model fit which variable should be the referent indicator when it has been determined that intercepts for all but two measures must be allowed to vary between groups. As presented in Step 3 in Table 5.4, it was completely ambiguous whether the intercepts of V1 or V2 should be constrained to be equal between groups for F1, and similarly whether the intercepts of V4 or V5 should be constrained to be equal between groups for F2. The modification indices, as well as the comparable chi-square difference tests, were identical for the between-group equality constraints on the intercepts of V1 and V2 and on the intercepts of V4 and V5. From this perspective, it is important to have at least two indicators per factor that are expected to have invariant parameters across groups.

The decision about which intercepts to constrain is an important one. In Table 5.5, we show that the choice among intercept constraints had a dramatic effect on the estimates of the differences in means for F1 and F2. We note that for these examples we set factor loadings to 1 for both groups for V2 and V5 and constrain factor loadings for V1 and V4 to be equal between groups. If the intercept constraints are imposed on V2 and V5 (Model A), the means on the academic and the social factors (F1 and F2, respectively) were greater in the day-care group than in the home-care group. However, the effect on F1 was small ($ES_{F1} = -.27$) and nonsignificant at the .05 level, whereas the effect on F2 was large ($ES_{F2} = -1.35$) and significant. In contrast, when constraints were imposed on V1 and V4 (Model B), the mean on the academic factor was higher in the home-care than the day-care group, whereas the mean on the social factor was higher in the day-care group. For both factors, the differences were significant and standardized effect sizes were moderate ($ES_{F1} = +.65$ and $ES_{F2} = -.61$). Although Models A and B produce very different estimates of differences in factor means, they are equivalent models and, therefore, the choice between them cannot be made based on model fit.

TABLE 5.5 SMM Results for Models with Various Between-Group Constraints on Intercepts for Measured Variables

	Group 1			Group 2			Between-Group Difference		
	Loadings			Loadings					
	F1	F2	Intercepts	F1	F2	Intercepts	Means	p value	Effect size
Model A with intercept constraints for V2 and V5									
V1	1.74		49.14	1.74		57.41			
V2	1.00		82.60	1.00		82.60			
V3	0.84		104.95	1.28		98.78			
V4		1.54	78.58		1.54	85.62			
V5		1.00	54.95		1.00	54.95			
V6		2.12	119.91		1.49	136.74			
F1			0			−1.38	−1.38	.143	−.27
F2			0			−8.38	−8.38	<.001	−1.35
Model B with intercept constraints for V1 and V4									
V1	1.74		49.14	1.74		49.14			
V2	1.00		82.60	1.00		77.85			
V3	0.84		104.95	1.28		92.70			
V4		1.54	78.58		1.54	78.58			
V5		1.00	54.95		1.00	50.39			
V6		2.12	119.91		1.49	129.97			
F1			0			+3.37	+3.37	<.001	+.65
F2			0			−3.82	−3.82	<.001	−.61

(continued)

TABLE 5.5 SMM Results for Models with Various Between-Group Constraints on Intercepts for Measured Variables (continued)

	Group 1			Group 2			Between-Group Difference		
	Loadings			Loadings					
	F1	F2	Intercepts	F1	F2	Intercepts	Means	p value	Effect size
Model C with intercept constraints for V1, V2, V4 and V5									
V1	2.28		49.64	2.28		49.64			
V2	1.00		81.07	1.00		81.07			
V3	0.94		104.95	1.53		93.70			
V4		1.32	79.44		1.32	79.44			
V5		1.00	53.96		1.00	53.96			
V6		1.95	119.91		1.42	132.92			
F1			0			−2.16	−2.16	<.001	−.49
F2			0			−6.09	−6.09	<.001	−.89

Note: Fits for models A and B were identical: $\chi^2(18) = 19.26$, $p = .38$, CFI = .998, SRMR = .035, RMSEA = .019. For model C, $\chi^2(20) = 54.91$, $p < .001$, CFI = .942, SRMR = .060, RMSEA = .093.

Estimates of factor mean differences are dissimilar for Models A and B because the estimates for the two models rely on different measured variables. For simplicity, we focus on F1 to illustrate that the estimates of factor mean differences rely on the measures with constrained intercepts. For Model A, the intercepts for V2 are constrained to be equal, and thus the estimate of the difference in factor means is a function of V2. The equations for the reproduced means for V2 are identical in groups 1 and 2 except for the factor means of the two groups: $_{G1}\overline{V2}_{Reproduced} = (1.00)_{G1}\overline{F1} + 82.60$ and $_{G2}\overline{V2}_{Reproduced} = (1.00)_{G2}\overline{F1} + 82.60$. Thus, the difference in reproduced means of V2 is simply the difference in the factor means for the two groups: $_{G2}\overline{V2}_{Reproduced} - _{G1}\overline{V2}_{Reproduced} = _{G2}\overline{F1} - _{G1}\overline{F1}$. The model accurately reproduced the difference in sample means for V2 of -1.38 ($= 81.22 - 82.60$) by estimating the difference in factor means to be -1.38 ($= -1.38 - 0$).

In contrast, for Model B, the intercepts for V1 are constrained to be equal, and thus the estimation of the difference in factor means is reliant on V1. The equations for the reproduced means for V1 are identical for groups 1 and 2 except for the factor means of the two groups: $_{G1}\overline{V1}_{Reproduced} = (1.742)_{G1}\overline{F1} + 49.14$ and $_{G2}\overline{V1}_{Reproduced} = (1.742)_{G2}\overline{F1} + 49.14$. Thus, the difference in reproduced means of V1 is the factor loading of 1.742 times the difference in factor means between the two groups: $_{G2}\overline{V2}_{Reproduced} - _{G1}\overline{V2}_{Reproduced} = (1.742)(_{G2}\overline{F1} - _{G1}\overline{F1})$. The model accurately reproduced the difference in sample means for V1 of 5.87 ($= 55.01 - 49.14$) by estimating the difference in factor means to be 5.87 [$= 1.742(3.37)$]. In summary, the mean of F1 was in favor of Group 1 for Model A because the mean on V2 was greater in that group, whereas the mean of F1 was in favor of Group 2 for Model B because the mean on V1 was greater in that group.

For Models A and B, we rely on a single measure to estimate the difference in means for each factor. It seems risky to base an estimate on a single measure, given the possibility for misspecification and the dramatically different estimates of factor mean differences for the two models. An alternative might be to impose between-group constraints on intercepts for more than one measure. As shown in Model C, we could impose between-group constraints on the intercepts of V1 and V2 for F1 and intercepts of V5 and V6 for F2. For our data, Model C does not fit as well as Model A or B, and on these grounds might be abandoned. On the other hand, the differences in factor means for Model C are somewhere between those for Models A and B and therefore could be considered a compromise position.

In summary, estimates of mean differences on factors are biased if equality constraints on parameters are incorrectly imposed between groups. On the other hand, to the extent that the number of equality constraints is minimized, decisions about differences in factor means will be based on a very restricted number of measures that fail to broadly represent the underlying construct (i.e., poorer construct validity) and still might lead to biased estimates. Researchers can minimize these problems by carefully designing studies that include for each factor multiple measures, and these measures, as a group, have good construct validity. Researchers also can assess whether the bias due to differences in intercepts of measures counterbalance each other across measures (i.e., the sum of the freely estimated intercepts for measures is approximately equal to zero within each group) when these intercepts are constrained to be equal between groups.

ACCOMMODATING MORE COMPLEX RESEARCH DESIGNS

One-Way Designs with More Than Two Groups

Our examples have illustrated the MIMIC and SMM approaches for the two-group case with populations distinguished by a single grouping variable. Traditionally, in ANOVA/MANOVA language, such a design might be called a *one-way* design with two groups. Extending the application of either the MIMIC or SMM approaches to three or more groups that differ according to a single dimension is relatively straightforward.

With the MIMIC model, we specified group membership for the two-group case using a single dummy-coded variable and tested the coefficient relating the coded variable to a latent variable to infer whether the two populations differ on the latent means. More generally, we could use $J-1$ dummy-coded variables simultaneously to evaluate differences among J populations. The omnibus hypothesis that the population means for a factor are equal between groups may be tested by comparing the fit of a model in which the coefficients between all coded variables and the factor of interest are freely estimated with a model in which these parameters are constrained to zero.

If the omnibus hypothesis is rejected for any factor, tests of pairwise differences between means for this factor can be assessed in one of two ways, depending on whether they involve the reference group. The reference group has zeros in all dummy-coded variables, whereas each of the other groups has ones in a dummy-coded variable—designated that group's dummy-coded variable—and zeros in all other dummy-coded variables. A test of a mean difference between a group and the reference group is conducted by evaluating whether the path coefficient for the non-reference group's dummy-coded variable is equal to zero. A test of a mean difference between two non-reference groups is performed by assessing whether coefficients for the dummy-coded variables for these groups are equal. Researchers can control for Type I errors across the multiple pairwise comparisons for any one factor; we prefer the use of the Shaffer modified sequential Bonferroni method that incorporates the prior test of an omnibus hypothesis (1986).

The SMM approach described in Table 5.3 can be applied to evaluate differences in factor means among three or more groups. Evaluation of between-group equality constraints in Steps 2 and 3 is more complex under these conditions than for the two-group case because factor loadings and intercepts for a measure may vary between two or more groups. For these analyses, researchers might have to make many judgment calls concerning equality constraints given potentially the large number of tests of these constraints. These judgment calls require a good understanding of the factor structure of the measures for the groups of interest based on past research and the literature in general.

With the SMM approach, Step 4 involves initial tests of omnibus hypotheses that the population means are equal for each of the factors. These tests are conducted by comparing the fit of two models. With the less constrained model, the factor mean for one group is fixed to zero, and the factor means for all remaining groups are freely estimated (i.e., the model resulting from Step 3). With the more constrained model, the means for a factor are all fixed at zero. A chi-square difference test is conducted to evaluate an omnibus hypothesis.

If the omnibus hypothesis is rejected for any factor, tests of pairwise differences between means for this factor can be evaluated. The simplest approach is to conduct a chi-square difference test based on two nested models. With the less constrained model, the factor mean for one of the groups involved in the pairwise comparison is fixed to zero, and the factor means for all remaining groups are freely estimated. With the more constrained model, the factor means for both groups involved in the pairwise comparison are fixed to zero, and the factor means for all remaining groups are freely estimated. A method should be used to control for Type I error across pairwise comparisons for each factor (e.g., Shaffer, 1986).

Multiway Designs

Methods used in MIMIC and SMM analyses with a single grouping variable can be extended to assess main and interaction effects in designs with multiple grouping variables (i.e., multiway designs). For MIMIC models, we follow methods analogous to procedures used to conduct analysis of variance with regression methods; that is, coded variables are included in the model for each grouping variable as well as cross-product terms between sets, and tests are conducted for particular hypotheses by constraining appropriate parameters. Of course, researchers should keep in mind the restrictive assumption implied by the MIMIC model that model parameters must be identical across all population groups except for factor means.

We illustrate the SMM approach by considering a 2×3 ANOVA, and believe that other multiway designs are straightforward generalizations of this illustration. In conducting our analyses, we impose our constraints at the cell level. For our two-way design, cells 1 through 3 are associated with the 1st through 3rd levels of B within the 1st level of A, and cells 4 through 6 are associated with the 1st through 3rd levels of B within the 2nd level of A.

Initially we investigate equality of factor loadings and intercepts of measures across cells of the multiway design. In investigating these parameters, we use the same approach as we would with a one-way design, with the cells being the groups. In Step 3, the mean for any one factor (denoted as F) is fixed to zero for one cell, and the factor means for the other cells are freely estimated. For simplicity, we arbitrarily specify the factor mean for the 1st cell to be fixed to zero (i.e., $_{G1}a_F = 0$). The model resulting from Step 3 becomes the less-constrained model for tests of the main effect and interaction hypotheses for our two-way design.

In Step 4, we impose constraints on this less-constrained model that are consistent with the main effect or interaction null hypotheses of interest. Based on our knowledge of ANOVA, the null hypotheses for the main effect hypotheses (based on unweighted means) and the interaction hypothesis in terms of the cell means are as follows:

A main effect H_0:

$$\left(_{G1}\mu_F + _{G2}\mu_F + _{G3}\mu_F \right) / 3 - \left(_{G4}\mu_F + _{G5}\mu_F + _{G6}\mu_F \right) / 3 = 0, \qquad (5.10)$$

B main effect H_0:

$$\left(_{G1}\mu_F + _{G4}\mu_F \right) / 2 - \left(_{G2}\mu_F + _{G5}\mu_F \right) / 2 = 0$$
$$\left(_{G2}\mu_F + _{G5}\mu_F \right) / 2 - \left(_{G3}\mu_F + _{G6}\mu_F \right) / 2 = 0, \qquad (5.11)$$

and AB Interaction effect H_0:

$$\left(_{G1}\mu_F - _{G2}\mu_F \right) - \left(_{G4}\mu_F - _{G5}\mu_F \right) = 0$$
$$\left(_{G2}\mu_F - _{G3}\mu_F \right) - \left(_{G5}\mu_F - _{G6}\mu_F \right) = 0. \qquad (5.12)$$

For the more constrained model, constraints on the factor means are imposed on the less-constrained model that are consistent with the null hypothesis of interest (and given $_{G1}a_F = 0$ for the less-constrained model):

A main effect constraint:

$$\left(_{G2}a_F + _{G3}a_F \right) / 3 - \left(_{G4}a_F + _{G5}a_F + _{G6}a_F \right) / 3 = 0, \qquad (5.13)$$

B main effect constraints:

$$_{G4}a_F / 2 - (_{G2}a_F + _{G5}a_F) / 2 = 0$$

$$(_{G2}a_F + _{G5}a_F) / 2 - (_{G3}a_F + _{G6}a_F) / 2 = 0$$

(5.14)

and
AB Interaction constraints:

$$-(_{G2}a_F) - (_{G4}a_F - _{G5}a_F) = 0$$

$$(_{G2}a_F - _{G3}a_F) - (_{G5}a_F - _{G6}a_F) = 0.$$

(5.15)

A chi-square difference test comparing the less constrained model and the more constrained model consistent with hypothesis of interest is conducted. See Green and Thompson (2012) for additional detail about multiway designs.

Inclusion of Manifest and Latent Covariates

The MIMIC and SMM approaches can readily be adapted to assess differences in factor means after controlling for one or more covariates. SEM can accommodate covariates specified either as measured variables or as factors based on multiple indicators. For example, we could examine differences between the home-care and day-care groups on the academic readiness factor, controlling for differences in social readiness—a latent covariate. As discussed earlier, a factor is a theoretically error-free representation of a construct so might provide a better way in which to operationalize a covariate if multiple indicators of the covariate are available. Inclusion of covariates can reduce the residual variance of the factors within groups and adjust factor means based on differences on the covariate.

Within a MIMIC approach, a manifest or latent covariate is included in the model along with the group-coded variable(s) as predictors of the factors. A covariance is included between the exogenous covariate and the group-coded variable. Evaluation of the difference in the population factor means, adjusted for the covariate, can then be made using the same parameter as before—the path from the group-coded variable to the factor. The same general approach can be adapted to accommodate multiple covariates. For example, using a MIMIC approach, Aiken et al. (1994) evaluated post-treatment mean differences in daily life factors across two treatment groups of drug addicts, after controlling for initial differences in daily life factors. Group differences were assessed by evaluating post-treatment factor means, adjusting for pre-treatment differences specifically on the same factor. Adjustment for pre-treatment differences was also made at the indicator level by specifying error covariances for corresponding indicators across pre- and post-treatment assessments. The flexibility of SEM for specifying covariates to control for any or all of the factors and indicators is an advantage over traditional methods such as MANCOVA, which does not allow selective pairings of covariates and outcomes.

Similarly, either manifest or latent covariates can be included in models as exogenous variables within an SMM approach. The covariates are added as predictors of the factors on which mean differences are to be assessed. The means of latent covariates, as well as those of the factors of interest, must be set to zero in the reference group. Additionally, whereas the MIMIC approach implies the homogeneity of slopes assumption by allowing a single parameter between the covariate and latent variable to be estimated, this invariance constraint is explicitly specified and can be tested in the SMM approach. Evaluation of the difference in latent population means, adjusted for the covariate, is made based on whether the intercept for the factor of interest differs from zero. See Hancock (2004) for further details and illustrated examples of MIMIC and SMM models with latent covariates.

SUMMARY AND CONCLUSIONS

We summarize the key points of this chapter on evaluating between-group differences in latent means by addressing the following simple question: "What steps can be taken that lead to the best chance for a successful analysis of differences in latent means?"

First and foremost, when planning the study, researchers should "get in the mood for modeling" and specify thoughtfully the models to be used, including consideration of any nested or non-nested models that are consistent with hypotheses of interest. For latent variable systems, SEM should be used to allow the relation between the constructs and measures to be properly specified. In addition, SEM is more powerful than MANOVA for latent variable systems (e.g., Kaplan & George, 1995). SEM methods also permit the use of theoretically error-free representations of constructs and flexibility in the specification of the model. However, it is extremely important to include valid measures of constructs in SEM analyses. The most valid set of measures possible should be used to represent each construct to avoid reaching incorrect decisions about factor mean differences.

Second, researchers are confronted with the choice between SMM and MIMIC methods for conducting their analyses. Although this choice could be made using a variety of strategies, we have a preference for a data-driven approach that could incorporate both methods. Initial analyses would assess whether model parameters are invariant across groups. These analyses would potentially involve CFAs within groups, multiple-group CFAs with between-group constraints, and structured means modeling with between-group constraints. If the results of these analyses suggest that the model parameters (except factor intercepts) are similar across groups, then MIMIC analysis would be appropriate and would allow for a straightforward interpretation of results. If the model parameters are dissimilar, then the stepwise approach to SMM presented in Table 5.3 should be used to test factor means.

We should also consider Type I error rate and power for the MIMIC and SMM approaches. Simulations by Hancock et al. (2000) demonstrated that SMM controls adequately for Type I error across a range of conditions in which factorial invariance does or does not hold. The MIMIC approach controlled adequately for Type I error under the more limited conditions of equal sample sizes across groups and/or equal determinants of covariance matrices across groups. Type I error rates were inflated when the larger sample had weaker loadings, and were too conservative when the smaller sample had weaker loadings. The power of the Wald test to detect differences in latent means appeared to be similar across the MIMIC and SMM approaches. Kaplan and George (1995) found power to be somewhat greater for larger models and for equal sample sizes across groups. They found power was not strongly affected by noninvariance when sample sizes were equal. Researchers would be well served to design studies to have approximately equal sample sizes across groups whenever possible. See Hancock and French's chapter in this volume for discussion of approaches for conducting power analysis and determining sample size in SEM; simulation studies focusing on power to detect noninvariant parameters may be used to provide recommendations for optimal design of studies and for approaches to evaluating factorial invariance (e.g., Chen, 2007; Lo & Thompson, 2007; Yoon & Millsap, 2007).

We have attempted to provide a fairly comprehensive introduction to SEM analyses of between-group differences in latent means. Although these methods have been available for quite some time, applied studies using latent means analysis have appeared rather infrequently in the literature. Perhaps this is because few resource books on SEM have illustrated these techniques; exceptions include sections in books by Bollen (1989) and, more recently, Kaplan (2009) and Byrne (2012; see also similar books for EQS, LISREL, and AMOS users). Conceptual introductions to latent means methods with illustrative examples are presented in Hancock's introductory article (1997) and his somewhat more technical chapter (2004). Addressing more specific aspects of these methods, Byrne et al. (1989) and Vandenberg and Lance (2000) discuss methods for assessing partial factorial invariance, while Cole et al. (1993) offer a comparative treatment of MANOVA and SEM approaches for multivariate group comparisons. Finally, a well-explained application of multiple-group latent means analysis can be found in Aiken et al. (1994).

APPENDIX A
Syntax for SMM and MIMIC Analyses for Dataset 1

Programs are presented here for SMM and MIMIC analyses for Dataset 1. The SMM syntax specifies the most complex and constrained form of the multiple-group model we discuss; other models presented can generally be specified by removing intercepts and/or constraints. Instructions are included for removing these constraints following each program.

To adapt syntax for the partial invariance example for Dataset 2, first replace the data with the covariance matrix and means presented in Table 5.2. Second, follow the stepwise instructions under Model Specification in Table 5.3. For example, for the combined-groups model under Model Specification for Step 1, the intercepts of measures are freely estimated in both groups and the intercepts of factors are set to 0 in all groups.

Syntax for SMM Analyses for Dataset 1

EQS Program

```
/TITLE
   SMM FOR DATA SET 1 — GROUP 1
/SPECIFICATIONS
   VARIABLES=6; CASES=200; MATRIX=COV; ANAL=MOMENT; GROUPS=2;
/EQUATIONS
   V1 =  1F1 +   50*V999 + E1;
   V2 =  *F1 +   80*V999 + E2;
   V3 =  *F1 +  100*V999 + E3;
   V4 =  *F2 +   80*V999 + E4;
   V5 =  *F2 +   50*V999 + E5;
   V6 =  1F2 +  120*V999 + E6;
   F1 = 0V999 + D1;
   F2 = 0V999 + D2;
/VARIANCES
   E1 TO E6 = *; D1 to D2 = *;
/COVARIANCES
   D1,D2=*;
/PRINT
   FIT=ALL; COV=YES;
/MATRIX
138.00
 45.58    80.49
 35.19    23.56    56.34
 45.13    32.00    16.64    232.17
 35.33    10.49    10.56     79.74    149.16
 73.34    28.73    33.21    117.30     79.90    324.36
/MEANS
 50.40 79.60 98.88 74.06 49.12 120.10
/END
```

```
/TITLE
   SMM FOR DATA SET 1 — GROUP 2
/SPECIFICATIONS
   VARIABLES=6; CASES=200; MATRIX=COV; ANAL=MOMENT;
/EQUATIONS
   V1 =   1F1 +   50*V999 + E1;
   V2 =   *F1 +   80*V999 + E2;
   V3 =   *F1 +  100*V999 + E3;
   V4 =   *F2 +   80*V999 + E4;
   V5 =   *F2 +   50*V999 + E5;
   V6 =   1F2 +  120*V999 + E6;
   F1 = *V999 + D1;
   F2 = *V999 + D2;
/VARIANCES
   E1 TO E6 = *; D1 to D2 =* ;
/COVARIANCES
   D1,D2=*;
/MATRIX
127.61
 58.49     76.81
 29.09     19.72      54.29
 45.84     28.94      31.82     223.09
 20.33     13.27       5.81      62.25    135.72
 50.64     55.45      30.15     126.74     62.16    337.37
/MEANS
 53.70     81.32     101.82      77.69     51.62    123.37
/CONSTRAINTS
   (1,V2,F1)=(2,V2,F1);
   (1,V3,F1)=(2,V3,F1);
   (1,V4,F2)=(2,V4,F2);
   (1,V5,F2)=(2,V5,F2);
   (1,V1,V999)=(2,V1,V999);
   (1,V2,V999)=(2,V2,V999);
   (1,V3,V999)=(2,V3,V999);
   (1,V4,V999)=(2,V4,V999);
   (1,V5,V999)=(2,V5,V999);
   (1,V6,V999)=(2,V6,V999);
/END
```

Note: Start values are provided for the intercepts of measures. They are the values in front of the * in the equations for the measures. These values are the means of the measures (to the nearest tenths place) in the first group. The equality constraints are in bold print under /CONSTRAINTS. An equality constraint on the model parameters can be removed by deleting the relevant constraint statement from the program.

Mplus Program

```
Title:     One-Way SMM Analysis of Latent Means for Dataset 1
Data:      File is "TWO GROUP DATA 1.dat";
           TYPE is MEANS COVARIANCE;
           NOBSERVATIONS = 200 200;
           NGROUPS=2;
Variable:  Names are V1 V2 V3 V4 V5 V6;
           USEVARIABLES ARE V1 V2 V3 V4 V5 V6;
           Grouping = Group (1=G1 2=G2);
Analysis:  Type = Mgroup;
           Estimator = ML;
           Convergence = .000000001;
Model:     F1 by V1@1 V2* V3*;
           F2 by V4* V5* V6@1;
           F1*;
           F2*;
           [F1@0];
           [F2@0];
           [V1]; [V2]; [V3]; [V4]; [V5]; [V6];
           V1-V6*;
Model g2:  F1*;
           F2*;
           [F1*];
           [F2*];
           V1-V6*;
OUTPUT: standardized sampstat residual;
```

Note: The data for this program should be stored in a file that ends with ".dat" (named TWO GROUP DATA 1.dat in this program) and should be saved in the same folder as this program. The data are as follows: means for the first group in the first row, lower left triangle of the covariance matrix for the first group in the next six rows; means for the second group in the next row, and lower left triangle of the covariance matrix for the second group in the last six rows. Parameters included in statements only under the Model command are constrained to be equal between groups. If a parameter is also specified under the Model g2 command, the equality constraint on this parameter no longer holds.

SIMPLIS Program

```
! SMM FOR DATA SET 1
GROUP1: DAYCARE
OBSERVED VARIABLES:
V1 V2 V3 V4 V5 V6
COVARIANCE MATRIX:
138.00
 45.58    80.49
 35.19    23.56    56.34
 45.13    32.00    16.64    232.17
 35.33    10.49    10.56     79.74    149.16
 73.34    28.73    33.21    117.30     79.90    324.36
MEANS:
 50.40    79.60    98.88     74.06     49.12    120.10
SAMPLE SIZE:
200
LATENT VARIABLES:
F1 F2
RELATIONSHIPS:
V1 = CONST + 1*F1
V2 = CONST +   F1
V3 = CONST +   F1
V4 = CONST +   F2
V5 = CONST +   F2
V6 = CONST + 1*F2
ITERATIONS = 500
GROUP2:HOMECARE
COVARIANCE MATRIX:
127.61
 58.49    76.81
 29.09    19.72    54.29
 45.84    28.94    31.82    223.09
 20.33    13.27     5.81     62.25    135.72
 50.64    55.45    30.15    126.74     62.16    337.37
MEANS:
 53.70    81.32   101.82     77.69     51.62    123.37
SAMPLE SIZE:
200
RELATIONSHIPS:
F1 = CONST
F2 = CONST
END OF PROBLEM
```

Note: To remove the constraints on factor variances, factor covariances, and error variances, add the following statements before the END OF PROBLEM statement:

```
Set Error Variance of V1-V6 FREE
Set Covariance of F1 and F2 FREE
Set Variance of F1-F2 FREE
```

Syntax for MIMIC Analyses for Dataset 1

EQS Program

```
/TITLE
   MIMIC ANALYSIS FOR DATASET 1
/SPECIFICATIONS
   VARIABLES=7; CASES=400; METHODS=ML; MATRIX=COV;
/EQUATIONS
   V1 = 1F1 + E1;
   V2 = *F1 + E2;
   V3 = *F1 + E3;
   V4 = *F2 + E4;
   V5 = *F2 + E5;
   V6 = 1F2 + E6;
   F1 = *V7 + D1;
   F2 = *V7 + D2;
/VARIANCES
   E1 TO E6 =*; V7=*; D1 TO D2=*;
/COVARIANCES
   D1,D2=*;
/PRINT
   FIT=ALL; COV=YES;
/MATRIX
135.19
 53.33     79.21
 34.48     22.86     57.33
 48.37     31.96     26.83     230.35
 29.82     12.93      9.99      73.08     143.64
 64.53     43.40     34.00     124.69      72.90     332.72
  0.83      0.43      0.73       0.91       0.62       0.82     0.25
/END
```

Mplus Program

```
TITLE: MIMIC MODEL FOR DATASET 1
DATA:
    FILE IS "MIMIC model data for dataset 1.dat";
    TYPE = COVARIANCE;
    NOBSERVATIONS = 400;
VARIABLE:
    NAMES ARE V1 V2 V3 V4 V5 V6 V7;
ANALYSIS:
    ESTIMATOR = ML;
    ITERATION = 500;
MODEL:
    F1 BY V1@1 V2-V3*;
    F2 BY V4-V5* V6@1;
    F1 WITH F2*;
    F2 ON V7*;
    F1 ON V7*;
    V1-V6*;
    F1-F2*;
OUTPUT:
    standardized sampstat residual;
```

Note: The data for this program should be stored in a file that ends with ".dat" (named `MIMIC model data for dataset 1.dat` in this program) and should be saved in the same folder as this program. The data consist of the lower left triangle of the covariance matrix for the six measures and the dummy-coded variable.

SIMPLIS Program

```
!MIMIC ANALYSIS FOR DATA SET 1
OBSERVED VARIABLES:
  V1 V2 V3 V4 V5 V6 V7
COVARIANCE MATRIX:
  135.19
   53.33    79.21
   34.48    22.86    57.33
   48.37    31.96    26.83    230.35
   29.82    12.93     9.99     73.08    143.64
   64.53    43.40    34.00    124.69     72.90    332.72
    0.83     0.43     0.73      0.91      0.62      0.82    0.25
SAMPLE SIZE:
  400
LATENT VARIABLES:
  F1 F2
RELATIONSHIPS:
  V1 = 1*F1
  V2 = F1
  V3 = F1
  V4 = F2
  V5 = F2
  V6 = 1*F2
  F1 = V7
  F2 = V7
SET COVARIANCE OF F1 AND F2 FREE
END OF PROBLEM
```

APPENDIX B
The Comparative Fit Index
and Measurement Invariance

The comparative fit index (CFI) is used to assess the global fit of a hypothesized model compared to a highly constrained model, referred to as a null model. The sample CFI is an estimate of the population comparative fit index and thus incorporates the degrees of freedom in the comparison of the two models:

$$\text{CFI} = \frac{(\chi^2_{\text{Null Model}} - df_{\text{Null Model}}) - (\chi^2_{\text{Hypothesized Model}} - df_{\text{Hypothesized Model}})}{(\chi^2_{\text{Null Model}} - df^2_{\text{Null Model}})}.$$

If the resulting value is negative, the CFI is set to 0; if the resulting value is greater than 1, the CFI is set to 1. Assuming that a researcher is investigating a series of models, the null model is chosen so that it is nested within the most constrained hypothesized model of the series; that is, the null model is obtained by imposing constraints on this model (Widaman & Thompson, 2003). It is important to maintain the same null model across a series of hypothesized models; otherwise, it would be unclear whether differences in CFIs among hypothesized models are due to differences in fit of hypothesized models or differences in fit of null models.

Historically, the focus of SEM was on the analysis of the covariance matrix among measures and not on their means. In this context, CFI was computed using a null model specifying no relationship among measured variables; that is, all measures are simply a function of errors. This highly restrictive model is nested within most hypothesized models in the analysis of covariance matrices of measures (for exceptions, see Widaman & Thompson, 2003) and thus is an appropriate null model in assessing these models.

If analyses are conducted on means of measures as well as their variances and covariances, the choice of a null model becomes more complex. In our assessment of measurement invariance, the most highly constrained model in the series of tests involves between-group equality constraints on the intercepts. Accordingly, the null model should specify that each measure is a function of error plus an intercept that is constrained to be the same value across groups. In other words, the appropriate null model for these series of tests specifies no relationship among measures and equal means (i.e., intercepts) of each measure across groups.

Researchers must be aware of the null model that is used by a software program to compute a CFI. For example, Mplus computes a CFI based on a null model in which the mean structure of the model is saturated; that is, the sample means of measures must be perfectly reproduced by the model. This model is not nested within a model in which intercepts are constrained to be equal across groups. Thus researchers should not report CFIs computed by Mplus when conducting analyses investigating measurement invariance. Instead they should compute a CFI based on the fit of a null model that is nested within the most constrained hypothesized model in the series of tests.

For example, when we assessed the fit for configural invariance with our first dataset using Mplus, the results were as follows: $\chi^2(16) = 26.05$ for the model, $\chi^2(30) = 454.60$ for the null model (labeled "Baseline Model" in Mplus), and CFI = .976. The null model specifies uncorrelated measures, but allows for differences in intercepts (i.e., means) of measures between groups. The results for the Mplus null model can be reproduced by using the following model statements:

> **Model:** V1-V6; [V1-V6];
> **Model g2:** V1-V6; [V1-V6];

The model assessing configural invariance is one of a series of models in the evaluation of measurement invariance, including a model with equal intercepts for measures. Consequently, the appropriate null model for these series of models must constrain the intercepts to be equal between groups using the Mplus model statements:

> **Model:** V1-V6; [V1-V6];
> **Model g2:** V1-V6;

Based on analyses using these statements, the chi-square statistic for the appropriate null model is $\chi^2(36) = 495.321$. Therefore, the CFI is

$$\text{CFI} = \frac{(495.321 - 36) - (26.049 - 16)}{(495.321 - 36)} = .978.$$

In this example, the Mplus CFI and CFI based on the appropriate null model are almost identical: .976 versus .978. However, the differences can be more substantial in other applications.

In this chapter, all CFIs based on multiple-group analyses were recomputed based on a null model specifying no covariance among measures and equal means between groups. It should be noted that we reported the Mplus CFIs for all single sample analyses, including those for the MIMIC model.

REFERENCES

Aiken, L. S., Stein, J. A., & Bentler, P. M. (1994). Structural equation analyses of clinical subpopulation differences and comparative treatment outcomes: Characterizing the daily lives of drug addicts. *Journal of Consulting and Clinical Psychology, 62,* 488–499.

Bentler, P. (1995). *EQS structural equations program manual.* Encino, CA: Multivariate Software Inc.

Bollen, K. A. (1989). *Structural equations with latent variables.* New York, NY John Wiley & Sons.

Bollen, K. A., & Bauldry, S. (2011). Three Cs in measurement models: Causal indicators, composite indicators, and covariates. *Psychological Methods, 16,* 265–284.

Bollen, K. A., & Lennox, R. (1991). Conventional wisdom on measurement: A structural equation perspective. *Psychological Bulletin, 110,* 305–314.

Byrne, B. M. (2012). *Structural equation modeling with Mplus: Basic concepts, applications, and programming.* New York, NY: Routledge.

Byrne, B. M., Shavelson, R. J., & Muthén, B. (1989). Testing for the equivalence of factor covariance and means structures: The issue of partial measurement invariance. *Psychological Bulletin, 105,* 456–466.

Chen, F. F. (2007). Sensitivity of goodness of fit indexes to lack of measurement invariance. *Structural Equation Modeling: A Multidisciplinary Journal, 14,* 464–504.

Cheung, G. W., & Rensvold, R. B. (1999). Testing factorial invariance across groups: A reconceptualization and proposed new method. *Journal of Management, 25,* 1–27.

Cheung, G. W., & Rensvold, R. B. (2002). Evaluating goodness-of-fit indexes for testing measurement invariance. *Structural Equation Modeling: A Multidisciplinary Journal, 9,* 233–255.

Chou, C.-P., & Bentler, P. M. (1990). Model modification in covariance structure modeling: A comparison among likelihood ratio, Lagrange multiplier, and Wald tests. *Multivariate Behavioral Research, 25,* 115–136.

Cohen, J. (1988). *Statistical power analysis for the behavioral sciences* (2nd ed.). Hillsdale, NJ: Erlbaum.

Cohen, P., Cohen, J., Teresi, M., Marchi, M., & Velez, C. N. (1990). Problems in the measurement of latent variables in structural equations causal models. *Applied Psychological Measurement, 14,* 183–196.

Cole, D. A., Maxwell, S. E., Arvey, R., & Salas, E. (1993). Multivariate group comparisons of variable systems: MANOVA and structural equation modeling. *Psychological Bulletin, 114,* 174–184.

Fan, W., & Hancock, G. R. (2012). Robust means modeling: An alternative for hypothesis testing of independent means under variance heterogeneity and nonnormality. *Journal of Educational and Behavioral Statistics, 37,* 137–156.

Green, S. B., & Thompson, M. S. (2003a). Structural equation modeling in clinical research. In M. C. Roberts & S. S. Illardi (Eds.), *Methods of research in clinical psychology: A handbook.* London: Blackwell.

Green, S. B., & Thompson, M. S. (2003b, April). *Understanding discriminant analysis/MANOVA through structural equation modeling.* Paper presented at the annual meeting of the American Educational Research Association, Chicago, IL.

Green, S. B., & Thompson, M. S. (2012). A flexible SEM approach for analyzing means. In R. H. Hoyle (Ed.), *Handbook of structural equation modeling*. New York, NY: Guilford Press.

Hancock, G. R. (1997). Structural equation modeling methods of hypothesis testing of latent variable means. *Measurement and Evaluation in Counseling and Development, 20*, 91–105.

Hancock, G. R. (2001). Effect size, power, and sample size determination for structured means modeling and mimic approaches to between-groups hypothesis testing of means on a single latent construct. *Psychometrika, 66*, 373–388.

Hancock, G. R. (2003). Fortune cookies, measurement error, and experimental design. *Journal of Modern and Applied Statistical Methods, 2*, 293–305.

Hancock, G. R. (2004). Experimental, quasi-experimental, and nonexperimental design and analysis with latent variables. In D. Kaplan (Ed.), *The SAGE handbook of quantitative methodology for the social sciences* (pp. 317–334). Thousand Oaks, CA: SAGE Publications.

Hancock, G. R., Lawrence, F. R., & Nevitt, J. (2000). Type I error and power of latent mean methods and MANOVA in factorially invariant and noninvariant latent variable systems. *Structural Equation Modeling: A Multidisciplinary Journal, 7*, 534–556.

Howell, R. D., Breivik, E., & Wilcox, J. B. (2007). Reconsidering formative measurement. *Psychological Methods, 12*, 205–218.

Johnson, E. C., Meade, A. W., & DuVernet, A. M. (2009). The role of referent indicators in tests of measurement invariance. *Structural Equation Modeling: A Multidisciplinary Journal, 16*, 642–657.

Kano, Y. (2001). Structural equation modeling for experimental data. In R. Cudeck, S. du Toit, & D. Sörbom (Eds.), *Structural equation modeling: Present and future — Festschrift in honor of Karl Jöreskog* (pp. 381–402). Lincolnwood, IL: Scientific Software International, Inc.

Kaplan, D. (2009). *Structural equation modeling: Foundations and extensions* (2nd ed.). Thousand Oaks, CA: Sage Publications.

Kaplan, D., & George, R. (1995). A study of the power associated with testing factor mean differences under violations of factorial invariance. *Structural Equation Modeling: A Multidisciplinary Journal, 2*, 101–118.

Keselman, H. J., Algina, J., Lix, L., Wilcox, R. R., & Deering, K. (2008). A generally robust approach for testing hypotheses and setting confidence intervals for effect sizes. *Psychological Methods, 13*, 110–129.

Lo, W.-J., & Thompson, M. S. (April, 2009). *Using goodness-of-fit indices and alternative search procedures for identifying partial measurement invariance*. Paper presented at the annual meeting of the American Educational Research Association, San Diego, CA.

MacKenzie, S. B., Podsakoff, P. M., & Jarvis, C. B. (2005). The problem of measurement model misspecification in behavioral and organizational research and some recommended solutions. *Journal of Applied Psychology, 90*, 710–730.

Meredith, W. (1993). Measurement invariance, factor analysis and factor invariance. *Psychometrika, 58*, 525–544.

Millsap, R. (2011). *Statistical approaches to measurement invariance*. New York, NY: Routledge.

Muthén, L. K., & Muthén, B. O. (2010). Mplus user's guide (6th ed.). Los Angeles, CA: Muthén & Muthén.

Rensvold, R. B., & Cheung, G. W. (2001). Testing for metric invariance using structural equation models: Solving the standardization problem. In C. A. Schriesheim & L. L. Neider (Eds.), *Research in management*, (Vol. 1, pp. 25–50). Greenwich, CN: Information Age Publishing.

Shaffer, J. P. (1986). Modified sequentially rejective multiple test procedures. *Journal of the American Statistical Association, 81*, 826–831.

Treiblmaier, H., Bentler, P. M., & Mair, P. (2011). Formative constructs implemented via common factors. *Structural Equation Modeling: A Multidisciplinary Journal, 18*, 1–17.

Vandenberg, R. J., & Lance, C. E. (2000). A review and synthesis of the measurement invariance literature: Suggestions, practices, and recommendations for organizational research. *Organizational Research Methods, 3*, 4–70.

Widaman, K. F., & Thompson, J. S. (2003). On specifying the null model for incremental fit indices in structural equation modeling. *Psychological Methods, 8*, 16–37.

Yoon, M., & Millsap, R. E. (2007). Detecting violations of factorial invariance using data-based specification searches: A Monte Carlo study. *Structural Equation Modeling: A Multidisciplinary Journal, 14*, 435–463.

CHAPTER 6

CONDITIONAL PROCESS MODELING

Using Structural Equation Modeling to Examine Contingent Causal Processes

Andrew F. Hayes and Kristopher J. Preacher

There is much theoretical and applied value to research that attempts to establish whether there is a causal relation between two variables X and Y. But establishing a causal association is rarely sufficient for broad understanding and application. We better understand some phenomenon and can better use that understanding when we can answer not only whether X affects Y, but also *how* X exerts its effect on Y, and *when* X affects Y and when it does not, or does so strongly as opposed to weakly. The "how" question relates to the underlying psychological, cognitive, or biological process, mechanism, or causal chain of events that links X to Y, whereas the "when" question pertains to what might be considered the boundary conditions of the causal association—under what circumstances, or for which types of people, does X exert an effect on Y and under what circumstances, or for which type of people, does X not exert an effect?

Structural Equation Modeling: A Second Course (2nd ed.), pages 219–266
Copyright © 2013 by Information Age Publishing
All rights of reproduction in any form reserved.

If it is true, and we believe it is, that all causal associations exist through some kind of causal chain of events and that all effects have at least some boundary conditions, then any study that addresses only the how question or the when question, but not both, is incomplete in significant ways, just as is a study that answers only whether or not an effect exists. A more complete analysis of a phenomenon will address both questions simultaneously, so as to uncover how a sequence of causal events—in terms of the direction, magnitude, or existence of the effect or lack thereof—depends on contextual or individual difference factors. The purpose of this chapter is to introduce some of the important principles and procedures of an analytical approach that does just this, an approach called *conditional process modeling*. After first introducing conditional process modeling and its history, we define some of the important terms and concepts we will use throughout this chapter. We then describe how to convert a conceptual diagram of a conditional process into a statistical model, the parameters of which can then be estimated using a structural equation modeling (SEM) program. We then show how parameters of the model are pieced together to yield estimates of the conditional nature of the process being modeled and how inferences about those estimates are made. We end by discussing some extensions of the procedure we outline into the arena of latent variable modeling.

WHAT IS CONDITIONAL PROCESS MODELING?

Process modeling is undertaken when the goal is to understand, explore, and estimate the mechanism by which some putative causal variable affects an outcome through at least one intermediary variable. *Conditional* process modeling is undertaken when this process is thought to be contingent on additional variables. As described in detail throughout this chapter, it is used to estimate the *direct* and *indirect* pathways through which a variable transmits its effects, as well as to model how the size of those effects depend on (or are *conditional on*) the value(s) of one or more *moderators*.

The term *conditional processing modeling* represents a melding of two ideas both conceptually and analytically: *process modeling* and *moderation analysis*. Judd and Kenny (1981) described *process modeling* in their early seminal piece in *Evaluation Review* as an attempt to "specify the causal chain" (p. 602) through which interventions exert their effects on an outcome of interest. Process modeling is now better known as *mediation analysis*; the label *process modeling* never gained traction. Of course, "mediation analysis" did gain traction following the publication of Baron and Kenny's extraordinarily popular *Journal of Personality and Social Psychology* article (Baron & Kenny, 1986). We adopt "process modeling" here partly to honor the contributions that Charles Judd, David Kenny, and their early colleagues have made to the literature in this area, but also because mediation is a controversial term which can invite confusion depending on how it is used and defined (see, e.g., Mathieu & Taylor, 2006).

The *conditional* in "conditional process modeling" stems from moderation analysis and the concept of *interaction* (see, e.g., Aiken & West, 1991; Jaccard & Turrisi, 2003). If an antecedent variable X's effect on some consequent variable Y depends on a third variable W, we can say that X's effect is *moderated by* or is *conditional on* W, or that X and W *interact* in influencing Y. In such a case, it is not meaningful to talk about X's effect on Y without first conditioning that discussion on the value or values of the moderator, W. Typically, when a moderating effect is found in such an analysis, the interaction is probed in order to estimate *conditional effects*, sometimes called *simple slopes* or *simple effects*, which quantify the effect of X on Y at various values of W that have some kind of theoretical or practical (if not arbitrary) meaning (Bauer & Curran, 2005; Hayes & Matthes, 2009).

Although the term *conditional process modeling* is new (see Hayes, 2013), the concept is not. Some of the earliest literature on mediation analysis describes scenarios in which a causal process could be described by combining moderation and process analysis. Judd and Kenny (1981) contemplated the possibility that an experimental treatment could affect the magnitude of a mediation effect by influencing the size of the association between mediator and outcome. James and Brett (1984) showed how a variable's effect on some outcome might be mediated for some people but not others if one or more of the paths in a mediation model is moderated, what they termed *moderated mediation*. Like James and Brett, Baron and Kenny (1986) discussed how a manipulation's effect on a mediator might be moderated, but they also described how a moderated effect of a manipulation on an outcome could be mediated, a phenomenon called *mediated moderation.*

There are several examples of piecemeal approaches to assessing moderated mediation and mediated moderation processes in the literature prior to the turn of the century (e.g., Druley & Townsend, 1998; Tepper, Eisenbach, Kirby, & Potter, 1998). It was not until the publication of a handful of articles starting in 2005, however, that research articles truly and analytically integrating moderation and mediation analysis began to appear in scientific journals. Muller, Judd, and Yzerbty (2005) provided analytical models and corresponding equations for examining when "mediation is moderated and moderation is mediated" (from the article title). Almost simultaneously but in different journals, Edwards and Lambert (2007) and Preacher, Rucker, and Hayes (2007) published derivations and analytical tools for quantifying and testing hypotheses about what Preacher et al. (2007) called the *conditional indirect effect*—the indirect effect of one variable on another through a third expressed as a function of one or two moderator variables.

Edwards and Lambert also discussed how direct effects in a mediation model, like indirect effects, also can be mathematically modeled as a function of a moderator variable. Finally, a number of articles from David MacKinnon and his colleagues (e.g., Fairchild & MacKinnon, 2009; Morgan-Lopez & MacKinnon, 2006) provided advice and examples for how to appropriately model moderated mediation and mediated moderation effects.

The impact of these articles, gauged by the number of citations they have collectively received, has been large, to say the least. Analyses that simultaneously estimate both moderated and mediated effects in order to better illuminate the boundary conditions of a mediated causal process are now found in the literature with ease (recent examples include Antheunis, Valkenberg, & Peter, 2010; Parade, Leerkes, & Blankson, 2010; Van Dijke & de Cremer, 2010). Yet, as significant as these methodological developments are, to date they have been discussed and applied only in the context of fairly simple models, generally involving a single intervening variable or mediator, with a single indirect effect that is moderated by one or, in a few instances, two variables. In the pages that follow, after first introducing some important concepts and terms, we illustrate the steps involved in the estimation and interpretation of a conditional process model in the context of a fairly general model involving multiple pathways of influence and several moderators, some of which combine multiplicatively in their influence on the process being modeled. Although there is danger in starting with an overly complex framework, we believe that there is advantage in doing so, for illustrating the procedure with a complex model allows us to describe the various rules in constructing and estimating a conditional process model that will apply to both complex and simple models.

Before beginning, it is important to briefly comment on the position we take with respect to statistics, correlation, and causality. Conditional process modeling is a tool for understanding causal processes. Although the example we use here will include an experimental manipulation, many associations later in the sequence of the putative causal chain of events have various noncausal interpretations. Of course, no statistical tool can actually be used to ascertain whether an association is causal, for the inferences one can make stem not from the statistical procedures one uses but how one goes about collecting the data. Furthermore, inferences are products not of mathematics but of mind. We do not object to the modeling of data that stem from a design that affords only limited causal interpretation. However, the researcher should be aware of the limitations of data, and should take pains to ensure that the interpretation (causal or not) is sensible in light of theory or past research on the process being studied.

INDIRECT, DIRECT, AND CONDITIONAL EFFECTS

Before proceeding, we define some important terms and concepts that form the foundation of conditional process modeling. We start with *indirect effect*. Consider the simple causal model displayed in Figure 6.1(a, left),

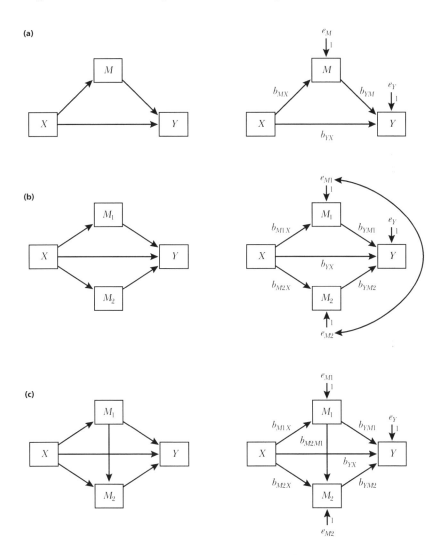

Figure 6.1 Conceptual (left) and statistical models (right) representing (a) simple mediation, (b) parallel multiple mediation, and (c) serial multiple mediation of the effect of X on Y. All variance and intercept parameters, and most covariance parameters, are omitted from this and subsequent figures to reduce clutter.

sometimes called a simple mediation model, represented in the form of a *conceptual model*. A conceptual model will be defined formally later. For now, suffice it to say that it is a visual representation of the process but not a formal path diagram, as one would use in SEM. Suppose M and Y are two observed, continuous measures, whereas X is either an observed continuous or dichotomous variable. Assuming linearity in all relations, the effect of X on both M and Y can be ascertained by estimating the parameters in the linear equations

$$M = i_M + b_{MX}X + e_M \qquad (6.1)$$

$$Y = i_Y + b_{YM}M + b_{YX}X + e_Y \qquad (6.2)$$

where i_M and i_Y are intercepts and e_M and e_Y are errors in prediction of M and Y, respectively. These coefficients come from the *statistical model* corresponding to the conceptual model of the process, depicted in Figure 6.1 (a, right), meaning the model as it would be specified in a SEM program. The parameters of these two equations, along with relevant variances and covariances, can be estimated using either SEM or two ordinary least squares regressions, the choice of which has little consequence on the results one will get. The indirect effect of X on Y through M is quantified as the product of the coefficients for the paths linking X to Y through M.[1] In this model, the path from X to M is b_{MX} and the path from M to Y is b_{YM}, so the indirect effect is $b_{MX}b_{YM}$. We can say that two cases differing by one unit on X are estimated to differ by $b_{MX}b_{YM}$ units on Y as result of how the difference in X causally influences M, which in turn causes differences in Y.

The indirect effect might not, and typically does not, completely quantify how differences in X map on to differences in Y, for X may affect Y *directly*, that is, independently of the indirect pathway through M. The *direct effect* of X on Y, quantified as b_{YX}, quantifies by how much two cases differing by one unit on X are estimated to differ on Y independent of M, or holding M constant. These two components of X's influence on Y, the indirect and direct effects, sum to the total effect of X on Y, $b_{MX}b_{YM} + b_{YX}$.

Early in the evolution of mediation or process analysis, great emphasis was placed on tests of significance for each coefficient in Equations 6.1 and 6.2 above as well as the total effect. The goal of early piecemeal approaches was to ascertain whether M meets various criteria outlined by Judd and Kenny (1981) and Baron and Kenny (1986) for establishing M as a mediator of X's effect either completely (when all but the direct effect is statistically different from zero) or partially (when the direct effect is statistically different from zero but closer to zero than the total effect). Although this approach to mediation analysis remains popular among researchers, it now is perceived by experts in mediation analysis as outdated and has fallen out of favor. In its place is a greater emphasis on the estimation of, and inferences about, the indirect effect, with much less concern about tests of significance for individual coefficients in the model, the relative sizes or significance of the direct and total effect, or labeling M as a complete or partial mediator (see, e.g., Cerin & MacKinnon, 2009; Hayes, 2009, 2013; Rucker, Preacher, Tormala, & Petty, 2011).

It is often the case, perhaps even typical, that X influences Y through more than one intermediary variable. Figure 6.1(b), depicts a *parallel multiple mediator model* (see, e.g., Preacher & Hayes, 2008) with two intervening variables. Again assuming linearity of all effects, this model requires three equations to estimate the effects of X:

$$M_1 = i_{M1} + b_{M1X}X + e_{M1} \tag{6.3}$$

$$M_2 = i_{M2} + b_{M2X}X + e_{M2} \tag{6.4}$$

$$Y = i_Y + b_{YM1}M_1 + b_{YM2}M_2 + b_{YX}X + e_Y. \tag{6.5}$$

There are two kinds of indirect effect in this model: specific and total. The *specific indirect effect* of X on Y through M_1 is the product of the paths linking X to Y through M_1, which here is $b_{M1X} b_{YM1}$. The specific indirect effect through M_2 is calculated similarly as $b_{M2X} b_{YM2}$. The sum of the specific indirect effects produces the *total indirect effect* of X: $b_{M1X} b_{YM1} + b_{M2X} b_{YM2}$. The direct effect of X is still b_{YX}, and the total effect of X is the sum of the direct and the total indirect effects: $b_{YX} + b_{M1X} b_{YM1} + b_{M2X} b_{YM2}$.

The model in Equations 6.3 through 6.5 assumes no causal association between M_1 and M_2, unlike the model in Figure 6.1(c). Such a causal association can be estimated by including M_1 in the model of M_2:

$$M_2 = i_{M2} + b_{M2X}X + b_{M2M1}M_1 + e_{M2}. \tag{6.6}$$

The specific indirect effects for M_1 and M_2 are still defined as above, but this addition to the model produces another specific indirect effect of X, this one through M_1 and M_2 serially, estimated as the product of the three constituent paths: $b_{M1X} b_{M2M1} b_{YM2}$. The total indirect effect of X is the sum of the three specific indirect effects ($b_{M1X} b_{YM1} + b_{M2X} b_{YM2} + b_{M1X} b_{M2M1} b_{YM2}$) and the total effect of X on Y is the sum of the direct effect (b_{YX}) and all indirect effects.

Whereas in mediation analysis, where focus is on estimating and interpreting the indirect effect X on Y through M, in *moderation analysis*, the goal is to ascertain whether X's effect on Y depends on some third variable W. When evidence of moderation is found, then interpretative focus is placed on estimating and interpreting the *conditional effect* of X on Y given W. The most basic moderation model is depicted graphically in Figure 6.2(a). The

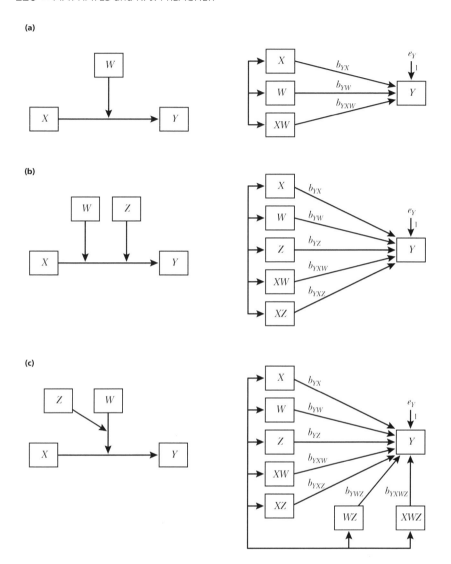

Figure 6.2 Conceptual (left) and statistical models (right) representing (a) simple moderation, (b) additive multiple moderation, and (c) multiplicative multiple moderation.

unidirectional arrow from W to the path linking X to Y represents 'depends on;' that is, the effect of X on Y depends on W. Although there are many different forms that moderation can take, the most commonly estimated form and the only form we will address in this chapter is linear moderation, or linear interaction. Assuming X and W are either continuous or dichotomous, the moderating effect of W on the effect of X on Y can be derived by estimating the parameters in the following model:

$$Y = i_Y + b_{YX}X + b_{YW}W + b_{YXW}XW + e_Y \qquad (6.7)$$

where XW is the product of W and X. In this model, X's effect on Y is estimated as a linear function of W, as can be seen by reexpressing Equation 6.7 in an algebraically equivalent form

$$Y = i_Y + (b_{YX} + b_{YXW}W)X + b_{YW}W + e_Y. \qquad (6.8)$$

In this model, the conditional effect of X on Y is $b_{YX} + b_{YXW}W$. It is interpreted as the amount by which two cases differing by a unit on X, given a specific value on W, are estimated to differ on Y. For instance, when $W = 2$, two cases who differ by one unit on X are estimated to differ by $b_{YX} + 2b_{YXW}$ units on Y. But when $W = 3$, two differing by one unit on X are estimated to differ by $b_{YX} + 3b_{YXW}$ units on Y. When b_{YXW} is different from zero, it is meaningless to talk about X's effect using a single point estimate, because its effect is not a constant. It is a function of W, meaning that it depends on W.

The effect of X on Y can be conditional on more than one variable. For the purposes of our discussion of conditional process modeling, there are two scenarios worth mentioning. Figure 6.2(b) depicts X's effect as additively conditional on both W and Z. Such a scenario would be modeled as

$$Y = i_Y + b_{YX}X + b_{YW}W + b_{YZ}Z + b_{YXW}XW + b_{YXZ}XZ + e_Y. \qquad (6.9)$$

This model can be written equivalently as

$$Y = i_Y + (b_{YX} + b_{YXW}W + b_{YXZ}Z)X + b_{YW}W + b_{YZ}Z + e_Y \qquad (6.10)$$

which shows how X's effect is an additive linear function of both W and Z. So the conditional effect of X on Y is $b_{YX} + b_{YXW}W + b_{YXZ}Z$. It depends on both W and Z. For instance, when $W = 2$ and $Z = 3$, the conditional effect of X is $b_{YX} + 2b_{YXW} + 3b_{YXZ}$.

Figure 6.2(c) depicts a situation where X's effect on Y is multiplicatively conditional on W and Z, modeled as

$$Y = i_Y + b_{YX}X + b_{YW}W + b_{YZ}Z + b_{YXW}XW + b_{YXZ}XZ + b_{YWZ}WZ + b_{YXWZ}XWZ + e_Y \qquad (6.11)$$

or, rewritten in equivalent form to reveal the conditional nature of X's effect on Y:

$$Y = i_Y + (b_{YX} + b_{YXW}W + b_{YXZ}Z + b_{YXWZ}WZ)X + \\ b_{YW}W + b_{YZ}Z + b_{YWZ}WZ + e_Y. \tag{6.12}$$

From Equation 6.12 it can be seen that X's effect on Y is a function of W, Z, as well as their product: $b_{YX} + b_{YXW}W + b_{YXZ}Z + b_{YXWZ}WZ$. So it is conditional on both W and Z. For instance, when $W = 3$ and $Z = 2$, the conditional effect of X is $b_{YX} + 3b_{YXW} + 2b_{YXZ} + 6b_{YXWZ}$.

If a causal effect can be conditional on a moderator, and indirect and direct effects quantify a variable's causal effect on some variable, then it follows that indirect and direct effects can also be conditional. Figure 6.3 depicts a simple model in which the direct effect of X on Y is moderated by W, as is the indirect effect of X on Y through M. This is what Edwards and Lambert (2007, p. 4) called a "direct effect and second stage moderation model." It is a conditional process model, albeit a rudimentary one, because it includes elements of both moderation and mediation. Of interest in such a model are the conditional direct and indirect effects of X, which can be derived from the coefficients of two linear models

$$M = i_M + b_{MX}X + e_M \tag{6.13}$$

$$Y = i_Y + b_{YX}X + b_{YW}W + b_{YM}M + b_{YXW}XW + b_{YMW}MW + e_Y. \tag{6.14}$$

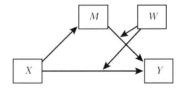

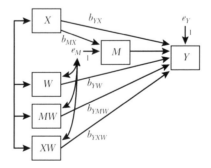

Figure 6.3 Conceptual (top) and statistical model (bottom) representing a form of conditional process model in which the indirect and direct effect of X is moderated by W.

By rewriting Equation 6.14 in an equivalent form,

$$Y = i_Y + (b_{YX} + b_{YXW}W)X + b_{YW}W + b_{YM}M + b_{YMW}MW + e_Y \qquad (6.15)$$

it can be seen that the direct effect of X on Y depends on W, for it is a linear function of W: $b_{YX} + b_{YXW}W$. So the direct effect is conditional on W.

Earlier it was shown that the indirect effect of X on Y through M is derived as the product of the coefficients linking X to Y going through M. From Equation 6.13, the effect of X on M is estimated as b_{MX}. But what is the effect of M on Y holding X constant? Unlike earlier, there is no single effect of M on Y, for its value depends on W, as can be seen by rewriting Equation 6.15 in an equivalent form:

$$Y = i_Y + (b_{YX} + b_{YXW}W)X + (b_{YM} + b_{YMW}W)M + b_{YW}W + e_Y. \qquad (6.16)$$

Here, the effect of M on Y is a linear function of W: $b_{YM} + b_{YMW}W$. Multiplying the effect of X on M by the conditional effect of M on Y given W yields the *conditional indirect effect* of X on Y: $b_{MX}(b_{YM} + b_{YMW}W)$. So, the indirect effect is not a single quantity defined by the parameter estimates but, instead, is a function of those estimates and that involves W. Substituting a different value of W into this formula for the indirect effect will yield a different estimate of X's effect on Y through M.

TRANSLATING A CONCEPTUAL MODEL INTO A STATISTICAL MODEL

Our hypotheses and beliefs about the process through which one variable affects another through a causal sequence of events is represented in the form of a diagram we call the *conceptual model*. This conceptual model is not the same as a formal path diagram or structural equation model and should not be interpreted as such. Rather, it visually represents the direct and indirect paths of influence from a causal or antecedent variable that is the focus of the study to the final outcome or consequent variable of interest, as well as which of those causal influences are moderated. The conceptual model uses a visual notation common for representing direct, indirect, and moderated causal effects. A unidirectional arrow represents the direction of causal flow or path from an antecedent variable (where the arrow begins) to a consequent variable (where the arrow ends), and a unidirectional arrow from one variable to another arrow or path represents moderation by that variable. Yet ultimately, this conceptual model must be translated into a formal statistical model—a path diagram or structural equation model—in order to test hypotheses and quantify the various effects of interest statistically. This section describes this translation process.

Thus far, our discussion of conditional process modeling has been abstract. To make it more concrete, we now introduce a simulated working example to be used throughout the rest of this chapter. This example, conceptually diagrammed in Figure 6.4, closely resembles the example used by Judd and Kenny (1981) in their seminal paper on process modeling. Participants in the study either are ($X = 1$) or are not ($X = 0$) exposed to a mass media campaign designed to lower risks of coronary heart disease, with exposure being determined by random assignment. It is hypothesized that exposure to the campaign will increase knowledge of the dietary causes of heart disease (M_1, measured such that a higher value corresponds to greater knowledge), and this knowledge will translate into dietary choices that are better for the heart (M_2, measured such that higher scores correspond to more healthy eating habits). Such choices, in turn, will result in better heart health (Y: measured with various physiological indicators of heart health). It is also proposed that those who use the mass media more often (W, mea-

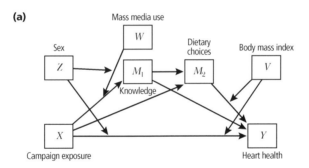

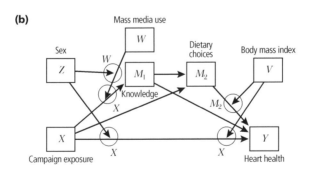

Figure 6.4 (a) The conceptual model for the Heart Health Campaign example; (b) points of moderation represented in the conceptual model following the completion of Step 2.

sured such that higher scores correspond to greater use) are more likely to learn from the mass media campaign, indicating moderation of the effect of the campaign on knowledge, but that this differential effect of exposure as a function of mass media use should depend on sex ($Z = 1$ for males, 0 for females). Further complicating the process, it is expected that the effect of dietary choices on heart health will depend on a person's body mass (V: quantified as BMI, a numerical function of weight and height, such that a higher score represents greater body mass). Finally, it is acknowledged that the campaign could influence heart health through other means. For instance, the campaign could influence dietary choices independent of knowledge, and it could influence heart health through means other than diet. Direct effects of the mass media campaign, however, are expected to independently (i.e., additively rather than multiplicatively) depend on sex as well as body mass. For the sake of this illustration, assume that sex, mass media use, and body mass are measured at the beginning of the campaign, knowledge of the dietary causes of heart disease is measured three months after the start of the campaign, dietary choices three months after that, and, finally, heart health three months still later (i.e., nine months after the start of the campaign). All variables are measured as continua except for sex and exposure to the campaign, which are dichotomous.[2] The simulated means, variances, and covariances are reported in Table 6.1.

This elaborate process is diagrammed in Figure 6.4(a) as a conceptual model. At the core of this conceptual model is a serial multiple mediator model with three indirect effects of exposure to the campaign (X) on heart health (Y) (via knowledge [M_1], dietary choices [M_2], and knowledge and dietary choices serially), as well as a direct effect. The effect of the campaign on knowledge is diagrammed as moderated by mass media use (W) (represented with the arrow from mass media use to the path linking exposure to knowledge), with the extent of this moderation itself moderated by sex (Z) (represented with the arrow from sex to the arrow linking mass media use to the path linking campaign exposure to knowledge—a three way interaction). In addition, the direct effect of exposure to the campaign is additively moderated by sex and body mass (V) (represented with the arrows from these two variables to the path linking exposure to heart health—two-way interactions). Finally, the effect of dietary choices on heart health is moderated by body mass (i.e., the arrow from body mass to the path linking dietary choices to heart health, also a two-way interaction).

A conceptual model might be incomplete, depending on additional hypotheses one might propose linking variables in the system. For instance, it might be reasonable to expect that there would be sex differences in knowledge. This path of influence from sex to knowledge is missing from the conceptual diagram in Figure 6.4. Although it could be added, it turns out that

TABLE 6.1 Means and Variance-Covariance Matrix for Heart Health Campaign Example

	Mean	Variance	M_1	M_2	Y	W	Z	V	WX	ZX	ZW	ZWX	VM_2	VX	X
M_1	61.113	118.292	118.292	8.303	27.006	1.354	-0.212	0.873	7.045	0.529	-0.204	2.807	233.987	41.827	1.503
M_2	7.906	2.738	8.303	2.738	8.249	0.054	-0.061	0.210	0.532	0.015	-0.234	0.110	76.831	3.248	0.114
Y	25.077	57.496	27.006	8.249	57.496	0.335	-0.067	12.477	3.650	0.454	-0.075	2.026	328.844	30.315	0.852
W	3.954	0.982	1.354	0.054	0.335	0.982	0.015	0.002	0.615	0.026	0.534	0.319	1.363	1.002	0.039
Z	0.495	0.250	-0.212	-0.061	-0.067	0.015	0.250	0.062	-0.030	0.119	0.996	0.483	-1.253	-0.229	-0.009
V	27.272	13.415	0.873	0.210	12.477	0.002	0.062	13.415	-0.013	0.035	0.237	0.124	111.339	7.021	0.015
WX	1.992	4.524	0.432	0.055	0.116	-0.025	-0.001	-0.072	0.244	0.488	0.168	2.194	14.505	27.456	1.008
ZX	0.235	0.180	-0.111	-0.010	0.066	-0.001	0.000	-0.003	0.004	0.062	0.492	0.731	0.636	3.276	0.119
ZW	1.973	4.443	-0.037	-0.018	0.025	-0.012	0.000	-0.011	-0.008	0.010	0.245	2.212	-4.870	-0.405	-0.018
ZWX	0.956	3.184	0.190	0.017	0.078	-0.007	0.010	0.025	-0.003	0.000	-0.006	0.061	3.705	13.304	0.484
VM_1	215.817	3,023.412	0.653	0.492	5.242	-0.121	-0.072	-0.448	0.003	-0.052	-0.011	0.011	37.962	146.748	3.259
VX	13.480	192.934	0.401	0.043	0.923	-0.072	-0.003	-0.003	-0.001	0.016	0.025	-0.002	1.716	3.353	6.824
X	0.494	0.250	1.503	0.114	0.852	0.039	-0.009	0.015	0.001	0.000	-0.001	0.004	0.043	0.000	0.250

Note: In the analysis reported in this manuscript, X, W, Z, V, and M_2 were mean centered and so have means of 0. The *Mean* and *Variance* columns are for uncentered data. Diagonal elements are variances after mean centering. Lower triangular elements are covariances after mean centering. Upper triangular elements are covariances before centering. Centering was conducted before computation of products.

doing so is not necessary, for it will end up in the statistical translation of the conceptual model anyway, as will be seen. Although there is room for debate on this point, we recommend that conceptual diagrams depict only those relations—direct, indirect, and moderated—that pertain specifically to the effect of the causal variable that begins the chain of events (exposure to the campaign) leading to the final outcome variable of interest (heart health). Later, additional causal influences in the system can be added to the statistical model if desired prior to estimation of the coefficients in the model.

We do not estimate the conceptual model. Rather, once the conceptual model is constructed, that model must be translated into a statistical model in the form of various linear equations the parameters of which can then be estimated. For novices, this is probably the more difficult part of conditional process modeling. The recipe described below for doing so will seem at first to contain a lengthy list of ingredients, but with an understanding of the basic steps and some practice, the process will become more intuitive and the recipe will not be required in order to complete it.

Step 1: Derive the Number of Linear Models Necessary to Model the Process Statistically

The first step is perhaps the easiest of them all. A conditional process model requires p linear models, one for each consequent variable. Recall that a consequent variable is a variable that receives a causal path, meaning that at least one variable in the model sends an arrow to it. As can be seen in Figure 6.4, there are three consequent variables in this conceptual model, M_1, M_2, and Y. So we need $p = 3$ linear models to represent the process, one for each consequent variable.

Step 2: Label the Points of Moderation in the Conceptual Model

The next step is a visual aid that helps in the derivation of the necessary product terms in the p linear models. Using the diagram of the conceptual model, circle any arrowhead that points at another path and label that circle with the name of the variable that sends an arrow *straight through* it. Importantly, make sure that your circle encompasses only a single arrowhead. So when drawing your conceptual diagram, make sure that if a path is moderated by more than one variable, you visually separate the arrowheads pointing at the same path. As can be seen in Figure 6.4(b), there are 5 points of moderation (two for the path from X to Y, one for the path from

M_2 to Y, one for the path from X to M_1, and one for the path from W to the path from X to M_1).

Step 3: Construct Sequences of Variable Names for Each Consequent

The next step is perhaps the most complicated to explain, but it is simple to complete once understood. This step involves tracing all *valid pathways* from an antecedent variable to a consequent variable while constructing a sequence of variable names as the pathway is traced. A pathway from an antecedent to a given consequent is considered a valid pathway if you can trace from that antecedent to the consequent while never tracing in a direction opposite to the arrow and never passing through another antecedent variable. A sequence of variable names is constructed for a given pathway by "collecting" variable names as you trace along the path from the antecedent to the consequent. A sequence of variable names always begins with the name of the antecedent variable at the starting point, and as you pass through a circle, append the name of the variable with which the circle is labeled to the end of the sequence *unless that variable name already occurs in the sequence you are generating*, in which case you simply ignore it. When you reach a consequent variable, stop. Do not include the name of the consequent variable at which you stop in the sequence.

Consider consequent M_1. There are three valid pathways from an antecedent variable to M_1. The first starts at X and passes through a circle labeled X before ending at M_1. So the sequence for that pathway is "X". We do not count the second X because X already exists in the sequence at the point our tracing intersects the X circle en route to M_1. The second pathway starts at W and passes through a circle labeled W and then a circle labeled X before ending at M_1. So the sequence for this second pathway is "WX". We add X to this sequence after W because X did not exist in the sequence up to the point we reached the X circle, but because W already exists in the sequence once we reached the W circle, that W is not included in the sequence. The final pathway starts at Z and passes through a circle labeled W, and then one labeled X, before terminating at M_1. So the sequence for this last pathway to M_1 is "ZWX". There are no other sequences for consequent M_1 because there are no additional means of tracing from a variable to M_1 without violating one of the two rules above. For instance, although there are two ways to trace from V to M_1, (V to M_2 to M_1, and V to Y to M_1), doing so violates not just one but both rules—tracing opposite of the direction an arrow points (M_2 to M_1 or Y to M_1) as well as passing through another variable (M_2 or Y).

Using this same procedure, there are two valid pathways to consequent M_2, each with only one variable in the sequence: "M_1" and "X". Finally, there are six valid pathways to consequent Y. The sequences for these six pathways are "X", "ZX", "VX", "M_1", "M_2" and "VM_2". Clearly, this step can create a lot of information that is hard to keep in memory, so we recommend you implement it on a piece of paper (as in Table 6.2), keeping careful track of what sequences are generated for each consequent so that you can refer to them later.

Step 4: Expansion of Sequences with at Least Three Variable Names

For most conditional process models, Step 4 is not necessary, but it is important to describe this step because you might, on occasion, need to implement it. Let k be defined as the maximum number of variables in any sequence for a given consequent. So for consequent M_1, $k = 3$, whereas k=1 for consequent M_2 and $k = 2$ for consequent Y. For any consequent with $k > 2$, take each sequence with at least three variable names and generate all possible combinations of variable names containing at least one fewer

TABLE 6.2   The Outcome of Various Steps Converting the Conceptual Model into a Statistical Model

	Consequent		
	M_1	M_2	Y
Variable sequences generated at Step 3	X	X	X
	WX	M_1	ZX
	ZWX		M_1
			M_2
			VM_2
			VX
Variable sequences added at Step 4	ZX	None	None
	ZW		
Predictor variable list after completion of Step 5	X	X	X
	W	M_1	V
	Z		Z
	WX		M_1
	ZX		M_2
	ZW		ZX
	ZWX		VX
			VM_2

variable. In most conditional process models to which this step applies, k is likely to be no greater than 3, but we illustrate first with $k = 4$. Suppose a sequence contained the variable names ABCD. All possible combinations of variable names with at least one fewer variable would be ABC, ABD, ACD, BCD, AB, AC, AD, BC, BD, and CD. In the current example, one sequence for consequent M_1 contains the sequence ZWX. So all possible combinations with at least two variables would be ZW, ZX, and WX. As there are no more sequences for any consequent variable with a least 3 variables, we stop.

Once all combinations are generated, add only those combinations generated at this step to the list of sequences for that consequent variable that are not already present in that consequent variable's sequence list. Ignore the order of the variable names in a sequence, meaning, for example, that WX can be considered equivalent to XW. Because we implemented this step only for consequent M_1, we need to consider only the list of sequences already generated for M_1. Recall that list contains the sequences "X", "WX" and "ZWX". Step 4 generated "ZX" "ZW", and "XW" by expansion of "ZWX". Ignoring order, "XW" already appears in the sequence list (as "XW") so we ignore it, but "ZX" and "ZW" do not already appear, so we add these two combinations to the list of sequences for consequent M_1, yielding the list "X", "WX", "ZWX" "ZX", and "ZW". This completes Step 4. It may be helpful to return to Table 6.2 at this point to check understanding.

Step 5. Use the List of Sequences to Generate the Linear Models for Each Consequent

After completion of Step 4 (if applicable), you are ready to construct the linear models for each consequent. The rules for doing so are simple. First, any variable name that appears in any sequence for a given consequent variable should be a predictor variable in the model of that consequent. Then all remaining sequences in the list for a given consequent containing at least two variables should be predictors in the model of that consequent, represented as products of those variables in the sequence. Each linear model should also include an intercept and a random error. Use whatever convention you want for labeling the regression parameters for each predictor variable in each linear model. Here, we use the $b_{\text{TO-FROM}}$ notation used throughout much of this book.

Referring to Table 6.2, the model for consequent variable M_1 should include X, W, Z (because these variables appear at least once in the list of sequences for consequent M_1), as well as WX, ZX, ZW, and ZWX as predictors. Thus,

$$M_1 = i_{M1} + b_{M1 X}X + b_{M1 W}W + b_{M1 Z}Z + b_{M1 WX}WX +$$

$$b_{M1ZX}ZX + b_{M1ZW}ZW + b_{M1ZWX}ZWX + e_{M1}. \tag{6.17}$$

The model for consequent variable M_2 is much simpler, for there are only two sequences in the list and no sequences with two or more variables, so no products of variables. The model is

$$M_2 = i_{M2} + b_{M2X}X + b_{M2M1}M_1 + e_{M2}. \tag{6.18}$$

The model for consequent variable Y is somewhat more complex, with five variables that appear somewhere in one of the sequences: X, Z, M_1, M_2, and V. In addition, there are three products: ZX, VX, and VM_2. So the model is

$$Y = i_Y + b_{YX}X + b_{YZ}Z + b_{YM1}M_1 + b_{YM2}M_2 + b_{YV}V + \\ b_{YZX}ZX + b_{YVX}VX + b_{YVM2}VM_2 + e_Y. \tag{6.19}$$

Step 6: Fine-Tuning the Models

These three linear equations are a mathematical representation of the conceptual model. At this point, it is worth taking a look at the equations and examining whether something important might be missing. For example, notice that the model assumes that there is no effect of body mass (V) on dietary choices (M_2). The researcher might feel that it is important to allow body mass to affect diet, or to at least statistically control for body mass when estimating the effect of dietary knowledge (M_1) on dietary choices (M_2). In that case, add V as a predictor to the model of M_2 (Equation 6.18). Of course, if this effect had been represented in the conceptual model, then V would have ended up as a predictor of M_2. It is also common in some fields to partial out various demographics to adjust all coefficients for the possibility that the associations they represent are spurious. For instance, we could include age or income as a predictor variable to all three models, and all coefficients would be adjusted for the influence of age and income on all consequent variables and path estimates leading to them.

It is important, however, that no variables are *removed* from the model generated from the first five steps, at least not yet. Some of the predictor variables in the linear models generated from this procedure are there because they are analytically necessary in order to properly estimate the moderated effects. For instance, the three way interaction between X, W, and Z in the model of M_1 requires the inclusion of all three two-way interactions in that model as well as X, W, and Z, and the inclusion of the interactions between X and Z and X and V in the model of Y requires the inclusion of X, W, and V in the model. So, for example, even though we might have no basis for believing that there are likely to be sex differences in knowledge or heart health, Z must be in the model of M_1 and Y because Z is a component of an interaction involving Z in both models. Excluding Z from the model of M_1 and Y would be equivalent to constraining its coefficient in these models to zero. Unless the corresponding parameters being estimated actually are zero, excluding these variables from the model would result in a biased estimate of the interaction. Later, if some of the higher order interactions are found to be nonsignificant, they (and other terms in the model that are analytically necessary for proper estimation of those interactions) could be deleted.

We also believe that in models that include indirect effects, all corresponding direct effects should be included in initial model estimation regardless of whether those effects are expected theoretically or otherwise. In this example, the model being estimated includes all possible direct and indirect effects from campaign exposure to heart health. However, one could imagine scenarios in which one's conceptual model includes only some of the possible direct effects. For instance, suppose there were no a priori basis for believing that knowledge influences heart health directly. We recommend including this direct effect in the model anyway, for if this a priori belief is wrong, a failure to include this direct effect can bias the estimation of the conditional indirect effects from knowledge to heart health through dietary choices as well as the conditional direct effects of exposure to the campaign on hearth health. Testing causal processes requires the estimation of both direct and indirect effects, or at least an openness to the existence of both types of effects. Excluding a direct effect represents a constraint in the model—that the direct effect is zero. We believe it is better to let the data tell you whether a direct effect is zero rather than forcing it to be so. The loss in parsimony that results if you are right is a small price to pay when it means less biased estimation of indirect effects in the event you are wrong, and the decision to include these untheorized direct effects can always be undone if an initial estimation reveals no evidence of such effects. Of course, if you heed this advice, there would be no need to make this change when fine-tuning the model, for your conceptual model will include all direct effects and they will appear in the equations after following the steps above.

Once you develop a familiarity with how to represent a process or a set of hypotheses in the form of a conceptual model, this fine-tuning process will no longer be necessary. With experience working with conditional process modeling, your conceptual model will probably translate into a statistical model that, with the exception of some terms included by analytical necessity to model a moderation process, exactly corresponds to the causal process you intend to model, with all necessary terms in the linear equations to quantify and test hypotheses of interest.

MODEL ESTIMATION

With the linear models representing the conceptual model now derived, it is possible to start estimating the parameters of the full conditional process model itself. In this section we describe our approach to estimation. We recognize that there is some room for debate about the steps we describe here, and that different analysts might approach the task differently depending on their own personal philosophy about model fitting and data exploration.

The conceptual model is holistic, in that it is an omnibus proposal about a set of associations, some unmoderated and some moderated, that link various antecedent variables to consequent variables. Of course, it is possible that the data might not be consistent with parts of the model. For example, if there is no evidence after estimation of the full model that the effect of dietary choices on heart health is moderated by body mass, then it would be sensible to reconceptualize the model without this moderated path. Doing so would change the linear models derived earlier, of course.

We believe that prior to moving forward with estimation of a full structural equation model, it is sensible to first estimate the linear model for each of the consequents in ordinary least squares regression or in an SEM program in order to ascertain whether the highest order interaction(s) that the process presumes exists actually are in evidence when subjected to empirical test. If the predicted interaction does emerge, then all is well with that component of the model. But if an interaction that the conceptual model and the theory or hypotheses that gave rise to it turns out not to have support in the data (as gauged by a hypothesis test on the corresponding parameter in the model), then remove it from the conceptual model and rederive the linear model for the consequent variable given the new conceptual model. For example, suppose an OLS regression analysis of the model for consequent M_1 (knowledge) revealed no statistically significant three-way interaction between X, W, and Z. In that case, it would be reasonable to reconceptualize the model to fix the interaction between X and W to be constant across values of Z. In terms of the conceptual model, this would involve removing the path from Z to the path from W moderating the effect of X on M_1. Following the steps described above, this would yield a new, much simpler linear model for M_1: $M_1 = i_{M1} + b_{M1X}X + b_{M1W}W + b_{M1WX}WX$. This model could then be estimated in OLS to support (or refute) that the effect of X on M_1 varies as a function of W. Suppose it does. Then one can proceed to the next consequent variable, and so forth, iteratively estimating and reestimating the model of each consequent until one settles on a model for each that is consistent with the data.

When we approached the first step of model estimation in this fashion, we found that the three-way interaction in the model of M_1 was statistically significant, implying that the moderation of X's effect by W was indeed moderated by Z. All other two-way interactions in this model are necessary for the proper estimation of the three-way interaction, so they must be retained in this model, as must the lowest order terms in the model (X, W, and Z) regardless. We also found that all four of the expected two-way interactions were statistically significant in the model of Y. So there was no need to reconceptualize the process by changing moderated paths to unmoderated ones.

A couple of comments are in order about this procedure. First, it will often be the case that this model modification stage, which we have described as occurring after the translation of the conceptual model into a set of linear models, actually will have occurred first. That is, an investigator may take a set of known moderated relations revealed from a set of initial and simpler analyses and piece them together in the form of a full conditional process model in order to estimate conditional direct and indirect effects. In fact, we believe this is probably more consistent with the way that researchers actually proceed. One's a priori beliefs about what the data will reveal are often tempered by the reality revealed in initial analyses, and those beliefs modified in light of what is now known before they are pieced together into a conditional process model.

Second, the procedure we just described sounds a lot like data mining, and there is always some danger to letting empirical criteria rather than theory or hypotheses overly influence the modeling process. At the same time, there is value to being sensitive to what preliminary analyses are suggesting and modifying one's original model as needed in order to bring it into closer alignment with what those preliminary analyses are telling you. In this case, because we are proposing making a model simpler by removing interactions that the data do not support, the dangers of overfitting described by, for example, MacCallum, Roznowski, and Necowitz (1992) are considerably reduced or eliminated. That is, the approach we describe here emphasizes parsimony over maximizing model fit, so it is somewhat more conservative relative to one which adds terms to a model because doing so improves fit. But because a null hypothesis can never be proven true, the possibility of model misspecification cannot be ruled out after such model pruning. The data analyst's own philosophy about how to balance parsimony with the dangers of misspecification will have to govern this stage of the modeling, along with knowledge of the substantive literature and theory guiding the investigation.

Once one has settled on a set of linear models that include moderated effects that are consistent with the data, those models can be pieced together and the parameters of the models estimated simultaneously in a structural equation model. If interest were solely on the parameter estimates and tests of significance for individual paths, then this would be unnecessary. Generally, little new is learned by estimating the paths simultaneously rather than in individual analyses of each consequent variable. However, the estimation and testing of conditional direct and indirect effects generally involves combinations or *functions* of parameters, as will be discussed in the next section, as well as the ability to conduct inferential tests on specific outputs from those functions. This requires an SEM program that can combine parameter estimates from the linear models and provide the information needed to compute conditional indirect effects.

Figure 6.5 depicts the conditional process model in the form of a formal path diagram. Each linear model is represented with unidirectional paths leading to the consequent from each of the predictor variables. We also explicitly include the covariances between variables that never serve the role of consequent (i.e., exogenous variables). In many SEM programs (such as Mplus), these covariances are estimated by default, but some programs (such as AMOS) require the user to explicitly estimate them by drawing them in the diagram or including them in the model code. It is important to become familiar with the chosen program's defaults before proceeding further.

We use Mplus to estimate the parameters of this model because the program has some powerful features for combining parameter estimates together into new estimates, something we will take advantage of when esti-

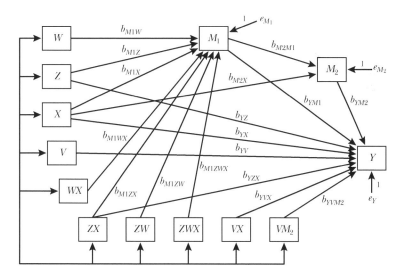

Figure 6.5 The statistical model for the Heart Health example in path diagram form. Not pictured, and also optional in full or in part (see the text), are 16 covariances involving model error covariances: $e_{M1} \leftrightarrow V$, $e_{M1} \leftrightarrow VX$, $e_{M1} \leftrightarrow VM_2$, $e_{M2} \leftrightarrow W$, $e_{M2} \leftrightarrow Z$, $e_{M2} \leftrightarrow V$, $e_{M2} \leftrightarrow WX$, $e_{M2} \leftrightarrow ZX$, $e_{M2} \leftrightarrow ZW$, $e_{M2} \leftrightarrow ZWX$, $e_{M2} \leftrightarrow VX$, $e_{M2} \leftrightarrow VM_2$, $e_Y \leftrightarrow W$, $e_Y \leftrightarrow WX$, $e_Y \leftrightarrow ZW$, $e_Y \leftrightarrow ZWX$.

mating conditional indirect effects and conducting inferential tests on their values given the data. The Mplus code we used can be found in Appendix A. A few comments on this code are in order. First, in the DEFINE section at the top, notice that we have chosen to mean center all variables that are involved in product terms, and we did so prior to computing products. There is some debate in the OLS regression literature as to whether such mean centering is necessary or desirable when estimating models with interactions (e.g., Echambi & Hess, 2007; Hayes, 2013; Kromrey & Foster-Johnson, 1998; Shieh, 2011). Our position is that in OLS regression the decision to center or not is one of personal choice and preference. But in structural equation modeling we recommend mean centering, as this can decrease the likelihood of convergence problems. Centering can either increase or decrease fit, depending on constraints in the model involving product terms. If the variables have means near zero to start with, such centering will generally have no effect. In this case, we found that the model diagrammed in Figure 6.5 (*without* the covariances noted in the figure caption) fit quite abysmally without centering [CFI = 0.254, RMSEA = 0.500, $\chi^2(16)$ = 3214.158, $p < 0.001$] but exceedingly well when we centered the variables involved in products [CFI > 0.999, RMSEA < .001, $\chi^2(16)$ = 10.392, $p = 0.845$] even though the parameter estimates and standard errors for the product terms were identical.

The linear models are spelled out in the MODEL section of the code, with consequent variables regressed ON predictors. Symbols in parentheses are labels attached to parameter estimates that are used later in the MODEL CONSTRAINT section for building estimates of conditional direct and indirect effects, to be discussed in the next section. The MODEL section in this code also includes covariances between consequent variable residuals and variables in the model that do not send a path to the corresponding consequent (there are 16 of these covariances: $e_{M1} \leftrightarrow V$, $e_{M1} \leftrightarrow VX$, $e_{M1} \leftrightarrow VM_2$, $e_{M2} \leftrightarrow W$, $e_{M2} \leftrightarrow Z$, $e_{M2} \leftrightarrow V$, $e_{M2} \leftrightarrow WX$, $e_{M2} \leftrightarrow ZX$, $e_{M2} \leftrightarrow ZW$, $e_{M2} \leftrightarrow ZWX$, $e_{M2} \leftrightarrow VX$, $e_{M2} \leftrightarrow VM_2$, $e_Y \leftrightarrow W$, $e_Y \leftrightarrow WX$, $e_Y \leftrightarrow ZW$, $e_Y \leftrightarrow ZWX$). Mplus will fix these at zero by default, so if they are desired, they must be explicitly stated in the model code. Freely estimating these yields a perfectly fitting model because the model is just identified. The parameter estimates for all paths can be found in Table 6.3.

The inclusion of these additional covariances is a matter of personal choice and modeling philosophy. A failure to include some or all of them can result in a misspecified model that may fit poorly. But a model can fit poorly for a number of reasons. In a model with observed variables only (i.e, no latent variables), poor fit can be the result of paths or covariances that are fixed to zero, or, for parsimony-adjusted measures of fit, the inclusion of paths that are unnecessary, meaning that they can be fixed to zero. Fixing a covariance between two variables to zero implies that the two variables are unassociated after accounting for the other effects in the

TABLE 6.3. Estimates of Structural Coefficients from Maximum Likelihood Estimation of the Model in Figure 6.5

		Consequent				
		M_1		M_2		Y
X	b_{M1X}	5.735 (0.732)	b_{M2X}	0.038 (0.108)	b_{YX}	2.086 (0.333)
W	b_{M1W}	1.236 (0.370)				
Z	b_{M1Z}	−0.828 (0.732)			b_{YZ}	0.320 (0.321)
V					b_{YV}	0.886 (0.044)
M_1			b_{M2M1}	0.069 (0.005)	b_{YM1}	−0.006 (0.017)
M_1					b_{YM2}	2.865 (0.109)
WX	b_{M1WX}	1.966 (0.740)				
ZX	b_{M1ZX}	−1.919 (1.464)			b_{YZX}	1.609 (0.641)
ZW	b_{M1ZW}	0.151 (0.740)				
ZWX	b_{M1ZWX}	3.150 (1.481)				
VX					b_{YVX}	0.180 (0.088)
VM_2					b_{YVM2}	0.104 (0.026)

model. It might be that this assumption is false, either because the variables share an unmodeled common cause, or one variable affects the other. One could, based on an examination of the size and significance of these covariances when they are freely estimated, modify parts of the model by including causal paths that are not a part of the original model. Alternatively, one could be agnostic about causal influence by modeling associations unaccounted for by the causal model through the estimation of covariance terms. Inclusion of a causal path merely because it improves model fit moves one into dangerous territory when it comes to the problem of capitalizing on chance that MacCallum et al. (1992) warned about. On the other hand, one should recognize that by excluding paths from a model of a given consequent variable (by fixing the path to zero) and freely estimating the covariance instead, there is the potential of introducing bias in the estimation of all other paths to that consequent variable. Some personal judgment is sometimes necessary when making the choice. The danger of overfitting can be counteracted by cross-validating the results one obtains on a new sample, or withholding half of the data for initial model fitting and then applying the empirically modified model to the other half for the sake of confirmation of the modified model.

In this example, and in the Mplus code in Appendix A, we do include all covariances listed above and in the caption in Figure 6.5. This model is saturated and so fits perfectly. However, the addition of these covariances did not significantly improve the model, $\chi^2(16) = 10.398$, and the estimates of model parameters were very similar compared to when these covariances were fixed to zero. But this should not be considered a general phenomenon. Model fit as well as parameter estimates can be different, even dramatically so, depending on the choice.

DERIVATION AND INFERENCES ABOUT (CONDITIONAL) DIRECT AND INDIRECT EFFECTS

With the parameters of the statistical model estimated, direct and indirect effects can be calculated, inferential tests about their size conducted, and their substantive meaning interpreted. In this section, we describe how to derive the expressions for the direct and indirect effects, estimate them for specific values of the moderators, and conduct inferential tests on their values.

Deriving Direct and Indirect Effects as Functions

The first step is to derive the mathematical expressions for the indirect and direct effects. In models with no moderators of any of the paths, this step is not necessary because the direct and indirect effects are constants (i.e., not functions of moderators) that are frequently provided in output from SEM programs. But when one or more of the direct or indirect effects is moderated, these effects become conditional and so are functions of parameters and moderators. SEM programs do not know what those functions are, so the user must specify what they are in order to obtain estimates of the conditional direct and/or indirect effects and inferential tests.

To facilitate the derivation process, it is helpful to label the paths from X to Y (directly or through an intermediary variable) in the diagram of the conceptual model with their corresponding parameters from the statistical model so that the direct and indirect effects can be built easily using the conceptual diagram. First, for those paths in the conceptual model that are unmoderated, label the paths with the corresponding parameter from the statistical model. In this example, there are three such unmoderated paths that are involved in the transmission X's effect to Y in the conceptual mod-

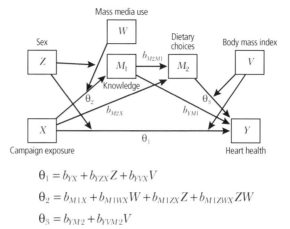

$$\theta_1 = b_{YX} + b_{YZX}Z + b_{YVX}V$$

$$\theta_2 = b_{M1X} + b_{M1WX}W + b_{M1ZX}Z + b_{M1ZWX}ZW$$

$$\theta_3 = b_{YM2} + b_{YVM2}V$$

Figure 6.6   The conceptual model with path labels for derivation of (conditional) direct and indirect effects.

el. The first is from X to M_2, which we labeled b_{M2X} in the statistical model. The second is from M_1 to Y, which is b_{YM1}. The third is from M_1 to M_2, or b_{M2M1} in the statistical model. These labels are attached the corresponding paths in Figure 6.6. Second, for each of the moderated paths (there are three in this example), there is no single parameter in the statistical model that defines it. Rather, these paths are functions of at least two parameters and one or more moderators, as discussed in the "Direct, Indirect, and Conditional Effects" section above. So pick any arbitrary symbol for the moderated paths and attach those labels to those paths in the conceptual diagram. In Figure 6.6, we use θ_1, θ_2, and θ_3 for the paths from X to Y, X to M_1, and M_2 to Y, respectively.

The next step is to define θ_1, θ_2, and θ_3 in terms of the parameters from the statistical model. In general terms, our goal is to derive the mathematical expression for an antecedent variable's path to a given consequent. This is done by extracting and summing the terms from the linear model of the consequent that involve the antecedent variable and then factoring the antecedent out of the sum. This requires a little algebraic manipulation of the linear model for the consequent to which the moderated path points.

Starting with θ_1, which is the moderated path from X to Y, we know from earlier (Equation 6.19) that the linear model for Y is

$$Y = i_Y + b_{YX}X + b_{YZ}Z + b_{YM1}M_1 + b_{YM2}M_2 + \\ b_{YV}V + b_{YZX}ZX + b_{YVX}VX + b_{YVM2}VM_2 + e_Y. \qquad (6.20)$$

To derive θ_1, first extract all terms involving the antecedent (X) and sum them, as such

$$b_{YX}X + b_{YZX}ZX + b_{YVX}VX \qquad (6.21)$$

and then, using the distributive law of multiplication, factor the antecedent out of the sum:

$$(b_{YX} + b_{YZX}Z + b_{YVX}V)X. \qquad (6.22)$$

The expression for the path of interest, θ_1 is the factor that doesn't contain the antecedent variable. So in this case,

$$\theta_1 = b_{YX} + b_{YZX}Z + b_{YVX}V. \qquad (6.23)$$

The same procedure is followed for deriving θ_2 and θ_3. θ_2 is the moderated path from X to M_1. Extracting and summing the terms involving X from the linear model for M_1 (Equation 6.17) yields

$$b_{M1X}X + b_{M1WX}WX + b_{M1ZX}ZX + b_{M1ZWX}ZWX \qquad (6.24)$$

which factors to

$$(b_{M1X} + b_{M1WX}W + b_{M1ZX}Z + b_{M1ZWX}ZW)X \qquad (6.25)$$

yielding

$$\theta_2 = b_{M1X} + b_{M1WX}W + b_{M1ZX}Z + b_{M1ZWX}ZW. \qquad (6.26)$$

For θ_3, M_2 is the antecedent and Y is the consequent. Extracting and summing the terms involving M_2 from the linear model for Y (Equation 6.19) and then factoring out M_2 leaves

$$\theta_3 = b_{YM2} + b_{YVM2}V. \qquad (6.27)$$

Having derived expressions for all of the paths in the conceptual model, we can now define the direct and indirect effects. The direct effect is simple, for it is merely the path from X to Y without going through any intermediary variables. We already derived this. It is just θ_1. Arbitrarily labeling the direct effect as δ,

$$\delta = \theta_1 = b_{YX} + b_{YZX}Z + b_{YVX}V. \qquad (6.28)$$

For the indirect effect(s), we trace all possible routes from X to Y going through at least one intermediary variable, multiplying products of paths as we go. As discussed earlier, there are three *specific* indirect effects. The first one goes from X to M_1 to Y. Arbitrarily using γ to denote an indirect effect,

$$\gamma_{M1} = \theta_2 b_{YM1} = (b_{M1X} + b_{M1WX}W + b_{M1ZX}Z + b_{M1ZWX}ZW)\, b_{YM1}. \tag{6.29}$$

Observe that γ_{M1} is a function of both W and Z. Thus, it is a conditional indirect effect. The second specific indirect effect goes from X to M_2 to Y, so

$$\gamma_{M2} = b_{M2X}\theta_3 = b_{M2X}\,(b_{YM2} + b_{YVM2}V). \tag{6.30}$$

This conditional indirect effect is a function of only V. The last indirect effect goes from X through both M_1 and M_2 before ending at Y. Multiplying all three paths yields

$$\gamma_{M1M2} = \theta_2 b_{M2M1}\theta_3 = (b_{M1X} + b_{M1WX}W + b_{M1ZX}Z + b_{M1ZWX}ZW)\, b_{M2M1}\,(b_{YM2} + b_{YVM2}V). \tag{6.31}$$

This conditional indirect effect is a function of all three moderators, W, Z, and V.

Quantification and Visualization of Conditional Direct and Indirect Effects

Equations 6.28 through 6.31 express the direct and indirect effects in terms of parameters of the model. But we can be more specific than these symbolic representations because the estimation of the model has yielded empirically-derived estimates of the corresponding parameters. Substituting the parameter estimates into their appropriate places yields

$$\hat{\delta} = 2.086 + 1.609Z + 0.180V \tag{6.32}$$

$$\hat{\gamma}_{M1} = (5.735 + 1.966W - 1.919Z + 3.150ZW)\,(-0.006) \tag{6.33}$$

$$\hat{\gamma}_{M2} = 0.038(2.865 + 0.104V) \tag{6.34}$$

$$\hat{\gamma}_{M1M2} = (5.735 + 1.966W - 1.919Z + 3.150ZW)\,(0.069)\,(2.865 + 0.104V). \tag{6.35}$$

With these functions for the direct and indirect effects expressed now in terms of moderators and parameter estimates, we can finally answer such questions as "What are the direct and indirect effects of the campaign on heart health?" But because the direct and indirect effects are moderated, the answer is "It depends." In order to answer this question, it has to be phrased more specifically, such as, "What are the direct and indirect effects of the campaign on heart health among men of moderate body mass but who use the mass media relatively little?" To answer this question, we can employ a procedure similar to the computation of simple slopes or conditional effects in moderated multiple regression. We simply select values of the moderators corresponding to those values on which we want to condition the estimates, and substitute them into Equations 6.32 through 6.35. For instance, if we arbitrarily define a person who uses the mass media relatively little as someone one standard deviation below the sample mean, and someone with moderate body mass as a person with average BMI, then the indirect and direct effects of the campaign for a male ($Z = 0.505$ after mean centering, as centering was done prior to computation of products) of average body mass ($V = 0$ after mean centering) and who uses the mass media relatively little ($W = -0.992$ after mean centering) are, according to Equations 6.32 through 6.35,

$$\hat{\delta} = 2.086 + 1.609(0.505) + 0.180(0) = 2.899$$

$$\hat{\gamma}_{M1} = [5.735 + 1.966(-0.992) - 1.919(0.505) + 3.150(0.505)(-0.992)]$$
$$(-0.006) = -0.007$$

$$\hat{\gamma}_{M2} = 0.038[2.865 + 0.104(0)] = 0.109$$

$$\hat{\gamma}_{M1M2} = [5.735 + 1.966(-0.992) - 1.919(0.505) + 3.150(0.505)(-0.992)]$$
$$(0.069)[2.865 + 0.104(0)] = 0.245$$

Because X is coded with a difference of 1 between those who did (-0.494) and did not (0.506) get exposure to the campaign, these direct and indirect effects can be interpreted as the average difference in heart health resulting from these effects. So among men of average body mass who use the mass media relatively little, those who were exposed to the campaign are $\hat{\delta} = 2.899$ healthy heart units higher than those not exposed to the campaign directly as a result of the exposure. They are $\hat{\gamma}_{M1M2} = 0.245$ units higher as a result of the effect of the campaign on knowledge which in turn affected dietary choices which then affected heart health. Furthermore, they are $\hat{\gamma}_{M2} = 0.109$ units higher as a result of the effect of the campaign on dietary choices independent of knowledge, which in turn affected heart health, but $|\hat{\gamma}_{M1}| = |-0.007| = 0.007$ units lower as a result of the effect of the campaign on knowledge, which in turn affected heart health independent of dietary choices.

In order to better grasp how the indirect and direct effects are contingent on the moderators, it is useful to complete these computations for various combinations of moderator values, simply by using the functions for the direct and indirect effects, substituting in the moderator values, and doing the computation. Table 6.4 displays the direct and indirect effects for people defined by combinations of sex, body mass, and use of mass media. For these computations, we use the convention of defining "relatively low," "moderate," and "relatively high" on continuous variables as one standard deviation below the sample mean, the sample mean, and one standard deviation above the mean, respectively. These are admittedly arbitrary, and other values could be used.

Another useful means of displaying the moderated nature of the direct and indirect effects is through a visual representation, such as a line graph (see, e.g., Bauer, Preacher, & Gil, 2006; Edwards & Lambert, 2007; Preacher et al., 2007). A graph for the indirect effect of the campaign through M_1 and M_2 can be found in Figure 6.7 using separate plots for levels of media use and separate lines for males and females. A comparable plot for the direct effect accompanies the plot of the indirect effect. Represented this way, it is clear that both direct and indirect effects increase with increasing body mass. That is, the campaign seems more effective, both directly and indirectly, among those with higher body mass. It is also apparent that the campaign seems to affect men directly more than indirectly through knowledge and diet choices, but also that this indirect effect is larger among men who use the mass media more. Among women, however, the indirect effect does not seem to depend on extent of mass media use. Finally, the difference in this indirect effect of the campaign between men and women varies by media use, with the larger sex differences among those who use the mass media relatively less.

Our discussion above, as well as Figure 6.7, neglects the specific indirect effects through knowledge only ($\hat{\gamma}_{M1}$) and through dietary choices only ($\hat{\gamma}_{M2}$). Comparable visual depictions of these effects could be generated. However, as can be seen in Table 6.4, these effects are rather small and, it turns out, not statistically different from zero regardless of the choice of moderator value(s). It is to the topic of statistical inference that we turn next.

Statistical Inference: Probing the Direct and Indirect Effects at Levels of the Moderator(s)

In this section thus far, focus has been on the derivation of the functions for the conditional direct and indirect effects, and using those functions given the parameter estimates and specific values of the moderator to quan-

TABLE 6.4. Direct and Indirect Effects of the Heart Health Campaign for Various Combinations of Sex, Mass Media Use, and Body Mass. (Bootstrap Interval Estimates in Parentheses)

Sex (Z)	Media (W)	BMI (V)	Indirect Effects				Direct Effect ($\hat{\delta}$)
			M_1 ($\hat{\gamma}_{M1}$)	M_2 ($\hat{\gamma}_{M2}$)	M_1 and M_2 ($\hat{\gamma}_{M1M2}$)	Total	
Female	Low	Low	−0.038 (−0.271, 0.177)	0.094 (−0.434, 0.606)	1.077* (0.561, 1.688)	1.133* (0.394, 1.906)	0.630*
Female	Moderate	Low	−0.040 (−0.275, 0.192)	0.094 (−0.434, 0.606)	1.146* (0.779, 1.589)	1.200* (0.555, 1.860)	0.630*
Female	High	Low	−0.043 (−0.299, 0.205)	0.094 (−0.434, 0.606)	1.215* (0.720, 1.809)	1.267* (0.544, 2.024)	0.630*
Female	Low	Moderate	−0.038 (−0.271, 0.177)	0.109 (−0.509, 0.690)	1.242* (0.645, 1.933)	1.313* (0.459, 2.185)	1.290*
Female	Moderate	Moderate	−0.040 (−0.275, 0.192)	0.109 (−0.509, 0.690)	1.322* (0.904, 1.814)	1.390* (0.652, 2.123)	1.290*
Female	High	Moderate	−0.043 (−0.299, 0.205)	0.109 (−0.509, 0.690)	1.401* (0.810, 2.045)	1.486* (0.624, 2.317)	1.290*
Female	Low	High	−0.038 (−0.271, 0.177)	0.123 (−0.580, 0.777)	1.407* (0.719, 2.187)	1.493* (0.521, 2.485)	1.949*
Female	Moderate	High	−0.040 (−0.275, 0.192)	0.123 (−0.580, 0.777)	1.497* (1.015, 2.068)	1.581* (0.748, 2.423)	1.949*
Female	High	High	−0.043 (−0.299, 0.205)	0.123 (−0.580, 0.777)	1.588* (0.937, 2.327)	1.669* (0.715, 2.635)	1.949*
Male	Low	Low	−0.007 (−0.122, 0.033)	0.094 (−0.434, 0.606)	0.212 (−0.263, 0.706)*	0.299 (−0.404, 0.986)*	2.239*
Male	Moderate	Low	−0.029 (−0.203, 0.135)	0.094 (−0.434, 0.606)	0.817* (0.481, 1.223)	0.883* (0.283, 1.524)	2.239

(continued)

TABLE 6.4. Direct and Indirect Effects of the Heart Health Campaign for Various Combinations of Sex, Mass Media Use, and Body Mass. (Bootstrap Interval Estimates in Parentheses) (continued)

			Indirect Effects				Direct Effect ($\hat{\delta}$)
Sex (Z)	Media (W)	BMI (V)	M_1 ($\hat{\gamma}_{M1}$)	M_2 ($\hat{\gamma}_{M2}$)	M_1 and M_2 ($\hat{\gamma}_{M1M2}$)	Total	
Male	High	Low	-0.050 (-0.349, 0.233)	0.094 (-0.434, 0.606)	1.422* (0.936, 1.427)	1.466* (0.762, 2.261)	2.239*
Male	Low	Moderate	-0.007 (-0.122, 0.033)	0.109 (-0.509, 0.690)	0.245* (-0.307, 0.806)	0.346* (-0.468, 1.133)	2.899*
Male	Moderate	Moderate	-0.029 (-0.203, 0.135)	0.109 (-0.509, 0.690)	0.942* (0.559, 1.389)	1.022* (0.326, 1.731)	2.899*
Male	High	Moderate	-0.050 (-0.349, 0.233)	0.109 (-0.509, 0.690)	1.640* (1.072, 2.312)	1.699* (0.884, 2.574)	2.899*
Male	Low	High	-0.007 (-0.122, 0.033)	0.123 (-0.580, 0.777)	0.277* (-0.347, 0.914)	0.393* (-0.528, 1.283)	3.558*
Male	Moderate	High	-0.029 (-0.203, 0.135)	0.123 (-0.580, 0.777)	1.067* (0.625, 1.564)	1.162* (0.376, 1.957)	3.558*
Male	High	High	-0.050 (-0.349, 0.233)	0.123 (-0.580, 0.777)	1.858* (1.205, 2.608)	1.931* (1.007, 2.931)	3.558*

* $p < 0.05$, assuming normality of the sampling distribution of the indirect effect. Bootstrap confidence intervals are bias corrected and based on 10,000 bootstrap samples.

Note: "Low," "moderate," and "high" media use and body mass were defined as one standard deviation below the sample mean, the mean, and one standard deviation above the mean, respectively. After mean centering, these correspond to -0.992, 0, and 0.992 for media use, and -3.665, 0, and 3.665 for body mass. Mean centered values of sex were used for males ($Z = 0.505$) and females ($Z = -0.495$). The total indirect effect is the sum of the three specific indirect effects.

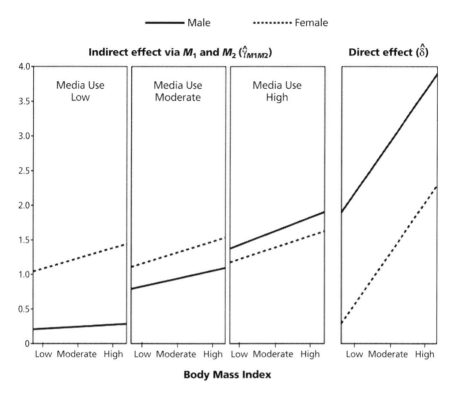

Figure 6.7 A visual representation of the conditional direct effect of the campaign and the conditional indirect effect through both knowledge and diet choices.

tify these effects. But our discussion has been purely *descriptive*. Although an estimate of, say, γ_{M1M2} for a given combination of moderator values may be descriptively different from zero, that does not mean that one can claim its population value is. There are inherent limitations of thinking dichotomously as "zero" or "not zero", but interpretation and discussion is often made simpler by focusing on those effects that are demonstrably different from zero while ignoring those that are not. Inferences for conditional direct and indirect effects can be conducted in a manner analogous to inferences about conditional effects or "simple slopes" in multiple regression with interactions. There are two approaches we will focus on, but only one that we can wholeheartedly endorse.

The first approach bases the inference on a ratio of the estimate of the effect for a given combination of moderator values to the standard error of that estimate. For example, to test the hypothesis that the population indirect effect of the campaign through knowledge and dietary choices (γ_{M1M2}) for females ($Z = -0.495$) relatively high in body mass ($V = 3.665$) and who use the mass media relatively little ($W = -0.995$), first derive the indirect effect using the parameter estimates from the full model (Equation 6.35):

$$\hat{\gamma}_{M1M2} = (5.735 + 1.966W - 1.919Z + 3.150ZW)(0.069)$$
$$(2.865 + 0.104V) = 1.322$$

Next, divide this ratio by an estimate of the standard error (obtained via the delta method and provided automatically by most SEM software; Sobel, 1982). If you assume the sampling distribution of the conditional indirect effect is normal, a p-value to test the null hypothesis that the population conditional indirect effect is zero for that combination of moderator values can be obtained from the standard normal distribution. In this case

$$Z = \frac{1.322}{0.231} = 5.723, p < .001.$$

Alternatively, a confidence interval can be constructed as the point estimate plus or minus about two standard errors, or (0.870, 1.776). Either way, we can claim that this conditional indirect effect is statistically different from zero. This process can be repeated for any or all direct and indirect effects of interest for various combinations of the moderator values. Doing so for the estimates of all effects reported in Table 6.4 reveals that the conditional indirect effect of the campaign through either knowledge (γ_{M1}) alone or dietary choices (γ_{M2}) alone is not different from zero for any of these combinations of moderator values. However, the conditional indirect effect through both knowledge and dietary choices (γ_{M1M2}) is positive and statistically different from zero except for males who use the mass media relatively little. Finally, the direct effect (δ) of the campaign is positive and statistically different from zero for all but women relatively low in body mass.

These computations can be automated in some SEM programs. We like to use Mplus in part because it makes this process simple. The "MODEL CONSTRAINT" section of the Mplus code in Appendix A constructs estimates of conditional indirect effects for various combinations of sex, mass media use, and body mass by implementing Equations 6.32 through 6.35 and generating the resulting estimates in the output.

We have not provided computational formulas for the standard errors of conditional direct and indirect effects here for four reasons. First, the standard error formulas for conditional *direct* effects can be found in sources on estimating and probing interactions in linear regression, and the same formulas apply here (see e.g., Aiken & West, 1991; Bauer & Curran, 2005; Hayes & Matthes, 2009). Second, the derivation and computation of standard errors for conditional direct and indirect effects can be exceedingly complex, involving linear combinations of parameters, moderator values, and covariances between parameter estimates. Preacher et al. (2007) provide formulas for the standard errors for five relatively simple conditional indirect effects, and even those are more complicated than anyone would want to try to implement by hand. Third, a good SEM program that can estimate functions of parameters will also provide standard errors for those functions, and often Z and p-values as well. Mplus, for example, generates standard errors using the delta method in the output that results from the MODEL CONSTRAINT section of the code.

But most important, we choose not to dwell on the computation of standard errors for conditional indirect effects because we do not recommend the use of this approach to inference. The problem with this approach is similar to the problem with the Sobel test when assessing indirect effects in mediation models without moderation. This test assumes the sampling distribution of the conditional indirect effect is normal. Conditional indirect effects are products of parameters that are themselves roughly normally distributed. But the sampling distribution of the product of random normal variables is not itself normal. Rather, the sampling distributions of products tend to be skewed and heavy tailed, meaning that p-values and confidence intervals that are based on the assumption of normality and symmetry can be inaccurate. Thus, we recommend bootstrap-derived confidence intervals as alternative approach to inference about conditional indirect effects that respects the nonnormality and asymmetry of the sampling distribution of the conditional indirect effect. Bootstrap confidence intervals for indirect effects have already been widely studied and recommended over the Sobel test in the context of the kinds of mediation models depicted in Figure 6.2 (see e.g., Hayes, 2009; Preacher & Hayes, 2004, 2008; Shrout & Bolger, 2002), as well as in conditional process models more rudimentary than those discussed here (Edwards & Lambert, 2007; Preacher et al., 2007). Furthermore, the evidence suggests that this approach is superior to the normal theory approach described above as well as many other tests for mediation that have been proposed in the mediation literature (Briggs, 2006; MacKinnon, Lockwood, & Williams, 2004; Williams & MacKinnon, 2008). There is no reason to believe these findings would not apply just as much to testing hypotheses about conditional indirect effects.

To construct a bootstrap confidence interval for a conditional indirect effect, the parameters of the model must be repeatedly estimated, each time after taking a random sample with replacement from the data equal in size to the original sample. In each resampling of the data, after estimation of the model parameters, the indirect effects are estimated as discussed above. Repeated many times (at least 1,000 times, but more is better; we recommend 5,000 to 10,000) this procedure empirically generates an approximation of the sampling distribution of the effect of interest. Interval estimates of the indirect (conditional or unconditional) effects are then constructed by finding the estimates in the distribution over repeated resampling that define the upper and lower 2.5th percentiles (for a 95% confidence interval) in the bootstrap sampling distribution of the effect. Inferences from such *percentile-based* bootstrap confidence intervals are made by ascertaining whether the confidence interval contains zero. If a 95% confidence interval does not contain zero, then the conclusion is that one can say with 95% confidence that the population conditional indirect effect is different from zero. However, if the confidence interval straddles zero, then one cannot claim that the population conditional indirect effect is different from zero, at least not with 95% confidence. Although not literally a hypothesis test, a failure for the confidence interval to contain zero can be interpreted much like one would interpret the outcome of a hypothesis test. One can reject the null hypothesis that the population conditional indirect effect is zero in favor of the alternative that it is different from zero.

Because the end points of the confidence interval are derived based on percentiles of an empirically generated distribution, there is no expectation or requirement that the endpoints are equidistant from the point estimate, as is true when constructing confidence intervals assuming the sampling distribution is normal. This reflects the reality that the sampling distribution of a product of normal variables is not typically symmetric. For this reason, interval estimates based on bootstrapping are often called "asymmetric confidence intervals." The endpoints can also be adjusted in various ways to produce bias corrected or bias corrected and accelerated confidence intervals (see, e.g., Efron & Tibshirani, 1993), although there is some debate in the literature about just how much of a difference these adjustments make to the validity and power of this approach to inference.

Of course, because it is so computationally intensive and repetitive, bootstrapping would be entirely impractical if it were not for the availability of statistical programs that implement this method. Fortunately, some SEM programs have bootstrapping routines built in. For instance, Mplus can produce percentile and bias corrected confidence intervals for model parameters as well as indirect effects. When combined with its MODEL CONSTRAINT function, both point and bootstrap confidence intervals for conditional indirect effects are easily calculated.

In the Mplus code in Appendix A, bootstrapping is implemented in the ANALYSIS and OUTPUT section of the code. Removing the comment symbol ("!") from "Bootstrap = 10000" tells Mplus to bootstrap all parameter estimates, including functions of parameters generated in the MODEL CONSTRAINT code, using 10,000 bootstrap samples. Removing the comment from "cinterval (bcbootstrap)" tells Mplus to produce bias corrected bootstrap confidence intervals for all those parameters and functions of parameters. For instance, earlier we calculated the point estimate for the conditional indirect effect of the campaign through knowledge and dietary choices for women high in body mass but low in mass media use as 1.322. Bootstrapping a confidence interval in Mplus yielded an interval estimate of (0.904, 1.814). So we can claim confidently that this indirect effect is positive and different from zero *without* making the assumption of normality of the sampling distribution. Bootstrap confidence intervals generated by Mplus using the code in Appendix A can be found in Table 6.4. As can be seen, these confidence intervals produce inferences that are largely identical to the normal theory tests, which will tend to occur with relatively large effects or in large samples such as this one.

EXTENSIONS TO LATENT VARIABLES

One of the primary benefits of SEM is that it grants researchers the ability to correct for the attenuating effects of measurement error by using latent variables with multiple observed indicators. Theoretically, latent variables are error-free representations of the constructs of interest. Because it is rarely possible to obtain error-free measurements of constructs, and because measurement error can drastically reduce the power for detecting an effect, it is usually to the researcher's advantage to substitute latent variables for measured variables when possible. Including latent variables in mediation models (without moderation) is straightforward, and indirect effects can be quantified and tested in exactly the same way as with measured variables (see e.g., Cheung & Lau, 2008; Coffman & MacCallum, 2005; Lau & Cheung, 2012).

If the ultimate outcome variable Y is latent, no fundamental changes are necessary to the procedures outlined in the preceding. The researcher merely needs to define a latent variable F_y with multiple observed indicators, then substitute F_y in the model for Y in all discussions and procedures described above. However, conditional process models present special challenges when the researcher wishes to use latent variables as antecedent or moderator variables. Recall that conditional process models require the researcher to compute products of antecedent variables. We achieved this earlier in Mplus by using a DEFINE statement to create additional variables that were equal to products of variables already present in the data set. The primary challenge here is that it is not possible to literally compute the product of two latent variables, or the product of a latent variable and an observed variable, for inclusion in a model—by definition, latent variables are unobserved.

Fortunately, methodologists have developed ways to estimate interaction effects in the latent variable context (see chapter by Marsh, Wen, Hau, & Nagengast in this volume, as well as Schumacker & Marcoulides, 1998, for overviews of several approaches). A method implemented in Mplus is termed the *latent moderated structural equations* (LMS) approach (Klein & Moosbrugger, 2000). Details behind exactly how this method proceeds are well beyond the scope of this chapter. Suffice it to say the LMS method is computationally intensive and requires numerical integration. On the other hand, LMS is easier to implement than competing methods because (1) it does not require the researcher to create products of measured indicators and (2) it is straightforward to invoke in Mplus. LMS results in unbiased, efficient estimates of interaction effects that are robust to departures from normality, and with unbiased standard errors.

We illustrate such a model with a brief example. In a recent study, Parker, Nouri, and Hayes (2011) explored the relation between perceptions of distributive justice (fairness of outcomes one receives at the company) and turnover intentions (likelihood of resigning) in a sample of 110 employees of accounting firms. Specifically, they hypothesized that perceptions of distributive justice (F_X: 4 indicators) influences turnover intentions (F_Y: three indicators) indirectly through promotion instrumentality (F_M: three indicators), which is the perception that good performance is rewarded through promotion. They further hypothesized that this indirect effect would be moderated because job performance (F_W: six indicators) was expected to influence the relation between promotion instrumentality and turnover intention (see Figure 6.8 for the conceptual model; this model corresponds to "Model 3" in Preacher et al., 2007 and the "second stage moderation model" in Edwards & Lambert, 2007). In Parker et al. (2011), all four constructs were measured by averaging responses to the items within each scale, but they could also be treated as latent variables with items as indicators. In the syntax provided in Appendix B, all four constructs are represented as latent variables with items as indicators (see Figure 6.8 for the statistical model in the form of a structural equation model). In Mplus, a latent product (here, $F_W F_M$) is created by use of the XWITH command, which also instructs Mplus to employ the LMS method. In this example, as in the published version using observed variables, the interaction between promotional instrumentality and job performance was statistically significant. Thus, it is sensible to estimate conditional indirect effects, which for this model would be $b_{F_M F_X}[b_{F_Y F_M} + b_{F_Y (F_W F_M)}]F_W$. When the moderator is latent, then conditional values may be chosen by using the latent moderator's mean (typically zero by default) and standard deviation (the square root of its estimated or fixed variance; here that is 1), just as with observed moderators. The conditional indirect effects of distributive justice on turnover intentions via promotion instrumentality at one SD below the mean (−1), the mean (0), and one SD above the mean (1) of latent job performance moderator were, respectively, 0.023, −0.355, and −0.734.

To determine whether these effects differ significantly from zero, we constructed 95% confidence intervals using a Monte Carlo method de-

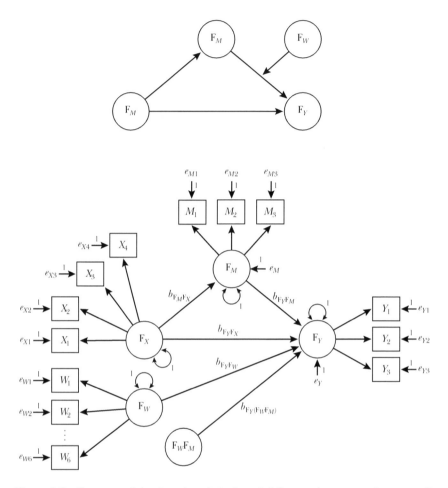

Figure 6.8 Conceptual (top) and statistical model (bottom) representing a conditional process model for the Parker et al. (2011) example involving latent variables in which the indirect effect of X is moderated by W. Not pictured in the statistical model to reduce visual clutter are freely estimated covariances among F_X, F_W, and $F_W F_M$ and between e_M and F_W.

scribed by MacKinnon et al. (2004) because Mplus does not provide bootstrap CIs if integration is used during model estimation. Using the Monte Carlo method, the conditional effects of distributive justice on promotion instrumentality and promotional instrumentality on turnover intentions were generated from normal distributions with means equal to their point estimates and standard deviations equal to their estimated *SE*s. Generating a large number of coefficient pairs (we used 100,000), along with their products, produces a Monte Carlo distribution of conditional indirect effects for each conditional value of job performance. The 2.5th and 97.5th percentiles of these distributions can serve as limits of 95% CIs. In the present case, these CIs were [0.431, –0.376] for low job performance, [–0.025, –0.766] for mean job performance, and [–0.175, –1.444] for high job performance.[3] Because the first CI includes zero while the latter two exclude zero, we conclude that the latter two conditional indirect effects differ significantly from zero at $\alpha = .05$. This pattern of significance mirrors the results reported by Parker et al. (2011). This method of confidence interval construction has the nice property that the indirect effects are not assumed to be normally distributed (or even symmetric).

CONCLUSION

The goal of this chapter was to describe the basic principles of conditional process modeling in an SEM framework. This analytical approach combines properties of moderation and mediation analysis with the goal of estimating and understanding the various pathways, direct and indirect, moderated and unmoderated, through which a putative causal agent influences an outcome. Our example was chosen in order to illustrate the fundamental steps in building a complex model, estimating the parameters in SEM software, constructing the effects of interest, and testing hypotheses. But these steps generalize to the estimation of simple models already discussed in the moderated mediation literature, and they can also be applied to latent variable models or models that mix observed and latent variables, as illustrated in the previous section. Although our focus has been exclusively on models with continuous consequent variables, in theory this approach can be extended to any analysis that can be parameterized in terms of a generalized linear model.

APPENDIX A

The Mplus code below estimates the path model in Figure 6.5, produces conditional direct and indirect effects from the resulting parameter estimates for various combinations of the moderators, and conducts inferential tests of those conditional effects. Removing the exclamation point (!) from the ANALYSIS and OUTPUT sections of the code implements bootstrapping for inference about conditional effects.

```
DATA:
     FILE IS C:\cpm2.txt;
     FORMAT IS free;
VARIABLE:
     NAMES ARE x w z v m1 m2 y;
     USEVARIABLES are x w z v m1 m2 y wx zx wz wzx vm2 vx;
DEFINE:
     !mean center variables involved in products;
     v = v-27.272;
     w = w-3.954;
     m2 = m2-7.906;
     z = z-0.495;
     x = x-0.494;
     !create product variables;
     wx = w*x;
     zx = z*x;
     wz = w*z;
     wzx = w*z*x;
     vm2 = v*m2;
     vx = v*x;
ANALYSIS:
     !bootstrap = 10000;
MODEL:
     m1 ON x (m1x)
           w
           z
           wx (m1wx)
           zx (m1zx)
           wz
           wzx (m1wzx);
     m2 ON x (m2x)
           m1 (m2m1);
     y ON  x (yx)
           v
           z
           m1 (ym1)
           m2 (ym2)
           vm2 (yvm2)
           zx (yzx)
           vx (yvx);
```

```
      M1 WITH v vx vm2;
      M2 WITH w z v wx zx wz wzx vx vm2;
      Y WITH w wx wz wzx;
      X;
MODEL CONSTRAINT:
      new (cdir1 cdir2 cdir3 cdir4 cdir5 cdir6
           cind1 cind2 cind3 cind4 cind5 cind6
           cind7 cind8 cind9 cind10 cind11 cind12
           cind13 cind14 cind15 cind16 cind17 cind18
           cind21 cind22 cind23 cind24 cind25 cind26
           cind31 cind32 cind33 ctind1 ctind2 ctind3
           ctind4 ctind5 ctind6 ctind7 ctind8 ctind9
           ctind10 ctind11 ctind12 ctind13 ctind14 ctind15
           ctind16 ctind17 ctind18
      zmale zfemale vlow vave vhigh wlow wave whigh);
      zmale = 0.505;
      zfemale = -0.495;
      vlow = -3.665;
      vave = 0;
      vhigh = +3.665;
      wlow = -0.992;
      wave = 0;
      whigh = 0.992;

      !cdir1-cdir6 are conditional direct effects of x on y at;
      !values of z and v;

      cdir1 = yx+yzx*zmale+yvx*vlow;
      cdir2 = yx+yzx*zfemale+yvx*vlow;
      cdir3 = yx+yzx*zmale+yvx*vave;
      cdir4 = yx+yzx*zfemale+yvx*vave;
      cdir5 = yx+yzx*zmale+yvx*vhigh;
      cdir6 = yx+yzx*zfemale+yvx*vhigh;

      !cind1-18 are conditional indirect effects via M1 and M2 at;
      !values of z, w, and v;

      cind1 = (m1x+m1wx*wlow+m1zx*zmale+m1wzx*wlow*zmale)*m2m1*
              (ym2+yvm2*vlow);
      cind2 = (m1x+m1wx*wlow+m1zx*zmale+m1wzx*wlow*zmale)*m2m1*
              (ym2+yvm2*vave);
      cind3 = (m1x+m1wx*wlow+m1zx*zmale+m1wzx*wlow*zmale)*m2m1*
              (ym2+yvm2*vhigh);
      cind4 = (m1x+m1wx*wave+m1zx*zmale+m1wzx*wave*zmale)*m2m1*
              (ym2+yvm2*vlow);
      cind5 = (m1x+m1wx*wave+m1zx*zmale+m1wzx*wave*zmale)*m2m1*
              (ym2+yvm2*vave);
      cind6 = (m1x+m1wx*wave+m1zx*zmale+m1wzx*wave*zmale)*m2m1*
              (ym2+yvm2*vhigh);
      cind7 = (m1x+m1wx*whigh+m1zx*zmale+m1wzx*whigh*zmale)*m2m1*
              (ym2+yvm2*vlow);
      cind8 = (m1x+m1wx*whigh+m1zx*zmale+m1wzx*whigh*zmale)*m2m1*
              (ym2+yvm2*vave);
```

```
 cind9 = (m1x+m1wx*whigh+m1zx*zmale+m1wzx*whigh*zmale)*m2m1*
         (ym2+yvm2*vhigh);
cind10 = (m1x+m1wx*wlow+m1zx*zfemale+m1wzx*wlow*zfemale)*m2m1*
         (ym2+yvm2*vlow);
cind11 = (m1x+m1wx*wlow+m1zx*zfemale+m1wzx*wlow*zfemale)*m2m1*
         (ym2+yvm2*vave);
cind12 = (m1x+m1wx*wlow+m1zx*zfemale+m1wzx*wlow*zfemale)*m2m1*
         (ym2+yvm2*vhigh);
cind13 = (m1x+m1wx*wave+m1zx*zfemale+m1wzx*wave*zfemale)*m2m1*
         (ym2+yvm2*vlow);
cind14 = (m1x+m1wx*wave+m1zx*zfemale+m1wzx*wave*zfemale)*m2m1*
         (ym2+yvm2*vave);
cind15 = (m1x+m1wx*wave+m1zx*zfemale+m1wzx*wave*zfemale)*m2m1*
         (ym2+yvm2*vhigh);
cind16 = (m1x+m1wx*whigh+m1zx*zfemale+m1wzx*whigh*zfemale)*
          m2m1*(ym2+yvm2*vlow);
cind17 = (m1x+m1wx*whigh+m1zx*zfemale+m1wzx*whigh*zfemale)*
          m2m1*(ym2+yvm2*vave);
cind18 = (m1x+m1wx*whigh+m1zx*zfemale+m1wzx*whigh*zfemale)*
          m2m1*(ym2+yvm2*vhigh);

!cind21-28 are conditional indirect effects via M1 only at;
!values of z, w;

cind21 = (m1x+m1wx*wlow+m1zx*zmale+m1wzx*wlow*zmale)*ym1;
cind22 = (m1x+m1wx*wlow+m1zx*zfemale+m1wzx*wlow*zfemale)*ym1;
cind23 = (m1x+m1wx*wave+m1zx*zmale+m1wzx*wave*zmale)*ym1;
cind24 = (m1x+m1wx*wave+m1zx*zfemale+m1wzx*wave*zfemale)*ym1;
cind25 = (m1x+m1wx*whigh+m1zx*zmale+m1wzx*whigh*zmale)*ym1;
cind26 = (m1x+m1wx*whigh+m1zx*zfemale+m1wzx*whigh*zfemale)*ym1;

!cind31-33 are conditional indirect effects via M2 only at;
!values of v;

cind31 = m2x*(ym2+yvm2*vlow);
cind32 = m2x*(ym2+yvm2*vave);
cind33 = m2x*(ym2+yvm2*vhigh);

!ctind1-ctind18 are total indirect effects at values of z, w,;
!and v;

ctind1 = cind1+cind21+cind31;
ctind2 = cind2+cind21+cind32;
ctind3 = cind3+cind21+cind33;
ctind4 = cind4+cind23+cind31;
ctind5 = cind5+cind23+cind32;
ctind6 = cind6+cind23+cind33;
ctind7 = cind7+cind25+cind31;
ctind8 = cind8+cind25+cind32;
ctind9 = cind9+cind25+cind33;
ctind10 = cind10+cind22+cind31;
ctind11 = cind11+cind22+cind32;
```

```
      ctind12 = cind12+cind22+cind33;
      ctind13 = cind13+cind24+cind31;
      ctind14 = cind14+cind24+cind32;
      ctind15 = cind15+cind24+cind33;
      ctind16 = cind16+cind26+cind31;
      ctind17 = cind17+cind26+cind32;
      ctind18 = cind18+cind26+cind33;
OUTPUT:
      !cinterval (bcbootstrap)
```

APPENDIX B

The Mplus code below estimates the statistical model in Figure 6.8, produces conditional direct and indirect effects from the resulting parameter estimates for various values of the moderator, and conducts inferential tests of those conditional effects.

```
DATA:
      FILE IS data.dat;
VARIABLE:
      NAMES ARE dj1 dj2 dj3 dj4 pi1 pi2 pi3
      jp1 jp2 jp3 jp4 jp5 jp6 to1 to2 to3;
      USEVARIABLES ARE dj1 dj2 dj3 dj4 pi1 pi2 pi3
      jp1 jp2 jp3 jp4 jp5 jp6 to1 to2 to3;
ANALYSIS:
      TYPE IS RANDOM;
      ALGORITHM IS INTEGRATION;
MODEL:
      dj BY dj1* dj2-dj4;
      pi BY pi1* pi2-pi3;
      to BY to1* to2-to3;
      jp BY jp1* jp2-jp6;
      dj@1 pi@1 to@1 jp@1;
      pijp | pi XWITH jp;
      pi ON dj (mx);
      to ON pi (ym);
      to ON dj jp;
      to ON pijp (ywm);
      pi WITH jp;
MODEL CONSTRAINT:
      NEW (eff1 eff2 eff3);
      eff1 = mx*(ym+ywm*(-1));
      eff2 = mx*(ym+ywm*0);
      eff3 = mx*(ym+ywm*1);
```

NOTES

1. We use the terms *path, coefficient,* and *regression coefficient* synonymously to refer to the weight for an antecedent variable in a model of a consequent variable. In all discussions here, these weights are in unstandardized rather than standardized form, although in some cases the mathematics applies equally to both forms.

2. The data we use here were fabricated for pedagogical purposes. None of the findings we report here should be interpreted as substantiated by actual empirical evidence.

3. R and SPSS code that accomplishes the Monte Carlo estimation are available from either author by request.

REFERENCES

Aiken, L. S., & West, S. G. (1991). *Multiple regression: Testing and interpreting interactions.* Thousand Oaks, CA: Sage Publications.

Antheunis, M. L., Valkenberg, P. M., & Peter, J. (2010). Getting acquainted through social network sites: Testing a model of online uncertainty reduction and social attraction. *Computers in Human Behavior, 26,* 100–109.

Baron, R. M., & Kenny, D. A. (1986). The moderator-mediator variable distinction in social psychological research: Conceptual, strategic, and statistical considerations. *Journal of Personality and Social Psychology, 51,* 1173–1182.

Bauer, D. J., & Curran, P. J. (2005). Probing interactions in fixed and multilevel regression: Inferential and graphical techniques. *Multivariate Behavioral Research, 40,* 373–400.

Bauer, D. J., Preacher, K. J., & Gil, K. M (2006). Conceptualizing and testing random indirect effects and moderated mediation in multilevel models. New procedures and recommendations. *Psychological Methods, 11,* 142–163.

Briggs, N. E. (2006). *Estimation of the standard error and confidence interval of the indirect effect in multiple mediator models.* Unpublished doctoral dissertation, The Ohio State University, Columbus, OH

Cerin, E., & MacKinnon, D. P. (2009). A commentary on current practice in mediating variable analysis in behavioural nutrition and physical activity. *Public Health Nutrition, 12,* 1182–1188.

Cheung, G. W., & Lau, R. S. (2008). Testing mediation and suppression effects of latent variables: Bootstrapping with structural equation models. *Organizational Research Methods, 11,* 296–325.

Coffman, D. L., & MacCallum, R. C. (2005). Using parcels to convert path models into latent variable models. *Multivariate Behavioral Research, 40,* 235–259.

Druley, J. A., & Townsend, A. L. (1998). Self-esteem as a mediator between spousal support and depressive symptoms: A comparison of healthy individuals and individuals coping with arthritis. *Health Psychology, 17,* 255–261.

Echambi, R., & Hess, J. D. (2007). Mean-centering does not alleviate collinearity problems in moderated multiple regression models. *Marketing Research, 26,* 438–445.

Edwards, J. R., & Lambert, L. S. (2007). Methods for integrating moderation and mediation: A general analytical framework using moderated path analysis. *Psychological Methods, 12*, 1–22.

Efron, B., & Tibshirani, R. (1993). *An introduction to the bootstrap.* New York, NY: Chapman & Hall.

Fairchild, A. J., & MacKinnon, D. P. (2009). A general model for testing mediation and moderation effects. *Prevention Science, 10*, 87–99.

Hayes, A. F. (2009). Beyond Baron and Kenny: Statistical mediation analysis in the new millennium. *Communication Monographs, 76*, 408–420.

Hayes, A. F. (2013). *Introduction to mediation, moderation, and conditional process analysis.* New York, NY: The Guilford Press.

Hayes, A. F., & Matthes, J. (2009). Computational procedures for probing interactions in OLS and logistic regression: SPSS and SAS implementations. *Behavior Research Methods, 41*, 924–936.

Jaccard, J., & Turrisi, R. (2003). *Interaction effects in multiple regression* (2nd ed). Thousand Oaks, CA: Sage Publications

James, L. R., & Brett, J. M. (1984). Mediators, moderators, and tests for mediation. *Journal of Applied Psychology, 69*, 307–321.

Judd, C. M., & Kenny, D. A. (1981). Process analysis: Estimating mediation in treatment evaluations. *Evaluation Review, 5*, 602–619.

Klein, A., & Moosbrugger, H. (2000). Maximum likelihood estimation of latent interaction effects with the LMS method. *Psychometrika, 65*, 457–474.

Kromrey, J. D., & Foster-Johnson, L. (1998). Mean centering in moderated multiple regression: Much ado about nothing. *Educational and Psychological Measurement, 58*, 42–67.

Lau, R. S., & Cheung, G. W. (2012). Estimating and comparing specific mediation effects in complex latent variable models. *Organizational Research Methods, 15*, 3–16.

MacCallum, R. C., Roznowski, M., & Necowitz, L. B. (1992). Model modification in covariance structure analysis: The problem of capitalizing on chance. *Psychological Bulletin, 111*, 490–504.

MacKinnon, D. P., Lockwood, C. M., & Williams, J. (2004). Confidence limits for the indirect effect: Distribution of the product and resampling methods. *Multivariate Behavioral Research, 39*, 99–128.

Mathieu, J. E., & Taylor, S. R. (2006). Clarifying conditions and decision points for meditational type inferences in Organizational Behavior. *Journal of Organizational Behavior, 27*, 1031–1056.

Morgan-Lopez, A., & MacKinnon, D. P. (2006). Demonstration and evaluation of a method for assessing mediated moderation. *Behavior Research Methods, 38*, 77–89.

Muller, D., Judd, C. M., & Yzerbyt, V. Y. (2005). When mediation is moderated and moderation is mediated. *Journal of Personality and Social Psychology, 89*, 852–863.

Parade, S. H., Leerkes, E. M., & Blankson, A. (2010). Attachment to parents, social anxiety, and close relationships of female students over the transition to college. *Journal of Youth and Adolescence, 39*, 127–137.

Parker, R., Nouri, H., & Hayes, A. F. (2011). Distributive justice, promotion instrumentality, and turnover intentions in public accounting firms. *Behavioral Research in Accounting, 23,* 169–186

Preacher, K. J., & Hayes, A. F. (2004). SPSS and SAS procedures for estimating indirect effects in simple mediation models. *Behavior Research Methods, Instruments, and Computers, 36,* 717–731

Preacher, K. J., & Hayes, A. F. (2008). Asymptotic and resampling methods for assessing and comparing indirect effects in multiple mediator models. *Behavior Research Methods, 40,* 879–891.

Preacher, K. J., Rucker, D. D., & Hayes, A. F. (2007). Assessing moderated mediation hypotheses: Theory, methods, and prescriptions. *Multivariate Behavioral Research, 42,* 185–227.

Rucker, D. D., Preacher, K. J., Tormala, Z. L., & Petty, R. E. (2011). Mediation analysis in social psychology: Current practices and new recommendations. *Social and Personality Psychology Compass, 5/6,* 359–371.

Schumacker, R. E., & Maroulides, G. A. (1998). *Interaction and nonlinear effects in structural equation modeling.* Mahwah, NJ: Lawrence Erlbaum Associates, Inc.

Shieh, G. (2011). Clarifying the role of mean centering in multicollinearity of interaction effects. *British Journal of Mathematical and Statistical Psychology, 64,* 462–477.

Shrout, P. E., & Bolger, N. (2002). Mediation in experimental and nonexperimental studies: New procedures and recommendations. *Psychological Methods, 7,* 422–445.

Sobel, M. E. (1982). Asymptotic confidence intervals for indirect effects in structural equation models. In S. Leinhardt (Ed)., *Sociological methodology 1982* (pp. 290–312). Washington, DC: American Sociological Association.

Tepper, B. J., & Eisenbach, R. J., Kirby, S. L., & Potter, P. W. (1998). Test of a justice-based model of subordinates' resistance to downward influence attempts. *Group and Organization Management, 23,* 144–160.

Van Dijke, M., & de Cremer, D. (2010). Procedural fairness and endorsement of prototypical leaders: Leader benevolence or follower control? *Journal of Experimental Social Psychology, 46,* 85–96.

Williams, J., & MacKinnon, D. P. (2008). Resampling and distribution of the product methods for testing indirect effects in complex models. *Structural Equation Modeling: A Multidisciplinary Journal, 15,* 23–51.

CHAPTER 7

STRUCTURAL EQUATION MODELS OF LATENT INTERACTION AND QUADRATIC EFFECTS

**Herbert W. Marsh, Zhonglin Wen,
Kit-Tai Hau, and Benjamin Nagengast**

Estimating interaction effects is an important concern in psychology and the social sciences more generally. In educational psychology, for example, it is often hypothesized that the effect of an instructional technique will interact with characteristics of individual students in determining learning outcomes (e.g., a special remediation program developed for slow learners might not be an effective instructional strategy for bright students). Developmental psychologists are frequently interested in how the effects of a given variable vary with age in longitudinal or cross-sectional studies. Furthermore, many psychological theories explicitly hypothesize interaction effects. For example, some forms of expectancy-value theory (Nagengast, Marsh, et al., 2011) hypothesize that resultant motivation is determined by the interaction between expectancy of success and the value placed on the outcome by the individual (e.g., motivation is high only if both probability of success and the value placed on the outcome are high). In self-concept

Structural Equation Modeling: A Second Course (2nd ed.), pages 267–308
Copyright © 2013 by Information Age Publishing
All rights of reproduction in any form reserved.

theory following William James (Marsh, 2008), the relation between an individual component of self-concept (e.g., academic, social, physical) and global self-esteem is hypothesized to interact with the importance placed on a specific component of self-concept (e.g., if a person places no importance on physical accomplishments, then these physical accomplishments—substantial or minimal—are not expected to be substantially correlated with self-esteem). More generally, a variety of weighted-average models posit—at least implicitly—that the effects of a given set of variables will depend on the weight assigned to each variable in the set (i.e., the weight assigned to a given variable interacts with the variable to determine the contribution of that variable to the total effect).

Analyses of quadratic (and other nonlinear) effects are also important. A strong quadratic effect may give the appearance of a significant interaction effect. Thus, without proper analysis of the potential quadratic effects, the investigator might easily misinterpret, overlook, or mistake a quadratic effect as an interaction effect (Klein, Schermelleh-Engel, Moosbrugger, & Kelava, 2009; Lubinski & Humphreys, 1990). Whereas quadratic trends and other nonlinearity can complicate interpretations of interaction effects, in other situations the nonlinearity of relations might be the focus of the analyses. Common examples include:

- nonlinear effects might be hypothesized between strength of interventions (or dosage level) and outcome variables such that benefits increase up to an optimal level, and then level off or even decrease beyond this optimal point;
- the effects of an intervention might be expected to be the highest immediately following the intervention and to decrease non-linearly over time;
- at low levels of anxiety, increases in anxiety might facilitate performance but at higher levels of anxiety, further increases in anxiety might undermine performance;
- self-concept might decrease with age for young children, level out in middle adolescence, and then increase with age into early adulthood;
- levels of workload demanded by teachers might be positively related to student evaluations of teaching effectiveness for low to moderate levels of workload but might have diminishing positive effects or even negative effects for possibly excessive levels of workload (Marsh, 2001).

Although nonlinear relations can become very complex, we are restricting our discussion to the relatively simple case of quadratic effects for ease of presentation.

In this chapter, we explore methods for the analysis of products of latent variables—latent interactions and latent quadratic effects. Although our focus is primarily on latent interactions that are based on products of two different variables, the rationale generalizes to quadratic effects that are based on the product of the same variable. We begin by briefly summarizing traditional methods for analyzing interactions between observed (manifest) variables, so that readers can better understand the alternative approaches to latent interactions. We then move to alternative methods that attempt to estimate nonlinear effects using more familiar traditional strategies with observed variables. These methods include the replacement of the latent interaction with observed variables, and the formation of multiple groups. For example, we could use factor scores to replace the latent variables and then revert to observed variable analytical techniques (e.g., regression). In multiple group analyses, the structural model (without the moderating variable) is estimated separately in different subsamples, each representing a certain level of the moderator. The limitations of these methods will be discussed.

In the analyses of fully latent interactions, we describe and present concrete examples for three broad approaches, namely product-indicator approaches (Kenny & Judd, 1984), distribution-analytic approaches (Klein & Moosbrugger, 2000; Klein & Muthén, 2007), and Bayesian approaches (e.g., Lee, Song, & Tang, 2007). We start with the product-indicator models and differentiate the constrained (Jaccard & Wan, 1995; Jöreskog & Yang, 1996) and unconstrained approaches (Marsh, Wen, & Hau, 2004). The unconstrained approach is among one of the recommended methods because of its relative ease in application. We discuss various issues related to the formation of product indicators and item centering, which ultimately lead to the question of how the appropriate standardized solutions should be obtained.

We then move on to discuss two relatively recent approaches—distribution-analytic and Bayesian approaches. Both are promising developments with reasonably good performance. The potential drawbacks to both of these approaches are that they are mathematically more sophisticated and have not been implemented in widely available and easily accessible software (with the LMS approach, Klein & Moosbrugger, 2000, being the exception) and are thus still not easily accessible to applied researchers. In the final section, we extend latent interaction models to latent quadratic effects and discuss some future directions of research.

TRADITIONAL (NON-LATENT) APPROACHES TO INTERACTIONS BETWEEN OBSERVED VARIABLES

Consider the effect of observed variables X_1 and X_2 on outcome Y. Traditional approaches to the analysis of interactions between X_1 and X_2, as summarized below, depend on the nature of X_1 and X_2.

When the independent variables X_1 and X_2 are categorical variables that can take on a relatively small number of naturally occurring values, interaction effects can easily be estimated with traditional analysis of variance (ANOVA) procedures. When one independent variable is a naturally occurring categorical variable with a few levels and the other one is a continuous manifest variable, one possibility is to divide the total sample into separate groups representing the different levels of the true categorical variable and then conduct separate multiple regressions on each group. The interaction effect is represented by the differences between unstandardized regression coefficients obtained with the separate groups and tests of the statistical significance of this difference (Aiken & West, 1991; Cohen & Cohen, 1983). Alternatively, researchers can represent the categorical variable with dummy coding and apply multiple regression (see Equation 7.1).

When the independent variables X_1 and X_2 are reasonably continuous manifest variables, the following multiple regression with a product term is used to estimate both main and interaction effects (e.g., Aiken & West, 1991; Cohen & Cohen, 1983):

$$Y = \beta_0 + \beta_1 X_1 + \beta_2 X_2 + \beta_3 X_1 X_2 + \varepsilon, \tag{7.1}$$

where β_1 and β_2 represent the main effects, β_3 represents the interaction effect, and ε is a random disturbance term with zero mean that is uncorrelated with X_1 and X_2. In order to test whether the interaction effect is statistically significant, the test statistic is

$$t = \frac{\hat{\beta}_3}{\text{SE}(\hat{\beta}_3)},$$

where $\text{SE}(\hat{\beta}_3)$ is the standard error of estimated β_3.

In the above analyses, when the interaction is statistically significant, typically graphs are drawn to probe the nature of this interaction. It is customary to plot the relation between the outcome and the focal predictor at different levels of the categorical independent moderator variable or at 0, ± 1 SD (or ± 2 SD) for a continuous moderator (e.g., Preacher, Curran, & Bauer, 2006).

The common practice to transform reasonably continuous variables into a few discrete levels (e.g., using a median split) for ANOVA or multiple regression analyses is generally not recommended. It reduces reliability, power, and variance explained, and creates problems in interpretation and summarization of the strength of interaction and non-linear relations (MacCallum, Zhang, Preacher, & Rucker, 2002; Marsh, Hau, Wen, Nagengast, & Morin, in press).

The critical feature common to each of these non-latent approaches is that all of the dependent and independent variables are manifest (non-latent) variables inferred on the basis of a single indicator rather than latent variables inferred on the basis of multiple indicators. In applied research, however, even a priori interactions based on strong theoretical predictions are typically small, nonsignificant, or non-replicable; and the problem is usually more serious in non-experimental designs (McClelland & Judd, 1993). When there are multiple indicators of these variables, latent variable approaches described below provide a much stronger basis for evaluating the underlying factor structure relating multiple indicators to their factors, controlling for measurement error, potentially increasing power, and, ultimately, providing more defensible interpretations of the interaction effect.

LATENT APPROACHES TO INTERACTIONS

When at least one latent independent variable is inferred from multiple indicators, there are many methods to analyze the interaction effect (see Schumacker & Marcoulides, 1998). Generally, structural equation modeling (SEM) provides many advantages over the use of analyses based on observed variables, but is more complicated to conduct. We begin our discussion with two approaches, factor scores and multiple group approaches, that we consider only briefly. We then move on to fully latent variable approaches in which all of the dependent and independent variables are latent constructs inferred on the basis of multiple indicators.

Analyses of Factor Scores

Using factor scores of each latent variable to analyze a multiple regression equation (see Equation 7.1) is an intuitive approach to the analysis of interaction effects between latent variables (Yang, 1998). The underlying rationale is to replace the multiple indicators of each latent factor with a single factor score and then to use the factor scores as observed variables in subsequent analyses. The procedure has two steps. In the first step, the measurement model is estimated for the two independent variables and the one dependent variable, assuming these factors to be freely correlated. From this initial analysis, factor scores representing each latent variable are generated for each case (e.g., participants). In the second step, the structural equation model is estimated as if the latent variables were observed. The generation of factor scores is a standard option in many statistical packages, making this approach an attractive alternative. This method is very easy to understand and to conduct. However, like other methods in which latent variables are replaced by observed variables, measurement errors are ignored. In particular, because multiple indicators of the latent constructs are not incorporated into the analyses, there is an implicit assumption that the variables are measured without error so that parameter estimates are not corrected for measurement errors. The factor score approach has some of the advantages of fully latent variable approaches to assessing interaction effects (e.g., evaluating the factor structure relating multiple indicators and latent constructs). However, because the final analysis that includes the interaction effect is based on derived scores rather than latent constructs, the approach really is a non-latent variable approach that suffers from most of the problems associated with other non-latent approaches. On this basis, we do not recommend its routine usage except, perhaps, for purposes of preliminary analyses.

Multiple Groups Analysis

When one of the independent variables is a latent variable based on multiple indicators and one is an observed categorical variable that can be used to form a relatively small number of groups, the invariance of the effect of the independent latent variable over the multiple groups provides an effective test of interaction effects (Bagozzi & Yi, 1989; Rigdon, Schumacker, & Wothke, 1998; Vandenberg & Lance, 2000). This method is an extension of the multiple regression approach based on separate groups with observed variables. Specifically, the models with and without constraining the path from the independent latent variable (ξ_1) to the dependent factor η to be equal across the groups of the observed variables (X_2) are compared. Assuming χ_1^2 (with df_1) and χ_2^2 (with df_2) are the chi-square test statistics for the models with and without the imposed constraint, the effects of ξ_1 on η are said to be different, or an interaction exists, if model deterioration is substantial in that $\chi_1^2 - \chi_2^2$ with $df_1 - df_2$ is statistically significant. Before formally comparing these two models, the invariance of the corresponding factor loadings and factor variances across different groups should be examined. Indeed, although beyond the scope of this chapter, the evaluation of measurement invariance and, more generally, the evaluation of goodness of fit using the χ^2 difference test and subjective indexes of fit are important topic (see e.g., Marsh, Muthén, et al., 2009; Vandenberg & Lance, 2000).

When both of the interacting variables are continuous latent variables based on multiple indicators, the multisample approach is no longer appropriate. In particular, it is inappropriate to use values based on one of the continuous variables to form discrete, multiple groups that could be incorporated in the multisample approach. Potential problems include the artificiality or arbitrariness of the groups, the possibly small sample sizes of the different groups, the loss of information in forming discrete groups, the inability to control for measurement error in the grouping variable, the resultant increases in Type II error rates, and the problematic interpretations of the size and statistical significance of interaction effects (see, for example, Cohen & Cohen, 1983; MacCallum et al., 2002; Ping, 1998). On this basis, we do not recommend the multiple group procedure unless one of the interacting variables is an observed naturally occurring categorical variable that can appropriately be represented by a relatively small number of discrete groups.

Fully Latent Variable Approaches Using Structural Equation Analyses

SEM plays a very important role in analyses of interaction effects in which all independent variables are latent constructs inferred from multiple indicators. A variety of such approaches were demonstrated in the monograph edited by Schumacker and Marcoulides (1998). A new generation of approaches was subsequently developed, including the centered constrained approach (Algina & Moulder, 2001), the partially constrained approach (Wall & Amemiya, 2001), the latent moderated structural equation modeling (LMS) approach (Klein & Moosbrugger, 2000), the quasi-maximum likelihood (QML) approach (Klein & Muthén, 2007), the two-step method of moments (2SMM) approach (Wall & Amemiya, 2000), the unconstrained approach (Marsh et al., 2004), Bayesian approaches (Arminger & Muthén, 1998; Lee, 2008; Lee et al., 2007), Blom and Christoffersson's (2001) method based on the empirical characteristic function of the distribution of the indicator variables, Cudeck, Harring, and du Toit's (2009) marginal maximum likelihood estimation, and Mooijart and Bentler's (2010) method based on third-order moments of indicators. In this chapter, we broadly divide the more popular approaches into the product-indicator approaches, the distribution-analytical approaches, and the Bayesian approaches and will discuss each in detail, focusing on the unconstrained approach in the product-indicator methods. We will, however, not cover the other approaches and methods that are either psychometrically weaker (less efficient, biased estimates) or seldom used in applied research.

PRODUCT-INDICATOR APPROACHES

Constrained Approaches

Although there are many approaches to estimating interaction effects between two latent variables, the most typical starting point is to identify a latent product variable with products of the indicators of the original

274 ■ H. W. MARSH et al.

latent variables. Typically these constrained approaches required a large number of parameter constraints that limited their applications. Kenny and Judd (1984) initially proposed this method for a simple model in which the dependent variable y is based on a single observed variable (although we subsequently drop this assumption to allow a latent dependent variable based on multiple indicators). In order to estimate the interaction effects of latent variables ξ_1 and ξ_2 on y, they formulated a latent regression equation with a product term

$$y = \gamma_1 \xi_1 + \gamma_2 \xi_2 + \gamma_3 \xi_1 \xi_2 + \zeta, \tag{7.2}$$

where y is an observed variable with a zero mean, ξ_1 and ξ_2 are latent variables with zero means, $\xi_1 \xi_2$ is the latent interaction term between ξ_1 and ξ_2, and ζ is the disturbance. In Kenny and Judd's demonstration, there were two observed indicators for each latent factor (x_1 and x_2 for ξ_1, and x_3 and x_4 for ξ_2) such that

$$x_1 = \xi_1 + \delta_1 \quad \text{and} \quad x_2 = \lambda_2 \xi_1 + \delta_2;$$

$$x_3 = \xi_2 + \delta_3 \quad \text{and} \quad x_4 = \lambda_4 \xi_2 + \delta_4;$$

where λ_1 and λ_3 were fixed to 1 for purposes of identification. Kenny and Judd suggested using all possible pairs of indicators of each factor to form product variables $x_1 x_3$, $x_1 x_4$, $x_2 x_3$, and $x_2 x_4$ as indicators of $\xi_1 \xi_2$, together with y, x_1, x_2, x_3, and x_4, to estimate the model, although other strategies are possible and, perhaps, preferable (see subsequent discussion on strategies of forming product indicators). However, this approach imposed many nonlinear constraints based on very demanding assumptions about the normality of latent factors that are typically not met (see subsequent discussion of violation of normality assumptions; also see Marsh, Wen, & Hau, 2004). For example,

$$x_2 x_4 = \lambda_2 \lambda_4 \xi_1 \xi_2 + \lambda_2 \xi_1 \delta_4 + \lambda_4 \xi_2 \delta_2 + \delta_2 \delta_4, \tag{7.3}$$

where the loading of $x_2 x_4$ on $\xi_1 \xi_2$ is constrained to be $\lambda_2 \lambda_4$. Altogether, Kenny and Judd posited 15 latent variables (e.g., $\xi_1 \xi_2$, $\xi_1 \delta_4$) and imposed many constraints on factor loadings and variances. They demonstrated their proposed technique using the program COSAN (Fraser, 1980) because it allowed the researcher to specify many nonlinear constraints at a time when this option was not generally available in other SEM packages. Nevertheless, the specification of these constraints was such a tedious task and so prone to error that this approach was rarely used in applied research.

Jaccard and Wan (1995; also see Hayduk, 1987) introduced a latent variable η instead of observed variable y as the dependent variable in their model, thus making the constrained approach a fully latent-variable approach. That is, Equation 7.2 became

$$\eta = \gamma_1 \xi_1 + \gamma_2 \xi_2 + \gamma_3 \xi_1 \xi_2 + \zeta. \tag{7.4}$$

They constrained the residual variance for each observed product term such that, for example, Equation 7.3 was written as

$$x_2 x_4 = \lambda_2 \lambda_4 \xi_1 \xi_2 + \delta_{24}, \tag{7.5}$$

where, under the assumption that x_2 and x_4 are normally distributed,

$$\operatorname{var}(\delta_{24}) = \lambda_2^2 \operatorname{var}(\xi_1)\operatorname{var}(\delta_4) + \lambda_4^2 \operatorname{var}(\xi_2)\operatorname{var}(\delta_2) + \operatorname{var}(\delta_2)\operatorname{var}(\delta_4). \tag{7.6}$$

Figure 7.1 illustrates the interaction model using indicator-products as the indicators of the latent product variable where η has three indicators, ξ_1 and ξ_2 each have two indicators, and two paired indicator-products $x_1 x_3$ and $x_2 x_4$ are used as the indicators of $\xi_1 \xi_2$ (see subsequent discussion on different strategies to form paired indicator products). Note, however, that the necessary constraints on the parameters involving the product terms are not presented in Figure 7.1, but are described in subsequent discussion.

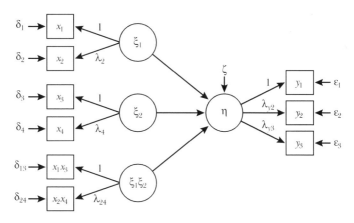

Figure 7.1 The interaction model using indicator products as the indicators of the product latent term. In this interaction model based on the constrained approach, indicator-products are used as the indicators of the latent product variable where η has three indicators, ξ_1 and ξ_2 each have two indicators, and two paired indicator products $x_1 x_3$ and $x_2 x_4$ are used as the indicators of $\xi_1 \xi_2$. ξ_1 and ξ_2 are allowed to correlate. The correlations between $\xi_1 \xi_2$ and its components are fixed to be zero in the constrained approach, but freely to estimated in the unconstrained approach.

Jöreskog and Yang (1996) provided a general model and thorough treatment of this constrained approach. For observed variables that are not mean-centered, they considered the Kenny-Judd model with intercepts in the structural and measurement models:

$$y = \alpha + \gamma_1 \xi_1 + \gamma_2 \xi_2 + \gamma_3 \xi_1 \xi_2 + \zeta \qquad (7.7)$$

where α is the constant term (i.e., the y-intercept of the regression line) and

$$x_1 = \tau_1 + \xi_1 + \delta_1 \quad \text{and} \quad x_2 = \tau_2 + \lambda_2 \xi_1 + \delta_2;$$

$$x_3 = \tau_3 + \xi_2 + \delta_3 \quad \text{and} \quad x_4 = \tau_4 + \lambda_4 \xi_2 + \delta_4. \qquad (7.8)$$

where τs are the item intercept terms. Under the supposition that ξ_1, ξ_2, ζ, and all δ terms are multivariate normal with mean of zero, and each is uncorrelated with the others (except that ξ_1 and ξ_2 are allowed to be correlated), they proposed a model with a mean structure. The mean vector and covariance matrix of ξ_1, ξ_2, and $\xi_1 \xi_2$ are, respectively (also see subsequent discussion of the violations of normality assumptions underlying the use of the constrained approach):

$$\kappa = \begin{bmatrix} 0 \\ 0 \\ \phi_{21} \end{bmatrix}, \quad \Phi = \begin{bmatrix} \phi_{11} & & \\ \phi_{21} & \phi_{22} & \\ 0 & 0 & \phi_{11}\phi_{22} + \phi_{21}^2 \end{bmatrix}. \qquad (7.9)$$

They further noted that even if ξ_1, ξ_2, and ζ are centered so as to have means of zero, $\kappa_3 = \phi_{21} = \text{cov}(\xi_1, \xi_2)$ will typically not be zero. Hence, a mean structure is always necessary, whether or not the independent observed variables include constant intercept terms (see Jöreskog, 1998; Yang, 1998). This implies that the Kenny-Judd (1984) model [or subsequent Jaccard-Wan (1995) model] without a mean structure is inappropriate unless $\text{cov}(\xi_1, \xi_2)$ is very small.

Algina and Moulder (2001) revised the Jöreskog-Yang model so that indicators are mean-centered as in Jaccard and Wan's model and latent variables have a mean structure as in the Jöreskog-Yang model (see Equation 7.9). Although their model is the same as the Jöreskog-Yang model except for mean centering, they found that the revised Jöreskog-Yang model (denoted here as the Algina-Moulder model) was more likely to converge. Even in cases when both models converged, their simulation results still favored their revised model. For this reason, we prefer the Algina-Moulder model among the variety of constrained approaches, and refer to it simply as the *constrained approach.*

The Algina-Moulder model can be characterized as follows: For the sake of simplicity, suppose that each of the latent variables η, ξ_1, and ξ_2 is associated with three indicators, namely, y_1, y_2, y_3; x_1, x_2, x_3; and x_4, x_5, x_6, respectively. Here we assume that we have three indicators per factor (in Figure 7.1 we assumed we had only two indicators per factor for simplicity sake), as it is typically advantageous to have at least three indicators per factor. We subsequently discuss more general strategies for forming product indicators, particularly when each factor has a different number of indicators. Further suppose that all of the independent indicators are centered so that they have zero means. The measurement equations are as below:

$$y_1 = \tau_{y1} + \eta + \varepsilon_1, \; y_2 = \tau_{y2} + \lambda_{y2}\eta + \varepsilon_2, \; y_3 = \tau_{y3} + \lambda_{y3}\eta + \varepsilon_3;$$

$$x_1 = \xi_1 + \delta_1, \; x_2 = \lambda_2\xi_1 + \delta_2, \; x_3 = \lambda_3\xi_1 + \delta_3;$$

$$x_4 = \xi_2 + \delta_4, \; x_5 = \lambda_5\xi_2 + \delta_5, \; x_6 = \lambda_6\xi_2 + \delta_6.$$

Indicators of the dependent variable (y_1, y_2, y_3) can either be mean-centered or not. The intercept terms τ_{y1}, τ_{y2}, and τ_{y3} of dependent indicators, however, must be included even if y_1, y_2, and y_3 are mean-centered, because η typically does not have a zero mean. The structural equation with the interaction term is (reiterating Equation 7.4):

$$\eta = \gamma_1\xi_1 + \gamma_2\xi_2 + \gamma_3\xi_1\xi_2 + \zeta, \tag{7.10}$$

where ξ_1, ξ_2, ζ, all δ terms, and all ε terms are multivariate normal (with mean zero), and are uncorrelated with each other (except that ξ_1 and ξ_2 are allowed to be correlated, and $\xi_1\xi_2$ is nonnormal even if ξ_1 and ξ_2 are normal). Note that when the measurement model for the y variables include the intercept terms (i.e., τ_y terms), the structural Equation 7.10 does not need an intercept term.

Interaction Effects: Constrained Approach

The LISREL specification using 3 matched-pairs for the constrained approach (Algina & Moulder, 2001) is:

$$
\begin{bmatrix} y_1 \\ y_2 \\ y_3 \end{bmatrix} = \begin{bmatrix} \tau_{y1} \\ \tau_{y2} \\ y_{y3} \end{bmatrix} + \begin{bmatrix} 1 \\ \lambda_{y2} \\ \lambda_{y3} \end{bmatrix} \eta + \begin{bmatrix} \varepsilon_1 \\ \varepsilon_2 \\ \varepsilon_3 \end{bmatrix},
$$

$$
\begin{bmatrix} x_1 \\ x_2 \\ x_3 \\ x_4 \\ x_5 \\ x_6 \\ x_1 x_4 \\ x_2 x_5 \\ x_3 x_6 \end{bmatrix}
=
\begin{bmatrix} 1 & 0 & 0 \\ \lambda_2 & 0 & 0 \\ \lambda_3 & 0 & 0 \\ 0 & 1 & 0 \\ 0 & \lambda_5 & 0 \\ 0 & \lambda_6 & 0 \\ 0 & 0 & 1 \\ 0 & 0 & \lambda_8 \\ 0 & 0 & \lambda_9 \end{bmatrix}
\begin{bmatrix} \xi_1 \\ \xi_2 \\ \xi_3 \end{bmatrix}
+
\begin{bmatrix} \delta_1 \\ \delta_2 \\ \delta_3 \\ \delta_4 \\ \delta_5 \\ \delta_6 \\ \delta_7 \\ \delta_8 \\ \delta_9 \end{bmatrix},
$$

$$
\eta = \gamma_1 \xi_1 + \gamma_2 \xi_2 + \gamma_3 \xi_1 \xi_2 + \zeta,
$$

$$
\kappa = \begin{bmatrix} 0 \\ 0 \\ \kappa_3 \end{bmatrix}, \quad
\Phi = \begin{bmatrix} \phi_{11} & & \\ \phi_{21} & \phi_{22} & \\ 0 & 0 & \phi_{33} \end{bmatrix}.
$$

$$
\Theta_\varepsilon = \mathrm{diag}(\theta_{\varepsilon 1}, \theta_{\varepsilon 2}, \theta_{\varepsilon 3}),
$$

$$
\Theta_\delta = \mathrm{diag}(\theta_{\delta 1}, \theta_{\delta 2}, \theta_{\delta 3}, \theta_{\delta 4}, \theta_{\delta 5}, \theta_{\delta 6}, \theta_{\delta 7}, \theta_{\delta 8}, \theta_{\delta 9}).
$$

The necessary constraints are:

(i) $\lambda_8 = \lambda_2 \lambda_5, \lambda_9 = \lambda_3 \lambda_6$;

(ii) $\kappa_3 = \phi_{21}$;

(iii) $\phi_{33} = \phi_{11}\phi_{22} + \phi_{21}^2$;

(iv) $\theta_{\delta 7} = \phi_{11}\theta_{\delta 4} + \phi_{22}\theta_{\delta 1} + \phi_{\delta 1}\theta_{\delta 4}, \theta_{\delta 8} = \lambda_2^2 \phi_{11}\theta_{\delta 5} + \lambda_5^2 \phi_{22}\theta_{\delta 2} + \theta_{\delta 2}\theta_{\delta 5}$,
$\theta_{\delta 9} = \lambda_3^2 \phi_{11}\theta_{\delta 6} + \lambda_6^2 \phi_{22}\theta_{\delta 3} + \theta_{\delta 3}\theta_{\delta 6}$.

It can be seen that these constraints are complicated and cumbersome to specify in the SEM syntax. More importantly, these constraints are based fundamentally on assumptions of the normality of the latent constructs that are seldom met (see subsequent discussion of the violations of normality assumptions).

Unconstrained Approach

Marsh et al. (2004) questioned whether all the constraints in the constrained approach were necessary or even appropriate. More specifically, they proposed an unconstrained approach in which the product of observed variables is used to form indicators of the latent interaction term, as in the constrained approach. Their approach, however, was fundamentally different in that they did not impose any complicated nonlinear constraints to define relations between product indicators and the latent interaction factor, only the mean of the latent product variable (i.e., κ_3) was constrained to be equal to the covariance of the two latent predictor variables $cov(\xi_1, \xi_2)$. They demonstrated that the unconstrained model is identified when there are at least two product indicators of the latent interaction effect.

In the unconstrained approach, the interaction model is the same as that in the constrained approach (when the first-order latent variables are normally distributed). Because the unconstrained approach does not constrain parameters based on assumptions of normality, it provides less biased estimates of the latent interaction effects than the constrained approach under widely varying conditions of non-normality. Importantly, the unconstrained approach is much easier for applied researchers to implement, eliminating the need for the imposition of complicated, nonlinear constraints required by the traditional constrained approaches. However, when the sample size is relatively small and normality assumptions are met, the precision of the unconstrained approach is somewhat lower than that of the constrained approach.

Robustness to Violations of Normality Assumptions in Product Indicator Approaches

There are three fundamentally different problems associated with violations in the assumption of normality that must be considered in the evaluation of the product indicator method to the estimation of interaction effects. Firstly, the maximum likelihood (ML) estimation typically used with each of these approaches is based on the assumption of multivariate normality of the indicators of ξ_1 and ξ_2. This problem is common to all CFA and SEM research based on ML estimation, and is not specific to the evaluation of latent interaction and quadratic effects. Secondly, even when the indicators of ξ_1 and ξ_2 are normally distributed, the distributions resulting from the products of these indicators are known to be non-normal (Jöreskog & Yang, 1996). Hence, this second problem is specific to the evaluation of latent interaction and quadratic effects with ML estimation. Thus, all product-indicator approaches suffer from both of these problems when ML estimation is used. Fortunately, ML estimation tends to be robust to violations of normality in terms of parameter estimates (e.g., Boomsma, 1983; Hau & Marsh, 2004). However, previous research suggests that the ML likelihood ratio test is too large and that ML standard errors are too small under some conditions of non-normality (Hu, Bentler, & Kano, 1992; West, Finch, & Curran, 1995). Although there are alternative estimators that do not assume multivariate normality (e.g., arbitrary distribution function, weighted least squares), simulation studies of latent interaction effects suggest that ML estimation outperforms these alternative estimation procedures under most conditions (Jaccard & Wan, 1995; Wall & Amemiya, 2001; also see Marsh et al., 2004). Nevertheless, it may be appropriate to impose adjustments on the standard errors and χ^2-statistics to correct for this bias (Marsh et al., 2004; Yang-Wallentin & Jöreskog, 2001).

The third problem associated with violations of the assumptions of normality is specific to the constrained approach, but does not apply to the unconstrained approach. As emphasized by Wall and Amemiya (2001) and by Marsh et al. (2004), the nature of the constraints imposed in the constrained approach depends fundamentally on the assumption that ξ_1 and ξ_2 are normally distributed. Importantly, estimates of the interaction effect based on these constraints are not robust in relation to violations of this assumption of normality; neither the size nor even the direction of this bias is predictable a priori, and the size of the bias does not decrease systematically with increasing N. Furthermore, this problem is not a function of the ML estimation procedure, but is inherent to the constrained approach. In contrast to the constrained approach, the unconstrained approach specifically avoids this problem in that it does not impose any constraints based on this assumption of normality.

There is no comprehensive evidence about how serious this third problem must be in order to threaten the validity of interpretations of the constrained approach. However, simulation studies by Wall and Amemiya (2001) and by Marsh et al. (2004) each provided examples in which the bias is substantial. Because most published evaluations of the constrained approach are based on simulated data that are specifically constructed to be multivariate normal, applied researchers have little research evidence upon which to justify the use of the constrained approach unless their data are multivariate normal. In contrast, the unconstrained approaches provide relatively unbiased estimates of the latent interaction effects under the widely varying conditions of non-normality considered by Marsh et al.

(2004). Furthermore, the small biases in non-normally distributed data that were observed became smaller as sample size increased. On this basis, we recommend that the constrained approach should not be used when assumptions of normality are violated.

Strategies in Forming Product Indicators

The determination of the optimal number and the best type of product indicators to be used is an important issue in the analysis of latent interactions using the product indicator models. Particularly when there are multiple indicators of each latent independent variable, there are a variety of different strategies that can be used in the formation of multiple product indicators used to infer latent interaction effects. Existing research, however, is typically based on ad hoc strategies that are, perhaps, idiosyncratic to a particular application, and has not systematically compared alternative strategies.

In a series of simulation studies, Marsh et al. (2004) compared three types of product indicators formed from the three indicators of ξ_1 and the three indicators of ξ_2: (1) all possible products (9 products in this example): $x_1 x_4$, $x_1 x_5$, $x_1 x_6$, $x_2 x_4$, $x_2 x_5$, $x_2 x_6$, $x_3 x_4$, $x_3 x_5$, $x_3 x_6$; (2) matched pair products (three matched pairs in this example): $x_1 x_4$, $x_2 x_5$, $x_3 x_6$; and (3) one pair: $x_1 x_4$. Different combinations of matched pairs and one pair were also compared. Their results showed that the precision of estimation for matched pairs was systematically better than other types of products. Based on their results, they posited two guidelines in the selection of the most appropriate strategies for the estimation of latent interactions:

(i) *Use all of the information*—all of the multiple indicators should be used in the formation of the indicators of the latent variable interaction factor;

(ii) *Do NOT reuse any of the information*—each of the multiple indicators should only be used once in the formation of the multiple indicators of the latent product variable to avoid creating artificially correlated residuals when the same variable is used in the construction of more than one product term.

Importantly, when the indicators are not reused, the variance-covariance matrix of errors is typically diagonal. This subsequently decreases the number of constraints in constrained-type analyses. Although intuitive and supported by their simulation research, more research is necessary to endorse their suggestions fully as guiding principles.

For situations where two latent variables have the same number of indicators (e.g., three indicators of each latent variable as in our simulations), matched pairs are easily formed with arbitrary combinations (or non-arbitrary combinations if there is a natural pairing of indicators of the two latent constructs). What if the number of indicators differs for the two first-order effect factors? Assume, for example, that there were five indicators for the first factor and ten for the second. In this situation, the application of the matched-pair strategy is not so clear-cut. The "do not reuse" principle dictates that we should only have five product indicators of the latent interaction factor, but our "use all of the information" principle dictates that we should have ten product indicators. We suggest several alternative strategies. One approach is to use five indicators and ten indicators, respectively, to define the two first-order effect factors, but to use the best (e.g., most reliable) five indicators of the second factor to form five matched-product indicators of the latent interaction effect. This suggestion suffers in that it does not use all of the information (i.e., only 5 of the 10 indicators of the second factor are used) and might require the researcher to select the best indicators based on the same data used to fit the model, but we suspect that it would work reasonably well so long as there were good indicators of both factors. A second approach is to use the items from the second factor to form item parcels for the calculation of product indicators. Thus, for example, the researcher could form five (item pair) parcels by taking the average of the first two items to form the first item parcel, the average of the second two items to form the second parcel, etc. The latent interaction variable would then be defined in terms of 5 matched-product indicators that each consisted of an item representing the first factor and an item-parcel representing the second factor. This strategy has the advantage of satisfying both of our principles. Simulation work by Jackman, Leite, and Cochrane (2011) showed that the parcelling approach and using as many matched product-indicators as there are items in the shorter scale yield comparable results. However, the use of item parcels instead of single indicators to represent the second factor is only justified when the underlying construct is strictly unidimensional and might cause further problems (see Marsh, Lüdtke et al., in press). Suggestions such as those outlined here are consistent with our empirical results and proposed guidelines, but further research is needed to evaluate their appropriateness more fully.

Indicator Centering and Mean Structure in Product Indicator Method

Centering the indicators and including the mean structure are important issues in the computation of the appropriate solutions in product indicator models. Here, without loss of generality, we use a simple example to illustrate how these strategies affect model and parameter specification in the case of the unconstrained approach. Assume that the endogenous latent variable η has three indicators (y_1, y_2, y_3), and the exogenous latent variables ξ_1 and ξ_2 have two indicators each $(x_1, x_2$ and x_3, x_4, respectively). The indicators of the product term $\xi_1\xi_2$ are the matched pairs (i.e., x_1x_3 and x_2x_4).

When uncentered indicators are used, the measurement equations of the product indicators are very complicated involving intercept, product, and error terms. Thus, for example,

$$x_2x_4 = (\tau_2 + \lambda_2\xi_1 + \delta_2)(\tau_4 + \lambda_4\xi_2 + \delta_4)$$

$$= \tau_2\tau_4 + \tau_4\lambda_2\xi_1 + \tau_2\lambda_4\xi_2 + \lambda_2\lambda_4\xi_1\xi_2 + \delta_{24},$$

where $\delta_{24} = \tau_4\delta_2 + \lambda_4\xi_2\delta_2 + \tau_2\delta_4 + \lambda_2\xi_1\delta_4 + \delta_2\delta_4$ denotes the error term of x_2x_4. The loadings of x_2x_4 on ξ_1 and ξ_2 are generally not zero. Importantly, a mean structure is always needed for the structural and measurement models. As pointed out by Jöreskog and Yang (1996), even if ξ_1, ξ_2, and ζ are centered so as to have means of zero, $\kappa_3 = \mathrm{E}(\xi_1\xi_2) = \mathrm{cov}(\xi_1,\xi_2)$ will typically not be zero.

Consequently, as in the case in the analyses of interaction in multiple regression, centering the observed variables is routinely carried out to simplify the model (Aiken & West, 1991; Algina & Moulder, 2001; Marsh et al., 2004). Let us denote the mean-centered variable of x as $x^C \left[x^C = x - \mathrm{E}(x) \right]$, the measurement models in Equation 7.8 become

$$x_1^C = \xi_1 + \delta_1, x_2^C = \lambda_2\xi_1 + \delta_2;$$
$$x_3^C = \xi_2 + \delta_3, x_4^C = \lambda_4\xi_2 + \delta_4. \tag{7.11}$$

When the indicators of the latent predictors are centered, the intercept terms in the measurement equations of the latent predictors and products indicators (i.e., $x_1^C x_3^C$ and $x_2^C x_4^C$) are no longer needed, whereas those of the outcome variables are still necessary even though they have been centered (see Algina & Moulder, 2001; Marsh et al., 2004). Moreover, the latent mean of $\xi_1\xi_2$ has to be included in the model because even if ξ_1 and ξ_2 are centered (means of zero), $\mathrm{E}(\xi_1\xi_2) = \mathrm{cov}(\xi_1,\xi_2)$ will typically not be zero (see Algina & Moulder, 2001; Marsh et al., 2004).

In most popular SEM software, though specification of the mean structure is not difficult, exploration of simpler approaches continues. One approach to remove the mean structure and thus simplify the model (Marsh et al., 2007) is to orthogonalize the interaction term by regressing the product term $\xi_1\xi_2$ on both ξ_1 and ξ_2 (Little, Bovaird, & Widaman, 2006; Marsh, Wen, Hau, Little, Bovaird, & Widaman, 2007). The product indicators $(x_1x_3$ and x_2x_4 in our example) based on the raw indicators are first regressed separately on the full set of indicators $(x_1, x_2, x_3, x_4$ in our example) and the residuals of this regression (denoted as $O\_x_1x_3$, $O\_x_2x_4$ respectively) are then used as indicators for the latent residualized product variable

$$O\_\xi_1\xi_2 = \xi_1\xi_2 - (\beta_0 + \beta_1\xi_1 + \beta_2\xi_2).$$

The orthogonalization method has a number of limitations in that it may affect the first-order effects, involve a cumbersome two-step procedure, and result in a non-random bias for each indicator of the interaction construct due to the consistency problem unless the third-order moments of the latent predictor variables are zero (Lin, Wen, Marsh, & Lin, 2010). Non-normal data are very common (Micceri, 1989; Pyzdek, 1995), and when the distribution of (ξ_1,ξ_2) is not bivariate normal, the third-order moments of (ξ_1,ξ_2) are not zero.

The problems can be solved by the double-mean-centering strategy proposed by Lin et al. (2010). The indicators of the latent variables are first mean-centered (Equation 7.11) before using them as indicators of the original latent variables and forming the matched product indicators for the latent interaction variable. The matched product indicators $x_1^C x_3^C$ and $x_2^C x_4^C$ are then mean-centered again (i.e., double-mean-centering) and denoted as $(x_1^C x_3^C)^C$ and $(x_2^C x_4^C)^C$, respectively. From Equation 7.11 we have

$$(x_2^C x_4^C)^C = x_2^C x_4^C - \mathrm{E}(x_2^C x_4^C) \tag{7.12}$$
$$= (\lambda_2\lambda_4\xi_1\xi_2 + \lambda_2\xi_1\delta_4 + \lambda_4\xi_2\delta_2 + \delta_2\delta_4) - \lambda_2\lambda_4\mathrm{E}(\xi_1\xi_2)$$
$$= \lambda_{24}[\xi_1\xi_2 - \mathrm{E}(\xi_1\xi_2)] + \delta_{24},$$

where

$$\delta_{24} = \lambda_2\xi_1\delta_4 + \lambda_4\xi_2\delta_2 + \delta_2\delta_4$$

has a zero mean. The latent interaction variable

$$\xi_1\xi_2 - \mathrm{E}(\xi_1\xi_2)$$

also has a mean at zero (see Equation 7.12). When these mean centered and double-mean-centered variables

$$x_1^C, x_2^C; \; x_3^C, x_4^C; \; (x_1^C x_3^C)^C, (x_2^C x_4^C)^C$$

are used as the indicators to the latent interaction model, actually it becomes the analysis of the following structural model

$$\eta = \gamma_1\xi_1 + \gamma_2\xi_2 + \gamma_3[\xi_1\xi_2 - \mathrm{E}(\xi_1\xi_2)] + \zeta, \tag{7.13}$$

suggesting that the mean structure is not necessary with mean centering of indicators of latent variables and double-mean-centering of indicators of latent interaction variable.

Noticing that some statistical packages (e.g., LISREL) automatically center the indicators if there is no mean structure in the model, Wu, Wen, and Lin (2009) confirmed that the single- and double-mean-centered indicators would produce identical results for the first-order and interaction effects if no mean structure were included in the model. Operationally, researchers adopting the product indicator approach should mean-center all indicators, form product indicators from these centered-indicators, and then fit the latent model without the necessity of imposing a mean structure. For some SEM software such as Mplus, in which a mean structure is included automatically, double-mean-centering of indicators is not necessary.

Appropriate Standardized Solution in Product-Indicator Models

In multiple regression analyses, standardization of the effects so that they represent relations between variables that have a mean of zero and a variance of unity (*z*-score) helps in the comparison of these effects where the original scales differ. Similarly in SEM, it is customary to report completely standardized solutions in which both latent factors and observed variables are standardized. The appropriate standardized solution for an interaction model with the product indicator approach, however, is not typically provided by commercial SEM package. Indeed, the so-called standardized solutions that are provided are typically wrong and should not be used. For simplicity, in the discussion below, we refer to the parameter estimates in the model as the *solution* to the set of standardized parameter estimates from the statistical package as *usual standardized solution*, and to the statistically appropriate standardized solution as the *appropriate standardized solution*.

For ease of understanding, let us examine the appropriate standardized solution with observed variables in multiple regression (Equation 7.1 repeated here for ease of reference):

$$Y = \gamma_0 + \gamma_1 X_1 + \gamma_2 X_2 + \gamma_3 X_1 X_2 + e \tag{7.14}$$

The usual standardized solution from the following equation,

$$Z_Y = \gamma_1' Z_{X_1} + \gamma_2' Z_{X_2} + \gamma_3' Z_{X_1 X_2} + e' \tag{7.15}$$

where Z_{X_1}, Z_{X_2}, and $Z_{X_1 X_2}$ are the standardized forms of X_1, X_2, and $X_1 X_2$, is not appropriate because $Z_{X_1 X_2}$ does not represent the product of Z_{X_1} and

Z_{X_2}. Friedrich (1982) provided a simple way to derive the appropriate standardized solution from the usual standardized solution in the interaction model. Specifically, (i) uncentered observed variables Y, X_1, X_2 are standardized to Z_Y, Z_{X_1}, Z_{X_2}, (ii) the product term $Z_{X_1}Z_{X_2}$ is formed, and (iii) the unstandardardized solution from the following equation is the appropriate standardized solution for the model in Equation 7.14:

$$Z_Y = \gamma_0'' + \gamma_1'' Z_{X_1} + \gamma_2'' Z_{X_2} + \gamma_3'' Z_{X_1} Z_{X_2} + e''. \tag{7.16}$$

In contrast to Equation 7.15, we have an intercept term γ_0'' in Equation 7.16, which is necessary. Generally this term is not equal to zero because even if Z_{X_1} and Z_{X_2} are zero, the mean of their product term is generally not zero unless the two terms are uncorrelated.

Wen, Marsh, and Hau (2010; also see Wen, Hau, & Marsh, 2008) derived a way to compute an appropriate standardized solution for latent interaction models. The appropriate standardized first-order effects γ_1' and γ_2' and the interaction effect γ_3' can be calculated from the respective usual standardized effects γ_1, γ_2, and γ_3 provided by many SEM software packages using the following equations:

$$\gamma_1' = \gamma_1, \ \gamma_2' = \gamma_2, \ \gamma_3' = \gamma_3 \frac{\sqrt{\phi_{11}\phi_{22}}}{\sqrt{\phi_{33}}} \tag{7.17}$$

where $\phi_{11} = \mathrm{var}(\xi_1)$, $\phi_{22} = \mathrm{var}(\xi_2)$, $\phi_{33} = \mathrm{var}(\xi_1\xi_2)$ are from the unstandardized solution (Wen et al., 2010). Whereas the main (first-order) effects of the appropriate standardized solution are identical to the usual standardized coefficients, the coefficient for the interaction term is not and has to be derived from Equation 7.17.

Related to the issue of obtaining the appropriate standardized solution is the appropriateness of the standard error of the appropriately standardized interaction effect that is used for computing confidence intervals as well as testing statistical significance. In SEM models without a latent interaction, the corresponding z-statistics of the unstandardized and standardized solutions are the same (Cudeck, 1989), as a parameter estimate and its standard error estimate are both rescaled by an identical factor. Thus, the statistical significance of the standardized effects can be conveniently determined using the unstandardized estimates.

To investigate whether the standard errors in models involving latent interactions are appropriate, Wen et al. (2010) used the bootstrap method and compared the appropriate standard errors (obtained from bootstrap) against the estimated standard errors (from SEM software). The estimated standard errors from SEM software and their associated z-values would be useful only if the standard errors from the two approaches (bootstrap and estimates from SEM software) are similar or identical. Their simulation showed minor differences between the z-values associated with the SEM software original estimates and those with the appropriate standardized estimates (using bootstrap). In most empirical studies, the usual standard errors obtained from SEM software are trustworthy. However, they should be accompanied by bootstrap standard errors when the associated z-values are close to the cut-off at the targeted significance level.

An Example Using the Unconstrained Approach

Here we use an example to illustrate the complete process in the unconstrained product-indicator method. In order to provide a context for the example, assume that η is mathematics achievement, ξ_1 is prior mathematics ability, ξ_2 is mathematics motivation, and $\xi_1\xi_2$ is the interaction of prior mathematics ability and mathematics motivation. The a priori prediction in this hypothetical example is that subsequent mathematics achievement is a function of not only math ability and math motivation, but also their positive interaction. That is, high ability contributes more to subsequent achievement for students who are also highly motivated or, equivalently, high motivation contributes more to subsequent achievement for students who are also have high ability.

In brief, the effect of the interaction between ξ_1 and ξ_2 on η is examined in the model:

$$\eta = \gamma_1\xi_1 + \gamma_2\xi_2 + \gamma_3[\xi_1\xi_2 - E(\xi_1\xi_2)] + \zeta$$

where the endogenous latent variable η is measured by y_1, y_2, and y_3, and the exogenous latent variables ξ_1 and ξ_2 are measured by x_1, x_2, x_3 and x_4, x_5, x_6, respectively. The sets of independent indicators x_1, x_2, x_3 and x_4, x_5, x_6 are mean-centered. We use a sample of simulated data generated from the population as described in Study 1 of Marsh et al. (2004).

All observed indicators y_1, y_2, y_3; x_1, x_2, x_3; and x_4, x_5, x_6 were first centered to have zero mean (but we still use the original notation for simplicity), then product indicators x_1x_4, x_2x_5, x_3x_6 were formed. The covariance matrix of the nine indicators and the three (matched-pair) product indicators are available from Appendix A with LISREL syntax to specify the unconstrained approach without mean structure.

The unstandardized estimates from the LISREL program are $\gamma_1 = 0.429$, $\gamma_2 = 0.319$, $\gamma_3 = 0.137$; $\hat{\phi}_{11} = 0.951$, $\hat{\phi}_{22} = 1.106$, and $\hat{\phi}_{33} = 1.223$, while the commonly provided standardized estimates from the same output are $\hat{\gamma}_1 = 0.423$, $\hat{\gamma}_2 = 0.338$, and $\hat{\gamma}_3 = 0.153$. Substituting these figures into Equation 7.17, we obtain the appropriate standardized effects $\gamma_1'' = 0.423$, $\gamma_2'' = 0.338$, and $\gamma_3'' = 0.142$, as seen in Figure 7.2. The appropriate standard errors and thus appropriate z-values (7.295, 5.460, and 2.395 for $\hat{\gamma}_1''$, $\hat{\gamma}_2''$, and $\hat{\gamma}_3''$, respectively) are obtained from 800 bootstrap samples each with sample size of 500 generated by PRELIS 2.72 (see the PRELIS syntax in Appendix B).

Substantively, the statistically significant first-order and interaction effects can be interpreted as follows: Students will have higher mathematics achievement when they have higher prior mathematics ability ($\gamma_1'' = 0.423$) or stronger mathematics motivation ($\gamma_2'' = 0.338$). The (appropriately standardized) effect of prior mathematics ability on mathematics achievement is $0.423 + 0.142 \times$ motivation, particularly stronger when students have stronger motivation. Alternatively, the interaction can also be interpreted such that the effect of mathematics motivation on mathematics achievement (i.e., $0.338 + 0.142 \times$ ability) is particularly stronger for students with better prior mathematics ability because $\gamma_3'' = 0.142$.

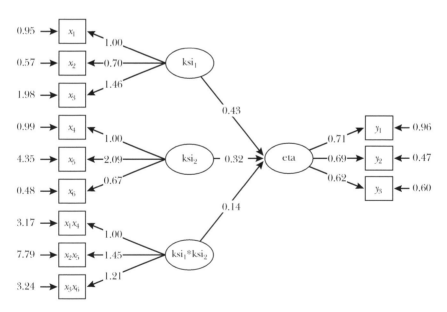

Figure 7.2 Path diagram and estimations of the unconstrained interaction model. Although not shown in the figure, the covariances between ξ_1, ξ_2, and $\xi_1\xi_2$ (i.e., ϕ_{21}, ϕ_{31}, and ϕ_{32}) are estimated freely. Values of ϕ_{31} and ϕ_{32} could be fixed to zero for known normal data.

DISTRIBUTION-ANALYTIC APPROACHES

So far, our main focus has been on methods that identified a latent product variable with multiple indicators that are cross-products of the original indicators of the main latent variables and issues related to this method. One advantage of these approaches is that they can be easily implemented in conventional SEM software.

In contrast to the product-indicator approaches, the distribution-analytic approaches (Cudeck et al., 2009; Klein & Moosbrugger, 2000; Klein & Muthén, 2007) are specialized alternatives for non-linear SEM models which explicitly model the non-normal distribution of the manifest indicator variables implied by the latent interaction effects and require specific software implementations. Currently two related methods are available. The Latent Moderated Structural Equations (LMS) approach (Klein & Moosbrugger, 2000) is implemented in Mplus (Muthén & Muthén, 1998–2010). The subsequent Quasi-Maximum Likelihood (QML) approach (Klein & Muthén, 2007) is still not incorporated in any commercial SEM package but a stand-alone program is available from its first author, Andreas Klein. Two other newly developed alternatives are based on the marginal maximum likelihood estimation (see Cudeck et al., 2009) and on modeling selected third-order moments with a method-of-moments approach (Mooijaart & Bentler, 2010). While marginal maximum likelihood estimation has not yet been implemented in widely available software, Mooijaart and Bentler's approach is available in EQS (Bentler, 2000–2008).

Distribution-analytic approaches directly estimate a structural equation model with a general quadratic form of latent independent variables as

$$\eta = \alpha + \Gamma\xi + \xi'\Omega\xi + \zeta, \tag{7.18}$$

where η is a latent dependent variable, α is an intercept term, ξ is a $(k \times 1)$ vector of latent predictor variables, Γ is a $(1 \times k)$ coefficient matrix, Ω is an upper triangular $(k \times k)$ coefficient matrix, and ζ is a disturbance variable. The usual interactions between two latent variables and the quadratic effect of one latent variable, discussed later, are special cases of Equation 7.18. For example, when

$$\Gamma = (\gamma_1, \gamma_2), \xi = (\xi_1, \xi_2)', \text{ and } \Omega = \begin{bmatrix} 0 & \gamma_3 \\ 0 & 0 \end{bmatrix}$$

then Equation 7.18 becomes Equation 7.4. Importantly, for structural models involving a latent interaction as presented in Equation 7.7 or other latent quadratic effects, both QML and LMS can estimate all model parameters without the need to use product indicators. More detailed description of both methods and current reviews are accessible (e.g., Kelava et al., 2011) and we provide only a brief summary here.

Distribution-analytic approaches estimate latent non-linear effects by directly modeling the nonnormality in the indicators of the outcome variables that is implied by latent interaction effects. They differ in the assumption they are making about this nonnormality, as we discuss below. The distribution-analytic approaches usually still assume that the indicators $\mathbf{x}$ of the latent predictor variables ξ, the structural disturbance term ζ, and all errors in the measurement model are normally distributed. In structural models without non-linear effects, the indicators $\mathbf{y}$ of the latent outcome variable are assumed to be unconditionally normally distributed in models. In contrast, in distribution-analytic approaches for models with non-linear effects, these indicators $\mathbf{y}$ of the latent outcome variable are assumed to be normally distributed only conditionally on the latent outcome variable η and the latent predictor variables ξ.

In LMS, the non-normal distribution of the latent outcome variable η is approximated with a finite mixture of normal distributions (Klein & Moosbrugger, 2000). The coefficients for the first-order and interaction effects, resting on the above assumptions, are then obtained with the expectation maximization algorithm (Dempster, Laird, & Rubin, 1977). This is achieved by a numerical integration technique which is needed to estimate the finite mixture distribution of the latent outcome variable and its indicators (Klein & Moosbrugger, 2000).

Simulation studies show that LMS provides efficient parameter estimators and a reliable model difference test, and that standard errors are relatively unbiased (Klein & Moosbrugger, 2000; Schermelleh-Engel, Klein, & Moosbrugger, 1998). Although models with multiple latent product terms can be analyzed with the LMS method, the method can become computationally very intensive for models with more than one product term or a large number of nonlinear effects in the structural equation and many indicators of latent variables.

The QML method for structural equation models with quadratic forms was developed for the efficient estimation of more complex nonlinear structural equation models which cannot be analyzed with the LMS method because of the computational burden involved in a more complex nonlinear model structure (Klein & Muthén, 2007). Though similar to the LMS approach that can directly model Equation 7.4 without constructing indicators for the latent interaction variable, it is based on a different and less restrictive set of distributional assumptions (Klein & Muthén, 2007). In contrast to the LMS approach that assumes that the distribution of the latent outcome variable η and its indicators can be approximated by finite—and potentially large—number of normal mixture components, QML simplifies and reduces the number of components representing the distribution of the indicators to two. It was based on an approximation of the nonnormal density function of the joint indicator vector by a product of a normal and a conditionally normal density. All other assumptions about the normality of the latent predictor variables and of residuals in the structural and measurement model still apply.

Operationally, the parameter estimates are obtained by the maximization of a quasi-log likelihood function using the Newton-Raphson algorithm. However, the QML approach still relies on the assumption that the first-order effect factors are normally distributed, which is generally violated when the measured variables are not normally distributed. However, recent simulation studies (Klein & Muthén, 2007) showed that QML effects and standard errors are sufficiently robust to violations of normality. Comparison studies between the LMS and QML methods (Kelava et al., 2011; Klein & Muthén, 2007) showed that QML is more robust against violations of the normal distribution of the indicator variables and residuals than LMS, but also slightly less efficient if the distributional assumptions of LMS are fulfilled. The biggest advantage of QML over LMS is its higher computational efficiency that makes more feasible the estimation of complex models with a larger number of interaction and non-linear effects.

In sum, except in cases where nonnormality is extreme, the tradeoff between the small biases in QML estimates appeared to be offset by its greater precision. However, at least for now, QML is not implemented in any commonly used structural equation modeling package, but is only available in a stand-alone software with limited capacities. Nevertheless, in the near future, the QML approach is likely to become an attractive choice among the various approaches for analyzing interaction and quadratic effects. An example analysis and syntax for both LMS and QML are available from the authors.

Comparison to Product-Indicator Approaches

As we have briefly discussed earlier, the product-indicator approaches and the distribution-analytic approaches have some substantial differences in their applications and assumptions:

1. In the product-indicator approaches, the latent interaction variable is constructed and identified with product indicators from indicators of the original latent variables. As outlined above such product indicators are not needed in the distribution-analytic approaches.
2. In contrast to product-indicator approaches which assume normality of the latent variables and their indicators, in the distribution-

analytic approaches the non-normality of the indicators and the dependent latent variable that arises due to the presence of interaction effects is explicitly taken into account in the maximization of the special fitting functions. Nevertheless, the models still rest on the assumption that the latent predictor variables and their indicators are normally distributed and on the assumption about the latent outcome variable and its indicators that are based on mixtures or conditionally normal distributions.

When QML's distributional assumptions are met, that is, when latent variable indicators and the residuals are normally distributed, the interaction effect estimates using QML are unbiased and are more precise than those using the unconstrained product-indicator approach (Klein & Muthén, 2007; Marsh et al., 2004). Even when the distributional assumptions are violated, the estimates from the currently available QML-program were slightly more efficient and the power to detect interaction effects with a likelihood ratio test was higher than with other methods. The standard error of the interaction by the QML approach tends to be slightly overestimated, while that by the unconstrained approach is slightly underestimated (Klein & Muthén, 2007). Results are similar in simulation studies that include both latent interaction and latent quadratic effects (Kelava et al., 2011).

The theoretical advantages of the distribution-analytic approaches, however, are somewhat offset by the need to use specialized software in estimation and the sometimes prohibitive computational demands. In contrast, the unconstrained approach can be implemented in most SEM packages, including the freely available R-packages (Fox, 2006; Kelava & Brandt, 2009) and OpenMx (Boker et al., 2010).

The distribution-analytic approaches do not routinely have a properly defined null model and thus will not provide any general fit statistics. (Kelava et al., 2011, discussed a new index of fit for the QML approach, but noted that it had not yet been investigated.) Nevertheless, chi-square-difference tests can be used for nested models while information criteria such as the AIC and the BIC can be used to compare non-nested models. Also noteworthy is that the QML-program provides a fully standardized solution, though without standard errors, by assuming all manifest and latent variables to be standardized. Within LMS, a fully standardized solution can be obtained when the latent predictor variables are identified by setting their variance to be equal to one and estimate all factor loadings freely. These estimates are equivalent to the appropriate standardized solutions discussed earlier in the product-indicator models.

BAYESIAN APPROACH

With the flexibility and accessibility of popular software (e.g., WinBUGS, see Spiegelhalter, Thomas, Best, & Nunn, 2004; Mplus, see Muthén & Muthén 1997–2010), Bayesian approaches to SEM are gaining popularity (see chapter by Levy and Choi, this volume). Importantly, the analyses of latent interaction from a Bayesian approach as developed by Lee et al. (2007) is fundamentally different from the likelihood based approaches discussed so far (also see Lee, 2008, for a thorough introduction into Bayesian SEM).

In the Bayesian approach, all parameters are assumed to be random variables and their distribution conditional on prior information and the data modeled (Gelman, Carlin, Stern, & Rubin, 2004). In contrast to the conventional frequentist approaches, the Bayesian approach can explicitly incorporate prior knowledge about parameter values into the model, though it is also possible to specify models without referring to prior information by using so-called non-informative priors. The prior information is then combined with the data, which is then summarized as the posterior distributions of all model parameters. The data have a greater influence on the final solution than the prior distribution when the sample size is large, while the choice of the prior distribution will have a bigger impact on the posterior parameter estimation in smaller samples. Logically, if only non-informative priors are provided, then the final parameter estimates will be a function of the data, in which case the Bayesian approaches converge into the conventional ML estimation methods.

As posterior distributions of the parameters are often analytically intractable, parameter estimates and their distribution cannot be obtained with conventional estimation methods. To solve the problem, the Markov chain Monte Carlo (MCMC) method (e.g., Geman & Geman, 1984; Hastings, 1970) is commonly used as an alternative for estimating Bayesian models in general, and for Bayesian structural equation models in particular. In these simulation-based procedures, large numbers of random draws from the corresponding conditional distributions are used to approximate the posterior distributions of the parameters. The central tendency (e.g., median or mean) and dispersion (e.g., SD) of these simulated distributions are used to make inferences about the parameters (Lee, 2008).

In contrast to the product-indicator model, but similar to the distribution-analytic approaches discussed earlier, latent interactions with the Bayesian approach do not involve the formation of product-indicators for the latent product variable. Operationally in the Bayesian approach, simulated values of the latent variables at each step of the MCMC-chain are drawn from their posterior distribution, which would then be used to calculate their cross-product. These procedures automatically take into account the nonnormal distribution of the latent outcome variable and its indicators.

Promising results have been obtained in simulation studies with Bayesian approach to non-linear structural equation models, especially in small samples (e.g., Lee et al., 2007). Inexperienced applied users, however, might find the specification of Bayesian models very demanding as they require sufficient statistical knowledge and careful consideration of the distributions of all model parameters and their priors. Particularly with small samples, it is advisable to pay extra attention and conduct sensitivity analyses with different priors.

The great flexibility of the Bayesian approach has its merits and disadvantages. On the one hand, theoretically the flexibility allows the applied researcher to extend the model not only to latent interactions, but also quadratic, higher-order interactions involving more than two latent variables, and other polynomial effects as well. These non-linear functions can be included and implemented as part of the structural model. On the other hand, the statistical knowledge for the specification of models and priors may be overwhelmingly difficult for most applied users. Furthermore, while Mplus (Muthén & Muthén, 1998–2010) provides a seemingly more user-friendly implementation of Bayesian SEM, currently it allows the specification of linear structural models only. The free WinBUGS software (Spiegelhalter et al., 2004), generally more statistically demanding to novice users, is the only widely available program that allows the implementation of the Bayesian approach for latent interaction. An example analysis and syntax using the Bayesian approach to latent interactions by Lee et al. (2007) as implemented in WinBUGS is available from the authors.

QUADRATIC AND NONLINEAR EFFECTS: UNCONSTRAINED AND OTHER APPROACHES

For observed variables, it is well known that unmodeled quadratic effects may be mistaken and misinterpreted as a significant interaction, particularly when predictors are correlated and have large measurement errors (Ganzach, 1997; Kromrey & Foster-Johnson, 1999; Lubinski & Humphreys, 1990; MacCallum & Mar, 1995). In structural equation models, starting with the early work of Kenny and Judd (1984), quadratic effects were often considered together with interaction effects because of the nonlinearity of latent factors, and this continues to be the case (e.g., Klein & Muthén, 2007; Klein et al., 2009; Laplante, Sabourin, Cournoyer, & Wright, 1998; Wall & Amemiya, 2001). Here we briefly introduce a simple latent quadratic model (see earlier discussion of substantive examples whereby this quadratic model might be used) and describe the application of the constrained, unconstrained, and other approaches to this issue. More generally, we stress that much less research has been conducted to evaluate the generalizability and robustness of approaches used to estimate interaction effects that might be confounded with polynomial effects.

In comparison, to test quadratic effects with observed variables is relatively easy. What is needed is the inclusion of a squared term of the targeted predictor. For SEM analyses involving latent quadratic terms, issues similar to those in latent interactions have to be considered. Similar to the latent interaction model without mean structure, the simple quadratic model is

$$\eta = \gamma_1\xi + \gamma_2[\xi^2 - E(\xi^2)] + \zeta \qquad (7.19)$$

with assumptions that all indicators are mean-centered and then the squared indicators are centered again (double-mean-centering).

For example, suppose that both latent variables η and ξ have three indicators, namely, y_1, y_2, y_3 and x_1, x_2, x_3, respectively. All indicators, y_1, y_2, y_3, x_1, x_2, x_3, are first centered to have zero mean, then squared indicators x_1^2, x_2^2, and x_3^2 are formed, and mean-centered again (this step is not necessary for LISREL because it will automatically center all indicators for a model without mean structure). The path diagram is shown in Figure 7.3. The LISREL syntax by unconstrained approach for the corresponding latent quadratic model is given in Appendix C.

We illustrate the unconstrained approach to quadratic effects with a simulated data example. In order to give some substantive context, consider the substantive issue of the relation between university students' evaluations of teaching effectiveness and student perceptions of course workload/difficulty (see Marsh, 2001). Let y_1, y_2, and y_3 be three rating items completed by students that assess the overall effectiveness of the teacher, whereas x_1, x_2, and x_3 are three indicators of students' perceptions of the course workload/difficulty. The means and SDs of simulated data are: y_1 (3.810, 0.833), y_2 (3.844, 0.813), y_3 (3.723, 0.834); x_1 (3.554, 1.038), x_2 (3.462, 0.962), x_3 (3.666, 0.991). The

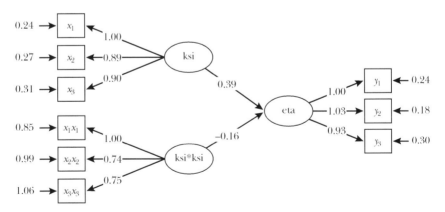

Figure 7.3 Path diagram and the estimations of the unconstrained quadratic model. Although not shown in the figure, the covariance between ξ and ξ^2 (i.e., ϕ_{21}) is estimated freely.

sample size N is 852. Independent indicators x_1, x_2, and x_3 are centered to have zero means.

The covariance matrix and mean vector of the six indicators and three squared-indicators are presented as part of the LISREL syntax (see Appendix C) and the path model and parameter estimates are shown in Figure 7.3. The estimated linear effect is $\hat{\gamma}_1 = 0.391$ (SE = 0.028, $z = 14.20$). The estimated quadratic effect is $\hat{\gamma}_2 = -0.164$ (SE = 0.023, $z = -7.21$). Both linear and quadratic effects are significant. The squared multiple correlation for the structural equation is 0.399. The model fits the data very well: $\chi^2(37) = 79.4$, RMSEA = 0.033, SRMR = 0.049, NNFI = 0.988, and CFI = 0.987.

In order to explain the model, we should leave all variables in their original metric. We then specified η to have the same units and mean as y_1 (i.e., $y_1 = \eta + \varepsilon_1$), and ξ as x_1 (i.e., $x_1 = \xi + \delta_1$). Thus, η and ξ in Equation 7.19 should be replaced by $\eta - 3.810$ and $\xi - 3.554$, respectively. Hence, we can write the estimated quadratic equation as:

$$\eta - 3.810 = 0.391(\xi - 3.554) - 0.164(\xi - 3.554)^2;$$

that is,

$$\eta = -0.164\xi^2 + 1.557\xi + 0.349.$$

The results indicate that both the linear and quadratic components of workload/difficulty are statistically significant. The linear effect is positive whereas the negative quadratic component represents an inverted U-shaped function. The hypothetical results (reflecting actual results from Marsh, 2001) indicate that student ratings of teaching effectiveness improve with increases in workload/difficulty for most of the range of workload/difficulty. However, using derivatives (i.e., setting $d\eta/d\xi$ to 0 and solving) we see that when

$$\xi = -\frac{1.557}{2\times(-0.164)} = 4.75$$

(near the top of the range of workload/difficulty) there is a maximum where the ratings plateau ($\eta = 4.04$), and then decrease slightly for courses that are perceived to be excessively difficult.

Analyses of quadratic effects with the distribution-analytic approaches are much simpler with both LMS and QML allowing the simultaneous estimation of interaction and quadratic effects in the same model (Klein & Moosbrugger, 2000; Klein & Muthén, 2007). The limited research suggests that the unconstrained product-indicator and distribution-analytic approaches all provide unbiased estimators for quadratic and interaction effects when normality assumptions for latent variable indicators are fulfilled (Kelava & Brandt, 2009; Kelava et al., 2011; Moosbrugger, Schermelleh-Engel, Kelava,

& Klein, 2009). Particularly for data that might violate the assumption of multivariate normality, following the advice for analyses with manifested variables (Ganzach, 1997), it may be desirable to routinely include latent quadratic terms in analyses of latent interaction effect (Klein et al., 2009).

Although our main focus has been on estimating interaction effects in this chapter, some additional concerns exist when these approaches are used to estimate quadratic effects. As is the case with the estimation of quadratic effects with manifest variables (e.g., Cohen & Cohen, 1983), there are substantial problems of multicollinearity in that the same variables are used to infer the linear and quadratic effects. In the latent variable approach, this is likely to result in less precise estimates and negatively affect the convergence behavior of the models used to estimate the quadratic effects. This problem of multicollinearity is countered substantially by mean-centering each of the indicators. Nevertheless, there has been insufficient evaluation of various approaches to the estimation of quadratic effects with either simulated data in Monte Carlo studies or real data in substantive applications. Hence, the applied researcher must apply appropriate caution in the interpretation of results based on these techniques.

For higher-order polynomials (such as cubic effects) or higher-order interactions that involve more than two latent variables, the adverse effects due to measurement error will be much larger than those with simple latent interaction because these errors aggregate multiplicatively. Research in these polynomials and higher-order interactions, however, is limited. Worth noting, currently the specifications of these polynomials and higher-order interactions are difficult with product-indicator models, impossible with the distribution-analytic methods, but possible and reasonably straightforward within the Bayesian framework.

Developments in semi-parametric approaches that model complex non-linear relations between latent variables as a latent variable mixture model offer a promising alternative to parametric models of non-linear relations (e.g., Bauer, 2005; Bauer & Curran, 2004; Pek, Sterba, Kok, & Bauer, 2009). Here, mixture components are used as a computational tool to approximate the non-linear regression function with linear within-group regression (see Bauer, 2005; Dolan & van der Maas, 1998) and the R program may be used to summarize and visualize the findings (Pek et al., 2009). Refinements to make it easier to use and extensions of this approach to cover additional, different non-linear models are still much needed.

SUMMARY AND LIMITATIONS

In Table 7.1 we briefly summarize the strengths, weaknesses, and assumptions of the various approaches that we have discussed. This leads us to offer the

TABLE 7.1 Procedures/Assumptions, Strengths and Weaknesses of Various Approaches Introduced in This Chapter

Method	Procedure/Assumption	Strength	Weakness
Non-Latent Approaches			
Factor Score Analyses	uses factor scores in regression; assumes error-free measurement	simple; easy to understand and implement; evaluates the factor structure of independent and dependent variables	measurement errors are ignored, standard errors not precise; does not analyze measurement and structural models simultaneously
Two-Stage Least Square method	instrumental variables are formed and used in the subsequent ordinary least square regression; does not assume normality of factors or indicators	easy to implement because does not involve SEMs (but more difficult than the factor score approach)	limited to a single-indicator outcome (without multiple indicators); results dependent on how latent variables are scaled; larger bias and lower power than fully latent approaches, which may increase with sample size, particularly when indicator reliability is low
Multiple-Group Analysis	multiple group SEM analysis	easy to understand and implement by using most SEM packages	applicable only when one of the independent variables can be represented by a small number of groups based on a single-indicator construct; requires a reasonable sample size for each group; strength of interaction not directly estimated but inferred on the basis of lack of invariance in parameter estimates across the multiple groups

(continued)

TABLE 7.1 Procedures/Assumptions, Strengths and Weaknesses of Various Approaches Introduced in This Chapter (continued)

Method	Procedure/Assumption	Strength	Weakness
Fully Latent Variable Approaches			
Product-Indicator Model: Constrained Approach	measurement and structural models estimated simultaneously, under the large number of constraints; normality assumed for factors, indicators, and constraints	appropriate consideration and treatment of measurement errors; less bias, better control of Type 1 error rate, higher power	complicated constraints required; imposed constraints based on multivariate normal indicators so that interaction effects are biased when variables are nonnormal
Product-Indicator Model: Unconstrained Approach	similar to the constrained approach, except constraints are not imposed	similar to those of the constrained approach, but without complicated constraints	somewhat less precise than constrained approach when indicators are normally distributed and small sample sizes
Latent Moderated Structural Equation (LMS)/Quasi-Maximum Likelihood (QML)	measurement and structural relation estimated simultaneously using special software; assumes latent variables and their errors are normal	more precise than constrained and unconstrained approaches	need special software, but simple to use, computational intensive in LMS; factors/indicators still assumed to be normal so that interaction effects are biased when indicators are not multivariate normal
Bayesian Method	parameter distribution conditional on prior information and data	does not involve setting up product-indicator; easily extend to quadratic, higher-order, other polynomial effects	complicated Markov Chain Carlo method, statistically demanding to specify the model and set priors; available in WinBUGS only

following advice, particularly for novice researchers. For investigators still new to CFA and SEM, it might be useful to begin with ANOVA (if both independent variables are categorical) or the multiple regression approach to ANOVA (if at least one of the independent variables is reasonably continuous) using scale scores instead of latent constructs. This should provide a preliminary indication of the expected results when more sophisticated approaches are applied. It may also be beneficial to use the factor score approach in which factor scores are used instead of scale scores. However, the use of scale scores and factor scores should give similar results for well-defined measures (i.e., the multiple indicators are all substantially related to the latent construct). When one of the independent variables is a categorical variable that can be represented by a relatively small number of naturally occurring groups, the multiple-group SEM approach is an attractive alternative. Importantly, researchers should be fully aware that these analyses using observed variables (or their aggregates like factor scores) are seriously limited by their low reliability, reduced power, and small variance explained.

Before proceeding to true latent variable approaches to assessing interactions, data should be evaluated for violations of univariate and multivariate normality. Particularly if there are violations of normality, then the various constrained approaches should be avoided. However, because of the simplicity of the unconstrained approach—compared to the constrained approach—we recommend the unconstrained approach instead of the constrained approach in most applications.

Applied researchers attempting to detect latent interactions should also be cautious in interpreting the goodness of fit of these models, which often is used as a very important criterion in assessing the hypothetical model of the relationships. Using goodness of fit for models with latent interaction is complicated by a number of issues. Firstly, as in other structural equation models, goodness of fit reflects the fit of the whole model and thus is not sensitive to any change related to the latent interaction path alone. Secondly, the saturated and null models for the product-indicator models and distribution-analytic approaches are not well defined, making fit indexes difficult to interpret in product-indicator models, and impossible to even obtain in distribution-analytic approaches. Note also that conventional fit indices are not sensitive to non-linear relations between latent variables (Mooijart & Satorra, 2009).

Finally, the use of highly sophisticated statistical tools such as SEM can mislead otherwise knowledgeable researchers into thinking that well-known problems that exist with less complicated approaches are no longer relevant. Although we have focused on important advantages to the use of latent variable approaches to these problems, the reader should not think that the latent variable approaches are a panacea for all of the complications in evaluating interaction and nonlinear effects. Theoretical and empirical research (e.g., Aiken & West, 1991) demonstrate that there are many inherent difficulties in estimating interaction effects, even when manifest (nonlatent) variable approaches are used. Misspecified models may be a problem. Also, even when the parameter estimates are well-defined, there are a range of approaches to visualize interaction effects (and their substantive importance) in manifest applications that have not been routinely implemented in latent-interaction studies. These include, for example, testing the statistical significance of simple slopes (i.e., whether the effect of one predictor variable is statistically significant at a certain value of the other predictor variable), regions of statistical significance (i.e., the range of the moderator values for which the relation between X_1 and Y are statistically significant), and the crossing point of the two regression lines when one predictor is categorical and the interaction is disordinal (see, e.g., Preacher et al., 2006). Furthermore, the analysis of interaction effects becomes more problematic when the relation between a given independent variable and the dependent variable is nonlinear or when the correlations among the interacting variables become increasingly large. There is no reason to suggest that problems and strategies such as these are not also relevant for latent variable approaches.

Most of the research conducted with fully latent approaches (e.g., the constrained and unconstrained approaches) has been based on simulated data with multivariate normal indicators. Whereas there has been some simulation research with nonnormal indicators (e.g., Kelava et al., 2011; Klein & Muthén, 2007; Marsh et al., 2004), these approaches have not been widely applied to substantive issues with real data (but see Nagengast, Marsh, et al., 2011). Hence, a more complete evaluation of the different procedures in actual practice is not possible. Because the robustness and convergence behavior of the approaches considered here have not been systematically evaluated with "real" data, researchers are encouraged to apply different approaches and to compare the results based on the different approaches. Particular caution is needed in interpreting polynomial effects, because the fully latent approaches discussed here have not been sufficiently evaluated—even with well-behaved, simulated data—for this application.

The estimation of latent interaction and quadratic effects is a rapidly developing area of research. Hence, applied researchers need to be aware that new procedures (e.g., QML, Bayesian) may be recommended over the approaches demonstrated here once efficient, user-friendly software becomes widely available or is incorporated into existing SEM packages such as LISREL, EQS, and Mplus.

APPENDIX A

LISREL Syntax for Latent Interaction Model Without Mean Structure by the Unconstrained Approach

```
Unconstrained Approach
All indicators were centered
ML ESTIMATION
DA NI=12 NO=500
CM
1.9362
0.6664  0.9363
0.5991  0.4335  0.9684
0.5106  0.2877  0.3067  1.8971
0.4827  0.2608  0.2560  0.6728  1.0402
0.8178  0.5522  0.4922  0.4394  0.9241  4.0012
0.4873  0.3966  0.2746  0.3660  0.3398  0.6222  2.0911
1.0671  0.7344  0.4734  0.7716  0.7487  1.1362  2.3343  9.2078
0.3688  0.2843  0.1974  0.2153  0.1965  0.3602  0.7304  1.5550  0.9738
0.0988  0.0370  0.1454  0.1564  0.0298  0.1404 -0.2260 -0.4774 -0.1191  4.3955
0.2236  0.1683  0.2383  0.0668  0.1214  0.3113 -0.0831 -0.2714 -0.0833  1.7310 10.3695
0.2675  0.1306  0.0688 -0.0493 -0.0973 -0.0883 -0.0241  0.1225 -0.0009  1.4920  2.1844  5.0454
MO NY=3 NE=1 NX=9 NK=3 LY=FU,FR PS=SY,FR TE=DI,FR TD=DI,FR
LE
eta
LK
ksi1 ksi2 ksi1*ksi2
PA PH
1
1 1
1 1 1
! PH 3 1 PH 3 2 could be fixed to zero for normal data
FI LY 1 1
FR LX 2 1 LX 3 1 LX 5 2 LX 6 2 LX 8 3 LX 9 3
VA 1 LY 1 1 LX 1 1 LX 4 2 LX 7 3
VA 0 LX 7 1 LX 7 2 LX 8 1 LX 8 2 LX 9 1 LX 9 2
FR GA 1 1 GA 1 2 GA 1 3
PD
OU ND=3 SC AD=off EP=0.0001 IT=500 XM
```

APPENDIX B
The PRELIS Syntax for Creating the Bootstrap Samples for Calculating Standard Errors for Appropriate Standardized Interaction Effects

```
SY='bs0.psf'
OU MA=CM RA=bs1.dat XM WI=11 ND=6 IX=111 BS=800 SF=100
```

APPENDIX C
LISREL Syntax for Latent Quadratic Model Without Mean Structure by the Unconstrained Approach

```
Unconstrained Approach
Independent indicators were centered
ML ESTIMATION
DA NI=9 NO=852
CM
 0.693
 0.472   0.661
 0.418   0.430   0.695
 0.335   0.352   0.339   1.075
 0.299   0.328   0.318   0.740   0.925
 0.301   0.313   0.311   0.749   0.662   0.980
-0.269  -0.265  -0.275  -0.117  -0.125  -0.131   2.187
-0.181  -0.191  -0.186  -0.109  -0.100  -0.066   1.002   1.729
-0.224  -0.240  -0.196  -0.122  -0.069  -0.123   0.993   0.740   1.805
LA
y1 y2 y3 x1 x2 x3 x1x1 x2x2 x3x3
MO NY=3 NE=1 NX=6 NK=2 LY=FU,FR PS=SY,FR PH=SY,FR TE=DI,FR TD=DI,FR
LE
eta
LK
ksi ksi*ksi
FI LY 1 1
VA 1 LY 1 1 LX 1 1 LX 4 2
FR LX 2 1 LX 3 1 LX 5 2 LX 6 2
FR GA 1 1 GA 1 2
PD
OU ND=3 SC AD=off EP=0.0001 IT=500 XM
```

REFERENCES

Aiken, L. S., & West, S. G. (1991). *Multiple regression: Testing and interpreting interactions.* Newbury Park, CA: Sage.

Algina, J., & Moulder B. C. (2001). A note on estimating the Jöreskog-Yang model for latent variable interaction using LISREL 8.3. *Structural Equation Modeling: A Multidisciplinary Journal, 8,* 40–52.

Arminger, G., & Muthén, B. O. (1998). A Bayesian approach to nonlinear latent variable models using the Gibbs sampler and the Metropolis-Hastings algorithm. *Psychometrika, 63,* 271–300.

Bagozzi, R. P., & Yi, Y. (1989). On the use of structural equation models in experimental designs. *Journal of Marketing Research, 26,* 271–284.

Bauer, D. J. (2005). A semiparametric approach to modeling nonlinear relations among latent variables. *Structural Equation Modeling: A Multidisciplinary Journal, 12,* 513–535.

Bauer, D. J., & Curran, P. J. (2004). The integration of continuous and discrete latent variable models: Potential problems and promising opportunities. *Psychological Methods, 9,* 3–29.

Bentler, P. M. (2000–2008). *EQS 6 structural equations program manual.* Encino, CA: Multivariate Software, Inc.

Blom, P., & Christoffersson, A. (2001). Estimation of nonlinear structural equation models using empirical characteristic functions. In R. Cudeck, S. Du Toit, & D. Sörbom (Eds.), *Structural equation modeling: Present and future.* (pp. 443–460). Lincolnwood, IL: Scientific Software.

Boker, S., Neale, M., Maes, H., Wilde, M., Spiegel, M., Brick, T., Spies, J., Estabrook, R., Kenny, S., Bates, T., Mehta, P., & Fox, J. (2010). OpenMx: The OpenMx statistical modeling package. R package version 0.3.0–1217. http://openmx.psyc.virginia.edu .

Boomsma, A. (1983). *On the robustness of LISREL against small sample size and nonnormality.* Doctoral dissertation, University of Groningen.

Cohen, J., & Cohen, P. (1983). *Applied multiple regression /correlational analysis for the behavioral sciences.* Hillsdale, NJ: Erlbaum.

Cudeck, R. (1989). Analysis of correlation matrices using covariance structure models. *Psychological Bulletin, 105,* 317–327.

Cudeck, R., Harring, J. R., & du Toit, S. H. C. (2009). Marginal maximum likelihood estimation of a latent variable model with interaction. *Journal of Educational and Behavioral Statistics, 34,* 131–144.

Dempster, A. P., Laird, N. M., & Rubin, D. B. (1977). Maximum likelihood from incomplete data via the EM algorithm. *Journal of the Royal Statistical Society—Series B, 39,* 1–38.

Dolan, C. V., & van der Maas, H. (1998). Fitting multivariate normal finite mixtures subject to structural equation modeling. *Psychometrika, 63,* 227–253.

Fox, J. (2006). Structural equation modeling with the SEM package in R. *Structural Equation Modeling: A Multidisciplinary Journal, 13,* 465–486.

Fraser, C. (1980). *COSAN User's Guide.* Toronto, Ontario: The Ontario Institute for Studies in Education.

Friedrich, R. J. (1982). In defense of multiplicative terms in multiple regression equations. *American Journal of Political Science*, 26, 797–833.

Ganzach, Y. (1997). Misleading interaction and curvilinear terms. *Psychological Methods, 3*, 235–247.

Gelman, A., Carlin, J. B., Stern, H. S., & Rubin, D. B. (2004). *Bayesian data analysis*. London: Chapman & Hall.

Geman, S., & Geman, D. (1984). Stochastic relaxation, Gibbs distribution, and the Bayesian restoration of images. *IEEE Transactions on Pattern Analysis and Machine Intelligence, 6*, 721–741.

Hastings, W. K. (1970). Monte Carlo sampling methods using Markov chains and their application. *Biometrika, 57*, 97–109.

Hau, K. T., & Marsh, H. W. (2004). The use of item parcels in structural equation modeling: Non-normal data and small sample sizes. *British Journal of Mathematical Statistical Psychology, 57*, 327–351.

Hayduk, L. A. (1987). *Structural equation modeling with LISREL: Essentials and advances*. Baltimore: Johns Hopkins University.

Hu, L., Bentler, P. M., & Kano, Y. (1992). Can test statistics in covariance structure analysis be trusted? *Psychological Bulletin, 112*, 351–362.

Jaccard, J., & Wan, C. K. (1995). Measurement error in the analysis of interaction effects between continuous predictors using multiple regression: Multiple indicator and structural equation approaches. *Psychological Bulletin, 117*, 348–357.

Jackman, M. G.-A., Leite, W. L., & Cochrane, D. J. (2011). Estimating latent variable interactions with the unconstrained approach: A comparison of methods to form product indicators for large, unequal numbers of items, *Structural Equation Modeling: A Multidisciplinary Journal, 18*, 274–288.

Jöreskog, K. G. (1998). Interaction and nonlinear modeling: issue and approaches. In R. E. Schumacker & G. A. Marcoulides (Eds.), *Interaction and nonlinear effects in structural equation modeling* (pp. 239–250). Mahwah, NJ: Erlbaum.

Jöreskog, K. G., & Yang, F. (1996). Nonlinear structural equation models: The Kenny-Judd model with interaction effects. In G. A. Marcoulides & R. E. Schumacker (Eds.), *Advanced structural equation modeling: Issues and techniques* (pp. 57–88). Mahwah, NJ: Erlbaum.

Kelava, A., & Brandt, H. (2009). Estimation of nonlinear latent structural equation models using the extended unconstrained approach. *Review of Psychology, 16*, 123-131.

Kelava, A., Werner, C. S., Schermelleh-Engel, K., Moosbrugger, H., Zapf, D., Ma, Y., Cham, H., Aiken, L. S., & West, S. G. (2011). Advanced nonlinear latent variable modeling: Distribution analytic LMS and QML estimators of interaction and quadratic effects. *Structural Equation Modeling: A Multidisciplinary Journal, 18*, 465–491.

Kenny, D. A., & Judd, C. M. (1984). Estimating the nonlinear and interactive effects of latent variables. *Psychological Bulletin, 96*, 201–210.

Klein, A. G., & Moosbrugger, H. (2000). Maximum likelihood estimation of latent interaction effects with the LMS method. *Psychometrika, 65*, 457–474.

Klein, A. G., & Muthén, B. O. (2007). Quasi-maximum likelihood estimation of strucural equation models with multiple interaction and quadratic effects. *Multivariate Behavioral Research, 42*, 647–663.

Klein, A. G., Schermelleh-Engel, K., Moosbrugger, H., & Kelava, A. (2009). Assessing spurious interaction effects. In T. Teo & M. S. Khine (Eds.), *Structural equation modeling in educational research: Concepts and applications* (pp. 13–28). Rotterdam, The Netherlands: Sense Publishers.

Kromrey, J. D., & Foster-Johnson, L. (1999). Statistically differentiating between interaction and nonlinearity in multiple regression analysis: A Monte Carlo investigation of a recommended strategy. *Educational and Psychological Measurement, 59*, 392–413.

Laplante, B., Sabourin, S., Cournoyer, L. G., & Wright, J. (1998). Estimating nonlinear effects using a structured means intercept approach. In R. E. Schumacker & G. A. Marcoulides, (Eds,), *Interaction and nonlinear effects in structural equation modeling* (pp. 183–202). Mahwah, NJ: Erlbaum.

Lee, S. Y. (2008). *Structural equation modeling. A Bayesian approach.* London, UK: Wiley.

Lee, S. Y., Song, X. Y., & Tang, N. S. (2007). Bayesian methods for analyzing structural equation models with covariates, interaction, and latent quadratic variables. *Structural Equation Modeling: A Multidisciplinary Journal, 14*, 404–434.

Lin, G. C., Wen, Z., Marsh, H. W., & Lin, H. S. (2010). Structural equation models of latent interactions: Clarification of orthogonalizing and double-mean-centering strategies. *Structural Equation Modeling: A Multidisciplinary Journal, 17*, 374–391.

Little, T. D., Bovaird, J. A., & Widaman, K. F. (2006). On the merits of orthogonalizing powered and product term: Implications for modeling interactions among latent variables. *Structural Equation Modeling: A Multidisciplinary Journal, 13*, 497–519.

Lubinski, D., & Humphreys, L. G. (1990). Assessing spurious "moderator effects": Illustrated substantively with the hypothesized ("synergistic") relation between spatial and mathematical ability. *Psychological Bulletin, 107*, 385–393.

MacCallum, R. C., & Mar, C. M. (1995). Distinguishing between moderator and quadratic effects in multiple regression. *Psychological Bulletin, 118*, 405–421.

MacCallum, R. C., Zhang, S., Preacher, K. J., & Rucker, D. D. (2002). On the practice of dichotomization of quantitative variables. *Psychological Methods, 7*, 19–40.

Marsh, H. W. (2001). Distinguishing between good (useful) and bad workload on students' evaluations of teaching. *American Educational Research Journal, 38*, 183–212.

Marsh, H. W. (2008). The elusive importance effect: More failure for the Jamesian perspective on the importance of importance in shaping self-esteem. *Journal of Personality, 76*, 1081–1121.

Marsh, H. W., Hau, K.-T., Wen, Z., Nagengast, B., & Morin, A. J. S. (in press). Moderation. In T. D. Little (Ed.), *The Oxford handbook of quantitative methods.* New York, NY: Oxford University Press.

Marsh, H. W,, Lüdtke, O., Nagengast, B., Morin, A. J. S., Trautwein, U., & Von Davier, M. (in press). Why item parcels are (almost) never appropriate: Two wrongs do not make a right—Camouflaging misspecification with item-parcels in CFA models. *Psychological Methods.*

Marsh, H. W., Muthén, B., Asparouhov, T., Lüdtke, O., Robitzsch, A., Morin, A. J. S., & Trautwein, U. (2009). Exploratory Structural equation modeling, integrating CFA and EFA: Application to students' evaluations of university teaching. *Structural Equation Modeling: A Multidisciplinary Journal, 16,* 439–476.

Marsh, H. W., Wen, Z., & Hau, K. T. (2004). Structural equation models of latent interactions: evaluation of alternative estimation strategies and indicator construction. *Psychological Methods, 9,* 275–300.

Marsh, H. W., Wen, Z. L., Hau, K. T., Little, T. D., Bovaird, J. A., & Widaman, K. F. (2007). Unconstrained structural equation models of latent interactions: Contrasting residual- and mean-centered approaches. *Structural Equation Modeling: A Multidisciplinary Journal, 14,* 570–580.

McClelland, G. H., & Judd, C. M. (1993). Statistical difficulties of detecting interactions and moderator effects. *Psychological Bulletin, 114,* 376–390.

Micceri, T. (1989). The unicorn, the normal curve, and other improbable creatures. *Psychological Bulletin, 105,* 156–166.

Mooijaart, A., & Bentler, P. M. (2010). An alternative approach for nonlinear latent variable models. *Structural Equation Modeling: A Multidisciplinary Journal, 17,* 357–373.

Mooijaart, A., & Satorra, A. (2009). On insensitivity of the chi-square model test to nonlinear misspecification in structural equation models. *Psychometrika, 74,* 443–455.

Moosbrugger, H., Schermelleh-Engel, K., Kelava, A., & Klein, A. G. (2009). Testing multiple nonlinear effects in structural equation modeling: A comparison of alternative estimation approaches. In T. Teo & M. Khine (Eds.), *Structural equation modeling in educational research: Concepts and applications* (pp. 103–136). Rotterdam, NL: Sense Publishers.

Muthén, L. K., & Muthén, B. O. (1997–2010). *Mplus user's guide.* Los Angeles, CA: Muthen & Muthen.

Nagengast, B., Marsh, H. W., Scalas, L. F., Xu, M. K., Hau, K.-T., & Trautwein, U. (2011). Who took the "×" out of expectancy-value theory? A psychological mystery, a substantive-methodological synergy, and a cross-national generalization. *Psychological Science, 22,* 1058–1066.

Pek, J., Sterba, S. K., Kok, B. E., & Bauer, D. J. (2009). Estimating and visualizing nonlinear relations among latent variables: A semiparametric approach. *Multivariate Behavioral Research, 44,* 407–436.

Ping, R. A., Jr. (1998). EQS and LISREL examples using survey data. In R. E. Schumacker & G. A. Marcoulides (Eds.), *Interaction and nonlinear effects in structural equation modeling* (pp. 63–100). Mahwah, NJ: Lawrence Erlbaum Associates.

Preacher, K. J., Curran, P. J., & Bauer, D. (2006). Computational tools for probing interactions in multiple linear regression, multilevel modeling, and latent curve analysis. *Journal of Educational and Behavioral Statistics, 31,* 437–448.

Pyzdek, T. (1995). Why normal distributions aren't [all that normal]. *Quality Engineering, 7,* 769–777.

Rigdon, E. E., Schumacker, R. E., & Wothke, W. (1998). A comparative review of interaction and nonlinear modeling. In R. E. Schumacker & G. A. Marcoulides (Eds.), *Interaction and nonlinear effects in structural equation modeling* (pp. 1–16). Mahwah, NJ: Lawrence Erlbaum Associates.

Schermelleh–Engel, K., Klein, A., & Moosbrugger, H. (1998). Estimating nonlinear effects using a latent moderated structural equations approach. In R. E. Schumacker & G. A. Marcoulides (Eds.), *Interaction and nonlinear effects in structural equation modeling* (pp. 203–238). Mahwah, NJ: Erlbaum.

Schumacker, R. E., & Marcoulides, G. A. (Eds.) (1998). *Interaction and nonlinear effects in structural equation modeling.* Mahwah, NJ: Erlbaum.

Spiegelhalter, D. J., Thomas, A., Best, N. G., & Lunn, D. (2004). *WinBugs user manual* (Version 1.4). Cambridge, UK: MRC Biostatistics Unit.

Vandenberg, R. J., & Lance, C. E. (2000). A review and synthesis of the measurement invariance literature: Suggestions, practices, and recommendations for organizational research. *Organizational Research Methods, 3,* 4–69.

Wall, M. M., & Amemiya, Y. (2000). Estimation for polynomial structural equation models. *Journal of the American Statistical Association, 95,* 929–940.

Wall, M. M., & Amemiya, Y. (2001). Generalized appended product indicator procedure for nonlinear structural equation analysis. *Journal of Educational and Behavioral Statistics, 26,* 1–29.

Wen, Z., Hau, K., & Marsh, H. W. (2008). Appropriate standardized estimates for moderating effects in structural equation models. *Acta Psychologica Sinica, 40,* 729–736.

Wen, Z., Marsh, H. W., & Hau, K. T. (2010). Structural equation models of latent interactions: An appropriate standardized solution and its scale-free properties. *Structural Equation Modeling: A Multidisciplinary Journal, 17,* 1–22.

West, S. G., Finch, J. F., & Curran, P. J. (1995). Structural equation models with nonnormal variables: Problems and remedies. In R. H. Hoyle (Ed.), *Structural equation modeling: Concepts, issues and applications* (pp. 56–75). Newbury Park, CA: Sage

Wu, Y., Wen, Z., & Lin, G. C. (2009). Structural equation modeling of latent interactions without using the mean structure. *Acta Psychologica Sinica, 41,* 1252–1259.

Yang, F. (1998). Modeling interaction and nonlinear effects: a step-by-step LISREL example. In R. E. Schumacker & G. A. Marcoulides (Eds.), *Interaction and nonlinear effects in structural equation modeling* (pp. 17–42). Mahwah, NJ: Erlbaum.

Yang-Wallentin, F., & Jöreskog, K.G. (2001). Robust standard errors and chi-squares for interaction models. In G.A. Marcoulides & R.E. Schumacker (Eds.), *New developments and techniques in structural equation modeling* (pp. 159–171). Mahwah, NJ: Erlbaum.

CHAPTER 8

USING LATENT GROWTH MODELING TO EVALUATE LONGITUDINAL CHANGE

Gregory R. Hancock, Jeffrey R. Harring, and Frank R. Lawrence

Researchers are frequently interested in understanding how some aspect of an individual changes over time. The focus of their investigation might be on general outcomes such as behavior, performance, or values; or it could be on more specific aspects such as substance abuse, depression, communication skills, attitudes toward disabled veterans, or advancement of math aptitude. Regardless of an investigation's focus, the real attraction in longitudinal studies is in understanding how change comes about, how much change occurs, how the change process might differ across individuals, and what the determinants of that change are.

Various methods can be used to analyze longitudinal data (see, e.g., Collins & Sayer, 2001; Gottman, 1995). Among the more traditional methods is analysis of variance (ANOVA), multivariate analysis of variance (MANOVA), analysis of covariance (ANCOVA), multivariate analysis of covariance (MANCOVA), and auto-regressive and cross-lagged multiple regression. No one method is necessarily superior, but each has strengths

Structural Equation Modeling: A Second Course (2nd ed.), pages 309–342
Copyright © 2013 by Information Age Publishing

and shortcomings that researchers should be aware of in order to select the analytic method best suited for the particular research context.

Selection of an appropriate analytic method revolves around two central issues. The first is the nature of the hypotheses being tested, while the second concerns the underlying assumptions necessary to apply a particular method. The method must be compatible with both the hypotheses being tested and the data, so as to be capable of providing evidence to tentatively support or refute the researcher's hypotheses. Unfortunately, traditional longitudinal data analytic techniques can present challenges on both fronts. With regard to the data, technical assumptions such as sphericity underlying repeated-measures ANOVA (see, e.g., Yandell, 1997) are rarely met in practice in the social sciences. Also, traditional methods tend to operate at the group-mean level, providing potentially interesting aggregated results but failing to address hypotheses regarding the nature and determinants of change at the level of the individual which are often the most important aspects of the data. Thus, when such incompatibilities of methods with data and hypotheses arise, conclusions from traditional methods can become somewhat circumspect (see, e.g., Rogosa & Willett, 1985).

To help overcome some of the limitations of more traditional analytic approaches to the assessment of change over time, two (highly-related and converging) classes of methods have emerged. The first, which is not the focus of the current chapter, operates from within a hierarchical linear modeling (HLM) framework, where data over time are treated as nested within the individual (see, e.g., Bollen & Curran, 2006; Raudenbush & Bryk, 2002; Singer & Willett, 2003). The second class of methods, relevant for this chapter, has emerged from the area of structural equation modeling (SEM). These methods, falling under the general heading of *latent growth modeling* (LGM), approach the analysis of growth[1] from a somewhat different perspective than the aforementioned methods. Specifically, LGM techniques can describe individuals' behavior in terms of reference levels (e.g., initial amount) and their developmental trajectories to and from those levels (e.g., linear, quadratic). In addition, they can determine the variability across individuals in both reference levels and trajectories, as well as provide a means for testing the contribution of other variables or constructs to explaining those reference levels and growth trajectories (Rogosa & Willett, 1985; Short, Horn, & McArdle, 1984). In doing so, LGM methods simultaneously focus on changes in covariances, variances, and mean values over time (Dunn, Everitt, & Pickles, 1993; McArdle, 1988), thus utilizing more information available in the measured variables than do traditional methods.

Consider Figure 8.1, a theoretical representation of five individuals' growth patterns over four equally spaced time intervals. A number of elements of this figure are noteworthy. First, the growth trajectory of all individuals displayed is linear; that is, for equal time periods a given individual

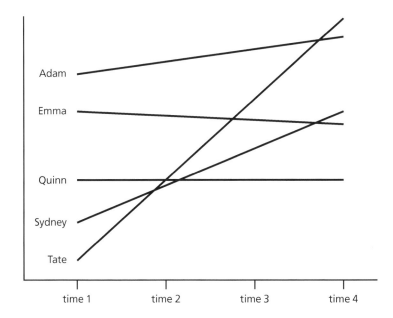

Figure 8.1 Linear growth for five individuals across four equally spaced time points.

is growing the same amount. Thus, Tate grows as much between time 1 and time 2 as he does between time 3 and time 4. Within the current chapter, we will begin by assuming that growth follows a linear trend. This linear form might or might not be a realistic representation of growth for a particular variable; growth might actually be of a different functional nature (e.g., quadratic or logarithmic). Given that growth processes are often measured over a fairly restricted portion of the span of development, even complex growth patterns can be well approximated by simpler models (Willett & Sayer, 1994). That said, after developing LGM concepts with this simple linear model, we will discuss the relaxation of the assumption of growth linearity and methods for evaluating other developmental patterns.

A second key point regarding Figure 8.1 is that, in contrast to what is often assumed in more traditional methods, the five individuals represented do not develop at the same rate. Tate, for example, starts below the others but finishes with the highest value because he progresses at such a rapid rate; Sydney starts slightly higher than Tate but finishes quite a bit lower. Adam, on the other hand, starts at the highest level but does not grow much from that point, while Quinn stays at the same level and Emma actually declines over the time period represented. Thus, all individuals differ in their initial levels of behavior as well as in their rates of growth from that initial level; assuming the same degree of change over time for all individuals is clearly unrealistic for these data. Fortunately, LGM methods typically only assume that all individuals' growth follows the same functional form, reflecting a deeper assumption that growth is grounded in some social and/or biological mechanism. In this case that mechanism is linear, that is, each individual's growth over this span is well represented by a line; as such, it may be summarized by, for example, a unique *intercept* and *slope*. An individual's intercept describes the amount of the variable possessed at the initial measurement point, while the slope captures information about how much that individual changes for each time interval following the initial measurement point.

Third, as a result of individual differences in these intercepts and slopes, changes occur in the relations among individuals' data across measurement occasions. For example, at time 1, Tate has the least amount of the variable of interest and Adam has the most; at time 4, Tate has the most and Quinn has the least. As a result of individuals' relative positions shifting across time, correlations between scores at different times are not identical. In addition, the variance of scores at different times also fluctuates; scores are more spread out at the first and last times, and less variable at the intermediate measurement points. These changes in correlation and variance combine to illustrate a violation of the sphericity assumption underlying repeated-measure ANOVA, a method commonly used in the evaluation of longitudinal data.

In short, then, a method is required that does not assume all individuals change at the same rate, thereby transcending the limitations of methods bound to this and other related assumptions. The purpose of the current chapter is to illustrate the use of LGM techniques, which overcome many traditional methods' limitations and thus serve as an extremely useful research tool for the explication of longitudinal data. Such methods have been used to analyze change in alcohol, cigarette, and marijuana use (Curran, Stice, & Chassin, 1997; Duncan & Duncan, 1994), tolerance of deviant behavior (Willett & Sayer, 1994), antisocial behavior (Patterson, 1993), client resistance during parent training therapy (Stoolmiller, Duncan, Bank, & Patterson, 1993), and family functioning (Willett, Ayoub, & Robinson, 1991). Our intent is to describe this methodological tool in enough conceptual and practical detail for readers with a basic understanding of SEM to be able to apply it. We start with the most common linear case and then expand to address nonlinearity, external predictors, and a host of variations and extensions to the basic model. Those readers wishing to explore LGM methodology further are encouraged to seek out the aforementioned references as well as some excellent texts (e.g., Bollen & Curran, 2006; Duncan, Duncan, & Strycker, 2006; Moskowitz & Hershberger, 2002; Preacher, Wichman, MacCallum, & Briggs, 2008).

LGM METHODOLOGY FOR
LINEAR DEVELOPMENTAL PATTERNS

To facilitate the introduction of LGM methods, consider the hypothetical situation in which high school students' mathematics proficiency has been assessed using the same instrument at the beginning of each school year. Thus, the researcher has data at an initial point (9th grade) and at three equally spaced subsequent intervals (as was depicted in Figure 8.1), in this case 10th, 11th, and 12th grade (note that data are gathered from individuals at the same measurement occasions, although this design constraint can be relaxed as will be discussed later). Having multiple data points such as these allows one to evaluate more accurately the functional form of the growth under examination. Stoolmiller (1995) recommended a careful consideration of the phenomenon under study to determine a suitable number and spacing of measurements, and we could not agree more. Predicting a precise window of time in which the phenomenon changes, coupled with concerns of how frequently observations must be gathered to adequately capture the behavior, can be quite challenging in practice.

Once data have been gathered, a preliminary assessment of several individuals' growth patterns,[2] coupled with a sound theoretical understanding of the nature of the behavior being investigated, often yields a reasonable assumption as to the functional form of the growth under investigation (linear, quadratic, exponential, etc.). In the current hypothetical case, assume that past research and theory dictate that math proficiency scores should follow a generally linear pattern over the time period studied, and that the initial point of measurement (time 1) is a useful point of developmental reference. As such, each individual's score can be expressed by the following functional form:

$$\text{score at time } t = \text{initial score} + (\text{change in score per unit time}) \times (\text{time elapsed}) + \text{error.}$$

As the initial score is simply an intercept, and the change in score per unit time is simply a slope, this expression could be expanded into a system of equations for each individual's score from time 1 (V1) through time 4 (V4).

time 1:	V1 = intercept + (slope) × (0) + E1
time 2:	V2 = intercept + (slope) × (1) + E2
time 3:	V3 = intercept + (slope) × (2) + E3
time 4:	V4 = intercept + (slope) × (3) + E4.

Thus, a student's observed math proficiency score at any time is believed to be a function of his or her own intercept and slope, as well as error. Note that the terms *intercept* and *slope* have the same interpretation here as they do in a simple regression model. That is, the intercept represents the expected value of the outcome at the point where the predictor variable is zero. Hence, the intercept is the expected value for V1 when elapsed time is zero. This can easily be seen from the system of equations. The first equation exemplifies this concept by fixing time to zero, thereby placing the intercept at that initial location. We can build on that interpretation by noting that the intercept appears in every subsequent equation. Thus, we can say that the intercept allows the initial amount of math proficiency to be 'frozen in' to the measured variable, while the slope describes change in proficiency beyond that initial score.

In linear LGM, these intercept and slope terms are treated as latent variables (factors) on which individuals may vary; as such these are not assumed to be measured directly. Certainly an individual's V1 measure provides an estimate of initial math proficiency, but, as the time 1 equation above indicates, this is not without error. Similarly, while one could estimate for each individual the slope of the best-fit line describing math proficiency over time, the rate of change possessed by an individual is treated as a latent trait. Observed scores, however, are not expected to be a perfect function of the latent growth constituents; the E terms in the equations represent error. There are many sources of error that appear in data and the modeling of data, including measurement error arising from instrument or rater unreliability and model misspecification arising from an incorrect assumptions regarding of the functional form of change. Thus, error may be regarded as the degree of deviation between the observed outcome and that which we would expect from the latent growth portion of the model. The system of equations represented above, which represents a theory about the data generation process for math proficiency, can be expressed in the model presented in Figure 8.2, shown with both covariance and mean structure parameters.

With regard to the covariance structure, note that in this model all paths to the variables (i.e., loadings) are fixed. Of particular theoretical importance is that all paths from the slope factor have been fixed.[3] This indicates that the researcher wishes to test if a model representing linear growth fits the data satisfactorily. This also leaves only variance and covariance parameters to be estimated within the covariance structure: all four error variances (c_{E1}, c_{E2}, c_{E3}, c_{E4}), which are usually only of peripheral interest[4]; both factor variances (c_{F1} and c_{F2}); and the factor covariance (c_{F1F2}). The intercept factor variance (c_{F1}) gives the researcher a sense of how much diversity exists in initial math proficiency scores, where a smaller variance would indicate that students start with rather similar math proficiency in 9th grade and a larger intercept variance indicates that students are more heterogeneous in terms of 9th grade math proficiency. The slope factor variance (c_{F2}) conveys the diversity in rates of change in math proficiency throughout high school, where a smaller variance would indicate that students grow at a fairly simi-

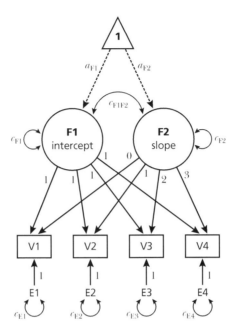

Figure 8.2   Linear latent growth model with covariance and mean structure.

lar rate throughout high school and a larger slope variance indicates that students change at more varied rates. Finally, the factor covariance (c_{F1F2}) captures the extent to which the rate of growth in math proficiency is related the initial amount of math proficiency. A positive covariance implies that more proficient students tend to grow more while less proficient students tend to grow less; conversely, a negative covariance implies that less proficient students tend to grow more while more proficient students tend to grow less.

While the covariance structure of the latent growth model contains information about individual differences in growth characteristics, the mean structure portion contains information about growth at the aggregate level. As detailed in the chapter by Thompson and Green (this volume), factor means can be estimated through the introduction of a pseudovariable that assumes a constant value of 1 for all subjects. The pseudovariable has no variance, and as such it cannot covary with (or have a meaningful causal effect on) any measured variable or factor. Nonetheless, it is placed into diagrams and structural models as though it were a variable, and its inclusion in a latent growth model as depicted in Figure 8.2 allows for intercept and slope factor means to be estimated.

Paths from the pseudovariable (depicted as a 1 inside a triangle) to F1 and F2 are denoted as a_{F1} and a_{F2}, respectively indicating the latent intercept and slope factor means. In general, a high value of a_{F1} implies that students start with high math proficiency in 9th grade *on average*, and a low value implies that students start with low math proficiency; it does not, however, communicate anything about individual differences, which was conveyed by c_{F1}. Further, a positive value of a_{F2} implies that students have a positive rate of growth in math proficiency across the high school years *on average*, while a negative value implies that students have a negative rate of growth on average; it does not say anything about individual differences in growth rate, which was conveyed by c_{F2} and which could imply (if sufficiently relatively large) that some students increase in math proficiency while others decline over time.[5]

The linear latent growth model in Figure 8.2, with both covariance and mean structure, has implications for the pattern of variances, covariance, and means that should be observed if the hypothesis of linear growth in math proficiency is correct. The model-implied covariance matrix and mean vector, $\hat{\Sigma}$ and $\hat{\mu}$, would be

$$
\hat{\Sigma} = \begin{bmatrix}
c_{F1} + c_{E1} & & & \\
c_{F1} + c_{F1F2} & c_{F1} + c_{F2} + 2c_{F1F2} + c_{E2} & & \\
c_{F1} + 2c_{F1F2} & c_{F1} + 2c_{F2} + 3c_{F1F2} & c_{F1} + 4c_{F2} + 4c_{F1F2} + c_{E3} & \\
c_{F1} + 3c_{F1F2} & c_{F1} + 3c_{F2} + 4c_{F1F2} & c_{F1} + 6c_{F2} + 5c_{F1F2} & c_{F1} + 9c_{F2} + 6c_{F1F2} + c_{E4}
\end{bmatrix}
$$

$$
\hat{\mu} = \begin{bmatrix} a_{F1} & a_{F1} + a_{F2} & a_{F1} + 2a_{F2} & a_{F1} + 3a_{F2} \end{bmatrix}.
$$

Thus, the model in Figure 8.2 would be fitted to the sample covariance matrix $\mathbf{S}$ and the vector of sample means $\mathbf{m}$, choosing estimates for the parameters in $\hat{\Sigma}$ and $\hat{\mu}$ so as to minimize the discrepancy of the observed moments $\mathbf{S}$ and $\mathbf{m}$ from the expected moments $\hat{\Sigma}$ and $\hat{\mu}$. As there are 14 pieces of data in $\mathbf{S}$ and $\mathbf{m}$ and 9 parameters to estimate in $\hat{\Sigma}$ and $\hat{\mu}$, the model will be overidentified with 5 degrees of freedom. The data-model fit is evaluated using the model-implied covariance matrix and mean vector, which represent the expected moments to be compared with the observed moments of our sample data. If the fit between the observed and expected moments is deemed poor, which would be signaled by unsatisfactory data-model fit indices (e.g., Standardized Root Mean Square Residual, SRMR; Root Mean Square Error of Approximation, RMSEA), then the model would be rejected. Generally speaking this could be due to a misspecification of the functional form of growth (which manifests itself in badness of fit in the mean and/or covariance structure), to a misspecification of the growth factor covariance structure, and/or to a misspecification of the error covariance structure. If model rejection is indeed warranted, then further theoretical and/or exploratory work is likely required. If, however, the data-model fit is deemed satisfactory, the researcher has gathered information supporting the hypothesis of linearity for growth in math proficiency (or, more accurately, the researcher has failed to disconfirm said hypothesis). Following the establishment of reasonable data-model fit, then, the interpretation of the parameters of interest becomes permissible, that is, of the intercept and slope factor variances, their covariance, and their latent means. An example of this linear latent growth model follows next.[6]

Example of a Linear Latent Growth Model

Continuing with the previous hypothetical context, imagine that a standardized test intending to measure math proficiency is administered to the same $n = 1,000$ girls in 9th, 10th, 11th, and 12th grades, yielding the following observed moments:

$$\mathbf{S} = \begin{bmatrix} 204.11 & & & \\ 139.20 & 190.05 & & \\ 136.64 & 135.19 & 166.53 & \\ 124.86 & 130.32 & 134.78 & 159.87 \end{bmatrix}$$

$$\mathbf{m} = \begin{bmatrix} 32.97 & 36.14 & 40.42 & 43.75 \end{bmatrix}.$$

As LGM is simply a constrained version of confirmatory factor analysis, with a mean structure, the model in Figure 8.2 may be imposed upon these data using any common SEM software. The Appendix contains appropriate syntax for this example for the EQS, SIMPLIS, and Mplus software packages, each of which yields output including data-model fit statistics, parameter estimates, and parameter estimate test statistics.

For this example, the data-model fit is excellent overall: SRMR = .010 and RMSEA = .027 (with 90% confidence interval of .000, .056). This implies that the hypothesis of linear growth in math proficiency over the high school years is reasonable, and allows further interpretation of specific model parameter estimates. For the mean structure, the estimate of the latent intercept mean is $\hat{a}_{F1} = 32.863$; this value is statistically significantly greater than zero ($z = 76.352$, $p < .001$), but this merely reflects the fact that the math proficiency scale itself contains all positive numbers. The estimate of the latent slope mean is $\hat{a}_{F2} = 3.654$; this value is also statistically significantly greater than zero ($z = 34.481$, $p < .001$), implying that growth in math proficiency throughout high school tended to be in the positive direction (i.e., increased). As the mean structure merely conveys average growth information for this linear model, we may consult the covariance structure parameters to gain a sense of individual differences in intercept and slope. The estimate of the intercept variance is $\hat{c}_{F1} = 148.714$, which is statistically significantly greater than zero ($z = 17.483$, $p < .001$). From this we may infer that a model positing that all 9th graders start with the same initial amount of math proficiency would be rejected, and that diversity in initial amount of math proficiency exists. The estimate of the slope variance is $\hat{c}_{F2} = 3.610$, which is also statistically significantly greater than zero ($z = 4.969$, $p < .001$). This implies that a model positing that all 9th graders have the same rate of change in math proficiency throughout high school would be rejected, and that diversity in growth rates does exist. Finally, the estimated covariance between the intercept and slope factors is $\hat{c}_{F1F2} = -7.217$, which is statistically significantly less than zero ($z = -3.953$, $p < .001$). Further interpretation of these parameter estimates follows.

Consider first the intercept parameter estimates of $\hat{a}_{F1} = 32.863$ and $\hat{c}_{F1} = 148.714$, the latter of which may be transformed to an estimated standard deviation for the intercept factor of $(148.714)^{1/2} = 12.195$. If one may reasonably assume normality for math proficiency in 9th grade, then the estimated latent mean and standard deviation could be used to create the distributional representation for 9th grade math proficiency shown in Figure 8.3. More interestingly, we could do the same for the slope parameter estimates of $\hat{a}_{F2} = 3.654$ and $\hat{c}_{F2} = 3.610$, the latter of which may be transformed to an estimated standard deviation for the slope factor of $(3.610)^{1/2} = 1.900$. Again assuming normality, the estimated latent mean and standard deviation could be used to create the distributional representation for growth in math proficiency also shown in Figure 8.3. We may infer from this second distribution that approximately 68% of students grow annually between 1.754 and 5.554 points along the latent math proficiency continuum, which has the same metric as the measured math proficiency variables. Assuming normality we may also estimate that approximately 3% of students either do not change over time, or in fact decline in math proficiency during high school.

Considering the relation between the intercept and slope factors, the estimated covariance of $\hat{c}_{F1F2} = -7.217$ corresponds to an estimated correlation between the intercept and slope factors of $-.311$. This implies that students who start with high math proficiency tend to grow at a slower rate, while students with lower initial math proficiency tend to grow at a faster rate. This result could be indicative of a ceiling effect of the math proficiency instrument itself, and/or a curricular emphasis targeted toward improving those students with lower initial math proficiency rather than challenging those students with higher initial math proficiency.

Latent intercept

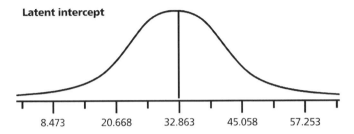

|8.473 | 20.668 | 32.863 | 45.058 | 57.253|

Latent slope

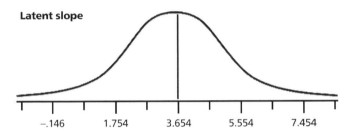

|−.146 | 1.754 | 3.654 | 5.554 | 7.454|

Figure 8.3 Hypothetical distributions for latent intercept and latent slope factors.

Variations on the Linear Latent Growth Model

Changes in growth metric. If a researcher wishes to assess the rate of change per an alternate unit of time, this may be accomplished simply by rescaling the slope factor loadings. For example, had semi-annual growth been of interest in the previous math proficiency example, the growth factor loadings for V1 through V4 could have been fixed to 0, 2, 4, and 6, respectively. This change would have no effect on data-model fit, but would merely rescale the parameters associated with the growth factor F2. Specifically, the estimated factor mean $\hat{a}_{F2}$, which was 3.654 math proficiency units per year, would become 1.827 math proficiency units per half year. As F2 now represents semi-annual growth, the estimated variance $\hat{c}_{F2}$, which was originally 3.610 for annual growth, would become one-fourth of its value (.902). The covariance of F2 with the intercept, $\hat{c}_{F1F2}$, would change from −7.217 to one-half of its value (−3.608).

Changes in reference point. The previous development and example had the initial amount of the variable of interest serving as the intercept. However, other reference points could also be chosen for the intercept factor. For example, F2 loadings of −3, −2, −1, and 0 would make the intercept factor F1 represent the amount of math proficiency in 12th grade rather than 9th. One could also choose F2 loadings of −1.5, −.5, .5, and 1.5, thus making the intercept factor F1 represent the amount of math proficiency in between 10th and 11th grade. Even a point outside of the measured range could be selected as the reference point for the intercept factor F1; the F2 loadings of 2, 3, 4, and 5, for example, would make 7th grade the temporal reference point for math proficiency. The researcher may choose any theoretically interesting developmental juncture as the temporal reference for F1, although caution certainly must be exercised when it lies outside of the range of observed measurements. Finally, note that changes in location of F1 will affect the estimated mean $\hat{a}_{F1}$, the estimated variance $\hat{c}_{F1}$, and the estimated covariance $\hat{c}_{F1F2}$ (for a more detailed account of the effects that choice of a reference point has on a LGM analysis, see, e.g., Biesanz, Deeb-Sossa, Papadakis, Bollen, & Curran, 2004; Hancock & Choi, 2006).

Unequally spaced time points. Although equally spaced times points have been used in the previous explanations for the linear model, a linear growth system could still be tested with unequally spaced time points. The only difference would be the choice of path coefficients from the slope factor to the measured variables. For example, if data were gathered at one year, three years, and six years after an initial measurement point, the paths from the F2 slope factor to V1 through V4 could simply be fixed to 0, 1, 3, and 6, respectively, to yield a growth rate in an annual metric (assuming the intercept factor F1 to be initial amount).

Balanced designs and complete data. The terms *balanced* and *complete* are frequently encountered in longitudinal studies. A *balanced* design describes a longitudinal study in which subjects are planned to be measured at the same time points, whereas an *unbalanced* design occurs when not all subjects are planned to be measured at the same time points. *Complete*, and its complement *incomplete*, are terms that refer to the data themselves. Complete data occur when there are no missing data; observations that were planned were realized (i.e., all data were collected). Incomplete data, on the other hand, indicates missingness: observations were planned but not fully realized. It is possible to have a balanced design with complete data, a balanced design with incomplete data, an unbalanced design with complete data, or an unbalanced design with incomplete data. One of the advantages that LGM techniques have over conventional methods for longitudinal analyses like ANOVA or MANOVA is that they do not require the longitudinal design to be balanced or the data to be complete. LGM offers a great deal of flexibility in terms of the longitudinal design as well as how it accommodates missing data. Generally, LGM can be used in cases where data collection schemes are unique at the individual level so that unequal numbers of time points and spacing of assessments across subjects are possible. A consequence of the latter is that this allows the observations to be anchored directly to time rather than to the order in which the observations were obtained. This is particularly important in developmental studies where the phenomenon under investigation (e.g., language acquisition) is tied to age and where age distinctions are important substantive features. More subtly, perhaps, is that because LGM does not require the spacing of assessments to be equal across subjects, the data collection protocol does not require obtaining observations of subjects at the same points in time, thereby adding a great deal of flexibility to the timing and number of measurements for each subject. So, for example, one subject might be measured at 24, 29, 33, 37, and 40 months, while another might be measured at 23, 28, 29, and 35 months (see, e.g., Bauer, 2003; Mehta & West, 2000).

Although many longitudinal studies are designed to collect data on every subject in the sample at each time point, many studies have some missing observations. In fact, missingness seems to be the rule, and not the exception, in longitudinal studies in the social and behavioral sciences. Non-response for a subject can occur at any occasion in the design or accrue at the end of the study period through *attrition* or *dropout*, where some subjects withdraw from the study before its intended completion. Missingness could also occur by design, such as in *cohort sequential* (or *accelerated*) designs (see, e.g., Duncan, Duncan, & Hops, 1996). For example, a researcher interested in examining annual growth from age 5 through age 9, which would take four years using a conventional design, could instead track simultaneously a cohort from age 5 through age 7 (missing ages 8 and 9), a cohort from age 6 through age 8 (missing ages 5 and 9), and a cohort from age 7 through age 9 (missing ages 5 and 6); this study would take only two years.

However missing data arise, by design or otherwise, they have some important implications for longitudinal analyses. First, there is necessarily some loss of information, which can cause a decrease in the precision with which growth factors' parameters (means, variances, covariances) can be estimated. Not surprisingly, this reduction in precision is dependent on the amount of missing data, with greater reduction generally associated with greater amounts of missing data. Second, under certain conditions missing data can introduce bias in the parameter estimates associated with the growth factors, thereby potentially compromising the ability to draw valid inferences. It is therefore critical that missingness in longitudinal designs be treated using state-of-the-art approaches, such as full information maximum likelihood (FIML) or multiple imputation (MI); readers are referred to the chapter by Enders (2013) in this volume, as well as more detailed treatments (e.g., Enders, 2010; Little & Rubin, 2002; Schafer & Graham, 2002).

Fitted functions for each individual. In addition to quantifying the variation in growth characteristics across individuals (e.g., variance and covariance of growth factors), individual differences can also be assessed by examining the linear trajectories for each individual case based on its raw data. For linear models, for example, this requires estimating each individual's intercept and slope,[7] producing a fitted trajectory, and superimposing this fitted function against an

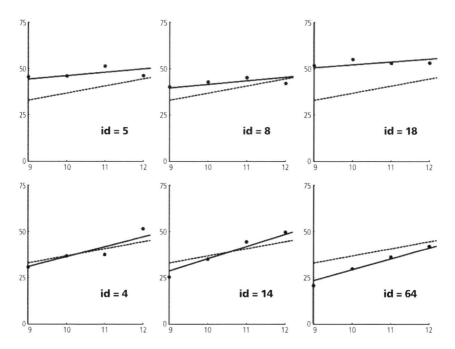

Figure 8.4 Fitted linear functions of six students' math proficiency scores, across grades 9 through 12.

individual's observed scores. These individual fitted linear functions across the sample could be organized in a graphical display known as a *lattice plot* (see, e.g., Cudeck & Harring, 2010). For example, the top row of plots in Figure 8.4 shows three students' fitted linear trajectories superimposed on their observed math proficiency scores, where those trajectories indicated growth rates that were less than the mean growth rate (shown as a dashed line). The bottom row of Figure 8.4, on the other hand, displays the fitted functions for three students whose growth rates were greater than the average.

LGM METHODOLOGY FOR
NONLINEAR DEVELOPMENTAL PATTERNS

Latent growth models certainly need not be restricted to linear forms. When theory dictates a specific nonlinear form, or if a researcher wishes to compare models with linear and nonlinear components (e.g., as done in trend analysis in ANOVA), several options exist. First, if one has a theoretical reason to suspect a specific nonlinear functional form for individuals' longitudinal change, then loadings for F2 may be chosen accordingly. For example, across a set of four equally spaced time points, if a researcher theorizes that change is decreasing with each interval, F2 loadings for V1 through V4 could be set to 0, 1, 1.5, and 1.75, respectively, such that each interval's change is half that of the immediately preceding interval. Similarly, F2 loadings for V1 through V4 could be set to 0, 1, $2^{1/2} = 1.414$, and $3^{1/2} = 1.732$, respectively, where growth in the outcome follows the square root of time.

Another option, which could be considered somewhat more exploratory, would be to assume that a single functional form underlies individuals' growth, but that such form is not known a priori. In this case the researcher may impose a latent growth model with an unspecified trajectory. For the four equally spaced measurements shown in Figure 8.2, the loadings from the growth factor F2 could be designed as 0, 1, b_{V3F2}, and b_{V4F2}, where the last two parameters are to be estimated rather than fixed to specific values. In such a model the amount of change for an individual in the initial interval from time 1 to time 2 becomes a yardstick against which other change is gauged. Thus, the values of b_{V3F2} and b_{V4F2} represent the amount of change from time 1 to time 3, and from time 1 to time 4, respectively, relative to the amount from time 1 to time 2. For example, consider estimated values of $\hat{b}_{V3F2} = 1.8$ and $\hat{b}_{V4F2} = 2.2$ which, although increasing numerically, indicate a general decline in the rate of change over time. The value of $\hat{b}_{V3F2} = 1.8$ implies that that change from time 1 to time 3 is 1.8 times that from time 1 to time 2, or that the growth from time 2 to time 3 is only .8 times the initial interval. Similarly, the value of 2.2 implies that that change from time 1 to time 4 is 2.2 times that from time 1 to time 2. This means that the growth from time 2 to time 4 is only 1.2 times the initial interval; further, because the growth from time 2 to time 3 was .8 times the initial interval, we may infer that growth from time 3 to time 4 is .4 times the initial interval.

For both the specified and unspecified alternatives for modeling non-linearity presented above, one may compare multiple models to assess competing hypotheses regarding growth's functional form. In each case, for example, one may fit a linear growth model as well as the nonlinear alternative. For the specified approach, with F2 loadings fixed to specific numerical values reflecting a hypothesized nonlinear trajectory, the linear and nonlinear models do not have a hierarchical relation; thus, comparison may proceed using data-model fit indices such as the Akaike Information Criterion (AIC) and/or Bayesian Information Criterion (BIC). For the unspecified strategy, the linear model is nested within that model; as such, the two models share a hierarchical relation and thus may be compared statistically using a χ^2 difference (likelihood ratio) test.

If a researcher suspects that growth is governed by multiple specific functional forms, such as having both linear and quadratic components, a model like that depicted in Figure 8.5 may be fitted to the data. In this model, which is one of many such variations, F1 is the intercept factor represent-

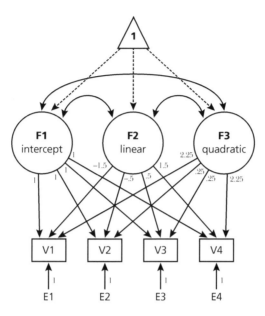

Figure 8.5 Latent growth model with linear and quadratic growth components.

ing the amount of the outcome at a reference point exactly between time 2 and time 3, F2 is the linear growth factor representing the expected rate of change per unit time, and F3 is the quadratic growth factor capturing the degree of quadratic curvature for each individual, with loadings equal to the squares of those for the linear factor. The additional parameters as a result of F3's inclusion include the latent mean a_{F3}, where a negative value indicates a general tendency for individuals' change to taper off over time, and a positive value indicates a general tendency for individuals' change to accelerate over time. The latent variance c_{F3} captures the diversity in magnitude of quadratic curvature, where a zero value would indicate a common degree of curvature (that reflected by a_{F3}) for all individuals. Lastly, the latent covariances c_{F1F3} and c_{F2F3} may be estimated, representing the degree to which quadratic curvature relates to individuals' amount of the outcome at the reference point and their linear rate of longitudinal growth, respectively.[8] Nonlinear models with multiple functional forms are discussed by Stoolmiller (1995), and an excellent example of this approach is illustrated by Stoolmiller et al. (1993).

Modeling with Nonlinear Functions

While quadratic models, and even higher order polynomials (e.g., cubic functions), fit easily into the LGM framework, there exist numerous research situations for which polynomials do not adequately summarize nonlinear change processes. Over a realistic time interval many behavioral processes either begin or end at a plateau while exhibiting differential rates of change throughout the span of development. Verbal facility, for example, increases rapidly in a child's formative years and reaches an optimal individual level in late adolescence. Behavior extinction may exhibit slow initial decline, followed by a stage of rapid drop off, and then decrease more slowly again as cessation of the behavior is attained. Many types of functions are available to describe this type of behavior (Seber & Wild, 1989), yet the final choice of a functional form is decided, in part, because its coefficients give some useful information about the underlying behavior that the function summarizes. Fortunately, the LGM framework can accommodate other nonlinear functions besides polynomials.

An exponential function, like that specified in Equation 8.1 and depicted in Figure 8.6, is frequently used where there is a rapid increase in response from a low initial starting value which then levels off at later measurements occasions:

$$V = F2 - (F2 - F1) \cdot e^{-\alpha t} + E. \tag{8.1}$$

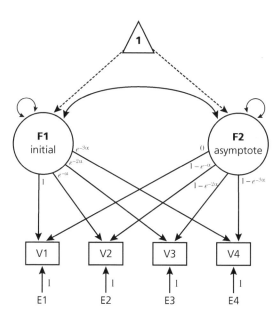

Figure 8.6 Exponential model for nonlinear growth that summarizes limiting or asymptotic behavior at later occasions of measurement.

In this model F1 is the intercept factor representing the initial amount of the outcome at time 1, F2 is the asymptotic growth factor representing potential performance at large values of time, and α is the parameter that controls how rapidly the function proceeds from F1 to F2. Larger values of α correspond to more rapid growth in the initial stages of the study period while smaller values of α reflect more gradual steady growth (Harring, Cudeck, & du Toit, 2006). One caveat worth pointing out is that the loadings, which were constants for linear and quadratic functions, are now functions of an exponential growth term α. In practice, this means that α is not allowed to vary (i.e., is fixed across individuals), nor does it covary with other growth factors. On the other hand, initial value and asymptotic behavior, represented by F1 and F2, are allowed to vary and covary across individuals. This type of function, in which the nonlinear aspects of the model (e.g., the exponential growth rate parameter α in Equation 8.1) are fixed across subjects while aspects that enter the function in a linear fashion vary/covary (e.g., the initial value factor F1 and asymptotic behavior factor F2 in Equation 8.1), is termed a *conditionally linear* model (Blozis & Cudeck, 1999), and can be analyzed in SEM software through the implementation of nonlinear constraints.[9] The Appendix contains appropriate syntax for fitting this conditionally linear exponential function to the math proficiency data using the Mplus software package.

Structured Latent Curve Models

Another type of latent growth model that captures change with a nonlinear function is the *structured latent curve model* (Browne, 1993). Formulation of this model begins with a specification of a function, referred to as the *target function* (e.g., logistic, exponential, Gompertz), to represent the mean response. Unlike individual trajectories of the conditionally linear model as described above, individuals need not follow the form of the target function in structured latent curve models. An individual's latent response is specified to be a linearly weighted combination of basis curves, a first-order Taylor polynomial taken with respect to the target function, which comprise the columns of the factor loading matrix. The basis curves for a particular application will be functions of the nonlinear parameters of the target function (Blozis, 2003). Another distinction between a conditionally linear latent growth model and a structured latent curve model is that the latter allows variation in the nonlinear parameter(s) to be estimated with other model parameters whereas the former model does not. Structured latent curve models, while dissimilar in orientation from conditionally linear latent growth models, may also be fitted using SEM software; the interested reader is referred to Preacher & Hancock (2012).

Piecewise Latent Growth Models

The concept of additional factors to capture nonlinearity may also be adapted to situations where different growth patterns are believed to occur during different time periods. Imagine, for example, five equally spaced time points in which a linear pattern is hypothesized to govern change from the first to third time points, and a potentially different linear pattern from the third through fifth. Reasons for such a model typically involve an event occurring, such as a transition from middle school to high school, or perhaps some kind of treatment intervention. The effect is to yield a discontinuity in the overall growth pattern, making it resemble two lines spliced together—a *spline*. For such *discontinuity* designs there are different options, depending upon the researcher's interest.

Consider the top of Figure 8.7, in which a single individual's linear trajectory is depicted as a solid line from time 1 to time 3 (phase 1), follow by a second linear trajectory from time 3 to time 5 as another solid line (phase 2). If the researcher wishes to view the second phase as entirely separate from the first, that is, as if time 3 were itself an intercept, then the dotted line shows the frame of reference for growth during the second phase. The corresponding model appears in the bottom of Figure 8.7, where F2 is intended to represent linear growth during the first phase and F3 is linear growth during the second phase. Notice that the paths from F2 stop changing in value at time 3, holding at a value of 2 for the remaining measurement points. This effectively makes the combination of F1 and F2 serve as the intercept for the second growth phase, freezing in the level of growth attained by time 3 as the point of departure for the second growth factor F3. During the first three time points F3 lay dormant, with paths of 0; at times 4 and 5, however, the linear growth from time 3 is modeled with the expected linear weights of 1 and 2. Because growth is treated in separate, although potentially related pieces, this approach is sometimes referred to as a *piecewise* growth model.

All of the expected latent parameters exist for this model for both the covariance structure (c_{F1}, c_{F2}, c_{F3}, and c_{F1F2}, c_{F1F3}, c_{F2F3}) and mean structure (a_{F1}, a_{F2}, a_{F3}; not depicted in Figure 8.7). Notice that a special case of this model is that in which a single linear process governs the entire time span; such a model would have variances $c_{F2} = c_{F3}$, covariances $c_{F1F2} = c_{F1F3}$ and $c_{F2F3} = (c_{F2}c_{F3})^{1/2}$ (i.e., F2 and F3 are perfectly correlated), and latent means $a_{F2} = a_{F3}$. Consequently, there is a hierarchical relation between the linear model and the piecewise model, where the former is nested within the latter. That said, as the linear model represents a boundary condition for the piecewise model (i.e., F2 and F3 correlating perfectly), comparing the models using data-model fit information (e.g., the AIC) might be a better

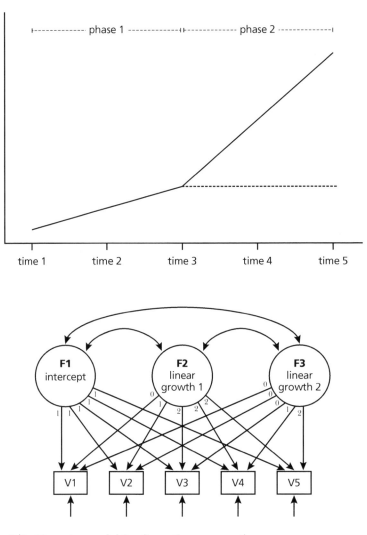

Figure 8.7 Piecewise model for discontinuous growth.

choice than the familiar χ^2 difference (likelihood ratio) test (e.g., Stoel, Garre, Dolan, & van den Wittenboer, 2006).

As a variation on the piecewise model, the researcher may wish to frame growth in the second phase somewhat differently. Specifically, one may consider the continuation of the first growth phase as one's baseline, and then model growth in the second phase as potentially *additive* to that process that is already underway. Consider the top of Figure 8.8, where the solid lines depict the same individual's two phases of growth as in Figure 8.7, but

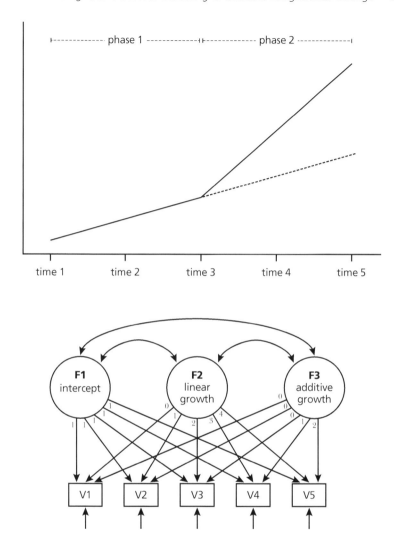

Figure 8.8   Model for additive growth.

where the dotted line now indicates the linear mechanism set into motion in the first phase. The corresponding model in the bottom of Figure 8.8 shows F2 to be the same as in a continuous growth model spanning all time points, and F3 to be a growth mechanism that does not engage until the second phase, just as in Figure 8.7. A key difference between F3 in Figures 8.7 and 8.8, however, is their frame of reference. Whereas in Figure 8.7 the F3 factor was essentially an entirely new linear growth process, in Figure 8.8 it is an additive process relative to the continuation of the first phase's growth mechanism. Thus, in Figure 8.7 if F3 were irrelevant then it implies that growth has stopped (flattened), while in Figure 8.8 an irrelevant F3 factor means growth continues just as it did in the first phase. This model again has all of the expected covariance and mean structure parameters, but with their interpretation adjusted according to the additive role for F3. Notice again that the single continuous process model is a special case of this model, and perhaps more obviously so than in the piecewise case of Figure 8.8. Quite simply, if variance $c_{F3} = 0$, covariances $c_{F1F3} = c_{F2F3} = 0$, and latent mean $a_{F3} = 0$, one has the traditional linear latent growth model.

Finally, three interesting variations on the piecewise models are worth mentioning briefly. First, rather than having a single transition point, one could have multiple transition points; in fact, one could have transitions at all time points, effectively modeling latent change between each pair of adjacent points. Variations of these *latent difference score models* are addressed in work by McArdle (2001) and Steyer and colleagues (e.g., Steyer, Eid, & Schwenkmezger, 1997; Steyer, Partchev, & Shanahan, 2000). Second, all of the above models, including those in Figure 8.7 and Figure 8.8, have the transition point from one phase to the next as a known value. It is possible, however, to model that transition point as a characteristic to be estimated within the model. That is, one might hypothesize that a transition occurs for the individuals under study, but not know where that occurs a priori. As Harring et al. (2006) demonstrated, such models may be estimated within any SEM software package that allows the imposition of nonlinear constraints. Lastly, the piecewise models reviewed here assume that the functions on either side of the transition point join at this point. There are instances when the functional pieces are not hypothesized to join and the change in response at the transition point is an important feature to model (see, e.g., Cudeck & Codd, 2012).

ACCOMMODATING EXTERNAL VARIABLES IN LATENT GROWTH MODELS

One of the unique and powerful advantages of utilizing LGM methodology over traditional methods is its ability to incorporate predictors of the latent growth factors, thereby attempting to explain individual differences in latent trajectories. For example, imagine a researcher was interested in the potential effect of mother's years of education on high school girls' initial math proficiency as well as on the degree to which the girls' math proficiency changes over time. Such questions can be addressed directly by introducing the appropriate predictor variable into the model. Consider the model in Figure 8.9, from which the mean structure has been omitted for simplicity.

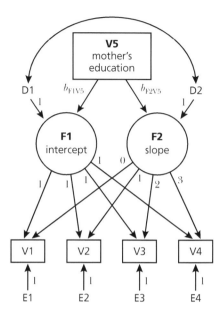

Figure 8.9 Latent growth model with predictor of intercept and slope factors.

Assuming the model fits satisfactorily to the covariance matrix relating V1 through V5 (i.e., including mother's education), one may interpret the new structural parameter estimates. In particular, the $\hat{b}_{F1V5}$ estimate describes the sign and magnitude of the relation between mother's education and initial math proficiency in the sample examined; the test statistic for this path facilitates inference regarding this relation in the population. As for the proportion of variability in F1 explained by mother's education, the researcher may compute this as one minus the squared value of the standardized path coefficient from D1 to F1 (which is part of standard SEM computer output). Similarly, the $\hat{b}_{F2V5}$ estimate describes the sign and magnitude of the relation between mother's education and growth in math proficiency in the sample examined, while the test statistic for this path facilitates inference at the population level. To determine the proportion of variability in F2 explained by mother's education, the researcher may compute one minus the squared value of the standardized path from D2 to F2.

In general, there are many interesting and highly useful variations to the inclusion of external predictors of latent growth factors. In the above example, a lone predictor of mother's education was included. Certainly researchers may wish to include multiple measured predictors in their models in order to assess their impact on growth factors, as is illustrated by numerous papers cited herein (e.g., Patterson, 1993; Stoolmiller et al., 1993). Particularly interesting are predictors within the piecewise and additive growth model framework, where the latter can be used to assess determinants of individual differences in responsiveness to an intervention. Researchers also have the option of using predictors that are latent in nature; that is, an exogenous factor (with its own measured indicators) may be included as a predictor of the growth factors. For example, imagine that a researcher wished to investigate the effect of 9th grade academic self-concept on initial math proficiency and growth in that proficiency. Rather than using a single measure of academic self-concept as a predictor, the researcher may define a self-concept factor from multiple indicators (e.g., questionnaire items), and build this factor as a causal element into the growth model.

Further examples of measured (as opposed to latent) external predictors in a growth model include group code (e.g., dummy) variables indicating group membership. For example, differences between males and females in initial math proficiency and proficiency growth could be assessed by creating a dichotomous sex variable as a predictor in a dataset containing high school students of both sexes. Statistically significant paths from that dummy variable to the intercept or slope factor would suggest sex differences in initial amount and/or growth in math proficiency in high school. As an extension, if differences among more than two populations are desired, a set of $k-1$ group code variables may be used as predictors to facilitate inference about differences among the k populations of interest.

Finally, the predictors discussed herein, whether measured or latent, are examples of *time-independent covariates*, whose values are assumed to be stable over time. In the case of mother's education, the assumption was that the mother received no additional education after her daughter's first math proficiency measurement in 9th grade. However, one can certainly imagine predictors whose values change over time; such variables are termed *time-dependent covariates* and are modeled as covarying predictors of the corresponding measured variables directly. In this manner the growth that is modeled is that above and beyond the variation accounted for by the covariate at each time point. The interested reader is referred to, for example, Bollen and Curran (2006), Duncan et al. (2006), and Preacher et al. (2008).

VARIATIONS, EXTENSIONS, AND CONCLUSIONS

While the models discussed in the current chapter were chosen to illustrate the basic principles of LGM methods, many additional variations and extensions of these models may be imposed upon one's data as well. For instance, in some cases investigators may obtain repeated measures on two or more variables, with interest in not only how each variable changes across the study period but also in how characteristics of change in the different variables are related. Continuing with the previous proficiency context, the same group of individuals might provide measures for both math proficiency and reading proficiency, where the interest is to determine whether individuals who start with below average proficiency in reading also begin with lower math proficiency. There might also be interest in determining whether growth in one measure is related to growth in the other. Studying parallel processes in the same LGM framework is sometimes referred to as *multidomain latent growth curve modeling* (e.g., Willett & Sayer, 1996), and falls under the more general heading of *multivariate repeated measures analysis* (see, e.g., Blozis, 2003; MacCallum, Kim, Malarkey, & Kiecolt-Glaser, 1997).

Note that multiple measures are sometimes present at each point in time because they are believed to be indicators of the same construct (e.g., multiple self-concept questionnaire items). Here the researcher is likely more interested in growth in the latent construct underlying those measures rather than in the individual measures themselves. In such cases one may analyze a *second-order latent growth model* (Hancock, Kuo, & Lawrence, 2001; Sayer & Cumsille, 2001), also known as a *curve-of-factors model* (McArdle, 1988), where the growth factors are modeled as second-order factors influencing the first-order constructs whose longitudinal change is of interest. Details of such models, including measurement invariance across time points and requirements of the mean structure, are presented elsewhere (see, e.g., Hancock et al., 2001; Ferrer, Balluerka, & Widaman, 2008). Also addressed is the interesting case where subjects' developmental levels necessitate that different indicators be used at different time points in the growth model (Hancock & Buehl, 2008).

Following along the lines of group comparisons, as discussed previously in this chapter with group code variables, researchers may actually build separate models for different groups of interest if sample size permits. For example, one may wish to compare the effect of two treatments over time. Constraining growth parameters to be equal across groups, and then examining the results of statistical tests of those equality constraints, would facilitate inference about longitudinal differences in the treatments under investigation (for a variation on this with additive growth models, see Muthén & Curran, 1997).

Latent growth modeling may even be used to compare groups whose membership is not known. *Latent growth mixture models* combine principles of LGM with those of finite mixture models (see Pastor & Gagné, this volume) to test hypotheses positing multiple latent growth classes. For example, a model with three classes might be found to fit the data better than models with one or two, suggesting, say, a class of individuals who start low change very little over time, another class that starts slightly higher and grows considerably, and a third class that starts high and stays high over time. These classes might differ in terms of one or more latent parameters, although differences in mean structure parameters might allow the greatest ability to detect distinct latent growth classes. More detail on such innovations may be found in seminal work by Muthén (e.g., 2001; 2004).

Other LGM innovations are both exciting and too numerous to address. New developments by leading LGM methodologists are constantly underway, further increasing researchers' options for addressing longitudinal change (see, e.g., Harring & Hancock, 2012). It is our hope that the introduction offered in the current chapter has clearly illustrated the benefits of LGM, and in turn will help researchers to utilize these methods in working toward a fuller explication of longitudinal growth in variables from their own areas of study.

NOTES

1. In the current chapter the term "growth" will be used in generic reference to change over time, regardless of whether that change is positive or negative.
2. Graphical displays are irreplaceable in judging whether a particular function fits individual growth patterns adequately. Spaghetti plots (e.g., Harring, 2009) in conjunction with exploring candidate functions vis-à-vis nonlinear least squares (Cudeck & Harring, 2007) are two visual approaches for determining an appropriate functional form.
3. Note that fixing the path from F2 to V1 to a value of 0 is equivalent to omitting the path altogether. Including the path in Figure 8.2 with a coefficient of 0 was done simply to illustrate the slope factor's impact on V1 relative to that on all other measured variables.
4. Note that part of the researcher's hypothesis regarding growth could involve specifying the equality of the variances of the variables' errors (i.e., $c_{E1} = c_{E2} = \ldots$). Such a hypothesis could be made if the researcher believed that measurement error in all variables was consistent across time points. The imposition of such an equality constraint creates a more restrictive, though conceptually simpler, explanation of the data (see, e.g., Dunn et al., 1993, p. 128; Patterson, 1993). Other types of error covariance structures are also possible (see, e.g., Blozis et al., 2008).
5. Two elements regarding this mean structure may seem inconsistent with latent means modeling in common practice (e.g., Thompson & Green, 2013, this volume). First, one notices the absence of paths from the pseudovariable to the measured variables, creating variable-specific intercept terms. Second, no factor mean needs to be fixed for identification purposes. In LGM, two properties make these requirements unnecessary. First, rather than each variable having its own intercept term, as in latent means models, one common intercept factor is built directly into the model (making variable-specific intercept terms superfluous). Second, because so many loadings are constrained, the model is identified and the factors are given the scale associated with all of the indicators simultaneously (which, in LGM, has the same units).
6. It is worth pointing out that a linear latent growth model with only three time points would have six parameters estimated in its covariance structure, and two

in its mean structure. While the model as a whole would have 1 degree of freedom, the covariance structure itself would be saturated. As such, the covariance structure alone does not pose a rejectable test of the hypothesis of linearity in individual growth; only the mean structure, from whence the single degree of freedom emanates, facilitates a test of linearity at the group (mean) level. Thus, three time points represents somewhat of a lower bound on fitting a latent growth model, while four time points represents a lower bound on rejectability of hypotheses regarding a single specified functional form.

7. Individuals' intercepts and slopes can be estimated after fitting the linear growth model. These predicted values are based on the joint multivariate normal distribution of the data and the growth factors and have the form of empirical Bayes estimates (Fitzmaurice, Laird, & Ware, 2011). For the ith individual, these are determined as $\hat{\boldsymbol{\eta}}_i = \hat{\boldsymbol{\Psi}}\hat{\boldsymbol{\Lambda}}_i'\hat{\boldsymbol{\Sigma}}_i^{-1}\mathbf{y}_i$, where the hats indicate that maximum likelihood estimates have been inserted into each parameter matrix. In this formulation for individual i, $\boldsymbol{\eta}_i$ are the linear growth factors, $\boldsymbol{\Psi}$ represents the matrix of variances and covariance of the linear growth factors, $\boldsymbol{\Lambda}_i$ is the factor loading matrix, $\boldsymbol{\Sigma}_i$ is the model implied covariance matrix, and $\mathbf{y}_i$ is the observed repeated measures. The i subscript on some matrices and vectors simply allows them to be of different dimension, not necessarily comprised of different elements.

8. Note that the nonlinear model depicted in Figure 8.5, with four time points, would have 10 parameters estimated in its covariance structure and three in its mean structure. Thus, while the model as a whole would have 1 degree of freedom, the covariance structure itself would be saturated and thus not pose a rejectable test of the hypothesis of nonlinearity. However, by comparing the linear and nonlinear models one may infer the reasonableness of including the hypothesized nonlinear (i.e., quadratic) component. More generally, as expected, the number of measured time points will limit the number of nonlinear components to the latent growth model.

9. Nonlinear constraints are relatively new and appear in most recent versions of Mplus and LISREL, SAS PROC CALIS, among others. The idea is to be able to use transcendental functions like $ln(\ )$, $exp(\ )$, etc. in the design matrices of LGMs. Three variations on this theme have been addressed in work by Choi, Harring, and Hancock (2009); Grimm and Ram (2009); and Harring et al. (2006).

APPENDIX
Software Syntax for Linear Latent Growth Model Example

EQS (6.2)

```
/TITLE
 LINEAR LATENT GROWTH MODEL FOR MATH PROFICIENCY
/SPECIFICATIONS
 CASES=1000; VAR=4; MA=COV; AN=MOMENTS;
/LABELS
 V1=MATHSC9; V2=MATHSC10; V3=MATHSC11; V4=MATHSC12;
 F1=INTERCEPT; F2=SLOPE;
```

```
/EQUATIONS
 V1 = 1F1 + 0F2 + E1;
 V2 = 1F1 + 1F2 + E2;
 V3 = 1F1 + 2F2 + E3;
 V4 = 1F1 + 3F2 + E4;
 F1 = *V999 + D1;
 F2 = *V999 + D2;
/VARIANCES
 E1 TO E4 = *;
 D1 TO D2 = *;
/COVARIANCE
 D1,D2 = *;
/MATRIX
 204.11
 139.20  190.05
 136.64  135.19  166.53
 124.86  130.32  134.78  159.87
/MEANS
  32.97   36.14   40.42   43.75
/PRINT
 FIT=ALL;
/END
```

SIMPLIS (8.8)

```
LINEAR LATENT GROWTH MODEL FOR MATH PROFICIENCY
OBSERVED VARIABLES
MATHSC9 MATHSC10 MATHSC11 MATHSC12
COVARIANCE MATRIX
 204.11
 139.20  190.05
 136.64  135.19  166.53
 124.86  130.32  134.78  159.87
MEANS
  32.97   36.14   40.42   43.75
SAMPLE SIZE = 1000
LATENT VARIABLES
INTERCEPT SLOPE
RELATIONSHIPS
 MATHSC9 = 1*INTERCEPT + 0*SLOPE
 MATHSC10 = 1*INTERCEPT + 1*SLOPE
 MATHSC11 = 1*INTERCEPT + 2*SLOPE
 MATHSC12 = 1*INTERCEPT + 3*SLOPE
 INTERCEPT = CONST
 SLOPE = CONST
PATH DIAGRAM
END OF PROBLEM
```

Mplus (6.12)

Data file: MATHPROF.DAT

```
  32.97    36.14    40.42    43.75
 204.11
 139.20   190.05
 136.64   135.19   166.53
 124.86   130.32   134.78   159.87
```

Input file, option 1 (general format)

```
TITLE:     LINEAR LATENT GROWTH MODEL
           FOR MATH PROFICIENCY

DATA:      FILE IS C:\MATHPROF.DAT;
           TYPE IS COVARIANCE MEANS;
           NOBS IS 1000;

VARIABLE: NAMES ARE y1-y4;

ANALYSIS: TYPE IS MEAN;

MODEL:     INTERCEP BY y1-y4@1;
           SLOPE BY y1@0 y2@1 y3@2 y4@3;
           INTERCEP* SLOPE*;
           [y1-y4@0 INTERCEP* SLOPE*];
           INTERCEP WITH SLOPE*;

OUTPUT:    SAMP;
```

Input file, option 2 (LGM-specific format)

```
TITLE:     LINEAR LATENT GROWTH MODEL
           FOR MATH PROFICIENCY

DATA:      FILE IS C:\MATHPROF.DAT;
           TYPE IS COVARIANCE MEANS;
           NOBS IS 1000;

VARIABLE: NAMES ARE y1-y4;

MODEL:     INTERCEP SLOPE | y1@0 y2@1 y3@2 y4@3;

OUTPUT:    SAMP;
```

Software Syntax for the Conditionally Linear Latent Growth Model Example Using an Exponential Function

Mplus (6.12)

Data file: MATHPROF.DAT
```
 32.97    36.14    40.42    43.75
204.11
139.20   190.05
136.64   135.19   166.53
124.86   130.32   134.78   159.87
```

Input file, option 1 (general format)

```
TITLE:     CONDITIONALLY LINEAR LATENT GROWTH
           MODEL FOR MATH PROFICIENCY USING AN
           EXPONENTIAL FUNCTION

DATA:      FILE IS C:\MATHPROF.DAT;
           TYPE IS COVARIANCE MEANS;
           NOBS IS 1000;

VARIABLE: NAMES ARE y1-y4;

ANALYSIS: TYPE IS MEAN;

MODEL:     INT BY y1@1
               y2(C1)
               y3(C2)
               y4(C3);
           ASY BY y1@0
               y2(C4)
               y3(C5)
               y4(C6);

!C1-C6 in parentheses constrain loadings defined in the model
!constraint command below

           INT* ASY*;
           [y1-y4@0 INT* ASY*];
           INT WITH ASY*;

MODEL CONSTRAINT:
           NEW (ALPHA); !Define nonlinear parameter alpha
           C1 = EXP(-1*ALPHA);
           C2 = EXP(-2*ALPHA);
           C3 = EXP(-3*ALPHA);
           C4 = 1 - C1;
           C5 = 1 - C2;
           C6 = 1 - C3;

!Expressions for loadings C1-C6 correspond to those in Figure 8.6

OUTPUT:    SAMP;
```

REFERENCES

Bauer, D. (2003). Estimating multilevel linear modesl as structural equation models. *Journal of Educational and Behavioral Statistics, 28,* 135–167.

Biesanz, J. C., Deeb-Sossa, N., Papadakis, A. A., Bollen, K. A., & Curran, P. J. (2004). The role of coding time in estimating and interpreting growth curve models. *Psychological Methods, 9,* 30–52.

Blozis, S. A. (2003). Structured latent curve models for the study of change in multivariate repeated measures. *Psychological Methods, 9,* 334–353.

Blozis, S. A., & Cudeck, R. (1999). Conditionally linear mixed effects models with latent covariates. *Journal of Educational and Behavioral Statistics, 24,* 245–270.

Blozis, S. A., Harring, J. R., & Mels, G. (2008). Using LISREL to fit nonlinear latent curve models. *Structural Equation Modeling: A Multidisciplinary Journal, 15,* 346–369.

Bollen, K. A., & Curran, P. J. (2006). *Latent curve models.* Hoboken, NJ: Wiley.

Browne, M. W. (1993). Structured latent curve models. In C. M. Cuadras & C. R. Rao (Eds.), *Multivariate analysis: Future directions 2* (pp. 171–197). Amsterdam: Elsevier Science.

Choi, J., Harring, J. R., & Hancock, G. R. (2009). Latent growth modeling for logistic response functions. *Multivariate Behavioral Research, 44,* 620–645.

Collins, L. M., & Sayer, A. G. (Eds.). (2001). *New methods for the analysis of change.* Washington, D.C.: American Psychological Association.

Cudeck, R., & Codd, C. L. (2012). A template for describing individual differences in longitudinal data, with applications to the connection between learning and ability. In J. R. Harring & G. R. Hancock (Eds.), *Advances in longitudinal methods in the social and behavioral sciences* (pp. 3–24). Charlotte, NC: Information Age Publishing.

Cudeck, R., & Harring, J. R. (2007). Analysis of nonlinear patterns of change with random coefficient models. *Annual Review of Psychology, 58,* 615–637.

Cudeck, R., & Harring, J. R. (2010). Developing a random coefficient model for nonlinear repeated measures data. In S.-Y. Chow, E. Ferrer, & F. Hsieh (Eds.), *Statistical methods for modeling human dynamics: An interdisciplinary dialogue* (pp. 289–317). New York, NY: Routledge.

Curran, P. J., Stice, E., & Chassin, L. (1997). The relation between adolescent alcohol use and peer alcohol use: A longitudinal random coefficients model. *Journal of Consulting and Clinical Psychology, 65,* 130–140.

Duncan, S. C., & Duncan, T. E. (1994). Modeling incomplete longitudinal substance use data using latent variable growth curve methodology. *Multivariate Behavioral Research, 29,* 313–338.

Duncan, S. C., Duncan, T. E., & Hops, H. (1996). Analysis of longitudinal data within accelerated longitudinal designs. *Psychological Methods, 1,* 236–248.

Duncan, T. E., Duncan, S. C., & Strycker, L. A. (2006). *An introduction to latent variable growth curve modeling: Concepts, issues, and applications* (2nd ed.). Mahwah, N.J.: Lawrence Erlbaum Associates..

Dunn, G., Everitt, B., & Pickles, A. (1993). *Modelling covariances and latent variables using EQS.* London: Chapman & Hall.

Enders, C. K. (2010). *Applied missing data analysis.* New York, NY: The Guilford Press.

Enders, C. K. (2013). Analyzing structural equation models with missing data. In G. R. Hancock & R. O. Mueller (Eds.), *Structural equation modeling: A second course* (2nd ed.). Charlotte, NC: Information Age Publishing.

Ferrer, E., Balluerka, N., & Widaman, K. F. (2008). Factorial invariance and the specification of second-order latent growth models. *Methodology, 4,* 22–36.

Fitzmaurice, G. M., Laird, N. M., & Ware, J. H. (2011). *Applied longitudinal analysis* (2nd ed.). New York, NY: Wiley.

Gottman, J. M. (Ed.). (1995). *The analysis of change.* Mahwah, NJ: Lawrence Erlbaum Associates, Publishers.

Grimm, K. J., & Ram, N. (2009). Nonlinear growth models in M*plus* and SAS. *Structural Equation Modeling, A Multidisciplinary Journal, 16,* 676–701.

Hancock, G. R., & Buehl, M. M. (2008). Second-order latent growth models with shifting indicators. *Journal of Modern Applied Statistical Methods, 7,* 39–55.

Hancock, G. R., & Choi, J. (2006). A vernacular for linear latent growth models. *Structural Equation Modeling: A Multidisciplinary Journal, 13,* 352–377.

Hancock, G. R., Kuo, W., & Lawrence, F. R. (2001). An illustration of second-order latent growth models. *Structural Equation Modeling: A Multidisciplinary Journal, 8,* 470–489.

Harring, J. R. (2009). A nonlinear mixed effects model for latent variables. *Journal of Educational and Behavioral Statistics, 34,* 293–318.

Harring, J. R., Cudeck, R., & du Toit, S. H. C. (2006). Fitting partially nonlinear random coefficient models as SEMs. *Multivariate Behavioral Research, 41,* 579–596.

Harring, J. R., & Hancock, G. R. (Eds.) (2012). *Advances in longitudinal methods in the social and behavioral sciences.* Charlotte, NC: Information Age Publishing.

Little, R. J. A., & Rubin, D. B. (2002). *Statistical analysis with missing data* (2nd ed.). Hoboken, NJ: Wiley.

McArdle, J. J. (1988). Dynamic but structural equation modeling with repeated measures data. In J. R. Nesselroade & R. B. Cattell (Eds.), *Handbook of multivariate experimental psychology* (Vol. 2, pp. 561–614). New York, NY: Plenum Press.

McArdle, J. J. (2001). A latent difference score approach to longitudinal dynamic structural analyses. In R. Cudeck, S. du Toit, & D. Sorbom (Eds.), *Structural equation modeling: Present and future—A Festschrift in honor of Karl Jöreskog* (pp. 342–380). Lincolnwood, IL: Scientific Software International, Inc.

MacCallum, R. C., Kim, C., Malarkey, W. B., & Kiecolt-Glaser, J. K. (1997). Studying multivariate change using multilevel models and latent curve models. *Multivariate Behavioral Research, 32,* 215–253.

Mehta, P. D., & West, S. G. (2000). Putting the individual back into individual growth curves. *Psychological Methods, 5,* 23–43.

Moskowitz, D. S., & Hershberger, S. L. (Eds.). (2002). *Modeling intraindividual variability with repeated measures data: Methods and applications.* Mahwah, NJ: Lawrence Erlbaum.

Muthén, B. (2001). Latent variable mixture modeling. In G. A. Marcoulides & R. E. Schumacker (Eds.), *New developments and techniques in structural equation modeling* (pp. 1–33). Lawrence Erlbaum Associates, Publishers.

Muthén, B. (2004). Latent variable analysis: Growth mixture modeling and related techniques for longitudinal data. In D. Kaplan (Ed.), *The SAGE Handbook of*

Quantitative Methodology for the Social Sciences (pp. 345–368), Thousand Oaks, CA: Sage Publications.

Muthén, B., & Curran, P. (1997). General longitudinal modeling of individual differences in experimental designs: A latent variable framework for analysis and power estimation. *Psychological Methods, 2,* 371–402.

Pastor, D., & Gagné, P. (2013). Mean and covariance structure mixture models. In G. R. Hancock & R. O. Mueller (Eds.), *Structural equation modeling: A second course* (2nd ed.). Charlotte, NC: Information Age Publishing.

Patterson, G. R. (1993). Orderly change in a stable world: The antisocial trait as a chimera. *Journal of Consulting and Clinical Psychology, 61,* 911–919.

Preacher, K. J., & Hancock, G. R. (2012). On interpretable reparameterization of linear and nonlinear latent growth curve models. In J. R. Harring & G. R. Hancock (Eds.), *Advances in longitudinal methods in the social and behavioral sciences* (pp. 25–58). Charlotte, NC: Information Age Publishing.

Preacher, K. J., Wichman, A. L., MacCallum, R. C., & Briggs, N. E. (2008). *Latent growth curve modeling.* Thousand Oaks, CA: Sage Publications.

Raudenbush, S. W., & Bryk, A. S. (2002). *Hierarchical linear models: Applications and data analysis methods* (2nd ed.). Newbury Park, CA: Sage Publications.

Rogosa, D., & Willett, J. B. (1985). Understanding correlates of change by modeling individual differences in growth. *Psychometrika, 50,* 203–228.

Sayer, A. G., & Cumsille, P. E. (2001). Second-order latent growth models. In L. Collins & A. G. Sayer (Eds.), *New methods for the analysis of change* (pp. 177–200). Washington, DC: American Psychological Association.

Schafer, J. L., & Graham, J. W. (2002). Missing data: Our view of the state of the art. *Psychological Methods, 7,* 147–177.

Seber, G. A. F., & Wild, C. J. (1989). *Nonlinear regression.* New York, NY: Wiley.

Short, R., Horn, J. L., & McArdle, J. J. (1984). Mathematical-statistical model building in analysis of developmental data. In R. N. Emde & R. J. Harmon (Eds.), *Continuities and discontinuities in development* (pp. 371–401). New York, NY: Plenum Press.

Singer, J. D., & Willett, J. B. (2003). *Applied longitudinal data analysis: Modeling change and event occurrence.* New York, NY: Oxford University Press.

Steyer, R., Eid, M., & Schwenkmezger, P. (1997). Modeling true intraindividual change: True change as a latent variable. *Methods of Psychological Research Online, 2,* 21–33.

Steyer, R., Partchev, I., & Shanahan, M. J. (2000). Modeling true intraindividual change in structural equation models: The case of poverty and children's psychosocial adjustment. In T. D. Little, K. U. Schnabel, & J. Baumert (Eds.), *Modeling longitudinal and multilevel data: Practical issues, applied approaches, and specific examples* (pp. 109–126). Mahwah, NJ: Lawrence Erlbaum Associates, Publishers.

Stoel, R. D., Garre, F. G., Dolan, C., & van den Wittenboer, G. (2006). On the likelihood ratio test in structural equation modeling when parameters are subject to boundary constraints. *Psychological Methods, 11,* 439–455.

Stoolmiller, M. (1995). Using latent growth curve models to study developmental processes. In J. M. Gottman (Ed.), *The analysis of change* (pp. 103–138). Mahwah, N.J.: Lawrence Erlbaum Associates, Publishers.

Stoolmiller, M., Duncan, T., Bank, L., & Patterson, G. R. (1993). Some problems and solutions in the study of change: Significant patterns in client resistance. *Journal of Consulting and Clinical Psychology, 61*, 920–928.

Thompson, M. S., & Green, S. B. (2013). Evaluating between-group differences in latent variable means. In G. R. Hancock & R. O. Mueller (Eds.), *Structural equation modeling: A second course* (2nd ed.). Charlotte, NC: Information Age Publishing.

Willett, J. B., & Sayer, A. G. (1994). Using covariance structure analysis to detect correlates and predictors of individual change over time. *Psychological Bulletin, 116*, 363–381.

Willett, J. B., & Sayer, A. G. (1996). Cross-domain analysis of change overtime: Combining growth modeling and covariance structure analysis. In G. A. Marcoulides & R. E. Schumacker (Eds.), *Advanced structural equation modeling. Issues and techniques* (pp. 125–157). Mahwah, NJ: Lawrence Erlbaum Associates, Publishers.

Willett, J. B., Ayoub, C. C., & Robinson, D. (1991). Using growth modeling to examine systematic differences in growth: An example of change in the functioning of families at risk of maladaptive parenting, child abuse, or neglect. *Journal of Consulting and Clinical Psychology, 59*, 38–47.

Yandell, B. S. (1997). *Practical data analysis for designed experiments.* London: Chapman and Hall.

CHAPTER 9

MEAN AND COVARIANCE STRUCTURE MIXTURE MODELS

Dena A. Pastor and Phill Gagné

Mixture modeling is becoming an increasingly useful tool in applied research settings. At the most basic level, such methods might be used to determine whether a single univariate data set arose from a homogeneous population or from a heterogeneous population consisting of unknown or latent groups (often called *classes*) that differ in their distributional parameters (e.g., means and/or variances). More advanced applications of mixture modeling are used to assess whether a population consists of a mixture of groups that have different multivariate distributions (e.g., mean vectors and/ or covariance matrices). Mixture modeling can also be used in conjunction with a variety of different latent variable models, such as factor models and latent growth curve models. For instance, factor mixture models can be used to assess whether the population consists of groups that differ in their factor model parameters (e.g., factor variances, loadings) and growth mixture modeling can be used to investigate whether unknown groups exist in the population that differ in how their members' change over time.

Structural Equation Modeling: A Second Course (2nd ed.), pages 343–393
Copyright © 2013 by Information Age Publishing
343

We start our introduction to mixture modeling by briefly revisiting several non-mixture analyses that should be quite familiar. Such a review will help to set up our discussion of mixture analyses and to contrast the mixture procedures with their non-mixture analogs. From there, we will provide a general definition of a mixture model and then move into a conceptual and mathematical discussion of select mixture models. Examples of how to estimate a selection of these models in Mplus (v6.11; Muthén & Muthén, 1998–2010) will be provided. The chapter will conclude with several issues to consider in applying mixture modeling.

Group Membership Known

Among the simplest of statistical analyses are those for making comparisons of group means. Analysis of variance (ANOVA) is commonly used to test the null hypothesis of equal means among $K \geq 2$ groups measured on a single variable, with the group membership of each observation in a sample known a priori. The extension of ANOVA to two or more dependent variables is multivariate analysis of variance (MANOVA), which is used to determine whether there are differences in the centroids (i.e., vectors of means) of two or more groups defined a priori.

Somewhat more complex are the analyses of relations among observed variables. Pearson zero-order correlations are used to test relations between two observed variables with no causal inferences made. Multiple regression features the construction of an equation for predicting a manifest criterion variable from one or more manifest predictor variables in a sample of data. In samples comprised of two or more groups, with group membership defined a priori, multiple regression can be used to test the equality of parameters (e.g., slopes) across groups. A straightforward application is to investigate whether the nature/extent of the relation between the one predictor and the criterion variable differs across groups. Another application of multiple regression is to evaluate a possible difference in the group means on the criterion variable, while taking the effects of the predictor variable(s) into account, a process that is equivalent to analysis of covariance (ANCOVA).

In confirmatory factor analysis (CFA), relations among observed variables are explained by positing that they are caused by one or more continuous latent variables. A CFA model, generally speaking, represents relations in one population from which all of the observations are assumed to have originated, but CFA can be readily extended to the comparison of multiple groups when group membership is known a priori. Such a test might be conducted, for example, to determine whether the pattern of factor loadings is invariant across groups (see, e.g., Byrne, Shavelson, & Muthén, 1989; Cheung & Rensvold, 1999; Meredith, 1993; Thompson & Green, this volume). In a special case of CFA called latent growth curve modeling (LGCM; McArdle & Epstein, 1987; Meredith & Tisak, 1990), measurements are taken of the same variables from the same respondents at multiple time points, with characteristics of change over time (e.g., intercept and slope) modeled as latent variables that influence the observed variables. LGCM can be extended to compare the means and variances of the intercept factor and the slope factor across multiple groups when group membership is known a priori (Muthén & Curran, 1997).

Group Membership Unknown

We have touched on several analyses that test the equality of various parameters in multiple groups with the benefit of information about group membership for each observation in the sample. For such analyses, obtaining and statistically comparing group parameters are relatively straightforward processes. When group membership is not known, however, or when it is not even known whether a mixture of groups exists within a population, similar statistical questions can be addressed, but the analyses are a bit more complicated.

Before detailing such analyses, we should establish a definition of the term *mixture analysis*. In a manner of speaking, any population that is made up of observations from two or more groups can be thought of as a mixed population. In ANOVA, for example, the full population is known to consist of multiple groups and group membership is known for each observation, so parameters for each group can be estimated and compared directly. Because group membership is known a priori, ANOVA is not formally considered a mixture analysis. When a population lacks information about group membership but there is a research question to be answered involving the parameter estimates of potentially multiple groups, a mixture analysis is called upon. A mixture analysis is therefore an analysis that estimates parameters for a given number of hypothesized groups, known as *classes*, in a single data set without the availability of a classification variable or other such a priori information about group membership with which to sort the data (see, e.g., Everitt & Hand, 1981; Lindsay, 1995; McLachlan & Peel, 2000; Titterington, Smith, & Makov, 1985).

As a simple example of the application of a mixture analysis, consider Figure 9.1, which illustrates two different univariate data sets. The observations contributing to the distribution in Figure 9.1a all came from the same normally distributed population; describing the data with a single mean and variance (i.e., a single distribution) is therefore appropriate. In Figure 9.1b, we see that the distribution as a whole seems to be bimodal. A

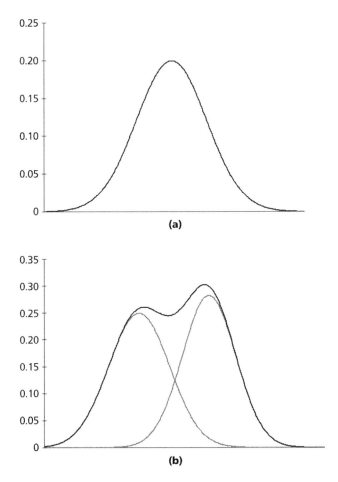

Figure 9.1 Distributions arising from a homogeneous and heterogeneous population: (a) Normal distribution for a homogeneous population; (b) Mixed distribution of two normally distributed groups in a heterogeneous population.

single normal curve would not fit as well as fitting two normal distributions with estimated means that correspond to the two peaks seen in the bimodal distribution and with estimated variances that may (but do not necessarily) differ between the two distributions.

If the data yielding Figure 9.1b had included group membership, then a statistical comparison of group means and/or variances would be fairly straightforward because parameters could be estimated for each group using the observations known to be from each group. Without a priori knowledge of group membership, however, a mixture analysis would be needed to determine the nature of the mixture and the parameter estimates of interest. Applied to the data in Figure 9.1b, a mixture analysis would, in a manner of speaking, "reverse engineer" the estimated group distributions by finding the two sets of parameter estimates (mean and variance of Class 1 and mean and variance of Class 2) that best fit the data. The 2-class model could then be compared to the 1-class model via a measure of data-model fit to determine which is the more desirable model.

The above example illustrates two very important decisions that must be made by the analyst in mixture modeling. First, conducting a mixture analysis typically involves fitting the data to multiple models that differ in the number of hypothesized groups, but the analyst must choose the number of groups for any one model. Second, in addition to a specific number of classes, the analyst needs to specify a distributional form for the potential class distributions. In the above example, the assumption is made that all potential class distributions are normal. Other distributions, however, can be applied in mixture analyses (see, e.g., Peel & McLachlan, 2000), and it is not necessary to assume that each hypothesized group has the same form. In this chapter, we limit our discussion to normal distributions as many statistical analyses assume normality within the classes.

The remainder of this chapter is spent describing specific mixture analyses. We will discuss scenarios as relatively simple as a univariate mixture with unknown group membership all the way up to testing for possible mixtures of factor models and latent growth curve models when group membership is not available a priori. The methods described in the previous section (i.e., ANOVA, multiple regression, CFA) are assumed to be familiar or at least easily refreshed. The analyses that follow, however, are assumed to be new to the reader, so much more detail is provided for each.

Examples using real data are provided for each type of mixture model described. Examples for the univariate mixture, multivariate mixture, and factor mixture models use data from Pastor, Barron, Miller, and Davis (2007). The example for the growth mixture model is based on data from Hulleman (2007). All variables in the examples are measures of college students' achievement goal orientations. Achievement goal orientation represents one's purpose for engaging in achievement-related behavior, as well as one's orientation towards evaluating his or her competence in the achievement activity. Information regarding the achievement goal orientation variables used in our examples is provided in Table 9.1. Although other chapters in this book have illustrated the use of several software programs for a given technique, our exposition is limited to Mplus 6.11 given that both LISREL and EQS are not currently capable of performing mixture modeling.[1] The syntax associated with each example is located in the Appendix and the Mplus output for each example is available from the first author. Because examples are provided for a limited number of models,

TABLE 9.1   Description of Variables Used in Examples

Variable	Description
MAP[a]	Mastery-approach
PAP[a]	Performance-approach
PAV[a]	Performance-avoidance
MAV[a]	Mastery-avoidance
MAP1[b]	Mastery-approach score obtained at time 1 (4th week)
MAP2[b]	Mastery-approach score obtained at time 2 (8th week)
MAP3[b]	Mastery-approach score obtained at time 3 (12th week)
MAP4[b]	Mastery-approach score obtained at time 4 (15th week)

Note: The possible range for all variables is 3–21. The four achievement goal orientations used in the examples are: mastery-approach, mastery-avoidance, performance-approach, and performance-avoidance. Students with mastery goal orientations are alike in that they are focused on mastering the material and developing their skills. The mastery-approach student, however, seeks to gain as much knowledge and skills as possible, whereas the mastery-avoidance student is focused on not losing the knowledge and skills they already have or misunderstanding the material. Similarly, students with performance goal orientations are alike in that both are concerned about their performance in relation to their peers. The performance-approach student, however, is focused on performing better than other students, whereas the performance-avoidance student is focused on not performing worse than other students. For more information on achievement goal orientation, see Elliot (2005).
[a] From Pastor et al. (2007); $N = 1868$
[b] From Hulleman (2007); $N = 310$

readers are encouraged to consult the literature for other Mplus examples as well as the Mplus manual.

ABOUT MIXTURE MODELS

Univariate Mixtures

As a simple case, suppose you gather a sample of observations and obtain data on a single continuous variable (X). Suppose also that you expect the population distribution to be normal, but instead, the distribution in your sample ends up being nonnormal. At this point you can test whether your data are more consistent with a homogeneous normally distributed population or a heterogeneous population consisting of a mixture of normally distributed groups.

To understand how to assess the fit between the sample data and a homogeneous normally distributed population, it is necessary to understand how the normal probability density function captures data-model fit. The normal probability density function

$$L_i = \left(1 / \sqrt{2\pi\sigma^2}\right) e^{(-.5)(x_i - \mu)^2 / (\sigma^2)}, \tag{9.1}$$

describes the likelihood (L_i) of obtaining value x_i on variable X from a normally distributed population with a mean and variance equal to μ and σ^2, respectively. In the absence of known population parameters, μ and σ^2 are set equal to estimates from the sample. The likelihood is a measure of fit, with larger likelihood values being indicative of superior data-model fit. For example, x_i values closer to the mean will have larger likelihood values and those further from the mean will have smaller likelihood values.

L_i captures fit for a single observation, but what is needed is a measure of fit for the entire sample. For a sample of N observations taken together, the assumption of independence of observations yields the joint likelihood equation

$$\prod_{i=1}^{N} L_i, \tag{9.2}$$

or the log-likelihood equation

$$LL = \ln\left(\prod_{i=1}^{N} L_i\right) = \sum_{i=1}^{N} [\ln(L_i)]. \tag{9.3}$$

The LL and $-2LL$, which is obtained by multiplying the LL by -2, are used to assess the fit of the sample data to the model, with values closer to zero indicating better data-model fit.

Example 1.1

In this example, Mplus is used to assess the fit between the normal distribution and the performance-approach (PAP) scores. The estimated mean and variance are equal to 15.34 and 16.79, respectively. The log-likelihood (Equation 9.3) is equal to –5,285.3, resulting in a value of 10,570.6 for the –2LL (see Table 9.2).

The values of the LL and $-2LL$ when the data are fit to only a 1-class model are not of much use by themselves, but will be important when the fit of the data to this model is compared to that of models specifying a different number of classes. For instance, suppose the population is instead considered to be comprised of two unknown groups or classes, with the classes differing in their distributional parameters. In this situation, a 2-class model is needed. In a 2-class model, the parameters for two separate distributions need to be estimated along with a *mixing proportion* (also known as a *class weight*) that quantifies the proportion of the population in each of the classes. The *i*th observation therefore has two likelihoods, one for each class, with each likelihood consisting of the class-specific parameters:

$$L_{i1} = \left(1/\sqrt{2\pi\sigma_1^2}\right)e^{(-.5)(x_i-\mu_1)^2/(\sigma_1^2)} \tag{9.4}$$

and

$$L_{i2} = \left(1/\sqrt{2\pi\sigma_2^2}\right)e^{(-.5)(x_i-\mu_2)^2/(\sigma_2^2)}, \tag{9.5}$$

with values of μ and σ^2 to be estimated for each class.

Because the full population is postulated to be a mixture of two unknown groups or classes, with class membership not known, the likelihood for the *i*th observation is a weighted sum of its likelihood in each class,

$$L_i = \varphi L_{i1} + (1-\varphi)L_{i2}, \tag{9.6}$$

where φ is the mixing proportion, directly interpreted as the proportion of the population in Class 1. The values of the means, variances, and the mixing proportion are estimated via an iterative process seeking to maximize *LL*, where (combining Equations 9.3 and 9.6)

$$LL = \sum_{i=1}^{N}[\ln(L_i)] = \sum_{i=1}^{N}\left[\ln\left(\varphi L_{i1} + (1-\varphi)L_{i2}\right)\right]. \tag{9.7}$$

It is important to note that the value of the mixing proportion, φ, is estimated along with the other parameters. Class membership is not known, so the proportion of the population in each class is also not known and could take on any value between 0 and 1 (but not including 0 or 1, because either of those would eliminate one of the two classes and thereby defeat the purpose of trying to fit a model that has two classes).

Although we have only introduced the 1-class and 2-class models, it is common for the data to be fit to models specifying various numbers of classes, beginning with a 1-class model and continuing by increasing the number of classes by one. Because the data are likely to be fit to models with more than two classes, it is instructive to write out the model more generally as a *K*-class model. In a *K*-class model, the likelihood for the *i*th observation is a weighted sum of its likelihood in each class *k*,

$$L_i = \sum_{k=1}^{K}\varphi_k L_{ik}, \tag{9.8}$$

with the likelihood of observing the data associated with the ith observation in each class k being

$$L_{ik} = \left(1 / \sqrt{2\pi\sigma_k^2}\right) e^{(-.5)\,(x_i - \mu_k)^2 / (\sigma_k^2)} \tag{9.9}$$

for a univariate normal distribution, and the log-likelihood calculated as

$$LL = \sum_{i=1}^{N} [\ln(L_i)] = \sum_{i=1}^{N} [\ln \sum_{k=1}^{K} \varphi_k L_{ik}]. \tag{9.10}$$

With a K-class model, there are $K - 1$ mixing proportions to be estimated and all mixing proportions are constrained to be non-negative and to sum to a value of one across classes.

Thus, mixture analysis can be applied to more than two distributions by incorporating additional terms into the equation for L_i. Furthermore, restrictions of statistical or theoretical interest can be made on the parameter estimates. For example, restricting the variances of all of the classes to be equal while not constraining the means to be equal is similar to a one-way ANOVA but without predetermined groups. Because it is actually quite difficult to estimate univariate mixture models with class-specific variances (McLachlan & Peel, 2000), it is quite common to constrain the variances to be equal across classes ($\sigma_1^2 = \ldots = \sigma_K^2$).

Example 1.2

To illustrate, the performance-approach scores are fit to a 2-class univariate normal mixture model. In this particular example, the variances are constrained to be equal across classes. The proportion of the population in the first class (φ) is estimated as 0.82, leaving the proportion of the population in the second class to be 0.18. The means of the performance-approach scores in the first and second class are estimated as 16.66 and 9.22, respectively, and the variance (constrained to be equal across classes) is estimated as 8.73. The log-likelihood associated with this model is −5,203.1, resulting in a value of 10,406.1 for the −2LL (see Table 9.2).

In our example, the log-likelihood values of the more complex 2-class model are closer to zero (indicating better fit) than the corresponding values of the 1-class model. However, the log-likelihood associated with the 2-class model cannot be further from zero than the log-likelihood from the 1-class model, so the question of fit is not, "Are the data more likely given two distributions than one?" but instead is, "Is the improvement in the likelihood worth the cost of having to estimate more parameters and thereby

TABLE 9.2 Fit Indices for Examples

Type of Mixture	Example	Classes	Number of Estimated Parameters	LL	−2LL	BIC	SABIC	LMR	Model Description
Univariate	1.1	1	2	−5,285.3	10,570.6	10,585.7	10,579.4	—	—
	1.2	2	4	−5,203.1	10,406.1	10,436.2	10,423.5	<0.001	ANOVA Model with Unknown Groups
Multivariate	2.1	1	14	−20,133.6	40,267.2	40,372.7	40,328.2	—	—
	2.1	2	29	−19,831.9	39,663.8	39,882.2	39,790.1	<0.001	Fully Unconstrained MVN Mixture
	2.2	2	19	−20,016.3	40,032.7	40,175.8	40,115.4	<0.001	MANOVA Model with Unknown Groups
	2.2	2	17	−20,136.3	40,272.6	40,400.7	40,346.7	<0.001	LPA Model
Factor	3.1	1	12	−20,248.3	40,496.7	40,587.1	40,548.9	—	Factor Model
	3.1	2	15	−20,165.0	40,330.1	40,443.1	40,395.4	<0.001	SP-FA Model
	3.2	2	21	−19,965.3	39,930.6	40,088.8	40,022.1	<0.001	Factor Mixture Model
Growth	4.1	1	9	−2,547.5	5,095.1	5,146.7	5,118.2	—	Latent Growth Model
	4.1	2	19	−2,512.3	5,024.7	5,133.7	5,073.4	0.1278	Growth Mixture Model

having a less parsimonious model?" When models are nested, this question is often addressed using a likelihood ratio test (LRT). Models are nested if the set of parameters in the simpler model is a subset of those in the more complex model. In mixture modeling, the Lo-Mendell-Rubin likelihood ratio test (LMR; Lo, Mendell, & Rubin, 2001) can be used to answer this question and is shown in Table 9.2. The small p-value implies that at least a 2-class model is needed to describe the data. Also shown in Table 9.2 are a variety of different information criteria (BIC, SABIC), which are all functions of the LL that impose various adjustments for model complexity based on different assumptions about the models. The values of the information criteria are all lower for the 2-class model, indicating that the 2-class model should be favored over the 1-class model. These statistical criteria (described in detail later in this chapter) are typically used along with theory and other practical considerations to choose among mixture models.

Multivariate Mixtures

Multivariate mixtures are needed when a collection of continuous variables are being modeled. For a vector $\mathbf{x}_i$ consisting of p continuous variables the multivariate normal probability density function can be used to obtain the likelihood of the ith observation in class k

$$L_{ik} = (2\pi)^{-p/2} |\Sigma_k|^{-1/2} e^{(-.5)(\mathbf{x}_i - \mu_k)' \Sigma_k^{-1} (\mathbf{x}_i - \mu_k)}. \tag{9.11}$$

As with the univariate mixture, the likelihood for the ith observation across classes is still a weighted sum of its likelihood in each class k as shown in Equation 9.8 and the LL is calculated as shown in Equation 9.10.

In a fully unconstrained multivariate normal mixture model with K classes, $K - 1$ mixing proportions are estimated along with K mean vectors and K covariance matrices. This model is called "fully unconstrained" because a unique set of means, variances, and covariances are estimated for each class.

Example 2.1

To illustrate, the four goal orientation scores (MAP, PAP, PAV, MAV) are fit to fully unconstrained 1-class and 2-class multivariate normal mixture models. The fit indices are shown in Table 9.2. On the basis of the fit indices alone, the 2-class model is championed over the 1-class model. Class 1 in the 2-class model consists of 29% of the population and is characterized by high means and low variances on the approach goal orientations and relatively lower means with higher variances on the avoidance goal orientations (see Table 9.3). Class 2 consists of 71% of the population and has relatively lower approach means compared to Class 1, but similar avoidance means. Unlike

TABLE 9.3 Parameter Estimates for Select Mixture Models

2-Class Multivariate Mixtures

Example 2.1: Fully Unconstrained

Class	φ		μ	Σ			
1	0.29	MAP	18.51	3.21	0.33	0.02	0.17
		PAP	18.99	1.01	2.96	−0.03	0.10
		PAV	12.06	0.19	−0.22	20.80	0.30
		MAV	13.04	1.47	0.87	6.52	23.43
2	0.71	MAP	15.38	10.28	0.11	0.27	0.04
		PAP	13.84	1.30	14.76	0.15	0.56
		PAV	11.66	3.11	2.09	12.62	0.33
		MAV	12.56	0.43	7.83	4.25	13.41

Example 2.2: MANOVA Model with Unknown Groups

φ		μ	Σ			
0.13	MAP	16.57	10.23	0.47	0.20	0.10
	PAP	8.61	4.82	10.15	0.11	0.24
	PAV	11.30	2.53	1.36	15.00	0.32
	MAV	9.22	1.19	2.87	4.71	14.41
0.87	MAP	16.25	10.23	0.47	0.20	0.10
	PAP	16.33	4.82	10.15	0.11	0.24
	PAV	11.85	2.53	1.36	15.00	0.32
	MAV	13.21	1.19	2.87	4.71	14.41

Example 2.2: LPA Model

φ		μ	Σ			
0.55	MAP	14.86	10.60	0.00	0.00	0.00
	PAP	13.05	0.00	14.82	0.00	0.00
	PAV	10.93	0.00	0.00	11.50	0.00
	MAV	11.45	0.00	0.00	0.00	12.29
0.45	MAP	18.05	4.17	0.00	0.00	0.00
	PAP	18.17	0.00	4.76	0.00	0.00
	PAV	12.83	0.00	0.00	17.39	0.00
	MAV	14.25	0.00	0.00	0.00	16.67

2-Class Factor Mixtures

Example 3.1: SP-FA Model

Class	φ		τ	Λ^a		Θ
1	0.46	MAP	15.49	**1.00**	(0.34)	8.97
		PAP	12.45	3.52	(0.97)	1.06
		PAV	11.42	0.46	(0.13)	14.77
		MAV	11.57	1.42	(0.39)	13.62
		κ = **0.00**				
		φ = 1.17				
2	0.54	MAP	15.49	**1.00**	(0.34)	8.97
		PAP	12.45	3.52	(0.97)	1.06
		PAV	11.42	0.46	(0.13)	14.77
		MAV	11.57	1.42	(0.39)	13.62
		κ = 1.47				
		φ = 0.36				

Example 3.2: Factor Mixture Model

φ		τ	Λ^a		Θ
0.27	MAP	17.72	**1.00**	(0.28)	5.12
	PAP	13.85	7.08	(0.90)	5.27
	PAV	10.55	−1.22	(−0.22)	12.70
	MAV	8.61	−0.01	(−0.00)	7.16
	κ = **0.00**				
	φ = 0.43				
0.73	MAP	15.75	**1.00**	(0.73)	5.12
	PAP	15.89	1.07	(0.73)	5.27
	PAV	12.24	0.61	(0.38)	12.70
	MAV	14.22	0.81	(0.59)	7.16
	κ = **0.00**				
	φ = 5.82				

2-Class Growth Mixture

Example 4.1: Growth Mixture Model

φ		Θ		κ	Φ	
0.46	MAP1	6.82	intercept	14.29	4.65	−0.04
	MAP2	4.03	slope	−0.25	−0.02	0.03
	MAP3	4.52				
	MAP4	4.70				
0.73	MAP1	1.36	intercept	16.91	5.92	−0.23
	MAP2	1.60	slope	−0.08	−0.08	0.02
	MAP3	0.86				
	MAP4	1.70				

Note: Bolded values indicate that the parameter was fixed to the shown value. In covariance matrices values along main diagonal are variances, below the main diagonal are covariances and above the main diagonal are correlations.
[a] Unstandardized (Standardized)

Class 1, the variances of all goal orientations in Class 2 are similar to one another and moderate in size. Many of the covariances among goal orientations differ across the two classes. In particular, the relation between PAP and MAV is stronger in Class 2 ($r = .56$) compared to Class 1 ($r = .10$).

A path diagram for a fully unconstrained multivariate mixture model is shown in Figure 9.2a. The latent variable called "C" in the path diagram is a latent categorical variable consisting of K categories in a K-class model. In the path diagram shown, all manifest variables are allowed to covary within class as indicated by the path from each variable to every other variable. In the fully unconstrained multivariate mixture model, where all parameters are allowed to vary across classes, the latent categorical variable can be conceptualized as a *moderating* variable, allowing the means and variances as well as the direction and magnitude of the relations among variables to vary across classes (Bauer & Curran, 2004; Jedidi, Jagpal, & DeSarbo, 1997).

As in the univariate case, the multivariate mixture model can be simplified by constraining certain parameters to be equal across classes. It is common for the elements in the mean vectors to remain freely estimated across classes, but for all or some of the elements in the covariance matrices to be

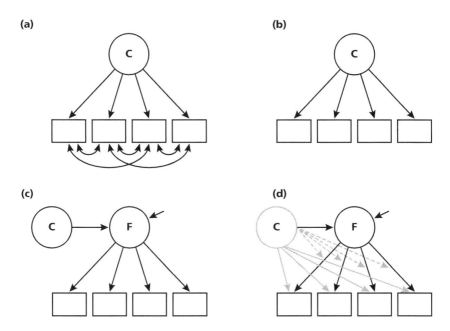

(a)   **(b)**   **(c)**   **(d)**

Figure 9.2 Path diagrams for selected mixture models. *Note:* To simplify the diagram, the paths for the residuals associated with each manifest variable in 2b – 2d are omitted. 2a = fully unconstrained MVN mixture model, 2b = LPA model, 2c = SP-FA model, 2d = factor mixture model.

constrained.[2] For example, if all elements of the covariance matrix are constrained to be equal across classes, then a MANOVA model with unknown groups is obtained. Another common constraint is to set all within-class covariances to zero, which results in the model used in latent profile analysis[3] (LPA; Gibson, 1959; Lazarfeld & Henry, 1968). Because LPA is commonly used, it is important to understand how it models relations among variables and what role the latent categorical variable plays in this model.

Although the covariances among variables are constrained to be zero within classes in LPA, non-zero relations among variables in the population are anticipated and considered to be the result of mixing groups together in the population that have differing means on the manifest variables. As described by Loken and Molenaar (2008), such a situation would occur if hair length were correlated with body height in a population comprised of both males and females. Within each gender group, hair length and body height would be uncorrelated, but when the groups are mixed, the differences between groups in their averages on the two variables create a negative correlation between the two variables. In this trivial example the categorical grouping variable of gender is observed, whereas in LPA the categorical grouping variable is unobserved.

In LPA, the latent categorical variable can be conceptualized as an *explanatory* variable. Because the latent categorical variable is considered to explain the relations among variables, the relations among manifest variables should be zero (a) once controlling for the latent categorical variable and (b) for persons in the same category (i.e., class) of the latent categorical variable. In other words, when a latent categorical variable is specified as an explanatory variable, manifest variables are assumed to be locally independent (i.e., within-class correlations are zero). This is identical to the assumptions made about a factor, which is a latent continuous variable, in factor analysis. The similarities[4] between LPA and factor analysis are even more apparent when one considers the path diagram for an LPA model, which is shown in Figure 9.2b. Figure 9.2b differs from Figure 9.2a in that the paths among manifest variables are constrained to zero and are therefore absent. Figure 9.2b only differs from a path diagram for a factor model in that a latent categorical variable "C" is being used as opposed to a latent continuous variable or factor, typically denoted "F".

Example 2.2

In this example, the four goal orientation scores (MAP, PAP, PAV, MAV) are fit to two additional 2-class multivariate normal mixture models. The first model differs from the fully unconstrained model by restricting the covariance matrices to be equal across classes and the second model differs from the unconstrained model by forcing all covariances within class to zero. The first model is a MANOVA model with unknown groups and the second model is a LPA model. The fit indices are shown in Table 9.2 and the parameter estimates in Table 9.3. Up to this point, we have only compared the LL, $-2LL$ and information criteria across models that differ in their number of classes. However, these fit indices can be compared across any mixture model based on the same set of manifest variables. For instance, a comparison of the information criteria across all 1-class and 2-class multivariate normal mixture models indicates superior fit of the 2-class fully unconstrained model. Although modifications of the LRT (e.g., LMR) are used to compare models with a different number of classes, the traditional LRT can be used to compare nested models with the same number of classes. For example, the traditional LRT can be used to compare the fit of the remaining 2-class models against the fully unconstrained model because these models are nested within the fully unconstrained model. These LRTs are statistically significant, indicating that the fully unconstrained model has superior data-model fit compared to the other 2-class models.

From this small example alone, it should be apparent that a variety of different specifications are possible when multivariate mixture models are used. Not only can models differ in the number of classes being specified, but they can also differ in the constraints being made to the within-class covariance matrices.

FACTOR MIXTURES

As mentioned previously, the LPA model and the factor model are similar in that both are used to explain the relations among variables, but differ in that the former uses a latent categorical variable (C) and the latter uses a latent continuous variable (F). These models use a single kind of latent variable—either a latent categorical or latent continuous variable. Other models, known as *latent variable hybrids*, use both kinds of latent variables (Muthén, 2006, 2008). Latent variable hybrid models include structural equation mixture models (Arminger, Stein, & Wittenburg, 1999; Dolan & van der Maas, 1998; Jedidi et al., 1997) and factor mixture models (Lubke & Muthén, 2005; Yung, 1997). Our focus here is on the factor mixture model where the incorporation of a latent categorical variable into a traditional factor model allows factor model parameters to vary across unknown groups.[5]

The multivariate mixture model can be simplified into a factor mixture model by structuring the mean vector and covariance matrix within each class according to a factor model. In such a model for class k, the ith observation's vector of values $\mathbf{x}_{ik}$, on the p manifest variables indicating the m factors, is the function

$$\mathbf{x}_{ik} = \tau_k + \Lambda_k \mathbf{F}_{ik} + \delta_{ik}, \tag{9.12}$$

where τ_k is a $p \times 1$ vector of variable intercept terms, values for the ith observation on the theoretical latent variable F hypothesized to cause the manifest variables are contained in the $m \times 1$ vector $\mathbf{F}_{ik}$, the factor loadings (i.e., the unstandardized slopes for the theoretical regression of $\mathbf{x}$ on F) are contained in the $p \times m$ matrix Λ_k, and δ_{ik} is a $p \times 1$ vector of residuals for the ith observation. For this general factor model, the first and second moments implied by the model for class k are

$$\mu_k = E[\mathbf{x}_{ik}] = \tau_k + \Lambda_k \kappa_k,$$
$$\Sigma_k = \Lambda_k \Phi_k \Lambda'_k + \Theta_k,$$

(9.13)

where κ_k is the $m \times 1$ vector of factor means (κ_k is a scalar if there is only one factor), Φ_k is the $m \times m$ factor variance-covariance matrix and Θ_k is the $p \times p$ variance-covariance matrix of residuals (δ_{ik}). Assuming MVN, we can obtain the density in class k by substituting Equation 9.13 into Equation 9.11. As with the other mixtures we have described thus far, the likelihood for the ith observation across classes is a still a weighted sum of its likelihood in each class k as shown in Equation 9.8. For simplicity, we focus our attention here on factor mixture models with a single factor (where $m = 1$), although as indicated in the notation in Equations 9.12 and 9.13, factor mixture models can easily accommodate multiple factors (where $m > 1$).[6]

A variety of different specifications of factor mixture models are possible (Clark, 2010; Masyn, Henderson, & Greenbaum, 2010). One model, known as the semi-parametric factor analysis model (SP-FA; Masyn et al., 2010), constrains the measurement model parameters (e.g., intercepts τ_k, loadings Λ_k and error variances Θ_k) to be invariant across classes and allows some combination of the factor means, factor variances or factor covariances to vary across classes. The SP-FA model can be thought of as a multiple group factor model (Hancock, 1997; Thompson & Green, this volume) with unknown groups. A general path diagram for this model is shown in Figure 9.2c. The manifest variables are modeled as a function of the factor and the factor itself is modeled as a function of a latent categorical variable. The latent categorical variable allows the parameters of the factor distribution to vary across classes. With a single normally distributed factor, classes can be allowed to differ in their factor means and/or variances.

Example 3.1

In this example, the four goal orientation scores (MAP, PAP, PAV, MAV) are fit to two models of the form shown in Figure 9.2c. The first model is a 1-class model and is equivalent to a unidimensional factor model. The second model is a 2-class SP-FA model which constrains the measurement model parameters to be equal across classes, but allows the factor mean and variance to vary across classes. For identification purposes, the mean of the factor is set equal to zero in Class 1 and the loading associated with a manifest variable (here MAP) is set to 1. The fit indices are shown in Table 9.2 and the parameter estimates in Table 9.3. On the basis of the fit indices alone the 2-class SP-FA model is championed over the 1-class factor model. In the 2-class SP-FA model almost half of the population is in each class. Class 2 has higher and less variable scores on the factor compared to Class 1. Inspection of the standardized factor loadings indicate that the factor is strongly defined by performance approach goal orientation scores.

The overall factor score distribution obtained by combining the factor scores across classes in the 2-class SP-FA solution is clearly nonnormal as shown in Figure 9.3. Although the factors *within class* are assumed to follow a multivariate normal distribution in the SP-FA model[7] (Masyn et al., 2010), the factor distribution across classes can take on any form.

The results of this example might be interpreted to imply that the overall population's distribution is nonnormal because of the presence of two unobserved groups, each with a normal distribution on the factor, that differ in their factor means and variances. An alternative explanation is that the

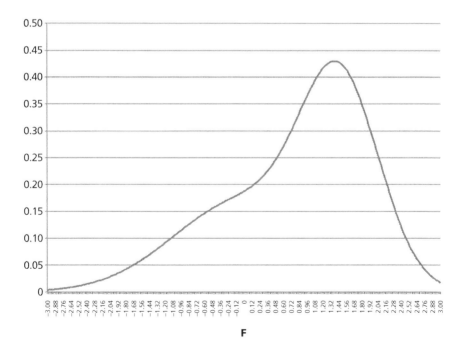

Figure 9.3 Overall distribution of factor across two classes in 2-class SP-FA model (Example 3.1).

population is homogeneous and that the overall distribution of the factor is simply nonnormal. SP-FA modeling is still considered useful in this situation because it allows a variety of different distributional forms to be modeled in the overall population. The distinction between *direct* uses of mixture modeling (where the technique is used to identify unknown groups in a heterogeneous population) and *indirect* uses (where the technique is used to model nonnormal distributions in a homogeneous population) applies to any type of mixture model (e.g., factor mixtures, growth mixtures) and is discussed later in the chapter.

With the factor mixture models presented thus far, measurement model parameters are held fixed across classes meaning that measurement invariance is assumed. Other factor mixture models allow some or all of the measurement model parameters (e.g., intercepts τ_k, loadings Λ_k) to be freely estimated across classes. In fact, the term "factor mixture models" is often reserved for such models (Masyn et al., 2010). Factor mixture models can be used to test the assumption that the measurement model is invariant across latent classes. For instance, the assumption of measurement invariance can be relaxed in the SP-FA model by allowing the intercepts of all manifest variables to vary across classes. If the fit of this model is significantly worse than the SP-FA model, then measurement invariance cannot be assumed.[8] More rigorous tests of measurement invariance can be obtained by fitting a model in which all intercepts *and* loadings are allowed to vary across classes. A graphical portrayal of this factor mixture model is shown in Figure 9.2d. The models in Figure 9.2c and Figure 9.2d are both considered "hybrid" models due to the presence of both latent categorical and latent continuous variables. The model in Figure 9.2d differs from that in Figure 9.2c by including direct effects from the latent categorical variable to the manifest variables (allowing the intercepts of these variables to vary) and the paths between the manifest variables and factor (allowing loadings to vary). The presence of these direct effects allows the measurement model parameters to vary across classes, whereas their absence constrains the parameters to be fixed across classes.

Example 3.2

Although a series of models are needed to thoroughly assess measurement invariance (see, e.g., Clark, 2010), we limit our example to the 2-class factor mixture model of the form shown in Figure 9.2d. The four goal orientation scores (MAP, PAP, PAV, MAV) are only fit to a 2-class model given that the corresponding 1-class model is equivalent to the 1-class model in Example 3.1. Intercepts τ_k and loadings Λ_k are freely estimated across classes while error variances Θ_k are constrained to be equal.[9] For identification purposes, the mean of the factor is set equal to zero in both classes[10] and the loading for a manifest variable (here MAP) is set to one in both classes. The fit indices are shown in Table 9.2 and the parameter estimates in Table 9.3. On the basis of the fit indices alone the 2-class factor mixture model is championed over the 1-class model and the 2-class SP-FA model. Because the 2-class SP-FA model is nested within the 2-class factor mixture model, a traditional LRT can be used to compare the fit of the models to one another. The results of the LRT indicate that the 2-class factor mixture model fits significantly better than the 2-class SP-FA. In the 2-class factor mixture model 27% of the population is in Class 1 and 73% is in Class 2. There are large differences between the intercept and loading estimates in each class, indicating a substantial lack of measurement invariance. There are also large differences in the factor variance estimates (0.43 and 5.82 in Classes 1 and 2, respectively). Although it is tempting to conclude that Class 2 is more variable than Class 1 on the factor, the large departure from measurement invariance precludes a comparison of classes on the factor. In fact, there is evidence to suggest that the meaning of the factor differs across classes. The standardized factor loadings indicate that the factor in Class 1 is strongly defined by only one goal orientation (PAP) and the factor in Class 2 is strongly defined by three of the four goal orientations (PAP, MAP, and MAV).

We have used seven different multivariate and mixture models in Examples 2.1 through 3.2 to model the average values, variances and covariances of the four goal orientation scores. Despite the fact that the models differ from one another in the perspective they are using to explain the data, the relative fit of these seemingly disparate models can be compared using the information criteria. For instance, when comparing the information criteria for all seven models, the most complex model, which is the 2-class fully unconstrained multivariate normal mixture, has the lowest information criteria. Although the data can be fit to a variety of different models and data-model fit compared using information criteria, it is important that a theoretical rationale support any model pursued and that evidence in addition to information criteria be provided to advocate for one model over another. Further information regarding model selection is provided later in the chapter.

Growth Mixtures

Growth mixture models (see, e.g., Li, Duncan, Duncan, & Acock, 2001; Muthén & Muthén, 2000; Muthén & Shedden, 1999) differ from factor mixtures in that the p-variables no longer represent different variables used to measure a factor, but instead represent the same variable collected at different time points for the purpose of assessing change over time. Although a variety of different forms of growth can be specified, for simplicity we focus our attention here on a linear growth mixture model.[11]

The model in Equation 9.12 can be adapted to serve as a linear growth mixture model by considering the values in $\mathbf{x}_{ik}$ to be those associated with the same manifest variable collected at each of p time points and the values in $\mathbf{F}_{ik}$ to be associated with the m latent growth factors. In a linear growth mixture model $m = 2$ given that only an intercept and linear slope factor are required. To specify a linear form of growth, the values in the $p \times m$ matrix $\mathbf{\Lambda}_k$ are fixed to certain values with the first column in $\mathbf{\Lambda}_k$ consisting of 1s and the second column fixed to values that specify a linear progression (e.g., values of 0, 1, 2, 3 for $p = 4$). When considering Equation 9.12 as a growth mixture model, the $p \times 1$ vector of variable intercept terms $\mathbf{\tau}_k$ is constrained to 0 and $\mathbf{\delta}_{ik}$ becomes a $p \times 1$ vector of time-specific residuals for the ith observation.

The first and second moments implied by the growth mixture model are the same as those in Equation 9.13. The $m \times 1$ vector $\mathbf{\kappa}_k$ consists of factor means for class k, which now correspond to those associated with the intercept and linear slope factors. When $\mathbf{\kappa}_k$ is allowed to differ across classes, class-specific intercepts (which often represent the average level of the variable at the initial measurement occasion) and slopes (which capture the average change over time on the variable) are obtained. The values in the $m \times m$ factor variance-covariance matrix $\mathbf{\Phi}_k$ include the variances for the intercept and slope factors as well as their covariance. Permitting $\mathbf{\Phi}_k$ to differ across classes allows the interindividual variability in intercepts and slopes and the relation between intercepts and slopes to differ across classes.[12] $\mathbf{\Theta}_k$ is the $p \times p$ residual variance-covariance matrix, which can take on a variety of different specifications, although it is common to constrain all residual covariances to zero and to allow time-specific residual variances to vary across classes.

If the growth mixture model is limited to a single class ($K = 1$), then the model reduces to a traditional latent curve model (McArdle & Epstein, 1987; Meredith & Tisak, 1990). The growth mixture model can also be conceptualized as a multiple group latent curve model (Muthén & Curran, 1997), but with unknown groups.

Example 4.1

A path diagram for a linear growth mixture model is shown in Figure 9.4 for mastery-approach goal orientation scores collected from college students in an introductory psychology class at four different occasions (4 weeks, 8 weeks, 12 weeks, and 15 weeks) during a semester. All paths from the intercept factor to MAP1 through MAP4 are set to 1 and the paths from slope factor to MAP1 through MAP4 are set to 0, 4, 8, and 11, respectively, to capture a linear relation between MAP and time and to ensure that the intercept can be interpreted as the value of MAP at the initial measurement occasion. The path from the latent categorical variable C to the intercept and slope factors indicates that the factor means, variances and covariance are allowed to vary across classes. The data are fit to two growth mixture models of the form in Figure 9.4. Both models estimate a residual variance for each time point and allow these variances to be class-specific. The first model is a 1-class model, which is equivalent to a traditional latent growth curve model and the second model is a 2-class model. The fit indices are shown in Table 9.2 and the parameter estimates in Table 9.3. Although the LMR detects no statistically significant difference, other indices indicate that the 2-class model is favorable to the 1-class model. In the 2-class model almost half of the population is in each class. The classes have similar factor variances but differ in their factor means and covariance. In Class 1, the average MAP score at the initial measurement occasion is 14.29 and the MAP scores decrease, on average, by 0.25 points each week in the semester. Class 2 has somewhat higher average MAP scores at the initial measurement occasion (16.91) and a less steep decrease in MAP scores over time (0.08 points per week). There is essentially no relation between intercepts and slopes in Class 1 ($r = -0.04$) and a small negative correlation in Class 2 ($r = -.23$). Class 1 also has higher residual variances at each time point compared to Class 2.

As illustrated with Example 3.2, when normally distributed factors are combined across classes, a nonnormal overall factor distribution may emerge. This can occur as well with the growth factors and prompts two plausible explanations when a multiple class solution is favored. The first explanation is that multiple classes do indeed exist, with each class having normally distributed growth factors with class-specific distributional parameters. The other explanation is that the population does not consist

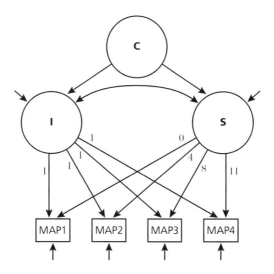

Figure 9.4 Path diagram for a linear growth mixture model.

of classes and that, instead, the distribution of the growth factors in this homogeneous population departs somehow from normality. These two explanations are provided here again to highlight the distinction between direct and indirect uses of mixture modeling and also as a reminder that a multiple-class solution does not necessarily mean that classes truly exist in the population.

PRACTICAL ISSUES IN MIXTURE MODELING

In this section we consider many practical issues in mixture modeling. Issues in estimation using maximum likelihood (ML) are addressed and a description of many of the popular indices used to choose among models is provided. We also address the importance of providing evidence that supports the final solution as being replicable and meaningful. Providing evidence that a mixture modeling solution is meaningful is particularly important given the many alternative explanations of the data that are plausible for a given model. The conclusion to this section describes some of the conditions under which spurious classes can emerge and makes the important distinction between indirect and direct uses of mixture modeling.

Estimation

A popular method for estimation in mixture modeling is maximum likelihood estimation via the expectation maximization (EM) algorithm. Conceptually in ML, several different sets of model parameter estimates are "tried out" with the data with each set associated with a *LL* value (see Equation 9.10). Because it is the intent of ML estimation to find the parameter values for which the sample data are most likely to occur, the final parameter estimates chosen are those associated with the highest likelihood value (a.k.a., the global optimum or global maximum). Unfortunately, it is quite common in mixture modeling for the estimation procedure to provide estimates that are not associated with this value. This is because the log likelihood distribution in mixture modeling has a tendency to be "bumpy," meaning that the distribution has multiple peaks. The bumpiness of this distribution is problematic in that convergence on a local optimum or maximum, which is any peak other than the highest peak, is likely to occur resulting in less than ideal parameter estimates and the possible adoption of the wrong model (Hipp & Bauer, 2006).

As noted by Hipp and Bauer (2006), many popular statistical models utilize ML for estimation, but local solutions are often not a concern because convergence on the global optimum is typically obtained regardless of the starting values used for the parameter estimates. In mixture modeling, however, the bumpiness of the likelihood surface makes convergence on the global optimum sensitive to the starting values. Mixture models are therefore estimated several times using different sets of starting values in order to differentiate global from local optima. The specific way in which this is accomplished in Mplus involves two stages. In the initial stage an attempt is made to obtain quality starting values and in the final stage the estimates from the best optimizations in the first phase are used as starting values in an additional series of model estimations. To obtain the starting values in the initial stage the model is estimated several times, and each estimation is based on a different set of randomly generated starting values for some of the model parameters. During this stage, the models are typically not fully optimized by limiting the number of iterations. The estimates from those models associated with the highest log likelihood values are then used as starting values in the final stage, where again the model is estimated several times, but this time the models are fully optimized.

The syntax used to execute this procedure in Mplus is provided in the Appendix. The first and second values after STARTS indicate the number of estimations to perform in the initial and final stages, respectively. The default values for the initial and final stages are 10 and 2, respectively. In order to arrive at the global solution, Hipp and Bauer (2006) recommended increasing the number of initial optimizations to at least 50 or 100 in order to ensure a more thorough examination of the likelihood surface. In the syntax shown in the Appendix, we are specifying that 500 estimations using randomly generated sets of starting values be performed in the initial stage.[13] We are also asking that the estimates associated with the 100 highest loglikehood values in the initial stage be used as starting values in the 100 estimations of the final stage. Only the solution in the final stage associated with the highest log likelihood value is provided in the output. The values we selected are not intended to serve as recommendations, but to illustrate a relatively cautious approach to avoiding local optima.

It is important that the range from which starting values are randomly sampled is sufficiently wide in order "to fully probe the parameter space" (Hipp & Bauer, 2006, p. 49). To accomplish this, it would be ideal to randomly generate starting values from sufficiently wide ranges for all parameters. However, which parameters are provided starting values and how much control the researcher has over the randomization process is dependent on the software being used (Hipp & Bauer, 2006). It is therefore of interest to know which starting values are randomized in Mplus in the initial stage optimizations and the manner in which they are randomized.

In Mplus random starting values are used for all parameters except the variances and covariances and the random starting value (v_i) for a parameter is a function of its original starting value (w_i). The original starting values are either the default starting values provided by Mplus (described in Chapter 14 of the user manual) or user-supplied starting values. If there is any question as to what original starting values are being used, they can easily be obtained using TECH1 on the OUTPUT command. To illustrate, consider the default starting values in Mplus shown in Table 9.4 for the 2-class fully unconstrained MVN mixture used in Example 2.1. The default in Mplus is to use starting values for the class weights such that an equal proportion of the population is in each class.[14] The starting values of the variances in each class are equal to half the observed variance of each variable and the starting values for all within-class covariances are set to zero. The original starting values for the variable means in each class are a function of the observed means for each variable (Ms) and their standard deviations (SDs). It appears that in a 2-class model the starting values for each variable mean in Class 1 are equal to $M - 1SD$ and in Class 2 are equal to $M + 1SD$.[15] As illustrated with this example, sometimes the default starting values for a model can depart widely from the model specifications. For example, although the default starting values for Σ_k in Table 9.4 are for a fully unconstrained MVN mixture, they are more in line with an LPA model. It is also interesting to note that the same default starting values shown in Table 9.4 were used in Mplus for the three very different 2-class multivariate mixtures in Examples 2.1 and 2.2.

TABLE 9.4 Descriptive Statistics for Goal Orientation Variables and Mplus Default Starting Values for a 2-Class Fully Unconstrained MVN Mixture

			Descriptive Statistics			
			M	*SD*	*VAR*	*VAR*.5
		MAP	16.28	3.20	10.25	5.12
		PAP	15.34	4.10	16.80	8.40
		PAV	11.78	3.88	15.04	7.52
		MAV	12.70	4.02	16.20	8.10

			Starting Values				
Class	φ		μ		Σ		
1	0.50	MAP	13.09	5.12	0.00	0.00	0.00
		PAP	11.24	0.00	8.40	0.00	0.00
		PAV	7.90	0.00	0.00	7.52	0.00
		MAV	8.68	0.00	0.00	0.00	8.10
2	0.50	MAP	19.49	5.12	0.00	0.00	0.00
		PAP	19.44	0.00	8.40	0.00	0.00
		PAV	15.66	0.00	0.00	7.52	0.00
		MAV	16.72	0.00	0.00	0.00	8.10

Of course, if researchers are concerned about the appropriateness of the default starting values, they can easily provide their own.

For many parameters, the random start values that are generated are centered on the values of original starting values. Because Mplus does not randomize the starting values for the variances and covariances, only the means of the variables within each class and the class weights are randomized in this particular model. In Mplus, random starting values[16] are obtained using the following equation:

$$v_i = w_i + sr_i b_i \qquad (9.14)$$

where w_i is either the user-supplied or default original starting value, r_i is a random number from a uniform distribution bounded by [−0.5, 0.5], b_i is a base parameter that places the perturbation on the proper scale for the parameter, and s is a scale variable that controls the strength of the perturbation. Although a specific range cannot be provided for start values in Mplus, the values of randomly generated starting values can be somewhat controlled by providing user-supplied start values and altering the value of the scale variable (s) using the STSCALE option. Setting s to a value larger than the default of 5 will increase the range of the random starting values, which may allow one to more thoroughly examine the likelihood surface, but may also lead to increases in nonconvergence. Note that at this time randomly generated starting values are not generated for the variance and covariances and that the value of the STSCALE option applies to all parameters for which starting values are randomly generated. Thus, being able "to fully probe the parameter space" may be somewhat limited using the STARTS option in Mplus. The extent to which these limitations are problematic has not fully been explored. Hipp and Bauer (2006) illustrated, however, that a study's conclusions may be very different depending on how thoroughly the parameter space is examined.

Although the use of a wider range of starting values increases the chances of converging on the global maximum, there is no guarantee that this will be the outcome. Often researchers inspect the *LLs* from the final stage optimizations in hopes of understanding their success in this endeavor and note how many local solutions exist. The number of local solutions are defined here as the number of different *LLs* in the final optimizations in Mplus, with the exception of the highest, which is hoped to be the global optimum. Although it may seem tempting to consider models with a large number of local solutions as incorrect models, a simulation study by Hipp and Bauer (2006) indicated that even when the data are fit to the correct model, multiple local solutions may exist. Thus, how many local solutions there are should *not* be used to choose among models and should be expected to increase as the data are fit to models with more classes. The simulation study by Hipp and Bauer

(2006) indicated, however, that when the data are fit to the correct model, the highest *LL* occurs frequently. Unfortunately, the highest *LL* also occurred frequently in some conditions in their simulation in which the data were fit to an incorrect model. Thus, the frequency of the highest *LL* should also *not* be used to choose among models, although researchers should avoid championing models where the highest *LL* occurs infrequently.

If statistical criteria and theory all suggest favoring a model in which the highest *LL* occurs infrequently, the researcher may want to compare the estimates associated with those solutions having the highest *LL*s to determine if they differ widely in value. This can be accomplished in Mplus by estimating the model again, but including only two lines of code ("TYPE = mixture;" and "OPTSEED= 318388;") under ANALYSIS, with the OPTSEED value set equal to the seed number associated with the *LL* of interest. If the estimates associated with the highest *LL*s do not drastically differ, then more confidence can be placed in the solution with the higher of the *LL*s.

It might be tempting to reject those models where either a larger number of solutions do not converge during optimization or improper solutions are obtained. Although this issue was not the focus of Hipp and Bauer's simulation, their results indicated that improper solutions and nonconvergence can occur even when the data are fit to the correct model. Thus, the rate of nonconvergence or improper solutions should *not* be used to choose among models. In fact, higher rates of nonconvergence should be expected when the data are fit to models with more classes.

Estimation problems should also be expected when using models having complex within-class specifications, such as a fully unconstrained MVN mixture. It is not uncommon with such models to encounter singularities, which can lead to nonconvergence (Hipp & Bauer, 2006). Singularities, which are spikes going to infinity on the likelihood surface, often occur "when the mean vector of a class becomes centered on the observed values of a single case and the variance (or determinant) of that class goes to zero" (Hipp & Bauer, 2006, p. 38). If random starts or reasonable starting values cannot be used to overcome this problem, a simpler model may need to be adopted. This solution is not without problems, however, because estimation problems may still exist and more importantly, the simplified model may not correspond with theory. As noted by Lubke (2010), choices made in specifying mixture models should be theory driven, but are oftentimes driven by such practical considerations as model convergence.

Choosing Among Models Differing in Their Number of Classes

Given that the number of local solutions and convergence rates should not be used to choose among models, how does one decide among the models differing in their number of classes? There are a variety of different types of statistical indices alluded to earlier in the chapter that can be used for this purpose. We have limited our presentation to those indices that are either easily available in Mplus or that performed well in selecting the model with the correct number of classes in the simulation studies we reviewed (Henson, Reise, & Kim, 2007; Nylund, Asparouhov, & Muthén, 2007; Tofighi & Enders, 2008; Yang, 2006). Similar to Henson et al., the statistical indices are classified below into four different groups: information criteria, likelihood ratio tests (LRTs), classification-based approaches, and distributional statistics.

Information Criteria (IC). Information-based measures of fit are commonly used in multiple regression and in structural equation modeling, and their application to mixture modeling is straightforward. The IC measures AIC (Akaike, 1973), consistent AIC (CAIC; Bozdogan, 1987), BIC (Schwarz, 1978), and sample-size adjusted BIC (SABIC; Sclove, 1987) can be computed rather easily for each model:

$$AIC = -2LL + 2q, \tag{9.15}$$

$$CAIC = -2LL + (\ln N + 1)q, \tag{9.16}$$

$$BIC = -2LL + \ln(N)q, \tag{9.17}$$

$$SABIC = -2LL + \ln\left(\frac{N+2}{24}\right)q \tag{9.18}$$

where q equals the number of parameters and N equals sample size. These IC all penalize the $-2LL$ for model complexity, although each does so in a unique way. For each IC, the values are compared across models differing in their number of classes, and the model with the smallest value is considered the preferred model. According to the simulations of Tofighi and Enders (2008), Yang (2006), and Henson et al. (2007), the IC most accurate at favoring the model with the correct number of classes is the SABIC. However, in the simulations of Nylund et al., the SABIC did not perform as favorably. The BIC and CAIC had mixed results, with a solid showing in Nylund et al., but a mediocre performance in Henson et al. All simulation studies reviewed showed the AIC to be the least effective in identifying the correct model.

Likelihood Ratio Tests (LRTs). With neighboring class models, which are models that only differ from one another in a single class (e.g., a $K-1$ versus a K-class model), the question of whether the increase in the likelihood is worth the cost of having to estimate more parameters arises. When models are nested this question is often answered using a likelihood ratio test (LRT). The LRT statistic is calculated as -2 times the difference between the LL values of the two models (LRT statistic $= -2[LL_{\text{parsimonious}} - LL_{\text{complex}}]$) and approximately follows a χ^2 distribution with degrees of freedom equal to the difference between models in the number of parameters being estimated. A significant LRT is taken as evidence that the more complex model should be adopted. In order for the LRT statistic to be used to compare the neighboring class models, the $K-1$ class model must be nested within the K class model. Indeed, mixture models differing in their number of classes are nested models. For instance, the $K-1$ class model can be obtained by setting a mixing proportion to zero in the K class model. However, because a parameter is being set to a value at the boundary of the parameter space to obtain the more parsimonious model, the regularity conditions needed to ensure that the LRT statistic is asymptotically χ^2 distributed do not hold (McLachlan & Peel, 2000).

The fit of $K-1$ versus K class models can still be compared using the Lo-Mendell-Rubin likelihood ratio test (LMR; Lo et al., 2001), which uses an approximation for the distribution of the LRT statistic. An adjusted LMR is also available in Mplus, however simulation studies indicate little difference between the LMR and its modification. Bootstrapping can also be used to create a distribution of the LRT statistic when comparing a K-class model to a model with $K-1$ classes (McLachlan & Peel, 2000). With the bootstrap LRT (BLRT), samples are generated using the parameters of the $K-1$ class model. The data for each sample are fit to both the $K-1$ and K-class models, with the resulting LLs used to calculate the LRT statistic for each sample. A distribution is then created of the LRT statistics across samples, to which is compared the LRT statistic of the original data. For both the LMR and BLRT, a p-value less than a nominal α level indicates the retention of at least a K-class model.

Nylund et al. (2007) is the only simulation study we reviewed that investigated the performance of the BLRT. Their results indicated that the BLRT performs incredibly well in selecting the correct model. The LMR has also been shown to perform well in those studies that have investigated its functioning (Henson et al., 2007; Nylund et al., 2007; Tofighi & Enders, 2008). In comparison to IC, Nylund et al. recommended the BLRT (their favored LRT) over the BIC (their favored IC). Both Tofighi and Enders and Henson et al. recommended their favored IC, the SABIC, over the LMR, which was the only LRT investigated in these studies.

Based on these results it is recommended that the SABIC, BIC, LMR, and BLRT be used to choose among models differing in their number of classes. Although the BLRT is recommended, it is not reported in Table 9.2 because of issues we encountered in its calculation, which are not uncommon. It is not unusual to obtain a warning that either the estimations are not converging or that the highest LL was not replicated in a particular solution. The Mplus software provides many options that can be used in an effort to circumvent these problems and make the BLRT more trustworthy. To avoid estimating the K-class model again in Mplus during the bootstrap LRT, it is recommended that the researcher first estimate the K-class model without the BLRT, then obtain the BLRT in a separate run using the OPT-SEED option so that the K-class solution with the highest LL in the initial run is obtained. It might also be beneficial to use the option K-1STARTS to increase the number of random starts being used in the initial and final stage optimizations when the data are fit to a $K - 1$ class model. To increase the number of random starts when the bootstrap samples are being fit to $K - 1$ and K-class models, the option LRTSTARTS can be used. Because the bootstrap samples are being generated according to a $K - 1$ class model, it is reasonable to assume that fewer random starts would be needed for the $K - 1$ class model than for the K-class model.

Classification-based approaches. When choosing among models it is also possible to utilize information regarding how well the model is able to classify persons into classes. Methods used to assess classification accuracy are based on posterior probabilities, which are calculated for each observation and capture the probability of membership in each of the K classes,

$$\hat{\phi}_{k|\mathbf{x}_i} = \frac{\hat{\phi}_k f_k(\mathbf{x}_i | \hat{\boldsymbol{\theta}}_k)}{\displaystyle\sum_{k=1}^{K} \hat{\phi}_k f_k(\mathbf{x}_i | \hat{\boldsymbol{\theta}}_k)}. \tag{9.19}$$

Equation 9.19 utilizes the final model parameter estimates (contained in $\hat{\boldsymbol{\theta}}_k$) as well as the individual's values on the manifest variables (contained in $\mathbf{x}_i$). Posterior probabilities can easily be acquired for each observation in Mplus by using SAVE=CPROBABILITES in the SAVEDATA command, followed by the name of a file created using FILE IS. The resulting file will consist of posterior probabilities for each observation as well as a variable indicating the class associated with the highest posterior probability for that observation. This variable captures class membership using modal assignment of observations to classes.

To the extent that a model yields one probability for each person that is larger than the other $K - 1$ probabilities for that person, the model's classes are more clearly separated. The classification table can be used to summarize this information. There are as many rows and columns in the classification table as there are classes. To illustrate, a hypothetical classification table is shown for a 4-class model in Table 9.5. The first row contains the averages based on only the 308 persons that were assigned to Class 1. Because these persons were assigned to the Class 1 using modal assignment, as anticipated the average posterior probability for these persons is highest for

TABLE 9.5   Example of a Classification Table for a 4-Class Mixture Model

Latent Class Modal Assignment	n	Latent Class Average Posterior Probability			
		1	2	3	4
1	308	**0.84**	0.00	0.00	0.17
2	285	0.00	**0.89**	0.02	0.09
3	248	0.01	0.01	**0.71**	0.27
4	1,027	0.13	0.00	0.06	**0.81**

Note: Values in bold represent the average posterior probability associated with the latent class to which persons were assigned.

Class 1. Similarly, in the second row the highest average for those 285 persons assigned to Class 2 is associated with Class 2. Note that these averages, the averages associated with the classes to which persons were assigned, are captured in the main diagonal of the classification table. For this reason, higher averages on the main diagonal of the classification table reflect greater accuracy in the assignment of persons to classes. The remaining averages can be examined to determine which particular classes may not be distinct from one another. For instance, the third row of the table contains the average posterior probabilities for persons assigned to Class 3. As expected, the highest average (.71) is associated with the class to which these persons were assigned, Class 3. The second highest average for persons assigned to Class 3 is associated with Class 4 (.27). Because the value of this average is sizeable, it indicates some overlap between Classes 3 and 4.

Classification utility can also be captured using a single statistic. This statistic is known as the entropy statistic and it ranges from 0 to 1 with higher values indicative of higher classification utility. The entropy statistic is calculated using the posterior probabilities from Equation 9.19, the overall sample size N and number of classes K.

$$\text{Entropy} = 1 - \frac{E}{N \ln K} \tag{9.20}$$

where

$$E = \sum_{i=1}^{N} \sum_{k=1}^{K} \left(-\phi_{k|\mathbf{x}_i} \ln \phi_{k|\mathbf{x}_i} \right). \tag{9.21}$$

McLachlan and Peel (2000) described indices for relative model fit that combine the likelihood-based approach and classification accuracy using formulas similar to the aforementioned information criteria but which include a term that is a function of classification clarity. The classification likelihood information criterion (CLC) and the integrated classification likelihood (ICL-BIC) have $-2LL$ and BIC, respectively, as the first term in their equations and two times E from Equation 9.21 as the second term

$$CLC = -2LL + 2E \qquad (9.22)$$

$$ICL - BIC = -2LL + \ln(N)q + 2E. \qquad (9.23)$$

Like with the information criteria, smaller values of these indices are more favorable. Although the CLC and ICL-BIC are not provided in Mplus, they can easily be calculated using the LL and entropy values[17] provided in the output.

The only simulation study we reviewed that examined the performance of these classification-based approaches was Henson et al. (2007) who found that CLC and ICL-BIC, but not entropy, performed exceptionally well. If classification accuracy is important, then we recommend that the CLC and ILC-BIC be consulted in addition to the SABIC, BIC, LMR and BLRT. It is important to note, however, that classification accuracy should not always be taken into consideration when deciding among models. Whether classification accuracy is important depends on whether one's purpose in using the mixture modeling is to accurately assign persons to classes. For instance, classification accuracy is not as important in indirect applications of mixture models, which use the different classes as a statistical tool for capturing nonnormality in the overall population. However, classification accuracy is very important in those direct applications of mixture modeling where the procedure is being used to assign class membership—particularly if the purpose in classifying subjects is to provide class-specific treatment or instruction. Even if classification accuracy is important to a researcher, it may be hard to obtain acceptable accuracy under certain conditions. For instance, classification accuracy might be hard to obtain in small classes, even when classes are well separated (Lubke, 2010).

Distributional statistics. All statistical indices described thus far are used to compare the fit of models relative to one another. Indices that assess the fit of the data to a particular model include the tests of multivariate skewness and kurtosis (SK) described in Muthén (2004). These tests can be obtained in Mplus using TECH13 on the OUTPUT command. For a given model, multiple data sets are generated according to the estimated model parameters. Values of multivariate skewness and kurtosis are then calculated for each data set and used to create distributions of these indices. The values of multivariate skewness and kurtosis in the observed sample are then compared to these distributions with the resulting p-values indicating how likely the observed values are given the values estimated by the model-generated data. High p-values associated with the SK tests are indicative of model fit; low p-values indicate that the data does not fit the model. The SK tests performed poorly in the simulation studies of Nylund et al. (2007) and Tofighi and Enders (2008).

Statistical indices for different kinds of model comparisons. When using statistical indices to compare models, which index one is able to use depends on the kind of models being compared. There are three different kinds of model comparisons. The first kind of model comparison involves models that have the same within-class parameterization, but that differ in the number of classes. The statistical indices have been described up to this point in the context of this kind of model comparison. We suggested the use of SABIC and BIC and because the LRT cannot be used for this kind of model comparison, the LMR and BLRT.

The second kind of model comparison involves models that differ in their number of classes, within-class specifications, or both. With the exception of the LMR and BLRT, which are only appropriate for comparing models that differ in the number of classes being specified, many of the other indices presented can be used for this kind of model comparison. For instance, the SABIC and BIC were used in Examples 2.1 through 3.2 to compare the relative fit of all the following models fit to multivariate data: a fully unconstrained MVN mixture, a LPA, and a variety of factor mixtures.[18] The SABIC and BIC can therefore be used to compare models that differ in number of classes, within-class specifications, or both.

The third kind of model comparison involves models with the same number of classes and nested within-class specifications (e.g., a 2-class fully unconstrained MVN mixture and a 2-class LPA). Information criteria can be used for such model comparisons as well as the traditional LRT, as illustrated in Examples 2.2 and 3.2.

Choosing a Final Solution. With so many quantities from which to choose from to evaluate the number of classes in a dataset, it is helpful that simulation studies have been conducted to test the accuracy of some of these indices. Although we have made recommendations based on simulation work, the reader should keep in mind that recommendations are likely to change as more simulation work is conducted. As noted by Lubke (2010), mixture modeling is a "rapidly evolving area" (p. 209) making it important to consult the most current research on the topic. As well, because the four simulation studies we reviewed vary in the models they test, the sample sizes they use, and in other design characteristics, there tends to be some disagreement when considering the results of multiple studies. Readers are therefore encouraged to consult the results of studies most relevant to their particular application.

Although statistical criteria can be used to choose among models, model selection should ultimately be based on theoretical expectations and substantive considerations. Responsible use of mixture modeling requires that a researcher consider the number and type of classes to expect prior to running the analysis. If the number and nature of the classes is strongly suggested by theory, a more confirmatory approach to mixture modeling, in which constraints are placed on parameters to reflect theoretical expectations, should be considered (Finch & Bronk, 2011). The researcher might also want to consider the uniqueness of additional classes in deciding among models. If the third class in a 3-class solution has parameter estimates similar to those in one of the remaining classes, then the more parsimonious 2-class solution may be favored.

It is also recommended that the size of classes be taken into consideration. For instance, a 2-class solution may be favored over a 3-class solution if solutions differ only in the addition of a third class consisting of a very small proportion of the population. Of course, this decision should be guided by theory. If theory can be used to argue for the relevancy of a class, it should be retained, regardless of size. However, if this situation occurs it should be kept in mind that unless the overall sample size is quite large, the estimates associated with small classes may be quite unstable. Researchers should also keep in mind that even though theory might support the presence of a small class, whether a model including such a class emerges and is championed as the "best" model depends on a variety of factors, including the particular data and models being used.

Although there may be reasons (e.g., small class size, similarity of class to other classes) to choose a more parsimonious solution over a more complex solution, researchers should keep in mind the situations that can lead to a more parsimonious solution being favored (perhaps mistakenly favored) over a more complex solution. As described by Lubke (2010), this may occur when the models being compared differ substantially in their number of parameters and information criteria are used to choose among models. Because information criteria often impose penalties for model complexity, it is quite likely that the simpler model will be favored, even though it may not be the correct model. Lubke (2010) also reminded researchers that more parsimonious models might be favored because convergence problems or unacceptable parameter estimates were encountered while estimating the more complex model.

Lubke (2010) provided several approaches to the analysis that may help prevent researchers from dismissing complex models. One recommendation is to fit the data to a series of models, beginning with simpler models and progressing to models with fewer and fewer constraints in each step. If convergence problems or improper solutions are encountered, the point at which they occur can be reported, providing a richer context for the results. Another recommendation is not to limit oneself to reporting the results of a single "best-fitting" model and instead report the results from several models yielding acceptable data-model fit.

Another important criterion for championing a model or set of models is whether the same solutions can be obtained in another sample. The replicability of solutions can be examined by implementing the mixture modeling procedure in an independent sample. Confidence in one's championed solutions is increased if the same solutions emerge as the favored solutions in the second sample.

Importance of Validity

If a direct application of mixture modeling is pursued, it is essential that validity evidence for the mixture modeling solution be acquired. Validity is established by providing evidence that the classes relate to other variables in ways anticipated by theory and previous research. We follow the lead of Marsh, Lüdkte, Trautwein, and Morin (2009) and refer to these "other" variables as *correlates* so as not to imply any causal ordering between the variables and the latent classes.

There are a variety of different methods that could be used to acquire validity evidence. One possibility includes classifying observations into classes using modal assignment (i.e., cases are assigned to the class in which they have the highest posterior probability) and then using the resulting grouping variable in other analyses (e.g., ANOVA, chi-square). A problem with this approach is that it ignores the classification accuracy of the model. Clark (2010) provided a compelling example illustrating this problem. They considered two individuals in a 2-class solution, both of whom have been assigned to the first class. The first individual had a posterior probability equal to 1.0 of being in Class 1 whereas the second individual had a posterior probability equal to .51 of being in Class 1. When class membership was used in an analysis such as an ANOVA or a chi-square, both individuals were considered to have a 1.0 probability of being in class 1. If classification accuracy of the mixture model is not perfect, the consequences in using this approach are possible incorrect conclusions about the validity of a solution.

In order to account for classification accuracy in the validity analyses, other approaches are needed. One approach involves the use of pseudo-class draws (Asparouhov & Muthén, 2010; Wang, Brown, & Bandeen-Roche, 2005), where individuals are assigned to classes based on random draws (Mplus uses 20 pseudo draws) taken from the distribution of their posterior probabilities. Using this approach, the class-specific statistics associated with a correlate are computed after each pseudo draw, with the final statistics needed to test the significance of the relation between class membership and the correlate obtained by averaging across all pseudo draws. This approach can be implemented for either categorical or continuous correlates in Mplus by adding (e) after correlates in the AUXILIARY command.

Another option for validity studies that also accounts for classification accuracy is a single-step approach, where the correlates are included in the mixture model. This approach requires one to provide and justify a causal ordering between the correlates and the latent class variable. For example, the correlates could be included in the mixture model as either predictors of the latent class variable (see Figure 9.5a) or as outcomes (see Figure 9.5b).

A disadvantage associated with the single-step approach is the fact that mixture modeling solution can *change* once correlates are incorporated into the model. For instance, consider the model in Figure 9.5b, where the correlates are considered outcomes of the latent class variable. Although correlates are considered as outcomes by the researcher, no distinction is made between manifest variables or correlates in the model—both are conceptualized to be endogenous variables. Thus, when correlates are incorporated as outcomes in a mixture model, the creation of latent classes will be based not only on the values of the manifest variables, but also on the values of the correlates. Despite their specification as exogenous variables (Figure 9.5a) in the model, the incorporation of correlates as predictors of the latent class variable can also change the mixture modeling solution. According to Marsh et al. (2009), this might occur because there are direct effects between the correlates and the other variables not specified in the model.

Correlates can be incorporated into the model in a variety of different ways; they are not limited to being specified as either predictors or outcomes. For instance, some correlates might be predictors, some might be outcomes and still others might have relations with the other variables in the model above and beyond their relation through the latent class variable. Because the model estimates and resulting conclusions are likely to differ depending on whether correlates are included and how they are included, researchers are advised to approach their decision about correlates carefully.

What particular steps should be taken to obtain validity evidence for a mixture modeling solution is an active area of research (see, e.g., Clark,

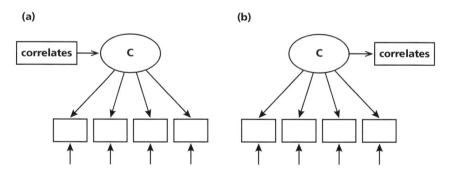

(a)   **(b)**

Figure 9.5 Path diagrams illustrating correlates as (a) predictors and (b) outcomes of the latent categorical variable.

2010; Marsh et al., 2009; Petras & Masyn, 2010) and guidance on the issue is likely forthcoming. In the meantime, if the goal is to gather validity evidence for a mixture modeling solution, then the safest approach might be to use the pseudo-class draws approach available using the AUXILIARY function in Mplus. This approach eliminates the need to specify a causal relation between the latent class variable and the correlates. In addition, the mixture modeling solution remains unchanged given that the correlates are not incorporated into the model.

Choosing from Among a More Comprehensive Set of Models

In Examples 2.1 through 3.2 information criteria was used to compare a variety of models that differed not only in number of classes, but also in the specification of their within-class covariance matrices. Fitting data to models that differ in both number of classes and their within-class specifications highlights an important relation in mixture modeling, which is the relation between within-class constraints and class enumeration. Put simply, when data are fit to a model with an overly constrained within-class specification relative to the true model, a model with more classes than the true model will yield superior fit. For instance, suppose the true model is a 2-class fully unconstrained MVN mixture, but data are only fit to LPA models with null within-class covariances matrices. The misspecification of the within-class component of the model will lead to the adoption of a model with more than two classes.

Similarly, if the data are fit to a model with an overly flexible within-class specification relative to the true model, a model with fewer classes than the true model will yield superior fit. More attention is paid to the situation where spurious classes emerge as a result of using an overly restricted within-class specification for two reasons. First, many researchers limit their analyses to models with constrained within-class components (e.g., LPA or factor mixtures), thus increasing the chances of adopting a model with too many classes. Second, when a model with multiple classes is championed, the researcher might be seduced into thinking that the classes are real and fail to consider alternative explanations.

One approach, which is advantageous in that a wide variety of alternative explanations are considered for the data, is the use of latent variable hybrids. Because latent variable hybrids utilize latent categorical variables and latent continuous variables (i.e., factors) models can differ not only in number of latent classes, but also in number of factors. Several researchers (Clark, 2010; Kuo, Aggen, Prescott, Kendler, & Neale, 2008; Lubke et al., 2007; Masyn et al., 2010; Muthén, 2006; Muthén & Asparouhov, 2006) have illustrated an approach to using these models that entails fitting the data to three different sets of models. The first set of models consist only of latent categorical variables (e.g., 1-class model, 2-class model, etc.), the second set of models consist only of latent continuous variables (e.g., 1-factor model, 2-factor model, etc.), and the third set of hybrid models consist of different combinations of latent classes and factors. Clark (2010) suggested that the largest number of classes and factors to use in the third of models be dictated by the best fitting models in the first and second set, respectively.

In addition to these models, researchers might also consider fitting the data to fully unconstrained multivariate mixtures that differ in their number of classes. In fact, Bauer and Curran (2004) suggested using the number of classes associated with the best fitting fully unconstrained multivariate mixture as the only number of classes to consider in latent variable hybrid models.[19] Their suggestion is motivated by the aforementioned relation between the within and between class components. Latent variable hybrid models, which model the within-class relations according to a factor model, might have an overly constrained within-class specification relative to the true model. If the number of classes in a latent variable hybrid model is limited to the number obtained with an unrestricted within-class model, there is less of a chance of adopting classes that exist solely to compensate for the misfit imposed by an overly restrictive within-class specification.

It is important for mixture modeling researchers to understand the relation between the within and between-class components of mixture models. Armed with this information, a researcher who adopts a particular model is aware of likely alternative explanations for the data. Being aware of this information may also encourage researchers to fit their data to a variety of types of models for the purposes of considering alternative explanations. According to a simulation study by Lubke and Neale (2006), fitting the data to a wider variety of models may also yield the benefit of increasing one's chances of selecting the correct model.

Other Alternative Explanations

As clearly demonstrated by Bauer (2007) and Bauer and Curran (2004), misspecification of the within-class component is just one reason why a model with the incorrect number of classes might be adopted. In order to understand these reasons a careful reading of Bauer and Bauer and Curran (2004) is highly recommended, along with Bauer and Curran (2003a), which spawned a spirited discussion in the same journal issue involving Rindskopf (2003), Muthén (2003), Cudeck and Henly (2003), and Bauer and Curran (2003b). Examples are provided below to illustrate some of the other conditions leading to selection of a model with the wrong number of classes as provided by Bauer and colleagues (a more thorough list of examples and conditions are found in Bauer (2007), Bauer and Curran (2003a), and Bauer and Curran (2004)).

One of the warnings made by Bauer and colleagues is that classes might emerge not because they reflect meaningful groups in the population, but because the wrong distributional form was utilized for the within-class model. To illustrate, consider Figure 9.3 where the overall population distribution of the factor scores is clearly nonnormal. By fitting the data to a mixture factor model and assuming normality of the factor within class, the overall population distribution can be reproduced fairly well by two classes. The dilemma, however, is whether these two classes represent meaningful, distinct groups or whether these two classes are just a statistically convenient way to represent the nonnormality in the overall population.[20] Although this example uses SP-FA modeling, the same principle applies to any mixture model assuming normality within-class (e.g., growth mixtures assuming normality of the growth factors within class as in Example 4.1). Thus, when multiple class solutions are championed the methodologist should keep in mind that the classes may only serve to capture idiosyncrasies in the overall distribution and that they may not represent meaningful latent groups in the population.

Another example in which spurious classes might emerge is when a linear model is used to capture a nonlinear relation. To illustrate with a simple example, consider the scatterplot in Figure 9.6, which captures the nonlinear relation between X and Y. As before, assume the population is homogeneous. Suppose that the data are fit both 1-class and 2-class fully unconstrained multivariate normal mixtures, which allow X and Y to covary within class, but assume the relation is linear. In this situation, the 1-class model would not recreate the relation between X and Y as well as the 2-class model (Figure 9.6), where the different covariances in each class together better approximate the overall relationship. Although a simplistic example was used here, the same principle applies to any mixture model assuming linearity of the relationships within-class (e.g., factor mixture models that assume a within-class linear relationships between items and factors; growth mixture models that assume within-class linear relationships between intercepts and slopes).

The reverse of these situations is also problematic, but not as likely to happen in practice. For example, if the overall population distribution is skewed as a result of mixing classes with different normal distributions and the data are fit to a model assuming skewed within class distributions, then too few classes will be retained. Because it is more common for the data to be fit to models assuming within-class normality and linearity, more attention is paid toward retaining a solution with too many classes. As well, more attention is devoted to situations in which spurious classes might be

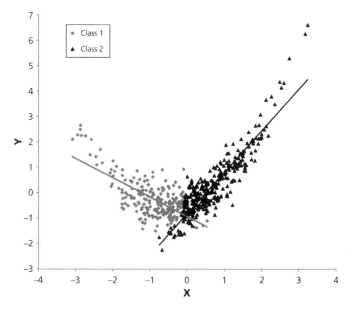

Figure 9.6 Scatterplot of nonlinear relation between X and Y captured by a 2-class mixture model where the linear relation between X and Y is allowed to vary across classes.

adopted to discourage researchers from prematurely treating these classes as being "real" and failing to consider alternative explanations.

Indirect versus Direct Applications

The concern that researchers will hastily conclude that the classes are "real" and not a possible result of model misspecification stems from the fact that most researchers use mixture modeling for the purpose of finding meaningful classes. Such applications of mixture modeling are known as *direct* applications (Titterington et al., 1985). In direct applications, the data are fit to mixture models for the purpose of identifying the number and nature of meaningful groups in the population. In contrast, *indirect* applications of mixture models use the classes as a statistical device, oftentimes to approximate overall distributions not easily captured by traditional parametric models (Titterington et al., 1985). As an example, SP-FA modeling might only be pursued to capture the nonnormality of the overall factor distribution, as in Figure 9.3.

The distinction between indirect and direct applications is important because it highlights that, although one's purpose in using mixture modeling might be direct, only inferences appropriate for an indirect application may be warranted (Bauer, 2007). Given the many situations in which spurious classes are likely to emerge, a conservative approach would be to limit the use of mixture modeling to only indirect applications. Thus, when multiple class solutions are favored, the researcher is less likely to be lured into thinking the classes are "real" and more likely to consider the alternative explanations for the solution.

Although indirect applications may not seem as exciting at first glance, they can be incredibly useful and might provide more realistic representations of the data than their non-mixture counterparts. For instance, indirect applications can be used to capture nonnormal factors or nonlinear relations in both factor modeling and growth modeling. A particularly interesting indirect application is illustrated by Bauer and Shanahan (2007), who demonstrated how indirect mixture modeling (specifically, LPA) can be used to capture nonlinear relations (of no specific form) in regression as well as complex interactions.

Even if the purpose of using mixture modeling is purely indirect, there is the danger of implying (perhaps not intentionally) that groups exist or of introducing the potential for others to attribute meaning to the classes (Bauer, 2007). To avoid these issues, Bauer (2007) recommended using the estimates from indirect applications of mixture modeling to calculate expected values in the overall distribution. This approach focuses the reader's attention on the overall distribution, avoiding the possible misinterpretation of the role classes play.

CONCLUSION

The intent of this chapter was to introduce readers to a variety of different types of mixture models. Models were introduced that differed in their complexity, with simple univariate mixtures on one end of the continuum and the relatively more complex latent variable hybrids on the other. Models were also distinguished by the purpose behind their use, with direct mixture models being used to discover the number and nature of unknown groups and indirect models being used as mathematical devices to help characterize a single population. In both indirect and direct applications of mixture modeling care should be taken to avoid convergence on a local maximum and those fit indices shown to have promise in simulation work should be consulted when choosing among models. In direct applications of mixture modeling additional care is needed. Theory should guide model selection and evidence should be gathered to convey the validity of the favored solution and the extent to which it is replicable. Furthermore, when mixture modeling is used to find unknown groups, it is important to acknowledge and rule out (if possible) alternative statistical explanations for the data. Of course, these additional concerns are not unique to direct applications of mixture modeling—they apply to any *exploratory* statistical procedure (e.g., exploratory factor analysis). Thus, it is important for researchers to keep in mind that theory, replication, and validity play important roles in all exploratory analyses and that mixture modeling is no exception.

APPENDIX

General Syntax

The general Mplus syntax for all examples is shown below. The numerical value associated with the CLASSES option corresponds to the number of classes in the model.

```
TITLE:   Initial setup for mixture models;
DATA:    FILE IS example.dat;
VARIABLE: NAMES ARE MAP PAP PAV MAV;
         CLASSES = C(2);    !numerical value represents # classes;
ANALYSIS: TYPE = mixture;
         STARTS = 500 100;  !1st numerical value= # initial
                             optimizations;
                            !2nd numerical value= # final optimizations;

[Example specific syntax goes here]

OUTPUT:   TECH11 TECH1;        !TECH11 = LMR, TECH1 = starting values;
SAVEDATA: SAVE = CPROBABILITIES; !Saves posterior probabilities;
          FILE IS postprobs.dat; !to this external file;
```

TECH14 can be used on the OUTPUT command to obtain the BLRT. See the chapter and the Mplus manual for further details.

Example Specific Syntax

Examples 1.1 and 1.2

To limit the analysis to the PAP variable, the line "USEVARIABLES ARE PAP;" would need to be added to the VARIABLE command in the general syntax. For Example 1.1, the value for CLASSES would be one and for Example 1.2 the value would be two. Recall in Example 1.2 that means were allowed to vary across classes, but variances were constrained to be equal. Because the default settings allow means to vary freely across classes while constraining variances to be equal, only the general syntax is needed for these two examples.

Examples 2.1 and 2.2

The example specific syntax for the fully unconstrained multivariate mixture in Example 2.1 is:

```
MODEL: %overall%
       MAP with PAP; MAP with PAV; MAP with MAV;    !covariances;
       PAP with PAV; PAP with MAV; PAV with MAV;    !covariances;

       %c#1%
       MAP-MAV;                                     !variances;
       MAP with PAP; MAP with PAV; MAP with MAV;    !covariances;
       PAP with PAV; PAP with MAV; PAV with MAV;    !covariances;

       %c#2%
       MAP-MAV;                                     !variances;
       MAP with PAP; MAP with PAV; MAP with MAV;    !covariances;
       PAP with PAV; PAP with MAV; PAV with MAV;    !covariances;
```

The code associated with the `%overall%` label is needed to specify what is common about the model across classes and the code affiliated with the `%c#1%` and `c#2%` labels is to specify class specific features. In this example, the two lines under the label `%overall%` are needed to override the default of all variables being unrelated within class. The three lines under `%c#1%` and `%c#2%` are needed to allow the variances and covariances to vary across classes. Means are freely estimated across classes by default. The first model in Example 2.2 is the MANOVA model with unknown groups where the covariance matrices are constrained to be equal across classes. This model can be obtained by eliminating `%c#1%` and `%c#2%` and their associated syntax. The second model in Example 2.2 is the LPA model, which constrains all within class covariances to zero and allows means and variances to vary across classes. The LPA model can be obtained by eliminating `%overall%` and its associated syntax and only including the lines for the variances under `%c#1%` and `%c#2%`.

Examples 3.1

The example specific syntax for the 1-factor 2-class SP-FA model, which constrains measurement model parameters to be equal across classes, is:

```
MODEL: %overall%
       f by MAP PAP PAV MAV;    !loadings;

       %c#1%
       [MAP] (i1);[PAP] (i2);[PAV] (i3); [MAV] (i4);  !intercepts;
       [f@0];                                         !factor mean;
       f;                                             !factor variance;

       %c#2%
       [MAP] (i1); [PAP] (i2); [PAV] (i3); [MAV] (i4); !intercepts;
       [f];                                           !factor mean;
       f;                                             !factor variance;
```

In this example, the line under the label `%overall%` specifies the factor model and the loading associated with the first variable (here MAP) is set to 1.0 by default. By default, residual variances are constrained to be equal across classes. (i1) through (i4) in the first line under `%c#1%` and `%c#2%` are used to constrain the variables' intercepts to be equal across classes. The second and third lines under `%c#1%` and `%c#2%` correspond to the factor means and variances, respectively. The factor variances are freely estimated in each class while the factor mean in Class 1 is constrained to 0 and freely estimated in Class 2.

Examples 3.2

The example specific syntax for the 1-factor 2-class factor mixture model that allows all measurement parameters (with the exception of residual variances) and factor variances to be class-specific is:

```
MODEL: %overall%
        f by MAP PAP PAV MAV;          !loadings;
        [f@0];                          !factor mean;

        %c#1%
        [MAP-MAV];                      !intercepts;
        f by MAP@1 PAP PAV MAV;         !loadings;
        f;                              !factor variance;

        %c#2%
        [MAP-MAV];                      !intercepts;
        f by MAP@1 PAP PAV MAV;         !loadings;
        f;                              !factor variance;
```

The factor means were constrained to zero in each class using the second line under the `%overall%` option. Intercepts and loadings were freely estimated in each class (with the exception of setting the loading for MAP equal to 1 in each class) using the first and second lines under `%c#1%` and `%c#2%`. Factor variances were freely estimated in each class using the third lines.

In Examples 3.1 and 3.2, not all final stage optimizations converged. The actual syntax therefore provided starting values for all parameters, including the mixing proportion. After starting values were provided, almost all final stage optimizations converged. Starting values can be provided by following the parameter by an asterisk and the starting value.

Example 4.1

Because this example is based on a different data set with different variables, FILE IS would need to be altered and NAMES ARE refers to MAP1 MAP2 MAP3 MAP4. The example specific syntax for the linear growth model is:

```
MODEL: %overall%
        i s|MAP1@0 MAP2@4 MAP3@8 MAP4@11;      !loadings;

        %c#1%
         [i s];                                !factor means;
         i s;                                  !factor variances;
         i with s;                             !factor covariance;
         MAP1 MAP2 MAP3 MAP4;                  !residual variances;

        %c#2%
         [i s];                                !factor means;
         i s;                                  !factor variances;
         i with s;                             !factor covariance;
         MAP1 MAP2 MAP3 MAP4;                  !residual variances;
```

The %overall% syntax sets up the linear growth model with an intercept (i) and slope (s) factor and provides the loadings of each of the variables on the slope factor. The lines under %c#1% and %c#2% allow for growth factor means, variances and covariances to vary freely across classes. Residual variances are also specified to be class-specific.

ACKNOWLEDGMENTS

We thank Deborah Bandalos, Christine DeMars, Sara Finney, and Gregory Hancock for their helpful comments on this chapter.

NOTES

1. Other software programs for mixture modeling are described in the Appendix of McLachlan and Peel (2000). Other options include Latent GOLD, AMOS, Mx, and several packages in R.
2. A variety of different specifications for the covariance matrices can be found in Table 1 of Pastor et al. (2007). Steinley and Brusco (2011a) also provided various specifications in Table 1 using nomenclature that corresponds to an eigenvalue decomposition of the covariance matrix (Banfield & Raftery, 1993).
3. If the LPA model is further constrained by specifying all variances to be equal and the classes to be of equal size, then the model is similar to the K-means approach in cluster analysis (MacQueen, 1967; Steinley, 2006; Thorndike, 1953), particularly when the classification EM algorithm (Celeux & Govart, 1992) is used for estimation. Further information about the similarities and differences between LPA and a K-means cluster analysis can be found in Steinley and Brusco (2011a) as well as in the rejoinders by McLachlan (2011), Vermunt (2011), and Steinley and Brusco (2011b).
4. The similarities and differences between LPA and factor analysis are more fully described on pages 4–9 of Bauer and Curran (2004).

5. This model can also be conceptualized as a latent profile model (LPA) where a factor is incorporated into the model for the purposes of relaxing the assumption of local independence. Seen through this lens, the factor mixture model allows variables to relate to one another within classes and the manner in which they relate to one another to be structured according to a factor model. The fully unconstrained MVN mixture also relaxes the LPA assumption of local independence by allowing the variables to relate to one another within class, but does not model those relations according to a factor model.

6. With multiple factors a confirmatory approach to the factor structure is typically taken (i.e., the number of factors is specified by the researcher and elements in Λ_k and Φ_k are freed or constrained to specify which manifest variables load on which factors and which factors are related to one another). The number of factors is typically held constant across classes, although it is possible to specify a different number of factors in each class. An exploratory approach to factor mixture models is also possible (McLachlan & Peel, 2000).

7. If the SP-FA model is constrained even further by setting the factor variances within class to zero, then a model referred to by Masyn et al. (2010) as the non-parametric factor analysis (NP-FA) model is obtained. The SP-FA and NP-FA models differ from one another in that the overall population distribution of the factor is considered a mix of (multivariate) normal distributions in the SP-FA and a mix of bars (a.k.a. mass points) in the NP-FA. In NP-FA, each class is represented by a bar. The position of the bar along the factor continuum is defined by the factor mean and the height of the bar is equal to the class proportion. For more information, see Clark (2010), Muthén (2008), and Masyn et al. (2010).

8. See Clark (2010) for a procedure that can be used to isolate which intercepts are non-invariant across classes.

9. Error variances were only constrained to be equal across classes because a model allowing them to vary would not converge to a proper solution.

10. Identifying the model in this way does force any mean differences in the factor across classes to be expressed as intercept differences. However, if measurement invariance does not hold (as evidenced by loadings differing across classes or the magnitude of the class differences in intercepts varying across indicators), comparisons across classes on the factor are not very meaningful.

11. Although we limit our focus to linear growth mixture models, where the same form of growth is specified in each class, it is possible to allow the form of growth to differ across classes by structuring Λ_k in Equation 9.12 differently for each class. This would require, however, some knowledge as to what class-specific forms of growth to anticipate. If this knowledge is lacking, class-specific forms of growth can be derived from the data by using a latent basis parameterization of the model, which frees some elements in Λ_k and allows those estimates to vary across classes. An example of such a model is provided by Ram and Grimm (2009).

12. A common constraint used in growth mixture modeling is to set all elements in Φ_k to zero, which implies no within-class variation in individual trajectories, resulting in a model similar to Nagin (1999). In this more parsimonious version of the model, the same class-specific growth parameters are considered

to adequately characterize the growth of all persons within a class. This model is similar to the non-parametric factor analysis model (NP-FA; Masyn et al., 2010) in that the overall distribution of the growth factors (e.g., intercept and slope factors) are captured with a series of bars (i.e., mass points), one for each class. Thus, this model can be used to capture overall growth factor distributions that depart from normality.

13. The option STITERATIONS can be used in the ANALYSIS command of Mplus to specify the number of iterations used in the initial stage optimizations. The default number of iterations is 10.

14. It is actually the intercepts (α_k) of a multinomial regression equation that are estimated in Mplus, which are related to the class weights. If no correlates are regressed on the latent categorical variable, the multinomial regression equation consists only of intercepts. The multinomial regression equation can be used to obtain the class weights from the intercepts:

$$\phi_k = \frac{\exp(\alpha_k)}{\sum_{k=1}^{K} \exp(\alpha_k)},$$

where α_K, the intercept for the last class, is set equal to 0. Original starting values for the class specific intercepts are 0, which corresponds to an equal proportion of the population in each class.

15. Using TECH1, the starting values for the means in each class for models differing in their number of classes were obtained. In a 3-class model the starting values for each variable mean were $M - 1SD$, M, $M + 1SD$, for each of the three classes, respectively. Interestingly, in a 4-class model the starting values were $M - 1SD$, M, M, $M + 1SD$ and in a 5-class model the starting values were $M - 2SD$, $M - 1SD$, M, $M + 1SD$, $M + 2SD$ for each of the classes, respectively. Only the starting values for the variable means varied across models differing in their number of classes, the starting values for the class weights, covariances, and variances remained the same.

16. Randomization of the starting values in Mplus is described in the Technical Appendix for the software.

17. The following equation can be used to obtain E from entropy: $E = [N \ln(K)]$ $(1 - Entropy)$.

18. These indices can also be used to compare factor mixture models that specify a different number of factors.

19. Bauer and Curran (2004) did note two drawbacks with this approach, particularly when there is sizeable number of manifest variables. First, given that information criteria penalize for model complexity these indices may not be ideal for choosing among fully unconstrained mixtures. Second, it may be difficult to obtain stable solutions with such complex models.

20. At this point, the researcher should consider why the overall population distribution appears nonnormal (Bauer & Curran, 2004). It might be that the distribution truly is nonnormal; however, the nonnormality could also be a result of using an instrument with floor or ceiling effects. Another possibility

to consider is that the sampling scheme resulted in the sample distribution appearing different from the population distribution.

REFERENCES

Akaike, H. (1973). Information theory and an extension of the maximum likelihood principle. In B. N. Petrov & F. Csake (Eds.). *Second international symposium on information theory* (pp. 267–281). Budapest: Akademiai Kiado.

Arminger, G., Stein, P., & Wittenburg, J. (1999). Mixtures of conditional mean- and covariance-structure models. *Psychometrika, 64,* 475–494.

Asparouhov, T., & Muthén, B. (2010). *Wald test of mean equality for potential latent class predictors in mixture modeling.* From www.statmodel.com/download/Mean-Test1.pdf

Banfield, J. D., & Raftery, A. E. (1993). Model-based Gaussian and non-Gaussian clustering. *Biometrics, 49,* 803–821.

Bauer, D. J. (2007). Observations on the use of growth mixture models in psychological research. *Multivariate Behavioral Research, 42,* 757–786.

Bauer, D. J., & Curran, P. J. (2003a). Distributional assumptions of growth mixture models: Implications for overextraction of latent trajectory classes. *Psychological Methods, 8,* 338–363.

Bauer, D. J., & Curran, P. J. (2003b). Overextraction of latent trajectory classes: Much ado about nothing? Reply to Rindskopf (2003), Muthén (2003), and Cudeck and Henly (2003). *Psychological Methods, 8,* 384–393.

Bauer, D. J., & Curran, P. J. (2004). The integration of continuous and discrete latent variable models: Potential problems and promising opportunities. *Psychological Methods, 9,* 3–29.

Bauer, D. J., & Shanahan, M. J. (2007). Modeling complex interactions: Person-centered and variable centered approaches. In T. D. Little, J. A., Bovaird, & N. A. Card (Eds.). *Modeling contextual effects in longitudinal studies,* (pp. 225–284). Mahwah, NJ: Lawrence Erlbaum Associates.

Bozdogan, H. (1987). Model selection and Akaike's Information Criterion (AIC): The general theory and its analytical extensions. *Psychometrica, 52,* 345–370.

Byrne, B. M., Shavelson, R. J., & Muthén, B. (1989). Testing for the equivalence of factor covariance and mean structures: The issue of partial measurement invariance. *Psychological Bulletin, 105,* 456–466.

Celeux, G., & Govart, G. (1992). A classification EM algorithm for clustering and two stochastic versions. *Computational Statistics & Data Analysis, 14,* 315–332.

Cheung, G. W., & Rensvold, R. B. (1999). Testing factorial invariance across groups: A reconceptualization and proposed new method, *Journal of Management, 25,*1–27.

Clark, S. L. (2010). *Mixture modeling with behavioral data* (Doctoral dissertation). Available from ProQuest Dissertations and Theses database. (UMI No. 3405665)

Cudeck, R., & Henly, S. J. (2003). A realistic perspective on pattern representation in growth data: Comment on Bauer and Curran (2003). *Psychological Methods, 8,* 378–383.

Dolan, C. V., & van der Maas, H. L. J. (1998). Fitting multivariate normal finite mixtures subject to structural equation modeling. *Psychometrika, 63,* 227–253.

Elliot, A. J. (2005). A conceptual history of the achievement goal construct. In A. J. Elliot & C. S. Dweck (Eds.), *Handbook of competence and motivation* (pp. 52–72). New York, NY: Guilford Press.

Everitt, B. S., & Hand. D. J. (1981). *Finite mixture distributions.* London: Chapman & Hall.

Finch, W. H., & Bronk, K. C. (2011). Conducting confirmatory latent class analysis using M*plus*. *Structural Equation Modeling: A Multidisciplinary Journal, 18,* 132–151.

Gibson, W. A. (1959). Three multivariate models: Factor analysis, latent structure analysis, and latent profile analysis. *Psychometrika, 24,* 229–252.

Hancock, G. R. (1997). Structural equation modeling methods of hypothesis testing of latent variable means. *Measurement and Evaluation in Counseling and Development, 30,* 91–105.

Henson, J. M., Reise, S. P., & Kim, K. H. (2007). Detecting mixtures from structural model differences using latent variable mixture modeling: A comparison of relative model fit statistics. *Structural Equation Modeling: A Multidisciplinary Journal, 14,* 202–226.

Hipp, J. R., & Bauer, D. J. (2006). Local solutions in estimating growth mixture models. *Psychological Methods, 11,* 36–53.

Hulleman, C. S. (2007). *The role of utility value in the development of interest and achievement.* (Doctoral dissertation). Available from ProQuest Dissertations and Theses database. (UMI No. 3279019)

Jedidi, K., Jagpal, H. S., & DeSarbo, W. S. (1997). Finite-mixture structural equation models for response-based segmentation and unobserved heterogeneity. *Marketing Science, 16,* 39–59.

Kuo, P-H., Aggen, S. H., Prescott, C. A., Kendler, K. S., & Neale, M. C. (2008). Using a factor mixture modeling approach in alcohol dependence in a general population sample. *Drug and Alcohol Dependence, 98,* 105–114.

Lazarfield, P. F., & Henry, N. W. (1968). *Latent structure analysis.* New York, NY: Houghton-Mifflin.

Li, F., Duncan, T. E., Duncan, S. C., & Acock, A. (2001). Latent growth modeling of longitudinal data: A finite growth mixture modeling approach. *Structural Equation Modeling: A Multidisciplinary Journal, 8,* 493–530.

Lindsay, B. G. (1995). *Mixture models: Theory, geometry, and applications.* Hayward, CA: Institute of Mathematical Statistics.

Lo, Y., Mendell, N. R., & Rubin, D. B. (2001). Testing the number of components in a normal mixture. *Biometrika, 88,* 767–778.

Loken, E., & Molenaar, P. (2008). Categories or continua? The correspondence between mixture models and factor models. In G. R. Hancock & K. M. Samuelsen (Eds.), *Advances in latent variable mixture models* (pp. 277–297). Greenwich, CT: Information Age Publishing, Inc.

Lubke, G. (2010). Latent variable mixture models. In G. R. Hancock & R. O. Mueller (Eds.), *Quantitative methods in the social and behavioral sciences: A guide for researchers and reviewers* (pp. 209–219). New York, NY: Routledge.

Lubke, G., & Muthén, B. (2005). Investigating population heterogeneity with factor mixture models. *Psychological Methods, 10,* 21–39.

Lubke, G., Muthén, B. O, Moilanen, I. K, McGough, J. J., Loo, S. K., Swanson, J. M., Yang, M. H., Taanila, A., Hurtig, T., Jarvelin, M.-R., & Smalley, S. L. (2007). Subtypes vs. severity differences in attention deficit hyperactivity disorder in the northern Finnish birth cohort (NFBC). *Journal of the American Academy of Child & Adolescent Psychiatry, 46,* 1584–1593.

Lubke, G., & Neale, J. R. (2006). Distinguishing between latent classes and continuous factors: resolution by maximum likelihood? *Multivariate Behavioral Research, 41,* 499–532.

MacQueen, J. (1967). Some methods of classification and analysis of multivariate observations. In L. M. Le Cam & J. Neyman (Eds.), *Proceedings of the Fifth Berkeley Symposium on Mathematical Statistics and Probability* (Vol. 1, pp. 281–297). Berkeley, CA: University of California Press.

Marsh, H. W., Lüdtke, O., Trautwein, U., & A. J. S. Morin (2009). Classical latent profile analysis of academic self-concept dimensions: Synergy of person- and variable-centered approaches to theoretical models of self-concept. *Structural Equation Modeling: A Multidisciplinary Journal, 16,* 191–225.

Masyn, K. E., Henderson, C. E., & Greenbaum, P. E. (2010). Exploring the latent structures of psychological constructs in social development using the dimensional-categorical spectrum. *Social Development, 19,* 470–493.

McArdle, J. J., & Epstein, D. (1987). Latent growth curves within structural equation models. *Child Development, 58,* 110–133.

McLachlan, G. J. (2011). Commentary on Steinley and Brusco (2011): Recommendations and cautions. *Psychological Methods, 16,* 80–81.

McLachlan, G., & Peel, D. (2000). *Finite mixture models.* New York, NY: Wiley.

Meredith, W. (1993). Measurement invariance, factor analysis, and factorial invariance. *Psychometrika, 58,* 525–543.

Meredith, W., & Tisak, J. (1990). Latent curve analysis. *Psychometrika, 55,* 107–122.

Muthén, B. (2003). Statistical and substantive checking in growth mixture modeling: Comment on Bauer and Curran (2003). *Psychological Methods, 8,* 369–377.

Muthén, B. O. (2004). Latent variable analysis: Growth mixture modeling and related techniques for longitudinal data. In D. Kaplan (Ed.), *Handbook of quantitative methodology for the social sciences* (pp. 345–368). Newbury Park, CA: Sage Publications.

Muthén, B. (2006). Should substance use disorders be considered as categorical or dimensional? *Addiction, 101* (Suppl. 1), 6–16.

Muthén, B. (2008). Latent variable hybrids: Overview of old and new models. In G. R. Hancock & K. M. Samuelsen (Eds.), *Advances in latent variable mixture models* (pp. 1–24). Greenwich, CT: Information Age Publishing, Inc.

Muthén, B., & Asparouhov, T. (2006). Item response mixture modeling: Application to tobacco dependence criteria. *Addictive Behaviors, 31,* 1050–1066.

Muthén, B., & Curran, P. J. (1997). General longitudinal modeling of individual differences in experimental designs: A latent variable framework for analysis and power estimation. *Psychological Methods, 2,* 371–402.

Muthén, B. O., & Muthén, L. K. (2000). Integrating person-centered and variable-centered analyses: Growth mixture modeling with latent trajectory classes. *Alcoholism: Clinical and Experimental Research, 24,* 882–891.

Muthén, B., & Shedden, K. (1999). Finite mixture modeling with mixture outcomes using the EM algorithm. *Biometrics, 55,* 463–469.

Muthén, L. K., & Muthén, B. O. (1998–2010). *Mplus user's guide* (6th ed.). Los Angeles, CA: Muthén & Muthén.

Nagin, D. (1999). Analyzing developmental trajectories: A semi-parametric, group-based approach. *Psychological Methods, 4,* 139–157.

Nylund, K. L., Asparouhov, T., & Muthén, B. O. (2007). Deciding on the number of classes in latent class analysis and growth mixture modeling: A Monte Carlo simulation study. *Structural Equation Modeling: A Multidisciplinary Journal, 14,* 535–569.

Pastor, D. A., Barron, K. E., Miller, B. J., & Davis, S. L. (2007). A latent profile analysis of college students' achievement goal orientation profiles. *Contemporary Educational Psychology, 32,* 8–47.

Peel, D., & McLachlan, G. J. (2000). Robust mixture modelling using the *t* distribution. *Statistics and Computing, 10,* 335–344.

Petras, H., & Masyn, K. (2010). General growth mixture analysis with antecedents and consequences of change. In A. Piquero & D. Weisburd (Eds.), *Handbook of quantitative criminology* (pp. 69–100). New York, NY: Springer.

Ram, N., & Grimm, K. J. (2009). Growth mixture modeling: A method for identifying differences in longitudinal change among unobserved groups. *International Journal of Behavioral Development, 33,* 565–576.

Rindskopf, D. (2003). Mixture or homogeneous? Comment on Bauer and Curran (2003). *Psychological Methods, 8,* 364–368.

Schwarz, G. (1978). Estimating the dimension of a model. *Annals of Statistics, 6,* 461–464.

Sclove, S. L. (1987). Application of model-selection criteria to some problems in multivariate analysis. *Psychometrika, 52,* 333–343.

Steinley, D. (2006). *K*-means clustering: A half-century synthesis. *British Journal of Mathematical and Statistical Psychology, 59,* 1–34.

Steinley, D., & Brusco, M. J. (2011a). Evaluating mixture modeling for clustering: Recommendations and cautions. *Psychological Methods, 16,* 63–79.

Steinley D., & Brusco, M. J. (2011b). *K*-means clustering and mixture model clustering: Reply to McLachlan (2011) and Vermunt (2011). *Psychological Methods, 16,* 89–92.

Thorndike, R. L. (1953). Who belongs in the family? *Psychometrika, 18,* 267–276.

Titterington, D. M., Smith, A. F. M., & Makov, U. E. (1985). *Statistical analysis of finite mixture distributions.* Chichester, England: Wiley.

Tofighi, D., & Enders, C. K. (2008). Identifying the correct number of classes in growth mixture models. In G. R. Hancock & K. M. Samuelsen (Eds.), *Advances in latent variable mixture models* (pp. 317–341). Greenwich, CT: Information Age Publishing, Inc.

Vermunt, J. K. (2011). *K*-means may perform as well as mixture model clustering bbut may also be much worse: Comment on Steinley and Brusco (2011). *Psychological Methods, 16,* 82–88.

Wang, C-P., Brown, H., & Bandeen-Roche, K. (2005). Residual diagnostics for growth mixture models: Examining the impact of a preventive intervention on multiple trajectories of aggressive behavior. *Journal of the American Statistical Association, 100,* 1054–1076.

Yang, C.-C. (2006). Evaluating latent class analysis models in qualitative phenotype identification. *Computational Statistical & Data Analysis, 50,* 1090–1104.

Yung, Y-F. (1997). Finite mixtures in confirmatory factor-analysis models. *Psychometrika, 62,* 297–330.

CHAPTER 10

EXPLORATORY STRUCTURAL EQUATION MODELING

**Alexandre J. S. Morin, Herbert W. Marsh,
and Benjamin Nagengast**

In a seminal publication, Cohen (1968) presented multiple regression as a generic data-analytic system for quantitative dependent variables (outcomes) encompassing classical analyses of variance and covariance, interactive effects, and predictive non-linearity among quantitative and qualitative predictors. However, multiple regression and most of the General Linear Model procedures were later shown to represent special cases of the even more encompassing framework of canonical correlation analysis, allowing for the inclusion of multiple outcomes within the same model (Knapp, 1978). Similarly, Structural Equation Modeling (SEM) was proposed as an even more flexible framework (Bagozzi, Fornell, & Larker, 1981; Fan, 1997; Graham, 2008), covering any relation that could be studied with canonical correlation analysis, but also allowing for the simultaneous estimation of chains of direct and indirect effects (i.e., path analysis) based on latent variables that implicitly correct the estimated relations for measurement error (i.e., confirmatory factor analysis [CFA]). Then, Muthén (2002; also see Skrondal & Rabe-Hesketh, 2004) incorporated all of these methods into an even more generic framework (generalized SEM [GSEM]), allowing for

Structural Equation Modeling: A Second Course (2nd ed.), pages 395–436
Copyright © 2013 by Information Age Publishing

the estimation of relations between any type of quantitative or qualitative observed and latent variables. Although exploratory factor analyses (EFAs) have been around for more than a century (Spearman, 1904) and represent an important precursor of CFAs (Cudeck & MacCallum, 2007), and thus of SEM and GSEM, it was until recently excluded from these generalized frameworks.

It is thus not surprising that EFA is now seen as less useful, and even outdated, compared to the methodological advances associated with CFA/ SEM (e.g., estimation of structural relations among latent constructs adjusted for measurement error, higher order factor models, method factors, correlated uniquenesses, goodness of fit assessment, latent mean structures, differential item functioning/measurement invariance, and latent curve modeling). This perception is reinforced by the erroneous semantically-based assumption that EFA is strictly an *exploratory* method that should only be used when the researcher has no a priori assumption regarding factor structure and that *confirmatory* methods are better in studies based on a priori hypotheses regarding factor structure. This assumption still serves to camouflage the fact that the critical difference between EFA and CFA is that all cross loadings are freely estimated in EFA. Due to this free estimation of all cross loadings, EFA is clearly more naturally suited to exploration than CFA. However, statistically, nothing precludes the use of EFA for confirmatory purposes, except perhaps the fact that most of the advances associated with CFA/SEM were not, until recently, available with EFA.

In fact, many psychological instruments with well-defined EFA structures are not supported by CFAs. A classic example is the Big-Five personality factor structure that has been consistently identified and supported by an impressive body of research relying on EFAs (see McCrae & Costa, 1997). However, CFAs have failed to replicate these findings when based on analyses of the items forming Big-Five questionnaires. On the one hand, these discrepant findings led some scholars (e.g., Vassend & Skrondal, 1997) to question the factor structure of the NEO instruments—the most widely used big-five personality instruments—as well as other big-five personality instruments. On the other hand, many researchers questioned the appropriateness of CFA for Big-Five research in principle (see Borkenau & Ostendorf, 1993; Church & Burke, 1994; McCrae, Zonderman, Costa, Bond, & Paunonen, 1996; Parker, Bagby, & Summerfeldt, 1993). For instance, McCrae et al. (1996) concluded: "In actual analyses of personality data [...] structures that are known to be reliable showed poor fits when evaluated by CFA techniques. We believe this points to serious problems with CFA itself" (p. 568). Marsh et al. (2009) presented a similar argument based on the Students' Evaluations of Educational Quality (SEEQ) instrument, whose structure also received strong support from EFAs (Marsh, 1983, 1987; Marsh & Hocevar, 1991), whereas CFA models failed to replicate these results (e.g., Toland & De Ayala, 2005). In addressing this issue, Marsh (1991a, 1991b) showed that the reason CFA structures did not provide an adequate fit to the data was that many items had minor cross-loading on other factors.

More generally Marsh, Hau, and Grayson, 2005 (also see Marsh, 2007) made the claim that "it is almost impossible to get an acceptable fit (e.g., CFI, RNI, TLI > .90; RMSEA < .05) for even "good" multifactor rating instruments when analyses are done at the item level and there are multiple factors (e.g., 5–10), each measured with a reasonable number of items (e.g., at least 5–10/per scale) so that there are at least 50 items overall" (p. 325). Marsh (30 August 2000, SEMNET@BAMA.UA.EDU) placed this claim on SEMNET (an electronic network devoted to SEM) and invited the more than 2000 members to provide counter-examples. No one offered a published counter-example, leading him to conclude that there are probably many psychological instruments routinely used in applied research that do not even meet minimum criteria of acceptable fit according to current standards.

More recently, however, Marsh and colleagues (Marsh, Liem, Martin, Morin, & Nagengast, 2011; Marsh, Nagengast, et al., 2011; Marsh, Nagengast, & Morin, 2012; Marsh et al., 2009, 2010) suggested that the independent cluster model inherent in CFA (ICM-CFA)—in which items are required to load on one, and only one, factor, with non-target loadings constrained to be zero—could be too restrictive for many multidimensional constructs. They also noted that in many CFA applications, even when the ICM-CFA model fits well (Marsh, Liem, et al., 2011), factor correlations are likely to be inflated unless all non-target loadings are close to zero (Marsh, Lüdtke, et al., 2011). This in turn undermines the discriminant validity of the factors and results in multicollinearity in the estimation of relations with outcomes (Marsh et al., 2010). Recent simulation studies seem to support that EFAs are better at recovering true population latent correlations and that CFA-based latent correlations can be severely inflated by the presence of even a few small cross loadings erroneously fixed to zero (Asparouhov & Muthén, 2009; Marsh, Lüdtke et al., 2011). Although there are advantages in having "pure" items that load only on a single factor, this is not a requirement of a well-defined factor structure, nor even a requirement of traditional definitions of simple structure in which non-target loadings are ideally small relative to target loadings but not required to be zero (Carroll, 1953; McDonald, 1985; Thurstone, 1947). Further, strategies that are often used to compensate for ICM-CFA models' inadequacies (e.g., parceling, ex post facto modifications such as ad hoc correlated uniquenesses) tend to be dubious, misleading, or simply wrong (Browne, 2001; Marsh et al., 2009, 2010). Why then do researchers persist with CFA models even when they have been shown to be inadequate? In fact, the recent dominance of CFAs in the literature might have generated the mistaken belief that EFAs are no longer viable or even acceptable. In addition, as we pointed out earlier, many recent advances in latent variable modeling have been reserved for CFA/SEM. Hence, failure to embrace these new and evolving methodologies could have unfortunate consequences for applied research.

Fortunately, this extreme solution is no longer necessary with the development of exploratory structural equation modeling (ESEM) by Asparouhov and Muthén (2009; also see Marsh et al., 2009). More specifically, ESEM is an integration of EFA within the global GSEM framework. Thus, ESEM combines the benefits associated with EFA's flexibility along with access to typical CFA/SEM parameters and statistical advances: standard errors; goodness of fit statistics; comparisons of competing models through tests of statistical significance and fit indices; inclusion of correlated uniquenesses; inclusion of both CFA and EFA factors based on the same, different, or overlapping sets of items; estimation of method effects; multiple indicators multiple causes models (MIMIC) models; and tests of multiple group and longitudinal invariance. Currently, ESEM is available in the commercial Mplus package (Muthén & Muthén, 2010), starting from version 5.1. Before moving on, we should acknowledge that, although Asparouhov and Muthén (2009) were the first to manage the integration of EFA within the CFA/SEM framework, previous efforts in this direction were made by Dolan, Oort, Stoel, and Wicherts (2009; see also Hessen, Dolan, & Wicherts, 2006), who developed ways to test for multiple group invariance of EFA solutions and simultaneous rotation in multiple groups in the Mx package (Neale, Boker, Xie, & Maes, 2003). Also, standard errors for rotated EFA solutions along with goodness-of-fit statistics and model tests have been available in the CEFA program (Browne, Cudeck, Tateneni, & Mels, 2010), although without capabilities for testing measurement invariance across multiple groups.

Previous Applications of ESEM

To date, ESEM applications are still few. In the first of those, Marsh et al. (2009) used ESEM to analyze over 30,000 sets of class-average responses to the SEEQ instrument, designed to evaluate students' evaluation of university teaching effectiveness. Previous research provided strong support for the EFA structure, but ICM-CFAs of SEEQ responses did not fit the data. Marsh et al. found support for their claim that ICM-CFA was inappropriate for this instrument and resulted in substantially inflated factor correlations among the nine SEEQ factors (median $r = .34$ for ESEM and .72 for CFA). They also showed that the SEEQ ESEM factor model was reasonably invariant over time and extended the ESEM model to approximate latent growth models using linear and quadratic functions of time as a MIMIC-like predictor of the factors. Finally, they conducted an extended set of MIMIC analyses to test that the potential biases (workload/difficulty, class size, prior subject interest, expected grades) to SEEQ responses were small in size and varied systematically for different ESEM factors, supporting a construct validity interpretation of the relations.

Marsh et al. (2010) conducted similar multiple group and longitudinal analyses on the answers provided by a sample of over 1500 high school students to the 60-item NEO Five Factor Inventory. Previous research based on this instrument found that the a priori (i.e., confirmatory) EFA factor structure was well defined, but that ICM-CFA models provided a poor fit to the data. Results showed that ESEM factors were more differentiated (less correlated) than ICM-CFA factors and that the ESEM factor structure was reasonably invariant over gender and time. More recently, Marsh et al. (2012) pursued extended analyses of Big-Five data to cover a lifespan perspective using data from the nationally representative British Household Panel Study ($N = 14,021$ participants aged 15–99). These authors contrasted three hypotheses describing personality evolution over the lifespan (i.e., the maturity principle, the plaster hypothesis, and their new La Dolce Vita hypothesis). Once again, ESEM resulted in a better fit to the data and less differentiated factors than ICM-CFA. Analyses were based on a set of multiple group, MIMIC, hybrid multiple-group MIMIC models (designed to investigate the loss of information due to the categorization of continuous variables) and introduced ESEM-within-CFA (EWC), an extension designed to further increase the flexibility of ESEM.

Marsh, Liem, et al. (2011) and Marsh, Nagengast, et al. (2011) used ESEM to conduct extensive psychometric evaluations—including multitrait multi-method analyses of the construct validity—of the Motivation and Engagement Scale and the Adolescent Peer Relations Instrument (APRI). Interestingly, both of these studies showed that ESEM might provide a more appropriate and realistic representation of the data—with substantially deflated factor correlations—even when the comparative ICM-CFA model fits the data reasonably well in the first place. In particular, Marsh, Nagengast, et al. conducted a multitrait-multimethod analysis of the APRI and showed that latent ESEM factors resulted in better discriminant validity than corresponding ICM-CFA factors. Based on cross-lagged autoregressive models of ESEM factors over three occasions, they also found that bully and victim ESEM factors of the APRI were reciprocally related over time—bullies become victims and victims become bullies.

Investigating the psychometric properties of the Physical Self Inventory, Morin and Maïano (2011) similarly showed that ESEM factors were more differentiated than ICM-CFA factors and provided a better fit to the data. This result is important since a repeated observation in physical self concept research (Marsh & Cheng, 2012) is the presence of elevated correlations among global self-worth, physical self-worth, and perceived physical appearance. Morin and Maïano's results showed that these correlations were related to some indicators of perceived physical appearance that also played a key role in defining global and physical self worth. In addition, their results allowed them to pinpoint problems related to negatively worded items that were not apparent with CFA.

Confirmatory Versus Exploratory Factor Analysis

We conceive of ESEM as primarily a confirmatory tool that provides a viable option to the sometimes over-restrictive assumptions of the ICM-CFA model. Ideally, applied researchers should begin with a well-established, a priori factor structure model—consisting of at least the number of factors and the pattern of target and non-target loadings. Tests of such an a priori model are clearly confirmatory in nature. Nevertheless, some post-hoc modifications or exploratory research questions might be required when the model is extended (e.g., partial invariance in tests of measurement invariance or relations with a set of covariates). ESEM studies considered thus far fit into this confirmatory framework and demonstrate that ESEM frequently outperforms corresponding ICM-CFA models even in cases where there is a clear a priori factor model to be tested. However, in applied research settings, researchers sometimes do not have the luxury of a well-defined a priori model. One common approach has been to use exploratory EFA to 'discover' an appropriate factor structure and then incorporate this post hoc model into a CFA framework. Clearly this approach blurs the distinction between confirmatory and exploratory factor analysis in a way that may offend purists, but we do not automatically reject the appropriateness of such an approach so long as interpretations are offered with appropriate caution. However, we would like to note that the logical next step following the discovery of the 'best' factor structure based on exploratory EFA might be a confirmatory ESEM rather than CFA as illustrated by two recent studies that we summarise here.

In the first of these studies, Meleddu, Guicciardi, Scalas, and Fadda (2012) explored the factor structure of an Italian version of the Oxford Happiness Inventory. After contrasting factor solutions with one to seven factors based on exploratory EFA, they retained a five-factor solution and showed that it provided a better representation of the data than alternative ICM-CFA models. Then they incorporated this exploratory EFA solution into a confirmatory ESEM framework and showed the five-factor model to be reasonably invariant across gender. Next, they used the ESEM factor correlation matrix to explore the single-factor higher-order structure of the instrument. Similarly, Myers, Chase, Pierce, and Martin (2011) explored the factor structure of the coaching efficacy scale for the head coaches of youth sports teams. After contrasting solutions with one to six factors based on exploratory EFA of ordered categorical indicators, they retained the five-factor ESEM solution, showed its superiority over a comparative ICM-CFA solution, and used ESEM to confirm the invariance of this factor model according to coach gender.

Organization of the Examples Provided in This Chapter

From these illustrations, it is obvious that the main use of ESEM to date has been a psychometric one, as it should be the case since obtaining a well-defined factor structure is a prerequisite to any predictive SEM analyses. This observation can also be explained by the fact that ESEM came to fill an important gap in research, allowing researchers using one of many instruments with a well replicated (i.e., confirmatory) EFA factor structure that did not fit the restrictive ICM-CFA model the possibility to embark on a more rigorous process of scale validation. Indeed, based on a total of eight different long and short instruments measuring seven different constructs, these illustrations all showed the superiority of ESEM compared to traditional ICM-CFA. However, it is also obvious from these illustrations that the full scope of ESEM possibilities is as large, or even larger since it can be based on EFA factors, CFA factors, or a combination of both, as in SEM or GSEM. It would thus be unrealistic for a single chapter to even attempt such an extensive coverage. Rather, we believe that the core aspect of ESEM lies in the extensive set of psychometric applications (e.g., measurement invariance) that are made available within the ESEM framework. We believe that once the reader masters how to specify, constrain, and more generally model EFA factors within the ESEM framework, most predictive, multilevel, mixture, or other applications of ESEM can easily be deduced from reading the other chapters of this advanced SEM textbook (or any introductory SEM textbook) once the limitations imposed by ESEM on these statistical models are understood. For this reason, we will devote significant space to psychometric applications of ESEM, before moving to a shorter section on predictive applications.

More specifically, the ESEM modeling framework approach will first be presented. Then, psychometric applications of ESEM for the estimation of measurement models, multiple group measurement invariance and longitudinal measurement invariance will be illustrated. Predictive ESEM models will then be briefly presented and these models will build on the preceding psychometric models, with a special attention given to the main limitations of ESEM. Finally, we also present Marsh et al.'s (2012) EWC generalization of ESEM as a way to circumvent some of the limitations of the ESEM models and provide some illustration of these models. To accompany this chapter, extensive sets of supplemental materials are available online (http://www.statmodel.com/esem.shtml), including all of the input files used to estimate the models described in this chapter, the simulated data file that we used, as well as the input used to generate this data file. It should be noted that, given the scope of this book, we assume readers to be reasonably familiar with EFA, CFA, and SEM, and to have previously conducted and interpreted such analyses according to current best practices (for a refresher, we suggest Bollen, 1989; Brown, 2006; Byrne, 2011; Cudeck & MacCallum, 2007; Fabrigar, Wegner, MacCallum, & Strahan, 1999; Gorsuch, 1983; Henson & Roberts, 2006; Kahn, 2006; Thompson, 2004; Thompson & Daniel, 1996).

THE ESEM APPROACH

The ESEM Model and Rotational Indeterminacy

As the objective of the present chapter is not to provide a detailed technical presentation of ESEM, we only summarize selected ESEM features of particular relevance. More details are available in the online supplemental materials (http://www.statmodel.com/esem.shtml). In the ESEM model (Asparouhov & Muthén, 2009; Marsh et al., 2009), there are p dependent variables $\mathbf{Y} = (Y_1, \ldots, Y_p)$, q independent variables $\mathbf{X} = (X_1, \ldots, X_q)$, and m latent variables $\boldsymbol{\eta} = (\eta_1, \ldots, \eta_m)$, forming the following general ESEM model:

$$\mathbf{Y} = \nu + \Lambda\boldsymbol{\eta} + \mathbf{KX} + \varepsilon$$

$$\boldsymbol{\eta} = \alpha + \mathbf{B}\boldsymbol{\eta} + \Gamma\mathbf{X} + \zeta.$$

Standard assumptions of this model are that the ε and ζ residuals are normally distributed with mean 0 and variance-covariance matrix θ and ψ, respectively. The first equation represents the measurement model where ν is a vector of intercepts, Λ is a factor loading matrix, $\boldsymbol{\eta}$ is a vector of continuous latent variables, $\mathbf{K}$ is a matrix of $\mathbf{Y}$ on $\mathbf{X}$ regression coefficients, and ε is a vector of residuals for $\mathbf{Y}$. The second equation represents the latent variable model where α is a vector of latent intercepts, $\mathbf{B}$ is a matrix of $\boldsymbol{\eta}$ on $\boldsymbol{\eta}$ regression coefficients, Γ is a matrix of $\boldsymbol{\eta}$ on $\mathbf{X}$ regression coefficients, and ζ is a vector of latent variable residuals.

In ESEM, $\boldsymbol{\eta}$ can include multiple sets of ESEM factors defined either as EFA or CFA factors. More precisely, the CFA factors are identified as in traditional SEM where each factor is associated with a different set of indicators. EFA factors can be divided into blocks of factors so that a series of indicators is used to estimate all EFA factors within a single block, and a different set of indicators is used to estimate another block of EFA factors. However, specific items may be assigned to more than one set of EFA or CFA factors. Assignments of items to CFA and/or EFA factors is usually determined based of a priori theoretical expectations, practical considerations, or, perhaps, post-hoc based on preliminary tests conducted on the data.

In a basic version of the ESEM model including only CFA factors (and thus equivalent to the classical SEM model), all parameters can be estimated with the maximum likelihood (ML) estimator or robust alternatives using conventional identification constraints. However, when EFA factors are posited, a different set of constraints is required to achieve an identified solution (Asparouhov & Muthén, 2009; Marsh et al., 2009). In the first step, an unconstrained factor structure is estimated. Given the need to estimate all loadings, a total of m^2 constraints are required to achieve identification for EFA factors (Jöreskog, 1969). These constraints are generally implemented by specifying the factor variance-covariance matrix as an identity matrix and constraining factor loadings in the right upper corner of the factor loading matrix to be 0 (for each ith factor, $i-1$ factor loadings are restricted to 0). Regarding the ESEM mean structure, the identification is similar to typical CFA: all items intercepts are freely estimated and all latent factor means are constrained to 0 (due to rotational difficulties, the alternative CFA method of constraining one intercept per factor to 0 to freely estimate the latent means is not recommended in ESEM). All of these constraints are built in as the default in the Mplus estimation process; in addition, Mplus uses multiple random starting values in the estimation process to help protect against nonconvergence and local minima. For a detailed presentation of identification and estimation issues, the readers are referred to Asparouhov and Muthén (2009), Marsh et al. (2009, 2010), and Sass and Schmitt (2010).

In the second step, this initial, unrotated solution is rotated using any one of a wide set of orthogonal and oblique rotations (Asparouhov & Muthén, 2009; Sass & Schmitt, 2010). As in EFA, multiple orthogonal and oblique rotation procedures are available in ESEM. A review of these rotational procedures is beyond the scope of the present chapter and extended discussions of alternative rotation procedures are available elsewhere (Asparouhov & Muthén, 2009; Bernaards & Jennrich, 2005; Browne, 2001; Jennrich, 2007; Marsh et al., 2009, 2010; Sass & Schmitt, 2010). In traditional applications of EFA, researchers often choose the rotational criteria on the basis of whether the resulting factors are believed to be orthogonal or correlated (e.g., Fabrigar et al., 1999; Henson & Roberts, 2006). This is clearly unsatisfactory and best practices are evolving (see Sass & Schmitt, 2010), but orthogonal rotations are generally considered to be unrealistic in psychological research. The choice of the most appropriate rotation procedure is to some extent still an open research area in EFA, and even more so in ESEM. However, due to rotational indeterminacy, all forms of rotations have equivalent implications for the covariance structure and thus represent statistically equivalent models.

Here, we focus on Geomin rotation (Browne, 2001; Yates, 1987). It was specifically developed to represent simple structure as conceived by Thurstone (1947), in which cross loadings are ideally small relative to target loadings but not required to be zero as in ICM-CFA. Geomin rotations also incorporate a complexity parameter (ε) consistent with Thurstone's original proposal. As operationalized in Mplus' defaults, this ε parameter takes on a small positive value that increases with the number of factors (Asparouhov & Muthén, 2009; Browne, 2001). Asparouhov and Muthén (2009) recommended estimating ESEM models with varying ε values. Marsh et al. (2009, 2010) generally recommended using an ε value of .5 with complex measurement instruments as the most efficient way of deflating factor correlations, but recently noted that target rotation could be more efficient in some cases (Marsh, Lüdtke et al., 2011). Until more is known about the conditions of superiority of one method over another, we recommend comparison of results based on alternative rotational procedures or clear arguments for the selection of a rotational method.

Although we described the generic ESEM model as starting from an unconstrained factor structure, it is also possible to build in equality constraints in this initial solution that can then be submitted to constrained rotation. The main application for this procedure is for tests of measurement or latent mean invariance across meaningful subgroups of participants or across multiple time-points for the same group of participants in a longitudinal study, in order to ensure that the constructs are defined the same way in these different groups, or time points. Thus, in order to test for invariance of sets of ESEM factors, blocks of model parameters can be restricted to be invariant across groups or time points in the estimation of the unconstrained solution and the subsequent rotation. In this way, ESEM easily implements tests of the invariance of factor loadings, item intercepts, uniquenesses, latent variance-covariances, and latent means that were not available in conventional applications of EFA.

Goodness of Fit

In applied SEM research, many scholars are seeking universal "golden rules" allowing them to make objective interpretations of their data rather than being forced to defend subjective interpretations (Marsh, Hau, & Wen, 2004). Many fit indices have been proposed (e.g., Marsh, Balla, & McDonald, 1988), but there is even less consensus today than in the past as to what constitutes an acceptable fit. Some still treat the indices and recommended cut-offs as golden rules; others argue that fit indices should be discarded altogether; a few argue that we should rely solely on chi-square goodness-of-fit indices; and many (like us) argue that fit indices should be treated as rough guidelines to be interpreted cautiously in combination with other features of the data. Generally, given the known sensitivity of the chi-square test to sample size, to minor deviations from multivariate normality, and to minor misspecifications, applied SEM research generally focuses on indices that are sample-size independent (Hu & Bentler, 1999; Marsh, Balla, & Hau, 1996; Marsh et al., 2004; Marsh et al., 2005) such as the Root Mean Square Error of Approximation (RMSEA), the Tucker-Lewis Index (TLI), and the Comparative Fit Index (CFI). The TLI and CFI vary along a 0-to-1 continuum, with values greater than .90 and .95 typically reflecting acceptable and excellent fit to the data, respectively. Values smaller than .08 or .06 for the RMSEA respectively support acceptable and good model fit.

For the comparison of two nested models, the chi-square difference test can be used, but this test suffers from even more problems than the chi-square test for single models that led to the development of other fit indices (see Marsh, Hau, Balla, & Grayson, 1998). Cheung and Rensvold (2002) and Chen (2007) suggested that if the decrease in fit for the more parsimonious model is less than .01 for incremental fit indices like the CFI, then there is reasonable support for the more parsimonious model. Chen (2007) suggested that when the RMSEA increases by less than .015 there is support for the more constrained model. For indices that incorporate a penalty for lack of parsimony, such as the RMSEA and the TLI, it is also possible for a more restrictive model to result in a better fit than a less restrictive model. However, we emphasize that these cut-off values only constitute rough guidelines.

Given the lack of consensus about fit indices, it is not surprising that there is also ambiguity in their application in ESEM and to the new issues that ESEM raises. For example, because the number of factor loadings alone for the EFA factors included in ESEM applications is the product of the number of items and the number of factors, the total number of parameter estimates in ESEM applications can be massively more than in CFA. This feature might make problematic any index that does not control for parsimony (due to capitalization on chance), and yet might call into question the appropriateness of controls for parsimony in indices that do. In the meantime, we suggest that applied researchers use a multifaceted approach based on the integration of a variety of different indices, detailed evaluations of the actual parameter estimates in relation to theory, a priori predictions, common sense, and a comparison of viable alternative models specifically designed to evaluate goodness of fit in relation to key issues.

DATA GENERATION AND ESEM ANALYSES

For present purposes we simulated a multivariate normal data set. This data set includes six items (X_1 through X_6) serving as indicators of two correlated factors, with the first three having their primary loadings on the first factor and the last three having their primary loadings on the second factor, and with two items presenting significant cross loadings on the other factor. These data will be referred to as *Time 1* data. We also simulated a second set of items (Y_1 through Y_6), referred to as *Time 2* data, designed to represent a second measurement point for the X items and simulated to have similar properties to the Time 1 data. Two subgroups of participants ($n = 3,000$ and $n = 1,500$) were simulated. The simulated data set (ESEM.dat), the inputs codes used to generate the data (data-generation.inp), and all of the (annotated) input codes used in the present chapter are available in online supplemental materials (http://www.statmodel.com/esem.shtml). The population generating model is presented in Figure 10.1. All analyses

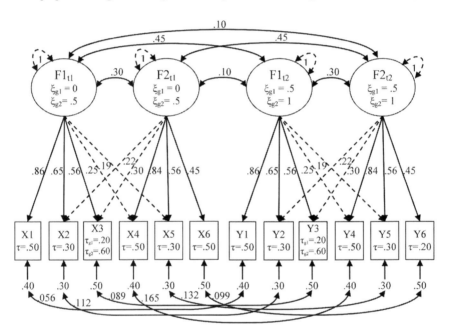

Figure 10.1 Graphical representation of the population generating model. *Note:* Ovals represent latent factors and squares represent observed variables; full unidirectional arrows linking ovals and squares represent the main factor loadings; dotted unidirectional arrows linking ovals and squares represent the cross-loadings; full unidirectional arrows placed under the squares represent the item uniquenesses; bidirectional full arrows linking the ovals represent factor covariances/correlations; bidirectional full arrows linking the squares represent the longitudinal correlated uniquenesses; bidirectional dashed arrows connecting a single oval represent factor variances; τ represents items intercepts; ξ represents the latent factor means; the subscripts g1 and g2 indicate that a parameter varies across both simulated subgroups.

were conducted with Mplus 6.1 (Muthén & Muthén, 2010), using the default maximum likelihood (ML) estimator—although other options might be preferable for other situations (e.g., multilevel, ordered-categorical, or missing data) that can be implemented easily in conjunction with ESEM. Following Marsh et al. (2009, 2010) we will rely on an oblique Geomin rotation (the default in Mplus) with an ε value of 0.5.

ILLUSTRATING ESEM: PSYCHOMETRIC APPLICATIONS BASED ON THE ESEM MEASUREMENT MODEL

Comparison of ESEM and CFA Models

The starting point for any ESEM investigation should be to test the a priori factor model and the hypothesis that the ESEM model provides a better fit to the data than a traditional ICM-CFA model.[1] Indeed, as emphasized by Marsh et al. (2009), the ESEM analysis is predicated on the assumption that ESEM performs noticeably better than the ICM-CFA model in terms of goodness of fit and construct validity of the interpretation of the factor structure. In the current example, CFA and ESEM models were first estimated separately on Time 1 and Time 2 data. The goodness of fit results from these models are reported at the top of Table 10.1. These results clearly indicate that the ICM-CFA model does not provide an acceptable fit to the data at either time point (significant χ^2; TLI < .95; RMSEA > .08). Conversely, the ESEM model provides an almost perfect fit to the data at both time points (non-significant χ^2; CFI and TLI > .95; RMSEA < .06). It is also instructive to compare parameter estimates based on the ICM-CFA and ESEM solutions, as reported in Table 10.2. The main difference—in addition to the ESEM cross-loadings—is that the CFA model results in highly inflated factor correlations due to the unrealistic assumption of 0 cross loadings (for Time 1 and Time 2 respectively, r = .728–.709 for CFA vs. .439–.429 for ESEM solution vs. .30 for the population model). Different forms of rotations were compared. These results are reported in the on-line supplementary materials (http://www.statmodel.com/esem.shtml) and support our decision to rely on Geomin rotation with ε = 0.5 with the current data set. However, the choice of the optimal rotation criteria remains an open question and different rotational procedures should generally be explored in ESEM studies. For the moment, we recommend that applied researchers should either explore alternative rotation procedures, or rely on a strong rationale for choosing a specific rotation procedure.

TABLE 10.1 Results from the Total Group and Multiple Group Cross Sectional Models

Model	χ^2	df	RMSEA	RMSEA 95% CI	CFI	TLI
Main models						
Time 1 data CFA model (CFA-Time1.inp)	339.062*	8	.096	.087–.105	.968	.940
Time 1 data ESEM model (ESEM-Geo.5-Time1.inp)	3.085	4	.000	.000–.020	1.000	1.000
Time 2 data CFA model (CFA-Time2.inp)	338.377*	8	.096	.087–.105	.969	.941
Time 2 data ESEM model (ESEM-Geo.5-Time2.inp)	7.864	4	.015	.000–.030	1.000	.999
Multiple Group Invariance Models (Time 1 data)						
Model 1: Configural (MG-inv-M1.inp)	9.064	8	.008	.000–.027	1.000	1.000
Model 2: Load. (MG-inv-M2.inp)	11.646	16	.000	.000–.014	1.000	1.001
Model 3: Load., uniq. (MG-inv-M3.inp)	17.730	22	.000	.000–.013	1.000	1.001
Model 4: Load., FVC. (MG-inv-M4.inp)	18.204	19	.000	.000–.018	1.000	1.000
Model 5: Load., int. (MG-inv-M5.inp)	246.682*	20	.071	.063–.079	.976	.964
Model 5p: Load., p.int. (MG-inv-M5p.inp)	16.163	19	.000	.000–.015	1.000	1.000
Model 6: Load., uniq., FVC. (MG-inv-M6.inp)	24.125	25	.000	.000–.016	1.000	1.000
Model 7: Load., int., uniq. (MG-inv-M7.inp)	256.920*	26	.063	.056–.070	.976	.972
Model 7p: Load., p.int., uniq. (MG-inv-M7p.inp)	22.342	25	.000	.000–.015	1.000	1.000
Model 8: Load., int., FVC. (MG-inv-M8.inp)	253.330*	23	.067	.059–.074	.976	.969
Model 8p: Load., p.int., FVC. (MG-inv-M8p.inp)	22.721	22	.004	.000–.018	1.000	1.000
Model 9: Load., int., uniq., FVC. (MG-inv-M9.inp)	263.316*	29	.060	.053–.067	.976	.975
Model 9p: Load., p.int., uniq., FVC. (MG-inv-M9p.inp)	28.732	28	.003	.000–.017	1.000	1.000
Model 10: Load., int., FMeans. (MG-inv-M10.inp)	641.270*	22	.112	.104–.119	.935	.912
Model 10p: Load., p.int., FMeans. (MG-inv-M10p.inp)	319.936*	21	.080	.072–.087	.969	.955

(continued)

TABLE 10.1 Results from the Total Group and Multiple Group Cross Sectional Models (continued)

Model	χ^2	df	RMSEA	RMSEA 95% CI	CFI	TLI
Model 11: Load., int., uniq., FMeans. (MG-inv-M11.inp)	654.880*	28	.100	.093–.106	.934	.930
Model 11p: Load., p.int., uniq., FMeans. (MG-inv-M11p.inp)	326.470*	27	.070	.063–.077	.969	.965
Model 12: Load., int., FVC., FMeans. (MG-inv-M12.inp)	641.699*	25	.105	.098–.112	.936	.923
Model 12p: Load., p.int., FVC., FMeans. (MG-inv-M12p.inp)	320.123*	24	.074	.067–.081	.969	.961
Model 13: Load., int., uniq., FVC., FMeans. (MG-inv-M13.inp)	655.453*	31	.095	.088–.101	.935	.937
Model 13p: Load., p.int., uniq., FVC., FMeans. (MG-inv-M13p.inp)	326.628*	30	.066	.060–.073	.969	.969
MIMIC Models						
Time 1 MIMIC null effect (MIMIC-null-Time1.inp)	634.162*	10	.118	.110–.126	.943	.881
Time 1 MIMIC saturated (MIMIC-satur-Time1.inp)	2.833	4	.000	.000–.019	1.000	1.001
Time 1 MIMIC invariant intercept (MIMIC-base-Time1.inp)	242.024*	8	.081	.072–.090	.979	.944
Time 1 MIMIC partially invariant intercept (MIMIC-DIF-Time1.inp)	7.440	7	.004	.000–.019	1.000	1.000
Time 2 MIMIC null effect (MIMIC-null-Time2.inp)	729.323*	10	.126	.119–.134	.936	.865
Time 2 MIMIC saturated (MIMIC-satur-Time2.inp)	7.785	4	.015	.000–.030	1.000	.998
Time 2 MIMIC invariant intercept (MIMIC-base-Time2.inp)	322.398*	8	.093	.085–.102	.972	.926
Time 2 MIMIC partially intercept invariant (MIMIC-DIF-Time2.inp)	8.655	7	.007	.000–.021	1.000	1.000

Note: Names of the input file in the supplementary materials are reported in parentheses; CFA: Confirmatory factor analysis; ESEM: Exploratory Structural Equation Modeling; χ^2: Chi square test of model fit; *df*: degrees of freedom; RMSEA: Root Mean Square Error of Approximation; RMSEA 95% CI: 95% confidence interval of the RMSEA; TLI: Tucker-Lewis Index; CFI: Comparative Fit Index; Load: Loadings invariance; Uniq: Uniquenesses invariance; FVC: Factor variance-covariance invariance; Int: Intercepts invariance; FMeans: Factor means invariance.

* $p < .05$

TABLE 10.2   Standardized Parameters from the CFA and ESEM Models Based on Time 1 Data

	CFA			ESEM geomin, $\varepsilon = .5$		
Item	F1	F2	Uniq.	F1	F2	Uniq.
Time 1	(CFA-Time1.inp)			(ESEM-Geo.5-Time1.inp)		
X1	.722		.478	.833	−.038	.333
X2	.843		.289	.690	.232	.330
X3	.742		.450	.536	.318	.461
X4		.835	.303	.139	.766	.300
X5		.774	.401	.166	.674	.420
X6		.510	.739	−.046	.566	.700
Correlations	.728			.439		
Time 2	(CFA-Time2.inp)			(ESEM-Geo.5-Time2.inp)		
Y1	.737		.457	.837	−.032	.322
Y2	.841		.292	.698	.222	.330
Y3	.751		.436	.560	.303	.449
Y4		.846	.284	.149	.762	.300
Y5		.776	.398	.140	.704	.400
Y6		.492	.758	−.054	.551	.719
Correlations	.709			.429		

Note: Names of the input file in the supplementary materials are reported in parentheses; All coefficients significant at the .05 level; CFA: Confirmatory factor analysis; ESEM: Exploratory Structural Equation Modeling; F1: standardized loadings on the first factor; F2: standardized loadings on the second factor; Uniq.: standardized uniquenesses.

Measurement Invariance: The Multiple Group Approach

Tests of measurement invariance evaluate the extent to which measurement properties generalize over multiple groups, situations, or occasions. Such tests are widely applied in SEM studies (Meredith, 1993; Vandenberg & Lance, 2000) and the evaluation of measurement structures more broadly. Measurement invariance is fundamental to the evaluation of construct validity and generalizability. More specific to our chapter, measurement invariance is an important prerequisite to any form of valid group-based comparison. Indeed, unless the underlying factors are measuring the same construct in the same way and the measurements themselves are operating in the same way (across groups or over time), mean differences and other comparisons might potentially be invalid. Before ESEM, the multigroup tests of invariance were seen as a fundamental advantage of CFA/SEM over EFA. Indeed, traditional EFA approaches were largely limited to descriptive comparisons of the factor loadings estimated separately in each group, while CFA approaches allowed researchers to constrain factor loadings to be the same in multiple groups.

Item intercepts play a key role in tests of measurement invariance, but not in traditional EFA approaches. In particular, an important assumption in the comparison of group means across multiple groups (or over time) is the invariance of a majority of item intercepts, a violation of which is also called *differential item functioning* (DIF; technically, monotonic DIF). For example, if group differences are not consistent in direction and magnitude across the items associated with a particular latent factor, then the observed differences on the latent factor depend on the mix of items considered. A subset of the items actually used or a new sample of items designed to measure the same factor might give different results.

Following Meredith (1993; also see Millsap, 2011; Vandenberg & Lance, 2000), Marsh et al. (2009, 2010) operationalized a taxonomy of 13 partially nested models varying from the least restrictive model of configural invariance with no invariance constraints to a model of complete invariance positing strict invariance of the factors loadings, items intercepts, and items uniquenesses as well as the invariance of the latent means and of the factor variance-covariance matrix (see Table 10.3). We illustrated this multigroup ESEM approach to invariance with our simulated data. As the results are highly similar across time points (and were specified to be similar in the population generating model), multiple group tests of invariance were only conducted on Time 1 data (i.e., variables X_1 through X_6) due to space limitations (readers are invited to conduct parallel tests with Time 2 data; files for doing so are contained in the supplemental materials at http://www. statmodel.com/esem.shtml). It should also be noted that when the multiple groups reflect a factorial combination of grouping variables (e.g., 3 age levels $\times$ 2 genders = 6 groups), it is possible to discern non-invariance due to the interaction of the two grouping variables as well as the "main" effects of each grouping variable considered separately. The results from the 13 steps of multiple group measurement invariance tests are reported in Table 10.1, shown previously. To conserve space, we will focus our presentation of results to the classical models of configural (Model 1), weak (Model 2), strong (Model 5), strict (Model 7) invariance as defined by Meredith's (1993) seminal paper, before exploring the invariance of the latent variance-covariance matrix and mean structure.

Configural invariance (Model 1). This taxonomy begins with a model having no invariance of any parameters across the groups, or *configural invariance* (Model 1). Like the traditional EFA approach, factor structures are freely estimated separately in each group, with only the number of factors being the same in all groups. This model is usually identified by using the latent standardization method of fixing the latent variances to 1 and the latent means to 0 in all groups so as to freely estimate all rotated factor loadings and item

TABLE 10.3   Marsh et al. (2009, 2010) Taxonomy of Invariance Tests

Model	Invariant parameters	Signification	Nesting
Model 1	None, apart from the number of factors.	Configural Invariance	none
Model 2	Factor loadings.	Weak measurement invariance	1
Model 3	Factor loadings, items' uniquenesses.		1, 2
Model 4	Factor loadings, factor variances-covariance.		1, 2
Model 5	Factor loadings, items' intercepts.	Strong measurement invariance	1, 2
Model 6	Factor loadings, items' uniquenesses, factor variances-covariance.		1, 2, 3, 4
Model 7	Factor loadings, items' intercepts, items' uniquenesses.	Strict measurement invariance	1, 2, 3, 5
Model 8	Factor loadings, items' intercepts, factor variances-covariance.		1, 2, 4, 5
Model 9	Factor loadings, items' intercepts, items' uniquenesses, factor variances-covariance.		1–8
Model 10	Factor loadings, items' intercepts, factor means.	Latent mean invariance	1, 2, 5
Model 11	Factor loadings, items' intercepts, items' uniquenesses, factor means.	Manifest mean invariance	1, 2, 3, 5, 7, 10
Model 12	Factor loadings, items' intercepts, factor variances-covariance, factor means.		1, 2, 4, 5, 6, 10
Model 13	Factor loadings, items' intercepts, items' uniquenesses, factor variances-covariance, factor means.	Complete factorial invariance	1–12

Note: Models with latent factor means freely estimated constrain intercepts to be invariant across groups, while models where intercepts are free imply that mean differences are a function of intercept differences. Nesting relations are shown such that the estimated parameters of the less general model are a subset of the parameters estimated in the more general model under which it is nested. All models are nested under model 1 (with no invariance constraints) while model 13 (complete invariance) is nested under all other models.

Source: Partially adapted from Table 1 of Marsh, Muthén, Asparouhov, Lüdtke, Robitzsch, Morin, & Trautwein (2009).

intercepts. Model 1 provides a baseline of comparison for the remaining models in the taxonomy that are nested under it (i.e., parameters in the subsequent, more restricted model are a subset of parameters of Model 1). This baseline model with no invariance constraints provides a satisfactory level of fit to the data (non-significant χ^2; CFI and TLI > .95; RMSEA < .06).

Weak measurement invariance (Model 2). Weak measurement invariance tests whether the factor loadings are the same in both groups. This is an important test, as all subsequent models in the taxonomy assume the invariance of factor loadings. When the loadings are fixed to equality across groups, the variances in all groups except in an arbitrarily selected reference group can be freely estimated. The test of weak measurement invariance is based on the comparison of Model 2 with Model 1. For our simulated data, this comparison results in a non-significant chi-square difference test, equivalent CFIs and TLIs (to two decimals), and a slight improvement in the RMSEA (that adjusts for model parsimony). These results provide strong support for weak measurement invariance. It should be noted that tests of weak measurement invariance are critical in ESEM as it is currently not straightforward to specify a model with partial invariance of factor loadings (but see the ESEM-within-CFA method, described later).

Strong measurement invariance (Model 5). Model 5 tests the strong measurement invariance of the model and requires that the indicator intercepts—in addition to the factor loadings—are invariant over groups. This is an important test since it justifies the comparison of latent means. When the intercepts are constrained to equality, the latent means can be freely estimated. The typical approach is to constrain the means in a selected referent group to be zero, and to use this as a basis of comparison for the remaining means that are freely estimated. Although the crucial comparison for strong measurement invariance is between Models 2 and 5, the invariance of items' intercepts can also be tested by the comparison of any pair of models differing only in regard to intercept invariance (Model 2 vs. Model 5; Model 3 vs. Model 7; Model 4 vs. Model 8; Model 6 vs. Model 9) and convergence or divergence of conclusions can be inspected for further information. For our simulated data, all of these comparisons resulted in significant chi-square difference tests, in substantial decreases in CFI and TLI, and an increase in RMSEA that exceeds the recommended cut-off. This shows that the hypothesis of strong measurement invariance does not hold for all items.

When tests of full invariance fail, it is typical to consider tests of partial invariance in which the invariance constraints for one or a small number of parameters are relaxed. For our simulated data, examination of the modification indices associated with these models suggests that the lack of invariance is limited to item 3 (consistent the population generating model). A series of partial invariance models were thus considered (those labeled "p" in Table 10.1) in which the invariance of the item 3 intercept was relaxed while invariance constraints were retained for the remaining intercepts. When the previously noted pairs of models were compared but relying on partial, rather than complete, invariance of the items intercepts, the results support that the remaining items are invariant across groups (non-significant $\Delta\chi^2$ and ΔTLI, ΔCFI, and ΔRMSEA < .01). When there are more than two groups, partial invariance could be limited to a subset of groups.

For instrument construction, DIF might be a sufficient basis for dropping an item (if there is a large pool of items with invariant factor loadings and intercepts from which to choose in the pilot version of the instrument). However, the post-hoc construction of models with partial invariance should generally be viewed with caution. When the sample size is small or non-representative, there is the danger of capitalizing on idiosyncrasies of the sample. When the number of target items measuring each factor is small, there is the danger that the invariance found for a subset of the items is not generalizable to the population of all possible items that could have been picked to assess this specific construct. Here, for example, the observed latent mean differences between groups are a function of the items that do have invariant intercepts and are assumed to generalize to other sets of appropriate items.

Strict measurement invariance (Model 7). Model 7 requires invariance of items' uniquenesses in addition to the invariance of factor loadings and intercepts. The invariance of uniquenesses is a prerequisite to the comparison of manifest scores based on factor scores but is also relevant and interesting in its own right as a test of the generalizability of measurement errors across groups. Still, tests of the invariance of items' uniquenesses are not as critical as other tests, as long as one relies on latent variable methodologies. Thus, if there is clear support for the invariance of uniquenesses, it is fine to impose this constraint; however, if the invariance of uniquenesses is not supported, comparisons of latent means or factor covariances are still justified without this constraint. Although the crucial comparison for strict measurement invariance is between Model 5 and Model 7 (or 5p and 7p in the case of partial invariance of items' intercepts), the invariance of items' uniquenesses can also be tested by the comparison of any pair of models differing only in regard to uniquenesses' invariance (Model 2 vs. Model 3; Model 4 vs. Model 6; Model 5 vs. Model 7; Model 8 vs. Model 9; Model 10 vs. Model 11; Model 12 vs. Model 13). Comparisons of each of these pairs of models result in a non-significant chi-square difference test and in equivalent CFI, TLI, and RMSEA values (to two decimals). Hence, there is support for partial strict measurement invariance—the partial label is determined by the non-invariant intercept.

Factor variance-covariance invariance (Model 4). Model 4 tests the invariance of the factor variance-covariance matrix—in addition to factor loadings. Although the invariance of each factor variance and covariance term can be tested separately in CFA, this is not routinely possible with ESEM (due to complications in rotation; but see the ESEM-within-CFA method described later). Tests of the invariance of the factor variance-covariance matrix (Model 4) are highly relevant to many substantively important issues, even more so than the comparison of latent means in some applications. Nevertheless, for purposes of testing group means, the invariance of the variance-covariance matrix is not necessary and does not represent an essential step in tests of measurement invariance. For our simulated data, the critical comparison is between Model 2 and Model 4 (or other pairs of models differing in respect to the invariance of the variance-covariance matrix). Although some of these comparisons result in statistically significant chi-square differences, none of them result in changes in CFI, TLI, or RMSEA that exceeds the recommended thresholds. These results thus provide reasonable support for the invariance of the factor variance-covariance matrix.

Invariance of latent means (Models 10 through 13). The final four models (Model 10 through Model 13) in the taxonomy all constrain mean differences between groups to be zero—in combination with the invariance of other parameters. In order for these tests to be interpretable, it is essential that there is support for the invariance of factor loadings and item intercepts (or at least partial invariance for a majority of items per factor, for further discussion see Byrne, Shavelson, & Muthén, 1989), but not item uniquenesses or the factor variance-covariance matrix. The most appropriate model will depend on the results of earlier invariance tests. In general, it is advisable to test the most constrained model that is justified by earlier tests (or to compare results of alternative models to determine whether there are substantively important differences). Hence, if there is support for the invariance of item uniquenesses and the factor variance-covariance matrix (as well as factor loadings and item intercepts) then Model 13 should be compared to Model 9. However, if there were invariance support for neither item uniquenesses nor the factor variance-covariance matrix (but there was support for factor loading and item uniquenesses), then the appropriate test would be the comparison of Model 10 and Model 5. For our simulated data, tests of Model 10 through Model 13 all lead to the conclusion that latent means differ systematically across groups (significant $\Delta\chi^2$ and ΔTLI, ΔCFI, and ΔRMSEA > .01). These comparisons show that, when group 1 latent means are constrained to be zero, group 2 latent means vary between .513 and .633 on factor 1 and .476 and .517 on factor 2—all close to the true population values of .50.

Summary. The multiple group approach provides a very general and elegant framework for tests of measurement invariance and latent mean differences when the grouping variable has a small number of discrete categories and the sample size for each group is reasonable. The extension of ESEM to incorporate this multiple group approach is one of the most important applications of ESEM. Nevertheless, this multiple group approach might not be practical for variables that are continuous (e.g., age), for studies that evaluate simultaneously many different contrast variables (e.g., age, gender, experimental/control) and their interactions, or when sample sizes are small. In such situations, a more parsimonious MIMIC approach (to be presented later) might be appropriate.

Measurement Invariance: The Longitudinal Approach

Essentially the same logic and the same taxonomy of models can be used to test the invariance of parameters across multiple occasions for a single group. One distinctive feature of longitudinal analyses is that they should normally include correlated uniquenesses (CUs) between responses to the same item on different occasions (see Jöreskog & Sörbom, 1977; Marsh, 2007; Marsh & Hau, 1996). When the same items are used on multiple occasions, the uniqueness component associated with each item from one occasion is typically positively correlated with the uniqueness component associated with the same item on another occasion. Failure to include these CUs generally results in biased parameter estimates. In particular, test-retest correlations among matching latent factors are systematically inflated, which can then systematically bias other parameter estimates and may even result in improper solutions such as a non-positive definite factor variance-covariance matrix or estimated test-retest correlations that exceed 1.0 (e.g., Marsh, Martin, & Hau, 2006; Marsh, Martin, & Debus, 2001). Importantly, the inclusion of CUs is another option within ESEM that was not typically possible in traditional EFAs.

For our simulated data, preliminary analyses were conducted to show that models including CUs (i.e., X_1 with Y_1, X_2 with Y_2, etc) provided a better fit to the data for both CFA and ESEM models (see first four models in Table 10.4). Interestingly, comparison of these four models demonstrated that the CFA model with CUs was acceptable according to some criteria (CFI, TLI > .95; RMSEA = .058)—even though all indices of fit were substantially better for the ESEM model (CFI, TLI > .99; RMSEA = .014). However, inspection of the factor correlations based on these four models, which are reported in Table 10.5, reveals that CFA factor correlations were systematically higher (and inflated relative to the known population parameters for our simulated data) when compared to the ESEM factor correlations. Of particular relevance, the test-retest correlations are systematically higher for models that do not contain CUs (and inflated relative to the known factor correlations for our simulated data) than for models that do. On this basis, we evaluated longitudinal invariance in relation to ESEM models that included CUs.

The results from the 13 steps of longitudinal measurement invariance tests are reported in Table 10.4. As the results from the longitudinal invari-

TABLE 10.4 Results from the Longitudinal Models

Model	χ^2	df	RMSEA	RMSEA 95% CI	CFI	TLI
Main models						
CFA model without CUs. (Longit-CFA.inp)	2605.798*	48	.109	.105–.112	.896	.857
CFA model with CUs. (Longit-CFA-CU.inp)	679.461*	42	.058	.054–.062	.974	.959
ESEM model without CUs. (Longit-ESEM.inp)	2002.267*	40	.104	.101–.108	.920	.869
ESEM model with CUs. (Longit-ESEM-CU.inp)	62.474*	34	.014	.008–.019	.999	.998
Longitudinal invariance models (with CUs)						
Model 1: Configural (Long-inv-M1.inp)	62.474*	34	.014	.008–.019	.999	.998
Model 2: Load. (Long-inv-M2.inp)	70.407*	42	.012	.007–.017	.999	.998
Model 3: Load., uniq. (Long-inv-M3.inp)	73.458*	48	.011	.005–.016	.999	.999
Model 4: Load., FVC. (Long-inv-M4.inp)	76.268*	45	.012	.007–.017	.999	.998
Model 5: Load., int. (Long-inv-M5.inp)	474.364*	46	.045	.042–.049	.983	.975
Model 5p: Load., p.int. (Long-inv-M5p.inp)	74.521*	45	.012	.007–.017	.999	.998
Model 6: Load., uniq., FVC. (Long-inv-M6.inp)	80.125*	51	.011	.006–.016	.999	.998
Model 7: Load., int., uniq. (Long-inv-M7.inp)	477.276*	52	.043	.039–.046	.983	.978
Model 7p: Load., p.int., uniq. (Long-inv-M7p.inp)	77.565*	51	.011	.005–.015	.999	.999
Model 8: Load., int., FVC. (Long-inv-M8.inp)	480.265*	49	.044	.041–.048	.983	.976
Model 8p: Load., p.int., FVC. (Long-inv-M8p.inp)	80.403*	48	.012	.007–.017	.999	.998

(continued)

TABLE 10.4. Results from the Longitudinal Models (continued)

Model	χ^2	df	RMSEA	RMSEA 95% CI	CFI	TLI
Model 9: Load., int., uniq., FVC. (Long-inv-M9.inp)	483.955*	55	.042	.038–.045	.983	.979
Model 9p: Load., p.int., uniq., FVC. (Long-inv-M9p.inp)	84.264*	54	.011	.006–.016	.999	.998
Model 10: Load., int., FMeans. (Long-inv-M10.inp)	1481.685*	48	.081	.078–.085	.942	.920
Model 10p: Load., p.int., FMeans. (Long-inv-M10p.inp)	1119.971*	47	.071	.068–.075	.956	.939
Model 11: Load., int., uniq., FMeans. (Long-inv-M11.inp)	1484.599*	54	.077	.073–.080	.942	.929
Model 11p: Load., p.int., uniq., FMeans. (Long-inv-M11p.inp)	1122.936*	53	.067	.064–.070	.957	.946
Model 12: Load., int., FVC., FMeans. (Long-inv-M12.inp)	1488.286*	51	.079	.076–.083	.942	.925
Model 12p: Load., p.int., FVC., FMeans. (Long-inv-M12p.inp)	1126.449*	50	.069	.066–.073	.956	.942
Model 13: Load., int., uniq., FVC., FMeans. (Long-inv-M13.inp)	1491.877*	57	.075	.072–.078	.942	.933
Model 13p: Load., p.int., uniq., FVC., FMeans. (Long-inv-M13p.inp)	1130.176*	56	.065	.062–.069	.956	.949

Note: Names of the input file in the supplementary materials are reported in parentheses; CFA: Confirmatory factor analysis; ESEM: Exploratory Structural Equation Modeling; χ^2: Chi square test of model fit; df: degrees of freedom; RMSEA: Root Mean Square Error of Approximation; RMSEA 95% CI: 95% confidence interval of the RMSEA; TLI: Tucker-Lewis Index; CFI: Comparative Fit Index; Load: Loadings invariance; Uniq: Uniquenesses invariance; FVC: Factor variance-covariance invariance; Int: Intercepts invariance; FMeans: Factor means invariance.

* $p \leq .05$

TABLE 10.5   Correlations Between Factors in the Longitudinal Models

	CFA models			ESEM models		
	Time 1 Factor 1	Time 1 Factor 2	Time 2 Factor 1	Time 1 Factor 1	Time 1 Factor 2	Time 2 Factor 1
Models without CUs						
Time 1 Factor 2	.728			.447		
Time 2 Factor 1	.545	.307		.559	.143	
Time 2 Factor 2	.342	.530	.709	.195	.551	.439
Models with CUs						
Time 1 Factor 2	.728			.438		
Time 2 Factor 1	.489	.304		.495	.166	
Time 2 Factor 2	.338	.432	.708	.215	.438	.430

ance tests closely parallel those from the multiple group invariance tests, they will not be described in detail. These results support the complete longitudinal invariance of the factor loadings, of the factor variance-covariance matrix, and of the items' uniquenesses. However, the results did not support the complete invariance of the items' intercepts. Examination of the modification indices associated with the various models allowing for the verification of items' intercept invariance suggests that this lack of invariance is limited to item 6 (consistent with the population generating model). A series of partial invariance models were thus estimated (labeled "p" in Table 10.4) in which the invariance of item 6 intercept was relaxed while the remaining intercepts were constrained to be invariant. Lastly, the final four models from the taxonomy all converged in rejecting the longitudinal invariance of the factor means. These models revealed that, when Time 1 latent means are constrained to be zero, Time 2 latent means vary between .465 and .508 on factor 1 and on .448 and .506 on factor 2, close to the population values of .50.

Summary. The longitudinal approach to invariance closely parallels the multiple group approach and provides a very flexible framework for the examination of the invariance assumptions inherent in longitudinal analyses. Due to space limitations, we do not pursue tests of multiple group longitudinal invariance, which would be the next logical step. Conducting these tests would involve multiplying by three the sequence of 13 steps proposed in Table 10.3 as these models allow for the testing of the invariance of the models parameters across groups, time period, and groups × time periods. In addition, new models could be added to this taxonomy (e.g., Marsh et al., 2010) to investigate the invariance of the longitudinal CUs across subgroups of participants. However, these tests can be conducted with our simulated data and we suggest that interested readers explore these possibilities.

ILLUSTRATING ESEM: PREDICTIVE APPLICATIONS BASED ON THE ESEM STRUCTURAL MODEL

The MIMIC Approach

The MIMIC model (Jöreskog & Goldberger, 1975; Marsh, Ellis, Parada, Richards, & Heubeck, 2005; Marsh, Tracey, & Craven, 2006; Muthén, 1989) is a multivariate regression model in which latent variables are regressed on observed predictors (see Figure 10.2). In addition, when the latent variables have multiple indicators, the MIMIC model can also be extended to test potential non-invariance of item intercepts, that is, DIF (technically, monotonic DIF). In that specific case, the MIMIC model has important advantages over the multiple group approach, but also some limitations. Particularly in applied research based on often modest sample sizes, the MIMIC model is much more parsimonious and does not require the separate estimation of the model in each group. Also, it allows researchers to consider multiple independent variables that would typically become unmanageable in multiple group analyses. A particularly important feature of the MIMIC model is that it allows researchers to consider continuous predictors (e.g., age, income, pretest scores) that cannot be evaluated in the multiple group approach without recoding them to form a small number of discrete groups—a strategy that we typically do not recommend. It is also possible to consider a combination of continuous and categorical predictors (and their interaction) in the

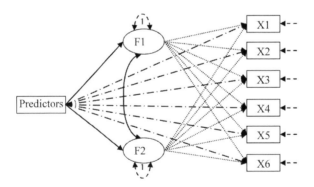

Figure 10.2 MIMIC model used to test differential item functioning. *Note:* Ovals represent latent factors and squares represent observed variables; dotted unidirectional arrows represent factor loadings; dashed unidirectional arrows represent items uniquenesses; full unidirectional arrows represent the paths needed to estimate the MIMIC model with invariant intercepts; dashed-and-dotted unidirectional arrows represent the paths needed to estimate the saturated MIMIC model; bidirectional full arrows linking the ovals represent factor covariances/correlations; bidirectional dashed arrows connecting a single oval represent factor variances.

same MIMIC model. However, while the MIMIC model is able to test DIF, it does not test the invariance of factor loadings (non-monotonic DIF), uniqueness, or the factor variance-covariance matrix (and implicitly assumes their invariance). We focus here on the use of the MIMIC model as a way to investigate monotonic DIF as it provides a more complete illustration of the full flexibility of the MIMIC approach, but note that this section more generally provides a generic example of how to relate ESEM factors to any number of observed variables (and their interactions).

In the MIMIC approach, monotonic DIF can be evaluated by the comparison of three models. The first (null effect) MIMIC model posits that the predictor variables have no effect on the latent variables and items intercepts (i.e., paths from predictors to latent factors and their indicators—the full lines and the dotted-dashed lines in Figure 10.2—are constrained to be zero). The second (saturated) MIMIC model has paths from each predictor variable to all item intercepts (i.e., the dotted-dashed lines in Figure 10.2), but not the latent factors. The third (invariant intercept) MIMIC model has freely estimated paths from the predictor variables to the latent factors (i.e., the full lines in Figure 10.2), but paths to item intercepts are all constrained to be zero. The comparison of Model 1 with Models 2 and 3 tests whether there are any effects of the predictors, the comparison of Model 1 and Model 3 tests whether the predictors have an effect on the latent variables, while the comparison of Model 2 and Model 3 tests whether the effects of the predictor variables on individual items can be fully explained in terms of effects on the latent factors. If Model 2 fits substantially better than Model 3, then there is evidence of monotonic DIF (i.e., non-invariance of intercepts). In this case, it might be appropriate to pursue partially invariant models in which the invariance constraint is relaxed for some item intercepts (but see previous discussion).

For our simulated data (see MIMIC Models in Table 10.1 and on-line materials), we again focus on the Time 1 results in order to conserve space (but note that Time 2 results are very similar and can easily be tested with the data set provided online). The MIMIC null effect model, in which the grouping variable is posited to be unrelated to the ESEM factors or the items, failed to provide an acceptable fit to the data (significant χ^2; TLI < .95; RMSEA > .08). This suggests that at least some effects of the predictor variable should be expected. Indeed, the saturated MIMIC model did provide a satisfactory fit to the data (non-significant χ^2; CFI and TLI > .95; RMSEA < .06) and a substantial improvement over the null effect model. The third (intercept invariant) MIMIC model (i.e., in which the grouping variable is only allowed to predict the latent factor scores but not the items) failed to provide an acceptable fit (significant χ^2; TLI < .95; RMSEA > .08), suggesting DIF. Examination of the modification indices associated with this model indicates that DIF was mainly associated with item 3. Allowing for direct effects of the predictor on item 3, in addition to its effects on the ESEM factors, results in a satisfactory fit to the data (non-significant χ^2; CFI and TLI > .95; RMSEA < .06) and in a fit that is comparable to the fit of the saturated MIMIC model [$\Delta\chi^2$ (df) = 4.607 (3), p > .05; ΔTLI and ΔCFI < .01]. Detailed results from this model reveal that participants' levels on factor 1 ($\hat{\beta}$ = .523, p < .001), factor 2 ($\hat{\beta}$ = .489, p < .001), and item 3 ($\hat{\beta}$ = .369, p < .001), tend to be higher in the second group.

Summary. The MIMIC approach to measurement invariance provides a generic framework for testing the relations between any number of observed continuous or categorical predictors, and their interactions, as well as a very powerful alternative to the multiple group approach for tests of DIF when the sample size is small, when the predictors are continuous or include many categories, when there are multiple predictors, when interactions among predictors are considered, and when the predictor variables themselves are latent variables based on multiple indicators. However, the MIMIC approach is limited in that it assumes the invariance of factor loadings and uniquenesses, but it does not allow for the verification of these assumptions. To further increase the flexibility of the MIMIC approach, Marsh et al. (2006) proposed a hybrid approach in which multiple group models and the MIMIC approach are combined for greater precision in the investigation of measurement invariance issues. So long as the two approaches converge to similar interpretations, there is support for the construct validity of these interpretations. More recently, Marsh, Nagengast, and Morin (2012) extended this approach to ESEM, including tests specifically designed to investigate the loss of information due to categorizing continuous variables where a MIMIC model is separately estimated in each of the separate groups.

Autoregressive Cross Lagged ESEM Models

An obvious extension of the MIMIC approach that we just presented would be to use latent variables to predict other latent variables. Rather than to present this simpler scenario, that can easily be pursued with the data set that is provided online, we present one final model to better highlight the flexibility of the ESEM approach in a way that might not be obvious to the reader. An important question that can be pursued in longitudinal research is related to the direction of the relations between two constructs over time (e.g., Marsh & Yeung, 1998; Morin, Maïano, Nagengast, Marsh, Morizot, & Janosz, 2011). These questions can be investigated in the context of autoregressive cross lagged models (Jöreskog & Sörbom, 1977; Marsh & Grayson, 1994), where each variable is expressed as an additive function of the preceding values on both variables (here Factors 1 and 2) and a random error. See Figure 10.3 for details of this model, as well as the syntax

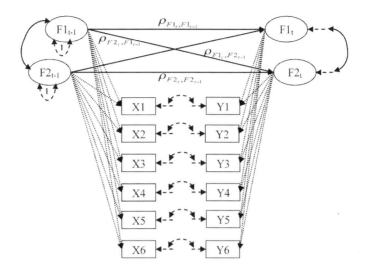

Figure 10.3 Autoregressive cross lagged ESEM model. *Note:* Ovals represent latent factors and squares represent observed variables; dotted unidirectional arrows represent factor loadings; dashed unidirectional arrows represent items uniquenesses and factor disturbances; full unidirectional arrows represent the autoregressive and cross lagged predictive paths; bidirectional full arrows linking the ovals represent time-specific factor covariances/correlations; bidirectional dashed arrows connecting a single oval represent factor variances; bidirectional dashed arrows linking the squares represent the longitudinal correlated uniquenesses.

used to estimate this model (ARCLM-from7p.inp) in the on-line materials. This model was built from the longitudinal Model 7p (i.e., partial strict measurement invariance) since measurement invariance of the constructs over time is a prerequisite to longitudinal analyses in order to ensure that the repeated measures estimate the same constructs over time. The results from this model reveal strong "horizontal" (i.e., test-retest) effects where each variable mostly predicted itself over time ($\hat{\beta}_{F1_t,F1_{t-1}} = .527$; $p < .001$; $\hat{\beta}_{F2_t,F2_{t-1}} = .425$, $p < .001$). Small longitudinal effects going from Factor 2 to Factor 1 were also apparent ($\hat{\beta}_{F1_t,F2_{t-1}} = -.069$, $p \leq .001$), but the reciprocal effects of Factor 1 to Factor 2 were nonsignificant ($\hat{\beta}_{F2_t,F1_{t-1}} = .027$, $p > .05$), suggesting that the direction of influence goes from Factor 2 to Factor 1 and is negative once the stability of each process is controlled for in the model.

This model can easily be extended to incorporate additional measurement points (see Marsh, Nagengast, et al., 2011) and can also serve as a starting point for the estimation of any predictive relationships between latent variables in the ESEM framework. It can also be combined with the MIMIC model so as to estimate the relations between observed and latent variables. For instance, we re-estimated this model while controlling for the effects of the grouping variable on the various latent factor, which is akin to estimating the previous autoregressive cross lagged model with the grouping variable used as a MIMIC-like predictor of the latent factors (see ARCLM-cont-from7p.inp). This model thus provides a straightforward extension of the previous one when applied researchers want to estimate predictive relationships above and beyond the effects of some confounding variables (e.g., gender) used as control. Not surprisingly given our population model, the overall pattern of results obtained from this example is not changed by the addition of this additional control ($\hat{\beta}_{F1_t,F1_{t-1}} = .494$; $p < .001$; $\hat{\beta}_{F2_t,F2_{t-1}} = .400$, $p < .001$; $\hat{\beta}_{F1_t,F2_{t-1}} = -.090$, $p \leq .001$; $\hat{\beta}_{F2_t,F1_{t-1}} = -.008$, $p > .05$) and show significant effects of the grouping variables on all latent factors ($\hat{\beta}_{F1_{t1},G} = .287$, $p < .001$; $\hat{\beta}_{F1_{t2},G} = .150$, $p < .001$; $\hat{\beta}_{F2_{t2},G} = .240$, $p < .001$; $\hat{\beta}_{F2_{t2},G} = .166$, $p < .001$).

EXTENDING ESEM: THE ESEM-WITHIN-CFA (EWC) APPROACH AND ILLUSTRATIONS

The ESEM approach is very flexible, but its current operationalization still presents some limitations when compared to CFA and SEM models. For instance, with standard ESEM models it is impossible, or at least very difficult, to: (a) fit a higher order factor on a set of first order ESEM factors, which also means that fully latent curve models cannot be estimated from longitudinal sets of ESEM factors, (b) evaluate mixture or factor mixture models, and (c) constrain ESEM latent means in multiple-group models, for example, to test linear and non-linear effects based on a single grouping variable such as age or interaction effects between two grouping variables such as age and gender interactions (see Marsh, Nagengast, & Morin, 2012). What is perhaps the most worrisome current limitation of ESEM, however, is that all of the factors forming a set of ESEM factors need to be simultaneously related or unrelated to other variables in the model. For instance, if we take the MIMIC model used previously, one cannot use the grouping variable to predict a single ESEM factor but not the other, or use a single ESEM factor to predict an outcome. Similarly, the invariance of the factor variance-covariance matrix can only be tested in an all-or-nothing manner. Applied researchers are not currently able to test the invariance of selected factor variances and covariances separately, or the relations between selected ESEM latent factors and other variables. Marsh, Nagengast, and Morin (2012) proposed a generalization of ESEM they called ESEM-Within-CFA (EWC) as a way to circumvent some of these problems.

EWC is based on an extension of an initial proposal by Jöreskog (1969; also see Muthén & Muthén, 2009, slides 133–146) that was designed to provide standard errors for EFA parameter estimates and greater flexibility in specifying factor structures—important limitations of EFA at that time. However, our extension of this earlier approach provides a possible solution to some of the aforementioned limitations of ESEM. The EWC model must contain the same number of restrictions as the ESEM model, that is, m^2 restrictions where m = number of factors (see earlier discussion). To achieve these restrictions, Jöreskog (1969) and Muthén and Muthén (2009) initially proposed that researchers select, based on preliminary EFAs, indicators with near-zero loadings on all but the latent constructs that they are designed to measure. The cross-loadings for these indicators were then constrained to be zero in a corresponding CFA model where each of the remaining loadings on all factors are freely estimated and the factors variances are constrained to be 1. Under appropriate conditions, the resulting CFA parameter estimates will approximate the corresponding ESEM model. In our EWC approach (Marsh et al., 2011) we offered an amendment to this approach for greater precision. The EWC model is estimated according to the following steps:

1. In preliminary analyses, compare ESEM and CFA models with regard to goodness of fit and parameter estimates. If the ESEM solution is not clearly superior, ESEM or EWC models should not be pursued and more parsimonious CFA models should be retained.
2. If preliminary ESEM models are better than CFA models, a complete ESEM analysis should be done to identify the best model with regard to goodness of fit and substantive interpretations.
3. If there is a need to conduct an additional analysis that cannot be easily implemented within the ESEM framework but can be estimated with CFA models, then all parameter estimates from the final ESEM solution should be used as starting values to estimate the EWC model.
4. Since a total of m^2 constraints need to be added for the EWC model to be identified, selected parameter estimates are fixed to the values obtained from the ESEM solution:
    (i) The m factor variances: by default these parameters are fixed to be 1 in a single-group ESEM solution or for the first group of a multiple group solution.
    (ii) A referent indicator is selected for each factor that has a large (target) loading for the factor it is designed to measure and small (non-target) cross-loadings. Then, for purposes of identification, these small cross loadings are fixed to their estimated values from the ESEM solution (i.e., do not allow these values to be freely estimated). Unlike the classical method of fixing these cross-loadings to zero, this approach routinely provides

an exact match to the ESEM solutions in terms of parameter estimates, and highly similar standard errors estimates.

(iii) For all other parameter estimates, the pattern of fixed and free estimates should be the same as in the selected ESEM solution. Thus, if the parameter is free in the ESEM solution it should be free in EWC and if the parameter is fixed or constrained in ESEM it should also be fixed or constrained in the same way in the EWC solution.

(iv) It should be noted that the mean structure from the EWC solution can be identified as in a standard CFA model (while using the ESEM start values when possible). This is usually done by freely estimating all items intercepts and constraining all factor means to zero (or the factor means from the first group of a multiple group solution).

The EWC solution will have the same degrees of freedom and, within rounding error, the same chi-square, goodness of fit statistics, and most importantly parameter estimates as the ESEM solution. Standard errors will also be highly similar, but might be slightly inflated, suggesting that caution still needs to be exerted in the interpretation of marginally non-significant results. In this sense, it is equivalent to the ESEM solution. Importantly, the researcher has more flexibility in terms of how to constrain or further modify the EWC model (as it is a true CFA model) than with the ESEM model upon which it is based. Interestingly, the EWC approach could be pursued in the present investigation. The fit statistics from the EWC model as estimated on Time 1 data are identical to the results from the ESEM model [χ^2 (df) = 3.085 (4), $p > .05$; CFI = 1.000; TLI = 1.000; RMSEA = .000]. The input (see EWCTime1.inp) is available in the on-line materials.

The EWC model is a convenient way of implementing a specific rotated ESEM solution within a conventional CFA model that allows more flexibility for further analysis than the original ESEM model. However, the initial ESEM model is needed to specify the EWC model. By allowing the researcher to import a well-tested and rotated ESEM measurement model to an even more flexible CFA framework for further analysis, EWC represents a useful complement to ESEM. For example, the EWC approach can be used to extend ESEM to applications that could not easily be tested with the traditional ESEM approach, such as conducting:

- Tests of the partial invariance of factor loadings (see EWC_partload.inp in the on-line materials for an illustration of how to specify this model);
- Tests of invariance separately for the factor variances and covariance (see EWC_var inv.inp in the on-line materials for an illustration);

- Tests of time × grouping variable latent mean differences, implementing contrasts to define main and interaction effects (see EWC_MG_L_7p.inp in the on-line materials). For an illustration of this method applied to more than two measurement points so as to represent nonlinear trajectories, we refer the interested reader to illustrations and syntax provided by Marsh, Nagengast, et al. (2011).

The above are relatively straightforward applications of existing SEM approaches that are easily implemented into the EWC framework so we decided not to further elaborate on them to conserve space. The interested reader can consult the online supplements for the annotated inputs files that may be used to estimate these models. Rather, we focus on the perhaps more common case of longitudinal mediation models, a final predictive example to illustrate how the EWC approach can help to circumvent some limitations of ESEM.

Mediation occurs when some of the effects of an independent variable (IV) on the dependent variable (DV) can be explained in terms of another mediating variable (MV) (Marsh, Hau, Wen, Nagengast, & Morin, in press). A mediator is thus an intervening variable accounting for at least part of the relation between a predictor and an outcome such that the predictor influences an outcome indirectly through the mediator. As such, the temporal ordering of the sequence IV → MV → DV is particularly important in mediation testing. Tests of longitudinal mediation based on latent variables have recently received increased scientific attention (Cole & Maxwell, 2003; MacKinnon, 2008; Selig & Preacher, 2009) as a method of choice to a fuller understanding of developmental processes. These tests are complex and are not all easily implemented in the regular ESEM framework where all factors from a set of ESEM factors need to be simultaneously related to other variables. For instance, consider the case where four predictors (defined as a set of EFA factors) are used to predict an outcome (defined as a single factor) indirectly via the action of two mediators (defined as a set of EFA factors) so that the first two predictors exert their effect via the first mediator and the last two predictors exert their effects via the second mediator. Mediation models taking this form are common in theoretically-grounded applied research (e.g., Loose, Régner, Morin, & Dumas, in press). However, these models cannot be estimated within the ESEM framework where all predictors need to be related to all mediators, which in turn need to be all related to the outcomes. EWC provides an interesting alternative to ESEM in such cases.

However, this is not the main limitation of ESEM for tests of mediation. Indeed, it is now well documented that bootstrapped confidence intervals represent the most efficient manner of testing the significance of indirect effects (represented as the product of the IV → MV and the MV → DV path coefficients) (e.g., Cheung & Lau, 2008; MacKinnon, Lockwood, & Williams, 2004). Unfortunately, bootstrapping still cannot be implemented in ESEM, but can easily be in EWC.

Moreover, the present data set was simulated with only two measurement points, creating another challenge when one wants to estimate longitudinal mediation models. In fact, with only two measurement points, mediation models fully ordered in time are not possible (unless, naturally, there is an inherent ordering of the variables—such as stable background variables like gender and ethnicity). The closest approximation would be to estimate difference scores between the two measurement points of one or both variables (MacKinnon, 2008; Selig & Preacher, 2009). For illustration purposes, suppose that we hypothesized that the relations between Time 1–Factor 2 (T1F2) and Time 2–Factor 2 (T2F2) would be mediated by changes in Factor 1 (CF1) occurring between Time 1 and Time 2. This model is illustrated in Figure 10.4. Note that this specific ordering of the effects is consistent with the results from the autoregressive cross-lagged models estimated in the

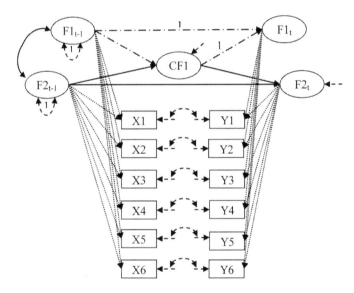

Figure 10.4 Longitudinal mediation EWC model with latent difference score for factor 1. *Note:* Ovals represent latent factors and squares represent observed variables; dotted unidirectional arrows represent factor loadings; dashed unidirectional arrows represent items uniquenesses and factor disturbances; full unidirectional arrows represent the longitudinal mediation paths; dashed-and-dotted unidirectional arrows are needed to estimate the CF1 latent difference variable; bidirectional full arrows linking the ovals represent time-specific factor covariances/correlations; bidirectional dashed arrows connecting a single oval represent factor variances; bidirectional dashed arrows linking the squares represent the longitudinal correlated uniquenesses.

previous section. Difference scores reflecting change occurring over time in specific latent variables can easily be implemented in a latent variable framework as latent difference scores, but represent a higher order factor that cannot be implemented with ESEM. So we estimated this model using EWC (see long-med-change-from7p.inp and long-med-change-from7p-boot.inp in the on-line supplemental materials) starting from longitudinal Model 7p. The obtained results show significant partial mediation in which T1F2 significantly predicts T2F2 ($\hat{\beta}_{T2F2,T1F2} = .540$; $p < .001$) and CF1 ($\hat{\beta}_{CF1,T1F2} = -.084$; $p < .001$), which in turns also predicts T2F2 ($\hat{\beta}_{T2F2,CF1} = .326$; $p < .001$). Not surprisingly, the indirect effect of T1F2 on T2F2 as mediated by CF1 is negative and significant as indicated from its bootstrapped 95% confidence interval that excludes 0 (indirect effect = $-.028$, 95%CI = $-.048/-.013$), showing that higher levels of T1F2 predict lower changes in F1 between time 1 and time 2 (CF1), while these changes predict higher levels in T1F2.

CONCLUSION

For pedagogical purposes, this chapter relied on a relatively simple simulated data set. This data set can be freely used for the initial stages of ESEM teaching and training, as the decisions to be made remain relatively simple and facilitated by knowing the real population parameter values. However, real data sets are often messier and involve more complicated decisions. As a next step, we recommend that interested readers consult previous applications of ESEM to real data sets that have been described at the start of this chapter to see how the range of possibility illustrated in this chapter can be implemented with real data sets in order to answer substantively important research questions.

Clearly, an important advantage of ESEM is to allow for a more appropriate representation of factor correlations due to the non-imposition of arbitrary zero cross-loadings—or at least a more systematic approach to the inclusion of cross-loadings into a measurement model than traditional CFA. ICM-CFA apparently systematically inflates the size of correlations among the latent factors when cross-loadings are present (Marsh, Lüdtke, et al., 2011). Indeed, typically the only way that these cross-loadings can be represented is by inflating the size of correlations. In relation to psychological and social science research more generally this can represent a particularly serious problem because it undermines support for: (a) the multidimensional perspective that is the overarching rationale for many psychometric instruments, (b) the discriminant validity of the factors that form these instruments, (c) the predictive validity of the factors due to multicollinearity, and (d) the usefulness of the ratings in providing diagnostic feedback to the persons being evaluated. We suggest that similar phenomena are likely to occur in most applications where ICM-CFA models are inappropriate. Conversely, allowing for cross-loadings when none are required, although it may result in the over parameterization of the model, is unlikely to result in a negative bias in factor correlations.

In this chapter, we compared factor solutions generated from ESEM relative to ICM-CFA. Guided by the taxonomy of invariance models, we also demonstrated the utility of ESEM for tests of multiple-group and longitudinal tests of measurement invariance based on the Marsh et al. (2009, 2010) 13-model taxonomy, which can also be applied in traditional ICM-CFA studies. However, to the extent that the ESEM solution provides an acceptable fit to the data and the CFA solution does not, then the appropriateness of the taxonomy for CFA models is dubious. In this respect we present the ESEM as a viable alternative to CFA, but do not argue that it should replace CFA. Still, ESEM should generally be preferred to ICM-CFA when the factors are appropriately identified by ESEM, the goodness of fit is meaningfully better than for ICM-CFA, and factor correlations are meaningfully smaller than for ICM-CFA. Furthermore, based on Marsh's (2007; Marsh et al., 2005) suggestion that almost no multidimensional psychological instruments widely used in practice provides an acceptable fit in relation to an a priori ICM-CFA structure, we suspect that ESEM is likely to generate better factorial solutions. In this situation, we suggest that advanced statistical strategies such as multi-group tests of measurement invariance, MIMIC models, and even latent growth models in many applications are more appropriately conducted with an ESEM approach than with a traditional ICM-CFA approach. To illustrate this point, we also showed that autoregressive cross lagged models and change-score based longitudinal mediation models could be estimated in ESEM, or within the complementary EWC method. While we believe that the results of existing research provide considerable support to ESEM, there have been only few large-scale applications of ESEM in applied research settings. Clearly there is need for further research based on the application of ESEM to other areas in psychology, education, and the social sciences more generally. Based on results from existing ESEM research we recommend that the psychometric evaluation of psychological assessment instruments should routinely apply ESEM and juxtapose the results with corresponding CFA models that are traditionally used.

NOTE

1. Although it is generally better to guide empirical research from well-supported, a priori hypothesis, in some cases exploratory applications cannot be avoided. An important issue when an ESEM model (then closer to classical EFA applications) is used for purely exploratory purposes is to determine the optimal number of factors required to best represent the data. A brief

presentation of criteria that could be used for this purpose is available in the supplemental materials at http://www.statmodel.com/esem.shtml. However, in this chapter we focus on the perhaps more common case where ESEM is used for confirmatory purposes based on a well-defined, a priori theoretical factor structure supported by theory, design, and empirical research.

ACKNOWLEDGEMENTS

The authors would like to thank Tihomir Asparouhov and Bengt Muthén for helpful comments at earlier stages of this research, and each of the co-authors of ESEM studies cited in this chapter (Tihomir Asparouhov, Rhonda Craven, Linda Hamilton, Gregory Arief Liem, Oliver Lüdtke, Christophe Maïano, Andrew Martin, Bengt Muthén, Roberto Parada, Alexander Robitzsch, and Ulrich Trautwein). This research was supported in part by a grant to the second author from the UK Economic and Social Research Council.

REFERENCES

Asparouhov, T., & Muthén, B. (2009). Exploratory structural equation modeling. *Structural Equation Modeling: A Multidisciplinary Journal, 16*, 397–438.

Bagozzi, R. P., Fornell, C., & Larcker, D. (1981). Canonical correlation analysis as a special case of a structural relations model. *Multivariate Behavioral Research, 16*, 437–454.

Bernaards, C. A., & Jennrich, R. I. (2005). Gradient projection algorithms and software for arbitrary rotation criteria in factor analysis. *Educational and Psychological Measurement, 65*, 676–696.

Bollen, K. A. (1989). *Structural equations with latent variables.* New York, NY: Wiley.

Borkenau, P., & Ostendorf, F. (1993). *NEO-Fünf-Faktoren Inventar nach Costa und McCrae.* [NEO Five Factor Inventory after Costa and McCrae]. Göttingen: Hogrefe.

Brown, T. A. (2006). Confirmatory factor analysis for applied research. New York, NY: Guilford.

Browne, M. W. (2001). An overview of analytic rotation in exploratory factor analysis. *Multivariate Behavioral Research, 36*, 111–150.

Browne, M.W ., Cudeck, R., Tateneni, K., & Mels, G. (2010*). CEFA: Comprehensive Exploratory Factor Analysis, version 3.04.* Available at: http://faculty.psy.ohio-state.edu/browne/software.php

Byrne, B. M. (2011). *Structural equation modeling with Mplus: Basic concepts, applications, and programming.* Mahwah, NJ: Routledge.

Byrne, B. M., Shavelson, R. J., & Muthén, B. (1989). Testing for the equivalence of factor covariance and mean structures: The issue of partial measurement invariance. *Psychological Bulletin, 105*, 456–466.

Carroll, J. B. (1953). An analytical solution for approximating simple structure in factor analysis. *Psychometrika, 18*, 23–38.

Chen, F. F. (2007). Sensitivity of goodness of fit indexes to lack of measurement invariance. *Structural Equation Modeling: A Multidisciplinary Journal, 14*, 464–504.

Cheung, G. W., & Lau, R. S. (2008). Testing mediation and suppression effects of latent variables: Bootstrapping with structural equation models. *Organizational Research Methods, 11*, 296–325.

Cheung, G. W., & Rensvold, R. B. (2002). Evaluating goodness-of-fit indexes for testing measurement invariance. *Structural Equation Modeling: A Multidisciplinary Journal, 9*, 233–255.

Church, A. T., & Burke, P. J. (1994). Exploratory and Confirmatory Tests of the Big 5 and Tellegens 3-Dimensional and 4-Dimensional Models. *Journal of Personality and Social Psychology, 66*, 93–114.

Cohen, J. (1968). Multiple regression as a general data-analytic system. *Psychological Bulletin, 70*, 426–443.

Cole, D. A., & Maxwell, S. E. (2003). Testing meditational models with longitudinal data: Questions and tips in the use of structural equation modeling. *Journal of Abnormal Psychology, 112*, 558–577.

Cudeck, R., & MacCallum, R. C. (Eds.). (2007). *Factor analysis at 100: Historical developments and future directions*. Mahwah, NJ: Erlbaum.

Dolan, C. V., Oort, F. J., Stoel, R. D., & Wicherts, J. M. (2009). Testing measurement invariance in the target rotated multigroup exploratory factor model. *Structural Equation Modeling: A Multidisciplinary Journal, 16*, 295–314.

Fabrigar, L. R., Wegener, D. T., MacCallum, R. C., & Strahan, E. J. (1999). Evaluating the use of exploratory factor analysis in psychological research. *Psychological Methods, 4*, 272–299.

Fan, X. (1997). Canonical correlation analysis and structural equation modeling: What do they have in common? *Structural Equation Modeling: A Multidisciplinary Model, 4*, 65–79.

Gorsuch, R. L. (1983). *Factor analysis* (2nd ed.). Hillsdale, NJ: Erlbaum.

Graham, J. M. (2008). The general linear model as structural equation modeling. *Journal of Educational and Behavioral Statistics, 33*, 485–506.

Henson, R. K., & Roberts, J. K. (2006). Use of exploratory factor analysis in published research: Common errors and some comment on improved practice. *Educational and Psychological Measurement, 66*, 393–416.

Hessen, D. J., Dolan, C. V, & Wicherts, J. M. (2006). Multi-group exploratory factor analysis and the power to detect uniform bias. *Applied Psychological Research, 30*, 233–246.

Hu, L., & Bentler, P. M. (1999). Cutoff criteria for fit indices in covariance structure analysis: Conventional criteria versus new alternatives. *Structural Equation Modeling: A Multidisciplinary Journal, 6*, 1–55 .

Jennrich, R. I. (2007). Rotation methods, algorithms, and standard errors. In R. Cudeck & R. C. MacCallum (Eds.), *Factor analysis at 100: Historical developments and future directions* (pp. 315–335). Mahwah, NJ: Erlbaum.

Jöreskog, K. G. (1969). A general approach to confirmatory maximum likelihood factor analysis. *Psychometrika, 34*, 183–202.

Jöreskog, K. G., & Goldberger, A. S. (1975). Estimation of a model with multiple indicators and multiple causes of a single latent variable. *Journal of the American Statistical Association, 10*, 631–639.

Jöreskog, K. G., & Sörbom, D. (1977). Statistical models and methods for the analysis of longitudinal data. In D. J. Aigner, & A. S. Goldberger (Eds.), *Latent variables in socio-economic models* (pp. 285–325). Amsterdam, NL: North-Holland Publishing.

Kahn, J. H. (2006). Factor analysis in counseling psychology research, training, and practice: Principles, advances, and applications. *The Counseling Psychologist, 34,* 684–718.

Knapp, T. R. (1978). Canonical correlation analysis: A general parametric significance testing system. *Psychological Bulletin, 85,* 410–416.

Loose, F., Régner, I., Morin, A. J. S., & Dumas, F. (in press). Are academic discounting and devaluing double-edged swords? Their relations to global self-esteem, achievement goals, and performance among stigmatized students. *Journal of Educational Psychology.*

MacKinnon, D. P. (2008). *Introduction to statistical mediation analysis.* Mahwah, NJ: Erlbaum.

MacKinnon, D. P., Lockwood, C. M., & Williams, J. (2004). Confidence limits for the indirect effect: Distribution of the product and resampling methods. *Multivariate Behavioral Research, 39,* 99–128.

Marsh, H. W. (1983). Multidimensional ratings of teaching effectiveness by students from different academic settings and their relation to student/course/instructor characteristics. *Journal of Educational Psychology, 75,* 150–166.

Marsh, H. W. (1987). Students' evaluations of university teaching: Research findings, methodological issues, and directions for future research. *International Journal of Educational Research, 11,* 253–388. (Whole Issue No. 3)

Marsh, H. W. (1991a). A multidimensional perspective on students' evaluations of teaching effectiveness: A reply to Abrami and d'Apollonia (1991). *Journal of Educational Psychology, 83,* 416–421.

Marsh, H. W. (1991b). Multidimensional students' evaluations of teaching effectiveness: A test of alternative higher-order structures. *Journal of Educational Psychology, 83,* 285–296.

Marsh, H. W. (2007). Application of confirmatory factor analysis and structural equation modeling in sport/exercise psychology. In G. Tenenbaum & R. C. Eklund (Eds.), *Handbook of sport psychology* (3rd ed., pp. 774–798). New York, NY: Wiley.

Marsh, H. W., Balla, J. R., & Hau, K. T. (1996). An evaluation of incremental fit indexes: A clarification of mathematical and empirical processes. In G. A. Marcoulides & R. E. Schumacker (Eds.), *Advanced structural equation modeling techniques* (pp. 315–353). Hillsdale, NJ: Erlbaum.

Marsh, H. W., Balla, J. R., & McDonald, R. P. (1988). Goodness-of-fit indexes in confirmatory factor analysis: The effect of sample size. *Psychological Bulletin, 103,* 391–410.

Marsh, H. W., & Cheng, J. H. S (2012). Physical self-concept. In G. Tenenbaum, R. Eklund, & A. Kamata (Eds.), *Measurement in Sport and Exercise Psychology* (pp. 215–226). Champaign, IL: Human Kinetics.

Marsh, H. W., Ellis, L., Parada, L., Richards, G., & Heubeck, B. G. (2005). A short version of the Self Description Questionnaire II: Operationalizing criteria for short-form evaluation with new applications of confirmatory factor analyses. *Psychological Assessment, 17,* 81–102.

Marsh, H. W., & Grayson, D. (1994). Longitudinal stability of latent means and individual differences: A unified approach. *Structural Equation Modeling: A Multidisciplinary Journal, 1,* 317–359.

Marsh, H. W., & Hau, K-T. (1996). Assessing goodness of fit: Is parsimony always desirable? *Journal of Experimental Education, 64,* 364–390.

Marsh, H. W., Hau, K-T., Balla, J. R., & Grayson, D. (1998). Is more ever too much? The number of indicators per factor in confirmatory factor analysis. *Multivariate Behavioral Research, 33,* 181–220.

Marsh, H. W., Hau, K-T., & Grayson, D. (2005). Goodness of fit evaluation in structural equation modeling. In A. Maydeu-Olivares & J. McArdle (Eds.), *Psychometrics. A festschrift to Roderick P. McDonald* (pp. 275–340). Hillsdale, NJ: Erlbaum.

Marsh, H. W., Hau, K. T., & Wen, Z. (2004). In search of golden rules: Comment on hypothesis testing approaches to setting cutoff values for fit indexes and dangers in over-generalizing Hu & Bentler's (1999) findings. *Structural Equation Modeling: A Multidisciplinary Journal, 11,* 320–341.

Marsh, H. W., Hau, K.-T., Wen, Z., Nagengast, B., & Morin, A. J. S. (in press). Moderation. In T. D. Little (Ed.), *Oxford handbook of quantitative methods.* New York, NY: Oxford University Press.

Marsh, H. W., & Hocevar, D. (1991). The multidimensionality of students' evaluations of teaching effectiveness: The generality of factor structures across academic discipline, instructor level, and course level. *Teaching and Teacher Education, 7,* 9–18.

Marsh, H. W., Liem, G. A. D., Martin, A. J., Morin, A. J. S., & Nagengast, B. (2011). Methodological-measurement fruitfulness of exploratory structural equation modeling (ESEM): New approaches to key substantive issues in motivation and engagement. *Journal of Psychoeducational Assessment, 29,* 322–346

Marsh, H. W., Lüdtke, O., Muthén, B. O., Asparouhov, T., Morin, A. J .S., & Trautwein, U. (2010). A new look at the big-five factor structure through Exploratory Structural Equation Modeling. *Psychological Assessment, 22,* 471–491.

Marsh, H. W., Lüdtke, O., Robitzsch, A., Nagengast, B., Morin, A. J. S., & Trautwein, U. (2011). *Two wrongs do not make a right: Camouflaging misfit with item-parcels in CFA.* Centre for Positive Psychology and Education, University of Western Sydney, Australia.

Marsh, H. W., Martin, A., & Debus, R. (2001). Individual differences in verbal and math self-perceptions: One factor, two factors, or does it depend on the construct? In R. Riding & S. Rayner (Eds.). *Self perception: International perspectives on individual differences* (pp. 149–170). Westport, CT: Ablex.

Marsh, H. W., Martin, A. J., & Hau, K-T. (2006). A multiple method perspective on self-concept research in educational psychology: A construct validity approach. In M. Eid & E. Diener (Eds.), *Handbook of multimethod measurement in psychology* (pp. 441–456). Washington, DC: American Psychological Association.

Marsh, H. W., Muthén, B., Asparouhov. T., Lüdtke, O., Robitzsch, A., Morin, A. J. S., & Trautwein, U. (2009). Exploratory structural equation modeling, integrating CFA and EFA: Application to students' evaluations of university teaching. *Structural Equation Modeling: A Multidisciplinary Journal, 16,* 439–476.

Marsh, H. W., Nagengast, B., & Morin, A. J. S. (2012). Measurement invariance of Big-Five factors over the life span: ESEM tests of gender, age, plasticity, maturity, and La Dolce Vita effects. *Developmental Psychology*. Advance online publication. doi:10.1037/a0026913.

Marsh, H. W., Nagengast, B., Morin, A. J. S., Parada, R. H., Craven, R. G., & Hamilton, L. R. (2011). Construct validity of the multidimensional structure of bullying and victimization: An application of exploratory structural equation modeling. *Journal of Educational Psychology, 103*, 701–732.

Marsh, H. W., Tracey, D. K., & Craven, R. G. (2006). Multidimensional self-concept structure for preadolescents with mild intellectual disabilities: A hybrid multigroup-mimic approach to factorial invariance and latent mean differences. *Educational and Psychological Measurement, 66*, 795–818.

Marsh, H. W., & Yeung, A. S. (1998). Top-down, bottom-up, and horizontal models: The direction of causality in multidimensional, hierarchical self-concept models. *Journal of Personality and Social Psychology, 75*, 509–527.

McCrae, R. R., & Costa, P. T. Jr. (1997). Personality trait structure as a human universal. *American Psychologist, 52*, 509–516.

McCrae, R. R., Zonderman, A. B., Costa, P. T., Jr., Bond, M. H., & Paunonen, S. (1996). Evaluating the replicability of factors in the revised NEO Personality Inventory: Confirmatory factor analysis versus Procrustes rotation. *Journal of Personality and Social Psychology, 70*, 552–566.

McDonald, R. P. (1985). *Factor analysis and related methods*. Hillsdale, NJ: Erlbaum.

Meleddu, M., Guicciardi, M., Scalas, L. F., & Fadda, D. (2012). Validation of an Italian version of the Oxford Happiness Inventory in adolescence. *Journal of Personality Assessment, 94*, 175–185.

Meredith, W. (1993). Measurement invariance, factor analysis and factorial invariance. *Psychometrika, 58*, 525–543.

Millsap, R. E. (2011). *Statistical approaches to measurement invariance*. New York, NY: Routledge.

Morin, A. J. S., & Maïano, C. (2011). Cross-validation of the Short Form of the Physical Self-Inventory (PSI-S) using Exploratory Structural Equation Modeling (ESEM). *Psychology of Sport and Exercise, 12*, 540–554.

Morin, A. J. S., Maïano, C., Nagengast, B., Marsh, H. W., Morizot, J., & Janosz, M. (2011). General growth mixture analysis of adolescents' developmental trajectories of anxiety: The impact of untested invariance assumptions on substantive interpretations. *Structural Equation Modeling: A Multidisciplinary Journal, 18*, 613–648.

Muthén, B. O. (1989). Latent variable modeling in heterogenous populations. *Psychometrika, 54*, 557–585.

Muthén, B. O. (2002). Beyond SEM: General latent variable modeling. *Behaviormetrika, 29*, 81–117.

Muthén, L. K., & Muthén, B. O. (2009). *Mplus short courses: Topic 1: Exploratory factor analysis, confirmatory factor analysis, and structural equation modeling for continuous outcomes*. Los Angeles CA: Muthén & Muthén. Retrieved from http://www.statmodel.com/course_materials.shtml

Muthén, L. K., & Muthén, B. O. (2010). *Mplus user's guide*. Los Angeles: Muthén & Muthén.

Myers, N. D., Chase, M. A., Pierce, S. W., & Martin, E. (2011). Coaching efficacy and exploratory structural equation modeling: A substantive-methodological synergy. *Journal of Sport and Exercise Psychology, 33,* 779–806.

Neale, M. C., Boker, S. M., Xie, G., & Maes, H. H. (2003). *Mx: Statistical modeling.* Richmond: Virginia Commonwealth University, Department of Psychiatry.

Parker, J. D. A., Bagby, R. M., & Summerfeldt, L. J. (1993). Confirmatory factor-analysis of the Revised Neo-Personality Inventory. *Personality and Individual Differences, 15,* 463–466.

Sass, D. A., & Schmitt, T. A. (2010). A comparative investigation of rotation criteria within exploratory factor analysis. *Multivariate Behavioral Research, 45,* 1–33.

Selig, J. P., & Preacher, K. J. (2009). Mediation models for longitudinal data in developmental research. *Research in Human Development, 6,* 144–164.

Skrondal, A., & Rabe-Hesketh, S. (2004). *Generalized latent variable modeling: Multilevel, longitudinal, and structural equation models.* New York, NY: Chapman & Hall/CRC.

Spearman, C. (1904). "General intelligence," objectively determined and measured. *American Journal of Psychology, 15,* 201–293.

Thompson, B. (2004). *Exploratory and confirmatory factor analysis: Understanding concepts and applications.* Washington, DC: American Psychological Association.

Thompson, B., & Daniel, L. G. (1996). Factor analytic evidence for the construct validity of scores: A historical overview and some guidelines. *Educational and Psychological Measurement, 56,* 197–208.

Thurstone, L. L. (1947). *Multiple factor analysis.* Chicago: University of Chicago.

Toland, M. D., & De Ayala, R. J. (2005). A multilevel factor analysis of students' evaluations of teaching. *Educational and Psychological Measurement, 65,* 272–296.

Vandenberg, R. J., & Lance, C. E. (2000). A review and synthesis of the measurement invariance literature: Suggestions, practices, and recommendations for organizational research. *Organizational Research Methods, 3,* 4–70.

Vassend, O., & Skrondal, A. (1997). Validation of the NEO Personality Inventory and the five-factor model. Can findings from exploratory and confirmatory factor analysis be reconciled? *European Journal of Personality, 11,* 147–166.

Yates, A. (1987). *Multivariate exploratory data analysis: A perspective on exploratory factor analysis.* Albany, NY: State University of New York Press.

PART III

ASSUMPTIONS

CHAPTER 11

NONNORMAL AND CATEGORICAL DATA IN STRUCTURAL EQUATION MODELING

Sara J. Finney and Christine DiStefano

Structural equation modeling (SEM) has remained a popular data analytic technique in education, psychology, business, and other disciplines (Austin & Calderón, 1996; MacCallum & Austin, 2000; Tremblay & Gardner, 1996). Given the frequency of its use, it is important to recognize the assumptions associated with different estimation methods, demonstrate the conditions under which results are robust to violations of these assumptions, and specify the procedures that should be employed when assumptions are not met. The importance of attending to assumptions and, consequently, selecting appropriate analysis strategies based on the characteristics of the data and the study's design, cannot be overstated. Put simply, violating assumptions can produce biased results in terms of data-model fit as well as parameter estimates and their associated significance tests. Biased results, in turn, might result in incorrect decisions about the theory being tested.

Structural Equation Modeling: A Second Course (2nd ed.), pages 439–492
Copyright © 2013 by Information Age Publishing
439

Although there are several assumptions underlying the popular normal theory (NT) estimators used in SEM, the two assumptions we focus on in this chapter concern the metric and distribution of the data. Specifically, the data required by NT estimators are assumed to be continuous and multivariate normally distributed in the population. We focus on these two assumptions because often the data modeled in the social sciences do not follow a multivariate normal distribution. For example, Micceri (1989) noted much of the data gathered from achievement and other measures are not normally distributed. This is disconcerting given that Gierl and Mulvenon (1995) found that most researchers do not examine the distribution of their data, but instead simply assume normality. In addition to the pervasiveness of nonnormal data, the applied literature is thick with examples of categorical data collected using ordinal scales (e.g., Likert-type scales).

Because of the prevalence of both nonnormal and categorical data in empirical research, this chapter focuses on issues surrounding the use of data with these characteristics. Specifically, we review the assumptions underlying NT estimators. We describe nonnormal and categorical data and review robustness studies of the most popular NT estimator, maximum likelihood (ML), in order to understand the consequences of violating these assumptions. Most importantly, we discuss three popular strategies often used to accommodate nonnormal and/or categorical data in SEM[1]:

1. Weighted least squares (WLS) estimation,
2. Satorra–Bentler (S–B) scaled χ^2 and robust standard errors, and
3. Robust diagonally weighted least squares (DWLS) estimation.

For each strategy, we present the following: (a) a description of the strategy, (b) a summary of research concerning the robustness of the χ^2 statistic, other fit indices, parameter estimates, and standard errors, and (c) a description of implementation across three software programs.

NORMAL THEORY ESTIMATORS

Assumptions of Normal Theory Estimators

As with most statistical techniques, SEM is based on assumptions that should be met in order for researchers to trust the obtained results. Central to SEM is the choice of an estimation method that is used to obtain parameter estimates, standard errors, and fit indices. The two common NT estimators are ML and generalized least squares (GLS), and both require the following set of assumptions (e.g., Bentler & Dudgeon, 1996; Bollen, 1989):

- *Independent observations*: Observations for different subjects are independent. This can be achieved through random sampling.
- *Large sample size*: All statistics estimated are based on an assumption that the sample is sufficiently large.
- *Correctly specified model*: The model being tested reflects the true structure in the population.
- *Multivariate normal data*: The observed scores have a (conditionally) multivariate normal distribution.
- *Continuous data*: The assumption of multivariate normality implies the data are continuous in nature. Categorical data, such as dichotomies or even Likert-type data, cannot, by definition, be normally distributed because they are discrete in nature (e.g., Kaplan, 2009). Therefore, it is often noted that NT estimators require *continuous, normally distributed* data from endogenous (i.e., dependent) variables.

If NT estimators are applied when the above conditions are satisfied, the parameter estimates have three desirable properties: asymptotic unbiasedness (neither over- nor under-estimate the true population parameter in large samples), consistency (parameter estimate converges to population parameter as sample size increases), and asymptotic efficiency (smallest asymptotic variance of all consistent estimators).

Defining Normal Theory Estimators

For both ML and GLS estimation methods, parameters are estimated using an iterative process. The final set of parameters minimizes the discrepancy between the observed sample covariance matrix, $\mathbf{S}$, and the model-implied covariance matrix calculated from the estimated model parameters, $\Sigma(\hat{\theta})$.[2] The fit function that is minimized, $\hat{F} = F[\mathbf{S}, \Sigma(\hat{\theta})]$, will equal zero if the model perfectly reproduces the elements in the sample covariance matrix. If the assumptions noted above are met, overall fit between the data and the model can be expressed as $T = \hat{F}(N-1)$, which follows a central χ^2 distribution with $\frac{1}{2}(p + q)(p + q + 1) - t$ degrees of freedom, where p is the number of observed exogenous variables, q is the number of observed endogenous variables, and t is the number of estimated parameters (Bollen, 1989). If the value of T is statistically significant, T follows a noncentral χ^2 distribution, and the hypothesis stating the population covariance matrix (Σ) equals the reproduced covariance matrix calculated from the population model parameters $[\Sigma(\theta)]$, $\Sigma = \Sigma(\theta)$, can be rejected and it is concluded that the model is misspecified.

The fit function for both ML and GLS estimators can be written in the same general form:

$$\hat{F} = \frac{1}{2} tr \left[\left([\mathbf{S} - \mathbf{\Sigma}(\hat{\mathbf{\theta}})] \mathbf{W}^{-1} \right)^2 \right]$$ (11.1)

where *tr* is the trace of a matrix (i.e., the sum of the diagonal elements), and $\mathbf{S} - \mathbf{\Sigma}(\hat{\mathbf{\theta}})$ represents a residual matrix of discrepancies between the elements in the sample covariance matrix and the elements in the model-implied covariance matrix. These residuals are weighted by the inverse of a weight matrix, $\mathbf{W}$. The weight matrix differs between the two NT estimators; GLS employs the observed sample covariance matrix, $\mathbf{S}$, as the weight matrix, whereas ML employs the model-implied covariance matrix, $\mathbf{\Sigma}(\hat{\mathbf{\theta}})$.[3] If all assumptions are met, the two weight matrices will be equivalent at the last iteration and the estimators will produce convergent results (Olsson, Troye, & Howell, 1999). However, if the model is misspecified, $\mathbf{W}$ at the last iteration will differ between the two techniques, even if all other assumptions are met. This difference in $\mathbf{W}$ results in different parameter estimates and fit indices across the estimators. Specifically, GLS has been found to produce overly optimistic fit indices and more biased parameter estimates than ML if the estimated model is misspecified. Seeing that most applied researchers are interested in the plausibility of a specified model and would, therefore, prefer fit indices sensitive to model misspecification, ML has been recommended over GLS (Olsson et al., 1999; Olsson, Foss, Troye, & Howell, 2000). We therefore limit subsequent discussion of NT estimators to ML.

NONNORMAL CONTINUOUS DATA

Assessing Nonnormality

In general, the effects of nonnormality on ML-based results depend on its extent; the greater the nonnormality, the greater the impact upon results. Therefore, researchers should assess the distribution of data from the observed variables prior to analyses in order to make an informed decision concerning estimation method. Three indices of nonnormality are typically used to evaluate the distribution: univariate skew, univariate kurtosis, and multivariate kurtosis. Unfortunately, there is not a clear consensus regarding an "acceptable" degree of nonnormality. Studies examining the impact of univariate normality on ML-based results suggest that problems might occur when univariate skewness and univariate kurtosis approach absolute values of 2 and 7, respectively (e.g., Chou & Bentler, 1995; Curran, West, & Finch, 1996; Muthén & Kaplan, 1985). In addition, there is no generally accepted cutoff value of multivariate kurtosis that indicates nonnormality. It has been suggested that data associated with a value of Mardia's normalized multivariate kurtosis (see Bollen, 1989, p. 424, equation 4 for formula) greater than 3 could produce inaccurate results when used with ML estimation (Bentler, 2004; Ullman, 2006). Future research should investigate the utility of such a cutoff value and the conditions under which it is relevant (e.g., model size, sample size).

Effects of Analyzing Nonnormal Continuous Data: Empirical Results

Given the abundance of nonnormal data analyzed in the social sciences, a question of significant interest concerns the robustness of ML estimation to nonnormality. Research examining the effects of nonnormality has typically focused on (a) the χ^2 statistic, (b) other model fit indices, (c) parameter estimates, and (d) standard errors. As detailed below, ML has been found to produce relatively accurate parameter estimates under conditions of nonnormality (e.g., Finch, West, & MacKinnon, 1997); however, the χ^2 statistic, approximate fit indices, and standard errors of the parameter estimates tend to exhibit bias as nonnormality increases (e.g., Bollen, 1989; Chou, Bentler, & Satorra, 1991; Finch et al., 1997).

Model chi-square and fit indices. When fitting a correctly specified model, the ML-based χ^2 does not follow the expected central χ^2 distribution if the multivariate normality assumption is violated. More specifically, research has shown that χ^2 tends to be biased under conditions of moderate nonnormality with values becoming more biased with increasing nonnormality (e.g., Chou et al., 1991; Curran et al., 1996; Hu, Bentler, & Kano, 1992; Yu & Muthén, 2002). Kurtotic distributions appear to have the greatest effect on χ^2 (e.g., Browne, 1984; Chou et al., 1991). Leptokurtic distributions (positive kurtosis) inflate the χ^2 statistic (Yuan, Bentler, & Zhang, 2005). The inflation of the χ^2 statistic might lead to an increased Type I error rate, which is a greater rate of rejecting a correctly specified model than expected by chance. Platykurtic distributions (negative kurtosis) attenuate the χ^2 statistic, which might lead to an increased Type II error rate when evaluating fit of the model.

In addition to the χ^2 statistic, the performance of other fit indices is important to understand given that most researchers are interested in the approximate fit of the data to the model instead of an exact fit evaluation determined solely by the χ^2 test (Bentler, 1990). As Hu and Bentler (1998) explained, "A fit index will perform better when its corresponding chi-square test performs well" (p. 427), meaning that because many fit indices (e.g., comparative fit index, CFI) are a function of the obtained χ^2, these too can be affected by the same factors that influence χ^2. Research has shown that if moderately to severely nonnormal data are coupled with a small sample size ($N \leq 250$), the ML-based Tucker-Lewis Index (TLI), CFI, and root mean-square error of approximation (RMSEA) tend to over-reject correctly specified models (Hu & Bentler, 1999; Yu & Muthén, 2002).

Parameter estimates and standard errors. Whereas parameter estimates are generally unaffected by nonnormality, their associated significance tests are incorrect if ML estimation is applied to nonnormal continuous data (e.g., Chou et al., 1991; Finch et al, 1997; Olsson et al., 2000). Specifically, the ML-based standard errors may underestimate or overestimate the true variation of the parameter estimates when data are nonnormal (Yuan et al., 2005). Leptokurtic distributions attenuate standard errors, which results in increased Type I error rates associated with statistical significance tests of the parameter estimates. Tests would imply that estimates of truly zero parameters were significantly different from zero, and thus important to include in the model. Platykurtic distributions have the opposite effect, in that standard errors are inflated and potentially contribute to a greater number of Type II errors.

Techniques to Address Nonnormal Continuous Data

In this section we will describe two methods to address problems encountered when modeling nonnormal data. The first involves an alternative method of estimation that does not require the distributional assumptions of ML estimation (Browne, 1984). The second involves adjusting the ML-based χ^2 and standard errors by a factor based on the level of multivariate kurtosis displayed in the observed data (Satorra–Bentler scaled χ^2 and robust standard errors). This method also involves adjusting approximate fit indices used to assess model fit (e.g., CFI, RMSEA).

Weighted Least Squares (WLS) estimator. Given the unrealistic assumption of multivariate normality and the lack of robustness of ML estimation to nonnormal data, Browne (1984) developed the WLS estimator (which is also called the asymptotic distribution free estimator when used with continuous nonnormal data). Unlike ML, WLS makes no assumption of normality; therefore, variables that are skewed and/or kurtotic should theoretically have no detrimental effect on the WLS χ^2 or standard errors. Thus, it would seem as though nonnormally distributed data could be easily accommodated by the WLS estimation technique, thus avoiding the problems encountered by NT estimators. But, as we will see, the WLS estimator only performs adequately under select conditions rarely encountered.

WLS estimator with nonnormal continuous data: Description. In order to understand the WLS estimator and some of its practical limitations, it is important to understand the form of the fit function, which is typically written as

$$F = (\mathbf{s} - \hat{\boldsymbol{\sigma}})' \mathbf{W}^{-1} (\mathbf{s} - \hat{\boldsymbol{\sigma}}), \tag{11.2}$$

where **s** represents a vector of the nonduplicated elements in the sample covariance matrix (**S**), $\hat{\sigma}$ represents a vector of the nonduplicated elements in the model-implied covariance matrix [$\Sigma(\hat{\theta})$], and $\mathbf{s} - \hat{\sigma}$ represents a residual vector of the discrepancies between the sample values and the model-implied values. These residuals are weighted by a weight matrix, **W**.

The weight matrix utilized with the WLS estimator is the asymptotic covariance matrix. The asymptotic covariance matrix of the observed sample variances and covariances provides us with information about the sampling variability associated with these sample estimates and the departure of the observed data from normality. That is, each element in the observed sample covariance matrix (each sample variance and covariance) has a variance associated with it and the square root of this variance is the sample estimate's standard error. The elements in the sample covariance matrix are not generally independent either; instead, the elements often covary. The asymptotic covariance matrix employed in WLS is a matrix of the variances (diagonal) and covariances (off-diagonal) of the elements in the observed sample covariance matrix. This information can then be used when estimating SEM parameters; those sample variances and covariances with less sampling variability receive more weight when estimating model parameters than those with more sampling variability.

Elements of the asymptotic covariance matrix, $W_{ij,kl}$, are calculated using the second and fourth-order moments (Bentler & Dudgeon, 1996),

$$W_{ij,kl} = s_{ijkl} - s_{ij}s_{kl}, \tag{11.3}$$

where s_{ijkl}, a quantity related to multivariate kurtosis, is defined as

$$s_{ijkl} = \frac{\sum_{a=1}^{N}(x_{ai} - \overline{x}_i)(x_{aj} - \overline{x}_j)(x_{ak} - \overline{x}_k)(x_{al} - \overline{x}_l)}{N}, \tag{11.4}$$

and s_{ij} and s_{kl} are the observed covariances of x_i with x_j and x_k with x_l, respectively. $W_{ij,kl}$ is the estimated asymptotic covariance between s_{ij} and s_{kl}. Thus, the elements in **W** reflect both departure from normality and sampling variability.

There are practical problems in implementing WLS estimation that are related to the weight matrix. Specifically, because the inverse of the weight matrix ($\mathbf{W}^{-1}$) needs to be calculated, a large weight matrix can make WLS estimation computationally intensive. Dimensions of **W** can be calculated as $\frac{1}{2}(p + q)(p + q + 1) \times \frac{1}{2}(p + q)(p + q + 1)$. For example, if a researcher has responses from a ten-item scale and wishes to employ confirmatory factor analysis, the dimensions of **W** would be 55×55, resulting in 3,025 elements in **W**.

As the number of observed variables increases, the number of elements in **W** increases rapidly. For example, if ten more items were added to the original ten-item measure, the dimensions of the weight matrix would be 210×210, resulting in 44,100 elements in **W**. Due to the computational intensity of the WLS technique, it requires very large sample sizes for results to converge to stable estimates. A minimum sample size of $1.5(p + q)(p + q + 1)$ has been suggested (Jöreskog & Sörbom, 1996), but much larger sample sizes may be needed to alleviate estimation and convergence problems.

WLS estimator with continuous nonnormal data: Empirical results. When analyzing nonnormal data with SEM, theoretically, the WLS estimator should produce parameter estimates with desirable properties and fit statistics that perform as expected (Browne, 1984). However, empirical research has shown otherwise. WLS tends to break down under common situations of moderate to large models (more than 2 factors, 8 items) and/or small to moderate sample sizes ($N < 500$). As discussed in detail below, the poor performance of the WLS estimator under many conditions makes it an unattractive option when modeling nonnormal continuous data.

Model chi-square and fit indices. With respect to data-model fit, WLS yields misleading results unless sample size is extremely large (e.g., Olsson et al., 2000). For example, when analyzing nonnormal continuous data, Hu et al. (1992) found that WLS estimation produced acceptable Type I error rates only when sample sizes reached 5,000. Similarly, Curran et al. (1996) found that the WLS-based χ^2 increased as sample size decreased and/or nonnormality increased, resulting in correctly specified models being rejected too frequently. Even more problematic is the WLS estimator's lack of sensitivity to model misspecification. Research has shown that WLS estimation produces overly-optimistic fit values when models are misspecified, which in turn could lead researchers to fail to reject an incorrectly specified model. The lack of sensitivity to specification errors becomes worse with increasing departures from normality (e.g., Curran et al., 1996; Foss, Jöreskog, & Olsson, 2011; Olsson, Foss, & Troye, 2003; Olsson et al., 2000).

Parameter estimates and standard errors. Empirical results concerning WLS-based parameter estimates and standard errors are also discouraging. Parameter estimates tend to be negatively biased unless the sample is large, with bias levels becoming more pronounced as kurtosis increases. In addition, WLS-based standard errors estimated from a correctly specified model under conditions of nonnormality have been found to be superior to ML-based standard errors only when the observed variables have an average univariate kurtosis larger than 3 and the sample size is greater than 400 (Hoogland & Boomsma, 1998).

WLS estimator with continuous data: Software implementation. Researchers who wish to utilize WLS as an estimation technique will find it easy to employ using LISREL, EQS, or Mplus. Unlike NT estimators, estimators that handle nonnormality require raw data. In LISREL (Jöreskog & Sörbom, 2007), WLS estimation requires the use of two programs (PRELIS and LISREL). As noted, WLS employs the asymptotic covariance matrix as the weight matrix. PRELIS (preprocessor for LISREL; Jöreskog & Sörbom, 1996) is used to produce both the asymptotic and observed covariance matrices from the raw data. These matrices are then employed by LISREL to estimate the model. Appendix A provides an example of the SIMPLIS command language (user-friendly language employed in the LISREL program) to specify WLS estimation. Notice that "WLS" needs to be specified on the OPTIONS line or else ML estimation will be employed, which will produce the Satorra–Bentler scaled χ^2 and robust standard errors (discussed below) when an asymptotic covariance matrix is used as input.

Similar to LISREL, the raw data file is necessary in order to construct the asymptotic covariance matrix in EQS (Bentler, 1985–2010) and Mplus (Muthén & Muthén, 1998–2010). Unlike LISREL, a preprocessor is not needed to construct this matrix in either EQS or Mplus; it is constructed and employed by specifying the estimator. Arbitrary GLS (AGLS) estimation is requested as the estimation method for EQS, whereas WLS is requested for Mplus. All three programs tend to produce similar parameter estimates, standard errors, and fit indices.

Satorra–Bentler scaled chi-square and robust standard errors. Another strategy employed to accommodate nonnormal continuous data involves adjusting the ML χ^2, fit indices, and standard errors by a factor based in part on the amount of multivariate kurtosis present in the data. Recall that when nonnormal continuous data are analyzed using ML, the parameter estimates are not affected. Therefore, the rationale behind this method is to adjust the fit indices and standard errors but not the ML-based parameter estimates.

Satorra–Bentler scaling method with nonnormal continuous data: Description. When data are normally distributed and the model is correctly specified, the expected value of the χ^2 is equal to the model degrees of freedom. Therefore, if a correctly specified model with 60 degrees of freedom was tested using multivariate normal data, a χ^2 value of approximately 60 would be expected. However, if data are moderately nonnormal, the ML-based χ^2 will be biased upward or downward, depending upon the distribution of the data (i.e., leptokurtic or platykurtic). This bias will occur even though the model is correctly specified. A scaling to correct the mean of the statistic, typically called the Satorra–Bentler (S–B) scaling method, uses the observed data's distributional characteristics to adjust the ML-based χ^2 in order to better approximate the theoretical χ^2 reference distribution:

$$\text{S–B scaled } \chi^2 = d^{-1}(\text{ML-based } \chi^2), \tag{11.5}$$

where d is a scaling factor that relates in part to the amount of multivariate kurtosis of the variables (Chou & Bentler, 1995; Satorra & Bentler, 1988, 1994). If no multivariate kurtosis exists, then the ML-based χ^2 equals the S–B scaled χ^2. However, as the level of multivariate kurtosis increases, the S–B scaled χ^2 becomes more discrepant from the ML-based χ^2.

The S–B scaling method is typically applied with ML estimation. Because ML is employed, computation problems experienced with the WLS estimator are avoided. Recall that WLS estimation requires the inversion of the full asymptotic covariance matrix of the sample variances and covariances (**W**). The S–B scaling applied to the ML χ^2 does not require inversion of the full **W** matrix. Instead, the matrices inverted in order to compute the scaling factor are of the smaller dimensions of $df \times df$, where df represents the degrees of freedom associated with the model (Satorra & Bentler, 1994).

It must be noted that the typical χ^2 difference test employed for nested model comparisons should not be calculated using the S–B scaled χ^2 values (i.e., simply subtracting the χ^2 value of the less parsimonious model from the χ^2 value of the more parsimonious model) because the difference between two S–B scaled χ^2 values is not distributed as a χ^2. Fortunately, calculations can be employed to correct the difference test in order to make nested model comparisons using the S–B scaled χ^2 values (Bryant & Satorra, 2012; Satorra & Bentler, 2001, 2010).

The standard errors are also corrected when employing the Satorra–Bentler method, alleviating some of the bias present when analyzing nonnormal data using ML estimation. Specifically, the ML standard errors are adjusted to approximate those that would have been obtained if the data were normally distributed. This adjustment to the NT standard errors involves the asymptotic covariance matrix of the sample covariance matrix. More specifically, robust standard errors are often called *sandwich* standard errors because in the formula information from the asymptotic covariance matrix of the sample variances and covariances is pre- and post-multiplied by the unadjusted (naïve) NT standard errors. Using the sandwich analogy, the unadjusted standard errors form the "bread" of the sandwich and the "meat" of the sandwich is the asymptotic covariance matrix of the sample variances and covariances (Bentler, 2004; Muthén, 1998–2004; Satorra & Bentler, 1994). Importantly, the asymptotic covariance matrix is not inverted in these computations.[4]

S–B Scaling Methods with Continuous Nonnormal Data: Empirical Results.

Model chi-square and fit indices. Studies using correctly specified models and continuous nonnormal data have shown that the S–B scaled χ^2 outperforms the ML-based χ^2, particularly as the degree of nonnormality increases (e.g., Chou et al., 1991; Curran et al. 1996; Hu et al., 1992; Yu &

Muthén, 2002). In addition, it performs better than the WLS-based χ^2 when estimating correctly specified models at all but the largest sample sizes (e.g., $N = 1000$, Curran et al., 1996; $N = 5000$, Hu et al., 1992). Recently, Foss, Jöreskog, and Olsson (2011) noted that the S–B scaled χ^2 may function poorly under conditions of model misspecification and nonnormal data. Specifically, they illustrated that the commonly noted concern of the insensitivity of the WLS χ^2 to model misspecification when modeling nonnormal data is also a concern with the S–B scaled χ^2. That is, similar to the WLS χ^2, the S–B χ^2 may suffer from an increasing loss of power as kurtosis increases.

If the S–B scaled χ^2 is employed to handle nonnormal data, it follows that the S–B scaled χ^2 would be incorporated into the calculation of fit indices (e.g., TLI, CFI). Nevitt and Hancock (2000) found that the S–B scaled RMSEA outperformed the unadjusted index for both correctly specified and misspecified models. Yu and Muthén (2002) also examined this index in addition to the S–B scaled TLI and CFI and found that, under conditions of moderate to severe nonnormality coupled with small sample size ($N \leq 250$), the S–B scaled versions of these three indices are preferred over the ML-based estimates for both correctly specified and misspecified models. Yu and Muthén suggested that values at or below .05 for the S–B scaled RMSEA and at or above .95 for the S–B scaled CFI indicate adequate fit, which is quite similar to the cutoff values recommended by Hu and Bentler (1999) for the unadjusted indices.

Parameter estimates and standard errors. Satorra–Bentler robust standard errors have also been found to outperform ML-based and WLS-based standard errors under conditions of nonnormality (Chou & Bentler, 1995; Chou et al., 1991). Similar to the ML-based and WLS-based standard errors, the robust standard errors showed some negative bias. However, they tended to be much closer to the expected values of the standard errors than those obtained from either WLS or ML estimation. Recall that the ML-based parameter estimates are not adjusted as they are not affected by nonnormality of the continuous data.

S–B scaling methods: Software implementation. Appendix B provides the syntax needed to employ the S–B scaling method. Similar to WLS estimation with continuous data, LISREL calculates the S–B scaled (or mean-adjusted) χ^2 and robust standard errors in two steps. The first step involves computing the asymptotic and observed covariance matrices from the raw data file using PRELIS. The second step involves specifying the model and estimation technique (ML) in the SIMPLIS program file.[5]

Similar to employing the WLS estimator, both EQS and Mplus read the raw data file directly into the program in order to construct the necessary matrices used to scale the χ^2 and standard errors. In addition to specifying ML as the estimator in EQS, the word "robust" is included to request the scaling method. In Mplus, one simply requests "MLM" as the estimator, which refers to the ML-mean-adjusted χ^2 and scaled standard errors.

In addition to the S–B scaled χ^2 (ML mean-adjusted χ^2), both Mplus and EQS can provide a mean- and variance-adjusted χ^2. Recall the mean adjustment results in the mean of the distribution of the test statistic better approximating the mean of the χ^2 reference distribution. The additional variance adjustment is designed to result in both the mean and the variance of the test statistic distribution approximating the mean and the variance of the χ^2 reference distribution (Asparouhov & Muthén, 2010; Satorra & Bentler, 1998). Mplus provides this χ^2 by specifying MLMV as the estimation method, whereas EQS provides both the mean- and the mean- and variance-adjusted χ^2 values if "robust, ML" is specified as the estimator.[6] The scaled standard errors are equivalent regardless if one requests the mean- or mean- and variance-adjusted χ^2.

Few studies have compared the performance of the mean-adjusted and mean- and variance-adjusted χ^2. Research suggests that the mean- and variance-adjusted χ^2 performs better (i.e., lower Type I error rate) than using only the mean-adjusted χ^2 when models are correctly specified (Herzog, Boomsma, & Reinecke, 2007), whereas the mean-adjusted χ^2 performs better (i.e., higher power) than the mean- and variance-adjusted χ^2 when models are misspecified (Fouladi, 2000; Nevitt & Hancock, 2004).

When employing the mean- and variance-adjustment, the difference in χ^2 values for nested models is not χ^2 distributed, thus the values cannot be simply subtracted to conduct χ^2 difference tests. The difference test is complicated and is different from the process used with mean-adjusted (S–B scaled) χ^2 values; however, the difference test can be conducted in Mplus using the DIFFTEST option (see Asparouhov & Muthén, 2006 for details on the computation of this test). Currently, EQS (6.1) does not have an option to compute the difference between mean- and variance-adjusted χ^2 values.

ORDERED CATEGORICAL DATA

As stated previously, ML estimation assumes the observed data are a sample drawn from a continuous and multivariate normally distributed population. In the social sciences, this type of data is not always collected. Frequently, researchers collect and analyze ordinal data, such as data obtained from the use of a Likert scale. Although researchers often treat ordinal data as continuous, the ordinal measurements are, as Bollen (1989, p. 433) noted, "coarse" and "crude." Even if the data appear to be approximately normally distributed (e.g., indices of skewness and kurtosis are almost zero or plots of the observed data appear to be normal), ordered categorical data are discrete in nature and, therefore, cannot be normally distributed by definition (Kaplan, 2009). Many studies have been conducted to examine the extent of bias present when applying a normal theory estimator such as ML to ordered categorical data, as detailed in the following sections.

Effects of Analyzing Approximately Normally Distributed Ordered Categorical Data Using ML Estimation: Empirical Results

When analyzing ordered categorical data, researchers often ignore the categorical nature of the data and apply ML estimation. By creating and analyzing a covariance matrix that represents relations among the observed categorical data, one is treating the ordinal data as if they were continuous. As the number of ordered categories increases, data more closely *approximate* continuous level data and, in turn, the observed relations are closer to their true values (Bollen & Barb, 1981). The fewer categories present, the more severe the attenuation in the observed relations and the greater the discrepancy between true and observed values. As discussed below, if few categories are used and ML estimation is employed, the model fit indices, parameter estimates, and standard errors can be biased.

Model chi-square and fit indices. In general, fit indices have been found to perform well if approximately normally distributed five-category ordinal data are treated as continuous (Babakus, Ferguson, & Jöreskog, 1987; Hutchinson & Olmos, 1998). Although the χ^2 was found to be robust when modeling data collected using four ordered categories, inflation occurred if fewer than four categories were used (Green, Akey, Fleming, Hershberger, & Marquis, 1997). In addition, slight underestimation of the goodness-of-fit index (GFI), adjusted GFI (AGFI), and Root Mean Square Residual (RMR) has been found if sample sizes are small and ordered categorical data with five categories are analyzed as continuous (Babakus et al., 1987). Researchers have suggested that when ordered categorical data are approximately normal and have at least five categories that the data may be treated as if they were continuous without great distortion in the fit indices (e.g., Bollen, 1989; Dolan, 1994; Muthén & Kaplan, 1985).

Parameter estimates and standard errors. When ordered categorical data have at least five categories and are approximately normal, treating data as continuous and applying the ML estimator produces slight underestimation in factor loading estimates (Babakus et al., 1987; Muthén & Kaplan, 1985; Rhemtulla, Brosseau-Liard, & Savalei, 2012). Standard errors have shown a greater sensitivity to categorization than the parameter estimates, exhibiting negative bias (Babakus et al., 1987; Muthén & Kaplan, 1985; West, Finch, & Curran, 1995). If standard errors are too small, tests of significance for parameter estimates may be inflated, resulting in Type I errors. As the number of ordered categories decreases, the underestimation in both the parameter estimates and standard errors becomes more severe, even if ordinal data are distributed symmetrically.

Effects of Analyzing Nonnormal Ordered Categorical Data using ML Estimation: Empirical Results

Ordered categorical data are considered by some researchers as inherently nonnormal (e.g., Muthén & Kaplan, 1985). However, as just described, if the observed data have many categories (e.g., at least five ordered categories) and are approximately normal, analyzing a covariance matrix using ML estimation does not result in severe levels of bias in fit indices, parameter estimates, or standard errors. Problems begin to emerge when the number of response options decreases or the observed item distributions diverge widely from a normal distribution. As the number of ordered categories is reduced, there are fewer response choices available for subjects to choose, leading to reduced variability in responses, and in turn a greater amount of attenuation in the observed covariances. Also, as the number of categories decreases, it becomes less likely that observed data could approximate a normal distribution.

Model chi-square and fit indices. Similar to results found when modeling continuous nonnormal data, fit indices are adversely affected when ordinal data follow nonnormal distributions. When modeling nonnormal ordered categorical data as continuous, ML-based χ^2 values (Green et al., 1997; Muthén & Kaplan, 1985; West et al., 1995) and RMR values were inflated, and values of the non-normed fit index (NNFI), GFI, and CFI were underestimated (Hutchinson & Olmos, 1998). These effects might lead to the rejection of a correctly specified model.

Parameters and standard errors. As univariate skewness and kurtosis levels of observed ordered categorical data become larger in absolute value, the negative bias observed with the ML-based parameter estimates and standard errors becomes more pronounced (Babakus et al., 1987). Bias levels increase with lower sample sizes, fewer categories, weaker relations between factors and indicators, or higher levels of nonnormality (e.g., Babakus et al., 1987; Bollen, 1989; Dolan, 1994).

Techniques to Address Ordered Categorical Data

In this section we will describe three methods commonly employed when modeling ordered categorical data. The first method involves applying the S–B scaling method to the ML-based χ^2 and standard errors, thus no adjustment is made to address bias in parameter estimates. The second and third methods involve alternative estimators that do not make the distributional assumptions of ML estimation: WLS and robust diagonally WLS. The latter of the two can be conceptualized as combining an alternative estimation method with the S–B scaling.

S–B scaling methods with ordered categorical data. A commonly employed strategy used to model ordered categorical data is to analyze the observed variable covariance matrix with ML and apply the S–B scaling to the fit indices and standard errors. That is, given that categorical data are nonnormal to some extent, and given the known effect of nonnormality on fit indices and standard errors, it would seem useful to employ an adjustment to these estimates. It is recognized that this method treats the categorical data as continuous, thus ignoring the metric level of the data. Hence, a caveat to this method is the ML-based parameters estimates are biased to some extent.

S–B scaling methods with nonnormal ordered categorical data: Empirical results. Although the functioning of the S–B correction has been usually examined with continuous nonnormal data, a few studies have evaluated how the correction performs with ordered categorical data. Often the focus of these studies is to assess the number of categories required for adequate performance of ML estimation using a covariance matrix coupled with S–B adjustments.

Model chi-square and fit indices. With regard to data-model fit, Green et al. (1997) found that S–B scaled χ^2 values were very close to the expected χ^2 values when modeling two-, four- or six-ordered category data that displayed symmetric, uniform, and negatively skewed distributions. The S–B scaled χ^2 did show positive bias when modeling variables that exhibited differential skew, with bias being the greatest in the two-category condition. Over all conditions, the S–B mean- and variance-adjusted χ^2 outperformed the ML-based χ^2. The mean- and variance-adjusted χ^2 displayed excellent Type I error rates when modeling two-, three-, and four-category data that were normal or kurtotic, where kurtosis levels ranged from approximately 4 to approximately 7 (Bandalos, in press). These Type I error rates were close to expected values even when sample sizes were small ($N = 150$ for 3 factor, 12 variable CFA model and $N = 500$ for 7 factor, 28 variable SEM). Moreover, the mean- and variance-adjusted χ^2 was found to perform better than the mean-adjusted χ^2 when modeling categorical data as continuous (Rhemtulla et al., 2012), with the former performing well when modeling six- and seven-ordered category data.

Parameter estimates and standard errors. As noted above, the S–B correction adjusts the standard errors and the χ^2. Thus, S–B-based parameter estimates will be equivalent to ML-based estimates. This implies no correction for the attenuation of the parameter estimates due to the categorical nature of the data. With respect to standard errors, research has found that robust standard errors exhibited greater precision than ML-based standard errors when nonnormally distributed ordered categorical data were analyzed (approximate skewness and kurtosis values of 2 and 6, respectively; DiStefano, 2002). The robust standard errors were found to be fairly accurate with six or seven ordered categories (Rhemtulla et al., 2012). Others have found the benefit of the robust standard errors when nonnormally distributed data have as few as three ordered categories (Bandalos, in press; DiStefano, 2003).

S–B scaling methods: Software implementation. As discussed, the S–B scaling method has been applied to ordered categorical data by treating the ordered categorical data as continuous in nature (i.e., analysis of covariances). This implies that the implementation of this method is the same across categorical or continuous data. All three software programs calculate the S–B scaled χ^2 and standard errors. Appendix B provides the syntax needed to employ the S–B scaling method when treating categorical data as continuous in nature.

Description of Categorical Variable Methodology (CVM). Alternative strategies for handling ordered categorical data consider the metric of the data by including this information in the estimation procedures. Specifically, Categorical Variable Methodology (CVM) incorporates the metric of the data into analyses by considering two components: (a) input for analyses that recognizes the ordered categorical variables and (b) the use of the correct weight matrix when employing a WLS estimator (Muthén, 1984; Muthén & Kaplan, 1985). Therefore, CVM is basically the use of a WLS estimator with specific input to accommodate ordered categorical variables.

Let us first discuss how to represent the categorical nature of the data. One way in which observed ordered categorical data are thought to occur is when a continuous, normally distributed latent response variable (y^*) is divided into distinct categories (e.g., Bollen, 1989; Muthén, 1993). This has been referred to as the *latent response variable formulation* (e.g., Muthén & Muthén, 1998–2010). The points that divide the continuous latent response variable (y^*) into a set number of categories (c) are termed thresholds (τ), where the total number of thresholds is equal to the number of categories less one ($c-1$). For example, if a Likert scale has five response choices, four threshold values are needed to divide y^* into five ordered categories. The observed ordinal data (y) are thought to be produced as follows:

$$y = \begin{cases} 1 & \text{if } y^* \leq \tau_1 \\ 2 & \text{if } \tau_1 < y^* \leq \tau_2 \\ 3 & \text{if } \tau_2 < y^* \leq \tau_3 \\ 4 & \text{if } \tau_3 < y^* \leq \tau_4 \\ 5 & y^* > \tau_4 \end{cases} \qquad (11.6)$$

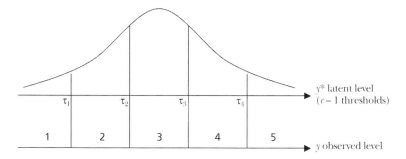

Figure 11.1 Relation between y^*, y, and thresholds.

As a result, $y^* \neq y$. Specifically, because subjects respond to the five-point Likert scale, the observed ordinal level data can only be reported as discrete values from 1 to 5. However, subjects' theoretical "true" levels of the latent response variable (y^*) are much more precise than allowed by the five-point response scale. Figure 11.1 illustrates the relation between the continuous, normally distributed latent response variable (y^*), observed level data (y), and the four ($c - 1$) threshold values for a variable with five ordered categories. This figure helps to illustrate that the observed ordinal data provide an approximation of the underlying continuous, normally distributed latent response variable.

Thus, with regard to metric, if data are continuous then data at the observed level are considered equivalent to the underlying latent response variable, that is, $y = y^*$. On the other hand, if data are from ordered categories, then $y \neq y^*$ and the difference between the two has two important consequences when modeling the data. First, unlike y^*, the standard linear measurement model ($y^* = bF + E$) does not hold when modeling y (that is, $y \neq bF + E$).[7] Second, and following from the first point, the hypothesis that the model reflects the true structure in the population, $\Sigma^* = \Sigma(\theta)$, does not hold when a covariance matrix of ordered categorical data is analyzed (Bollen, 1989). In brief, the population covariance matrix of the ordered categorical variables (Σ) will not equal the population covariance matrix of the continuous underlying latent response variables (Σ^*). Thus, if the covariance structure hypothesis is true when modeling y^*, that is, $\Sigma^* = \Sigma(\theta)$, and $\Sigma \neq \Sigma^*$ because $y \neq y^*$, then $\Sigma \neq \Sigma(\theta)$. Our interest is in estimating model parameters that describe the relations among the underlying latent response variables.

In order to avoid the consequences of modeling y using the standard linear confirmatory factor analysis model, y^* can be modeled (Bollen, 1989). Applying the linear factor model to the underlying latent response variable is illustrated in Figure 11.2. Notice the factor does not directly affect y, but instead, directly affects y^*. Because y^* is a continuous, normally distributed

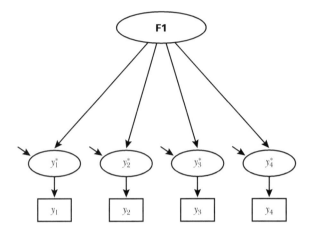

Figure 11.2 Latent response variable formulation.

(albeit latent) variable, the standard linear model ($y^* = bF + E$) can be used to estimate the relation between y^* and the factor.

Modeling the relation between F and y^* typically entails a three-step process: compute thresholds (τ), approximate the correlations among y^* values (i.e., latent correlations), and estimate model parameters from the thresholds and latent correlations using a WLS estimator. Because thresholds are used in the computation of latent correlations, these will be discussed first. Threshold values, which cut the underlying continuum into ordered categories (see Figure 11.1), might be estimated if the number of subjects that chose a certain category is known. These values are important to estimate not only because they are thought of as critical points that "move" a subject from one category to another but also because they are used to create marginal distributions of ordinal variables that assist with estimation procedures. Because the metric of the underlying latent response variable (y^*) is arbitrary, the mean and standard deviation are often set to 0 and 1, respectively. Using a mean of 0 and standard deviation of 1, item thresholds may be estimated by considering the cumulative area under the normal curve up to a given point (Bollen, 1989; Jöreskog & Sörbom, 1996) by:

$$\tau_i = \Phi^{-1}\left[\sum_{k=1}^{i} \frac{N_k}{N}\right], \quad i = 1, 2, \ldots, c-1 \tag{11.7}$$

where τ_i is a particular threshold, Φ^{-1} is the inverse of the normal distribution function, N_k is the number of subjects who selected category k, N is the total sample size, and c is the total number of categories. Equation 11.7 shows that thresholds are calculated using the proportions of subjects within each ordered category (N_k/N). The resulting threshold values (z-values) divide the underlying continuous, normal distribution into c categories and relate the discrete y values to the continuous y^* values.

The thresholds are then used to compute latent correlations, which in turn are used as input for estimating model parameters. Latent correlations (Σ^*) represent the theoretical relations among the underlying continuous, normally distributed latent response variables (y^*). For each pair of observed variables, a latent correlation can be estimated. If both observed variables are dichotomous, a *tetrachoric correlation* represents the relation between the y^* variables. If both variables are ordinal, a *polychoric correlation* represents the relation between the y^* variables. If one variable is ordinal and the other variable is continuous, a *polyserial correlation* represents the relation between the variables.

To illustrate how ordered categorical variables might be accommodated, consider an example employing Rosenberg's (1989) self-esteem scale. A sample of 120 college freshman responded to questions concerning their self-esteem using a scale with four ordered categories anchored at "Strongly Disagree" (1) to "Strongly Agree" (5). Responses to two of the questions, "On the whole, I am satisfied with myself" (SATISFIED) and "I feel that I have a number of good qualities" (QUALITIES) are provided in Table 11.1.

Threshold values are computed from the sample data using two pieces of information: (1) the proportion of students who responded to a certain category and (2) the area under the normal curve. Consider the first threshold for the variable, SATISFIED, which divides the underlying latent response variable into the two categories of "Strongly Disagree" (category 1) and "Disagree" (category 2). Students below this threshold will have responded "Strongly Disagree" to the statement. There are four students responding "Strongly Disagree" to the Satisfaction item. Using Equation 11.7, the cumulative probability through category 1 is .033 (4/120).

TABLE 11.1 Frequency of Responses to Self-Esteem Items

SATISFIED		QUALITIES	
Category	Frequency	Category	Frequency
1 (SD)	4	1 (SD)	4
2 (D)	16	2 (D)	27
3 (A)	50	3 (A)	42
4 (SA)	50	4 (SA)	47
N	120	N	120
Skew	−.812	Skew	−.514
Kurtosis	.154	Kurtosis	−.726

Note: SD = Strongly Disagree; D = Disagree; A = Agree; SA = Strongly Agree

TABLE 11.2   Threshold Values and Cumulative Area

	SATISFIED			QUALITIES		
	1	2	3	1	2	3
Threshold	1	2	3	1	2	3
Cumulative N	4	20	70	4	31	73
Cumulative Area	.033	.167	.583	.033	.258	.608
Threshold Value	−1.834	−0.967	0.210	−1.834	−0.648	0.275

Considering this as representative of the cumulative area under the normal curve, the threshold value is the z-value associated with .033, or a z-value of −1.83. The remaining thresholds may be calculated in a similar manner, as shown in Table 11.2.

To determine the polychoric correlation between two ordinal variables, a contingency table of students' responses to each pair of variables is needed. The frequency of responses to each option can be tabulated across the pair of variables. Table 11.3 shows the contingency table for the responses to the SATISFIED and QUALITIES variables. From the table, one can see a relation between the responses to the items. For example, those students who agreed or strongly agreed that they were satisfied with themselves generally agreed that they had a number of good qualities. However, some discrepancies are noticed. For example, six students who agreed that they possessed good qualities had some disagreement with the statement about having self-satisfaction.

The contingency table information reported in Table 11.3 and item thresholds are used to approximate the latent correlations among the ordinal variables (see Olsson, 1979). The estimated polychoric correlation represents the value with the greatest likelihood of yielding the observed contingency table, given the estimated thresholds. Here, the estimated polychoric correlation equals .649, reflecting the positive relation while acknowledging some inconsistency in student responses across the two items.

TABLE 11.3   Contingency Table Between SATISFIED and QUALITIES Variables

	QUALITIES				
SATISFIED	1 (SD)	2 (D)	3 (A)	4 (SA)	Total
1 (SD)	2	1	1	0	4
2 (D)	0	8	6	2	16
3 (A)	1	15	26	8	50
4 (SA)	1	3	9	37	50
Total	4	27	42	47	120

Note: SD = Strongly Disagree; D = Disagree; A = Agree; SA = Strongly Agree

Note that the polychoric correlation estimate is higher than the familiar Pearson product-moment correlation estimate (.551) because the polychoric correlation is disattenuated for the error associated with the variable's coarse categorization of the underlying latent variable's continuum. To illustrate the polychoric correlation graphically, the relation between the observed ordered categorical variables and the underlying latent responses variables are plotted in Figure 11.3.

WLS estimator with ordered categorical data: Description. One may question the use of the NT ML estimator when analyzing latent correlations. Research has shown the standard errors of parameter estimates and fit statis-

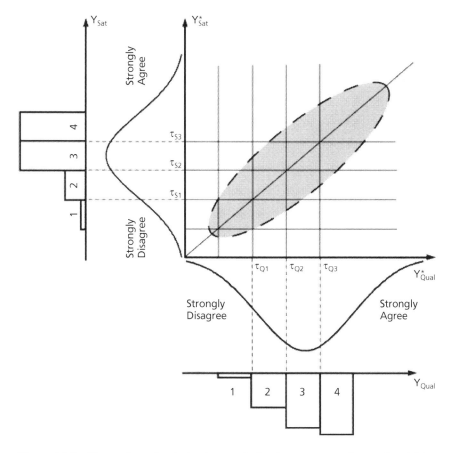

Figure 11.3 Illustration of polychoric correlation between two observed variables. *Note:* Sat = SATISFIED, Qual = QUALITIES. Threshold values are specific to the variable in question. For example, τ_{S1}, τ_{S2}, τ_{S3}, respectively, refer to thresholds 1 through 3 for the variable SATISFIED. Similarly, thresholds τ_{Q1}, τ_{Q2}, τ_{Q3}, respectively, refer to thresholds 1 through 3 for the variable QUALITIES.

tics are biased when employing ML estimation with polychoric correlations as input (Babakus et al., 1987; Dolan, 1994; Rigdon & Ferguson, 1991), hence the development and use of the WLS estimator when analyzing latent correlations. When applied to latent correlations, the WLS estimator has the following form:

$$F_{WLS} = (\mathbf{r} - \hat{\rho})' \mathbf{W}^{-1} (\mathbf{r} - \hat{\rho}) \qquad (11.8)$$

where $\mathbf{r}$ is a vector containing the sample latent correlations (and thresholds depending upon the software program), $\hat{\rho}$ is the corresponding vector from the model-implied matrix, and $\mathbf{W}$ is the asymptotic covariance matrix of $\mathbf{r}$, which is a matrix of the variances and covariances of the sample latent correlations (and thresholds depending upon the software program).

WLS with ordered categorical data: Empirical results. Although differences might exist in how various software programs employ WLS when analyzing ordered categorical data, the information presented here is made without specific reference to software packages (details concerning the implementation of CVM using LISREL, EQS, and Mplus are described below). In general, empirical studies have found that employing WLS estimation with ordered categorical data (i.e., thresholds and latent correlations) has both desirable and undesirable characteristics.

Model chi-square and fit indices. When approximately normally distributed ordered categorical data were fit to a correctly specified model using WLS estimation, values of χ^2 were close to expected values when small to moderate models were specified (15 parameters and less) or sample sizes were large ($N = 1000$; Flora & Curran, 2004; Muthén & Kaplan, 1985; Potthast, 1993). However, χ^2 values were greatly over-estimated, resulting in increased Type I error rates for large models (four factors with 16 variables), even when sample size was large ($N = 1,600$; Yang-Wallentin, Jöreskog, & Luo, 2010). The amount of inflation of the WLS-based χ^2 increased as sample size decreased, model size increased, or nonnormality of the data increased (DiStefano, 2002; Flora & Curran, 2004; Hutchinson & Olmos, 1998; Muthén & Kaplan, 1992; Potthast, 1993; Yang-Wallentin et al. 2010). RMSEA has been found to be somewhat robust to ordered categorical data analyzed using WLS estimation, and was not sensitive to sample size or model size when correctly specified models were estimated (Hutchinson & Olmos, 1998).

Parameter estimates and standard errors. Theoretically, CVM should result in unbiased parameter estimates when modeling ordered categorical data. The unbiased nature of the estimates is attributed to the analyses being conducted at the latent level, y^* (e.g., using polychoric correlations), rather than the observed level, y. Empirical studies have found that WLS parameter estimates tend to have little bias when estimating correctly specified models using a large sample, regardless of whether binary (Dolan, 1994) or nonnormal ordered categorical data (approximate values: skewness = 2.5, kurtosis = 6) were analyzed (DiStefano, 2002; Dolan, 1994; Potthast, 1993). However, with both correctly specified and misspecified models, parameter estimates became notably over-estimated when model size increased or sample size decreased (Flora & Curran, 2004; Yang-Wallentin et al., 2010). Research also suggests that estimated standard errors tend to be underestimated for both correctly specified and misspecified models. This negative bias became more pronounced as model size and nonnormality increased or sample sizes decreased (Bandalos, in press; DiStefano, 2002; Muthén & Kaplan, 1992; Potthast, 1993; Yang-Wallentin et al., 2010).

Importantly, there might be issues with convergence and admissibility of solutions when employing WLS with ordered categorical data (Bandalos, in press). These problems often stem from the size of the asymptotic covariance matrix. This matrix is difficult to compute and often cannot be inverted because it is non-positive definite. Specifically, when Flora and Curran (2004) used WLS to model a twenty-indicator CFA model using a sample size of 100, none of the asymptotic covariance matrices from the 500 replications used in the simulation study were positive definite. Increasing sample size to 200 resulted in the majority of asymptotic covariance matrices being positive definite, but the majority of solutions were inadmissible (Flora & Curran, 2004).

WLS with ordered categorical data: Software implementation. Software packages that employ WLS estimation with ordered categorical data typically use the multi-step approach described above (i.e., thresholds estimated, polychorics estimated, and WLS estimation applied to the polychorics). The syntax specific to LISREL, EQS, and Mplus can be found in Appendix A.

The LISREL approach to CVM requires both LISREL and PRELIS programs. PRELIS classifies variables with less than 16 distinct categories as ordinal unless otherwise specified by the researcher (Jöreskog & Sörbom, 1996). First, item threshold values are calculated from the sample data in PRELIS. Next, the threshold information and contingency tables between pairs of variables are used to compute the latent correlations. PRELIS then estimates the asymptotic covariance matrix of the latent correlations (threshold information is not included in the asymptotic covariance matrix; Jöreskog, 2005; Jöreskog & Sörbom, 1996; Yang-Wallentin et al., 2010). The latent correlations (**r** in Equation 11.8) and the asymptotic covariance matrix of the latent correlations are input into LISREL, and WLS estimation is employed.

A preprocessor is not needed for CVM when employing either EQS or Mplus. All of the necessary components are constructed and employed within the program. EQS uses a two-stage, not three-stage, process to analyze ordered categorical variables. In the first stage, partitioned maximum likelihood (PML) is used to estimate the thresholds and latent correlations simultaneously. PML partitions the model into submodels and estimates the relations among smaller parts of the model rather than the entire model at once. If both continuous and categorical variables are included in the same model, EQS estimates relations between continuous variables and categorical variables first (called the *polyserial partition*). Following the polyserial partition, relations among ordinal variables are estimated (called the *polychoric partition*). In the second stage, the asymptotic covariance matrix of the latent correlations (threshold information is not included in the asymptotic covariance matrix) is computed for use with the AGLS estimator, and estimates of model parameters and fit indices are computed.

The latent correlations and thresholds ($\mathbf{r}$ in Equation 11.8) are modeled when employing WLS estimation in Mplus (Muthén, 1984).[8] Thus, the asymptotic covariance matrix employed in Mplus as the weight matrix is formed using both the latent correlations and the thresholds. Moreover, Mplus has the capability to employ two different scalings of y^*, Delta and Theta, when undertaking CVM (Muthén & Asparouhov, 2002). Scaling is needed to provide a metric for the latent response distribution, y^*. Choosing a method to scale a variable is not new to those employing SEM, as researchers must choose among options for setting the scale of a factor (e.g., setting the variance of the factor to 1 vs. setting the metric of the factor using an indicator).

The Delta scaling fixes the total variance of y^* to 1 and obtains the residual variance of y^* as the remainder of $1 - \lambda^2\text{Var}(F)$. This scaling aligns with the conventions of latent correlations, which assume a variance of 1 for the underlying latent response variable, y^* (Kamata & Bauer, 2008). LISREL and EQS only employ this scaling of y^*, and this is the default for Mplus. Delta scaling coupled with fixing the factor variance to 1 result in parameters (factor pattern coefficients, residuals, and thresholds) in the familiar and commonly interpreted metric. That is, the factor pattern coefficients are interpreted as the amount of standard deviation change in y^* for every 1 standard deviation change in the factor; residuals indicate the proportion of variance in y^* not explained by the factor; and thresholds indicate the z-score corresponding to the cumulative area under the curve to the left of a particular category (i.e., probability of selecting any category up to category k).

The Theta scaling of y^* fixes the residual variance of y^* to 1. The total variance of y^* can then be computed as the sum of the residual variance and the common variance (i.e., $\lambda^2\text{Var}(F)$). Fixing the residual variance of y^* to 1 aligns with the conventions of a probit regression model (Kamata & Bauer, 2008; Muthén, 1998–2004). Use of the probit model produces unstandardized probit slopes (number of units probit y changes for every 1 unit change in the factor) and unstandardized probit thresholds (expected probit y when factor equals zero), which denote the relation between y^* and the factor.[9]

As shown in Table 11.4, the parameters based on Theta scaling are easily transformed to the commonly reported factor analytic parameters from the Delta scaling (Kamata & Bauer, 2008; Muthén & Asparouhov, 2002). In fact, when employing the Theta scaling in Mplus, the estimates from the Delta parameterization can be acquired by requesting the completely standardized solution (putting STDYX on the Output line).

Robust DWLS estimation with ordered categorical data: Description. Because the S–B scaling method does not adjust parameters for the metric of the data when modeling ordered categorical data, it may seem as though WLS estimation is a more attractive option. However, the computational demands of the WLS estimator make it an unrealistic option for handling ordered categorical data unless an extremely large sample size is available

---

TABLE 11.4 Mplus Scaling Options for y* and Conversion Formulas

	Theta Scaling	Delta Scaling
Fixed Parameters	• Factor variance = 1 • Factor mean = 0 • y^* residual variance = 1	• Factor variance =1 • Factor mean = 0 • y^* total variance = 1
Estimated Parameters	τ_T = threshold λ_T = factor loading	τ_D = threshold λ_D = factor loading
Values derived from model	$V(y^*)$ = total variance of y^*	θ_D = residual variance of y^*
Conversions formulas between scalings	$\tau_D = \dfrac{\tau_T}{\sqrt{V(y^*)}}, \lambda_D = \dfrac{\lambda_T}{\sqrt{V(y^*)}}$	$\tau_T = \dfrac{\tau_D}{\sqrt{\theta_D}}, \lambda_T = \dfrac{\lambda_D}{\sqrt{\theta_D}}$
Conversions to 2-PL IRT model from SEM b = IRT difficulty a = IRT discrimination	$b = \dfrac{\tau_T}{\lambda_T}, a = \lambda_T$	$b = \dfrac{\tau_D}{\lambda_D}, a = \dfrac{\lambda_D}{\sqrt{\theta_D}}$
Conversions to SEM framework from 2-PL IRT model	$\tau_T = ab, \lambda_T = a$	$\tau_D = \dfrac{ab}{\sqrt{1+a^2}}, \lambda_D = \dfrac{a}{\sqrt{1+a^2}}$

Note: These conversion formulas apply given the same distribution (normal or logistic) is used for the residual variance of y^* across the different scalings (Theta and Delta) and models (CFA vs. IRT). Assuming a normal distribution for the residuals, which is common for factor analytic models, results in IRT parameters in a probit, not logistic, metric. The probit IRT discrimination values can be converted to a logistic metric by multiplying by 1.7. The scaling of the latent factor can vary; however, we only present the parameterizations when the factor variance and factor mean are fixed to 1 and zero, respectively (see Kamata & Bauer, 2008, for additional scaling options for the latent factor). The total variance of y^*, $V(y^*)$, from the theta scaling can be computed using the scaling factor reported by Mplus: $V(y^*) = (1/\text{scaling factor})^2$.

(i.e., the full asymptotic covariance matrix is difficult to compute and invert). Diagonally WLS (DWLS) estimation was developed to overcome the limitations of full WLS estimation. Specifically, DWLS avoids the necessity of a large sample size by decreasing the computational intensity found with the full WLS estimator. In addition, DWLS can incorporate scaling similar to the S–B scaling method resulting in robust DWLS estimation.

DWLS follows the same general form of WLS (see Equation 11.8); however, what differs across the estimators is the weight matrix, **W**. Specifically, when estimating parameters using DWLS, **W** consists only of the diagonal elements of the full **W** matrix:

$$F_{\text{DWLS}} = (\mathbf{r} - \hat{\boldsymbol{\rho}})' (diag\mathbf{W})^{-1} (\mathbf{r} - \hat{\boldsymbol{\rho}}), \tag{11.9}$$

where the elements in $diag\mathbf{W}$ include the asymptotic variances of the latent correlation estimates (and thresholds depending upon the software program) and the off-diagonal elements are constrained to zero. Thus, the residuals $(\mathbf{r} - \hat{\boldsymbol{\rho}})$ are weighted by the asymptotic variances of the **r** elements (latent correlations or latent correlations plus thresholds, depending on the software program).

Recall that WLS estimation employs and inverts the full asymptotic covariance matrix in order to estimate parameters, standard errors, and χ^2. Employing only the diagonal of **W** to estimate parameters helps overcome convergence issues experienced with full WLS; however, if only the diagonal of the weight matrix is used to calculate standard errors and χ^2, these values would be biased. This is because, unlike the full asymptotic covariance matrix, the diagonal asymptotic covariance matrix of the polychorics is not a consistent estimate of the population asymptotic covariance matrix for the sampling distribution of the polychoric correlation estimates. Thus, information from the full asymptotic covariance matrix is needed to more accurately estimate the standard errors and χ^2 (Wirth & Edwards, 2007). Information from the full matrix can be incorporated into DWLS standard errors and χ^2 values in a different manner than with WLS, resulting in robust DWLS standard errors and χ^2 values. Specifically, to address the bias in the DWLS standard errors, a robust asymptotic covariance matrix for the parameter estimates is estimated and the robust DWLS standard errors are obtained from the diagonal elements of this matrix (Flora & Curran, 2004; Muthén, 1993; Muthén, du Toit, & Spisic, 1997). Computation of this robust asymptotic covariance matrix of the parameter estimates involves both the diagonal and the full asymptotic covariance matrix of the polychorics, but not the inverse of the full matrix (see formula A.24 in Jöreskog, Sörbom, du Toit, & du Toit, 2000; see formula 102 in Muthén, 1998–2004). That is, analogous to the S–B robust standard errors for nonnormal continuous data, these are sandwich standard errors. For DWLS, these sandwich standard errors consist of the DWLS standard errors as the bread of the sandwich and the full asymptotic covariance matrix of the polychorics as the meat of the sandwich. Again, this full asymptotic covariance matrix is not inverted in these computations.

Adjustments to the DWLS χ^2 are also accomplished using adjustments akin to the S–B corrections (i.e., mean-adjusted χ^2 or mean- and variance-adjusted χ^2; Asparouhov & Muthén, 2010; Yang-Wallentin et al., 2010). Similar to the robust DWLS standard errors, computation of the S–B robust DWLS χ^2 involves the full weight matrix but not its inverse.

Robust DWLS: Empirical results. Several simulation studies have been conducted to investigate the properties of robust DWLS parameter estimates, standard errors, and fit statistics. Given the issue of convergence associated with WLS, the convergence rates for robust DWLS are typically reported. With respect to convergence, robust DWLS estimators have been found to perform better than full WLS estimation with small sample sizes and large models (Flora & Curran, 2004; Yang-Wallentin et al., 2010). Robust DWLS estimation might have problems converging to an admissible solution under select situations. For example, convergence rates have been found to be especially low when sample size was 200 or less and variable-level skewness was 1.5 or greater (Forero, Maydeu-Olivares, & Gallardo-Pujol, 2009) and when non-normality produces empty cells between pairs of ordinal variables (DiStefano & Morgan, in press).

Model chi-square and fit indices. When modeling correctly specified models, robust DWLS χ^2 values were close to expected values, even with increasing levels of non-normality, decreasing sample size, and increasing model size (Bandalos, in press; Flora & Curran, 2004; Yang-Wallentin et al., 2010). Moreover, results have shown that with larger models (15 or 20 variables) and smaller sample sizes, robust DWLS estimators produced more accurate χ^2 values than WLS estimation (Bandalos, in press; Flora & Curran 2004; Yang-Wallentin et al., 2010). Robust DWLS χ^2 values do become slightly biased upward when models are large and samples are small (i.e., increased Type I error rate). Additionally, when comparing robust DWLS χ^2 values to unadjusted ML χ^2 values (when modeling ordered categorical data as continuous with ML), robust DWLS performed better (i.e., Type I error rate) when two- or three-category data were analyzed, whereas unadjusted ML χ^2 and robust DWLS χ^2 performed approximately equivalently when five- or six-category data were present (Beauducel & Herzberg, 2006).

With respect to power, robust DWLS might suffer from the same lack of power found previously with WLS estimation (Olsson et al., 2000) and the S–B scaling methods (Foss et al., 2011). That is, Bandalos (in press) found decreasing power with increasing levels of nonnormality and model misspecification for both the robust DWLS χ^2 and the mean- and variance-adjusted ML χ^2 (for the latter, the categorical data were treated as continuous). Similarly, when modeling five-category, ordinal data, robust DWLS χ^2 values were associated with less power as non-normality increased, sample size decreased, or model misspecification was less severe (Lei, 2009). However, when modeling dichotomous data (clearly nonnormal in nature), Yu and Muthén (2002) found high power associated with the robust DWLS χ^2 when sample size was 500 or higher. Given the limited study of this estimator with misspecified models, more work is needed to better understand its functioning.

Robust DWLS-based approximate fit indices have received limited attention. Yu and Muthén (2002) conducted one of the first studies of the performance of fit indices estimated using robust DWLS methods. Using both correctly specified and misspecified models with dichotomous data, the findings suggested that TLI, CFI, and RMSEA performed well with correctly specified models unless sample size was small ($N = 100$). Using a .95 cutoff for TLI and CFI and a .05 cutoff for RMSEA, these indices were associated with high model rejection rates when model misspecification was present. In a study comparing robust DWLS-based fit indices to unadjusted ML-based fit indices using a correctly specified CFA model and two to six response categories, Beauducel and Herzberg (2006) found both estimators produced fit values within appropriate ranges and none indicated poor model-data fit. Interestingly, when comparing the two estimators, they found slightly higher robust DWLS-based TLI and CFI values and lower RMSEA values when two- and three-category data were present (i.e., robust DWLS performed better), but this difference in estimators diminished for the CFI for five or six categories and actually reversed order for the TLI and RMSEA (i.e., ML performed better). When modeling normal or nonnormal two- through four-category data, Bandalos (2008) found that robust DWLS-based RMSEA and CFI values suggested good fit when models were misspecified (i.e., RMSEA values ≤ .05 and CFI ≥ .95), which could result in championing a misspecified model. Much more work is needed to assess the functioning of the DWLS-based fit indices.

Parameter estimates and standard errors. In general, robust DWLS has been found to provide fairly accurate parameter estimates (Bandalos & Webb, 2005; Flora & Curran, 2004; Oranje, 2003). When modeling correctly specified models, robust DWLS estimates were essentially unbiased (Flora & Curran, 2004; Yang-Wallentin et al., 2010). Although robust DWLS parameter estimates become more biased as sample size decreases, model size increases, factors have fewer indicators, indicator-factor relations weaken, and nonnormality increases, the amount of bias is slight when models are correctly specified (Flora & Curran, 2004; Forero et al., 2009; Yang-Wallentin et al., 2010). Of note, robust DWLS has been found to produce inadequate estimates of factor loadings when sample size is 200 or less, there are only three indicators per factor, indicator-factor relations are less than

.40, and skewness is 1.5 or greater (Forero et al., 2009). However, robust DWLS parameter estimates (i.e., factor loadings, factor correlations, and structural paths) were less biased than full WLS estimation, and were less vulnerable to decreasing sample size and increasing model size than full WLS estimates (Bandalos, in press; Flora & Curran, 2004; Potthast, 1993; Yang-Wallentin et al., 2010). When modeling misspecified models, the robust DWLS estimates became more biased as degree of misspecification increased and sample sized decreased; however, the bias was much less than that found for WLS estimates (Yang-Wallentin et al., 2010).

Robust DWLS standard errors tend to be fairly accurate when fitting both correctly specified and misspecified models (Yang-Wallentin et al., 2010). Bias tends to increase as sample size decreases and/or model size increases; however, the bias for robust DWLS is much smaller than that found for WLS (Bandalos, in press: Yang-Wallentin et al., 2010). For example, unlike WLS standard errors, very little bias in robust DWLS standard errors was observed for correctly specified large models (i.e., 2 factors with 20 indicators) coupled with large sample sizes ($N = 1,000$; Flora & Curran, 2004) or misspecified large models (i.e., 4 factors with 16 items) coupled with large sample sizes ($N = 800$; Yang-Wallentin et al., 2010). However, robust DWLS standard errors exhibited slight under-estimation (i.e., negative bias) as sample size decreased under 200 (Lei, 2009).

The greater bias in WLS parameter estimates, test statistics, and standard errors compared to robust DWLS parameter estimates, test statistics, and standard errors, especially with decreasing sample size and increasing model size, is attributed to the difference in the weight matrices used in the estimation process (Flora & Curran, 2004; Yang-Wallentin et al., 2010). Recall that WLS employs the elements from the full asymptotic covariance matrix of the polychorics which become inaccurate due to sampling error as sample size decreases and model size increases. Inversion of the full asymptotic matrix compounds these errors (inversion involves multiplying these error-prone elements), resulting in inaccurate elements in the inverted full matrix. Robust DWLS, however, only inverts the asymptotic variances, resulting in an inverted weight matrix less susceptible to sampling error and inaccuracies. Hence the robust DWLS parameter estimates, standard errors, and test statistics are not influenced by sampling instability to the same degree as are WLS values. This implies that, with small sample sizes, the off-diagonal elements of the *inverted* full asymptotic covariance matrix are not only not useful, but are potentially harmful, resulting in convergence problems and biased parameter estimates, standard errors, and fit indices.

Robust DWLS: Software implementation. Mplus and LISREL both offer robust DWLS estimation.[10] However, there are some differences between these two programs. One difference concerns the information included in the weight matrix. For Mplus, the elements in **r** (see Equation 11.9) consist of both the latent correlations and the thresholds. Given that the weight matrix is the diagonal of the asymptotic covariance matrix of **r**, the Mplus robust DWLS-based methods include both the latent relations and threshold information in the weight matrix. LISREL does not include threshold information in the weight matrix. That is, the weight matrix is the diagonal of the asymptotic covariance matrix of **r**, where **r** includes only the latent correlations.

In Mplus 6.0, there are two robust DWLS estimators: weighted least squares mean-adjusted (WLSM) and weighted least squares mean- and variance-adjusted (WLSMV). The WLSM produces a mean-adjusted χ^2. WLSMV differs from WLSM in that the χ^2 is both mean- and variance-adjusted (Asparouhov & Muthén, 2010). Thus, fit indices based on these two corrected χ^2 values differ; however, the robust standard errors and parameter estimates from WLSM and WLSMV are equivalent. Comparisons between these two corrected χ^2 values suggest that WLSMV outperforms WLSM, with WLSM showing higher Type I error rates (Asparouhov & Muthén, 2010; Muthén et al., 1997).

The syntax for using WLSM and WLSMV is very similar to the Mplus WLS syntax (see Appendix C). The categorical indicators are identified by including the heading "CATEGORICAL ARE:" when defining the "VARIABLES" command. Once the variables are defined to be categorical, CVM will be conducted. To request one of the robust estimation techniques, changes are made to the "ESTIMATOR IS:" heading under the "ANALYSIS" command. Insert "WLSM" to request mean-adjusted DWLS estimation or "WLSMV" to request the mean- and variance-adjusted DWLS estimation. Appendix C outlines the syntax to conduct these procedures.

Similar to MLM and MLMV, one cannot simply subtract the χ^2 values for two nested models estimated using WLSM or WLSMV; the difference in χ^2 values is not χ^2 distributed (Asparouhov & Muthén, 2006). The formula used to compute the difference test when employing MLM estimation is also used when employing WLSM estimation (Muthén, 1998–2004), whereas the DIFFTEST option employed in Mplus for MLMV estimation must also be used with WLSMV estimation (Asparouhov & Muthén, 2006; Muthén & Muthén, 1998–2010).[11]

In LISREL 8.80, the researcher has two options when applying DWLS with latent correlations: (1) inputting only the asymptotic variances of the latent correlations, or (2) inputting the full asymptotic covariance matrix of the latent correlations. Inputting only asymptotic variances will produce unadjusted DWLS χ^2 and standard errors values (Jöreskog, 1994). This option is not recommended. The second option employs but does not invert the full asymptotic covariance matrix to compute the robust standard errors (formula A.24 from Jöreskog et al., 2000) and the Satorra–Bentler scaled (mean-adjusted) χ^2 (Satorra & Bentler, 1988, equation 4.1). With respect to the χ^2, the unadjusted DWLS χ^2 is labeled the "Normal Theory Weighted Least Squares Chi-square" (C2) in the LISREL output. The χ^2 labeled the "Satorra-Benter Scaled Chi-square" (C3) is the Satorra–Bentler scaled DWLS χ^2; that is, "C3 is a correction to C2 which makes C3 have the correct asymptotic mean even under nonnormality" (Jöreskog et al. 2000, p. 196). Thus, it appears that robust DWLS in LISREL produces a scaled χ^2 similar to that from WLSM in Mplus (i.e., a mean-adjusted χ^2).

IRT Parameters via CVM

As has been noted by many methodologists (e.g., Muthén, 1984; Takane & de Leeuw, 1987), when analyzing ordered categorical data, factor analytic and item response theory (IRT) parameters are simple transformations of each other. If we focus on the use of binary data (0, 1) and a two-parameter IRT model, the difficulty (*b*) and discrimination (*a*) IRT parameters are similar to the threshold and factor loading parameters from a factor model, respectively. In the two-parameter IRT model, difficulty values, which can be derived from the factor model threshold values, indicate the value of the factor (often called *theta* in IRT) where there is a 50% probability of endorsing the item (i.e., responding 1 versus 0). Discrimination values are similar to factor loadings as they both represent the strength of the relation between items and factors. The last two rows in Table 11.4 provides the conversion formulas for the two-parameter IRT and factor analytic models. Currently, SEM software packages do not estimate what is equivalent to a three-parameter IRT model. That is, the *c* parameter, which is a non-zero lower asymptote often interpreted as a "guessing parameter," is estimated via IRT software but not with SEM software.

Conditional probability formulation. The simple conversion formulas between CFA and the two-parameter model are possible because both IRT and factor analytic parameters can be computed using a probit model. Recall that the latent response variable formulation described thus far for CVM specifies a linear relation between the factor and the underlying latent response variable (y^*). One way of conceptualizing the estimation of this relation is fitting a factor model to latent correlations (e.g., tetrachorics, polychorics) and thus far we have employed this conceptualization. Given that a latent correlation can be conceptualized as the covariance between two continuous latent response variables (y^*) with variances fixed to 1, this conceptualization aligns with the Delta scaling in Table 11.4 (Kamata & Bauer, 2008). However, an equivalent conceptualization aligns with the use of a probit model (often referred to as the conditional probability formulation), which estimates a linear relation between a predictor (a factor in this case) and the transformation of the categorical criterion (*y* in this case) to a continuous normally distributed criterion (y^*). The probit model fixes the residual variance of y^* to 1, thus, aligning with the Theta parameterization in Table 11.4. As shown earlier, the factor analytic parameters estimated using Delta or Theta scaling are easily converted to the other; that is, the parameters differ simply by the scaling of y^*.

Importantly, the use of a probit model to estimate the relation between a categorical indicator and a latent factor is equivalent to the two-parameter normal ogive IRT model. Specifically, the two-parameter normal ogive model estimates IRT difficulty and discrimination parameters which can be easily transformed to the Theta scaling factor analytic thresholds and factor loadings since both are employing the same scaling of y^* (residual variance fixed to 1) or to the Delta factor analytic parameters using slightly more complicated conversion formulas.

Figure 11.4 provides an example of the factor analytic and IRT parameters produced via Mplus 6.0 and the relations across the different scalings. A one-factor model was fit to the responses (0, 1) to 9 dichotomous items. First, note that WLSMV was selected as the estimator, employing the probit model to estimate both the factor analytic (thresholds and loadings) and IRT (difficulty and discrimination) parameters. Second, note that the factor variance was fixed to 1 (f1@1) and the factor mean was fixed to zero as default. Also, note that Theta scaling was requested on the analysis line. Parameters based on Delta scaling are produced by requesting STDYX on the Output line (i.e., fixes the total variance of both the factor and y^* to 1). The three differently scaled solutions can be translated to each other, showing their equivalence.

Benefits and limitations of conducting IRT in the SEM framework. As shown, the characteristics of ordered categorical variables can be estimated using either an IRT or SEM framework. Although these two perspectives are simple transformations of each other, there are some advantages to using the SEM framework to conduct IRT investigations. For example, IRT models are typically applied to data assumed to represent one unidimensional construct, which stems from most IRT software programs exclusively estimating unidimensional structure. If the goal is to specify and test a correlated multi-factor model with simple structure (each item represents only one of the multiple correlated factors), any SEM software program could be employed, whereas only one easily available IRT software package could be employed (i.e., NOHARM; Fraser & McDonald, 1988).

Likewise, complex items, which reflect multiple sources of common variance, are quite easy to model in SEM. For example, in SEM relations among item uniqueness terms can be easily identified through correlation residuals or modification indices and this information can be used to incorporate uniqueness correlations into the model (Hill et al., 2007). In contrast, IRT models typically assume there is no relation among items after controlling for

```
Data: File is dichitem.dat;
variable: names are p1 - p9;
categorical are p1-p9;
analysis: estimator = WLSMV; parameterization is Theta;
model: f1 by p1 - p9*;
f1@1;
Output: STDYX; RESIDUAL;

MODEL RESULTS
```

Theta parameterization constrains the residual variance of $y*$ to 1.

The asterisk in the "model" statement was used to override the fixing of the first factor loading to 1 to set the metric of the factor. Instead, the variance of the factor was fixed to 1.0.

	Estimate	S.E.	Two-Tailed Est./S.E.	P-Value
F1 BY				
P1	1.142	0.076	15.091	0.000
P2	0.925	0.065	14.231	0.000
P3	1.398	0.098	14.327	0.000
P4	1.122	0.071	15.719	0.000
P5	0.402	0.046	8.829	0.000
P6	1.195	0.076	15.821	0.000
P7	1.229	0.078	15.690	0.000
P8	1.223	0.078	15.584	0.000
P9	2.003	0.200	10.012	0.000
Thresholds				
P1$1	0.430	0.053	8.170	0.000
P2$1	-0.491	0.048	-10.319	0.000
P3$1	0.748	0.067	11.229	0.000
P4$1	-0.054	0.049	-1.099	0.272
P5$1	0.599	0.038	15.638	0.000
P6$1	-0.120	0.051	-2.353	0.019
P7$1	-0.057	0.052	-1.099	0.272
P8$1	-0.196	0.052	-3.744	0.000
P9$1	1.343	0.132	10.213	0.000
Variances				
F1	1.000	0.000	999.000	999.000

These values represent the slope relating the probit of each item to the factor (F1).

Given the factor is standardized, the value 1.142 represents the number of units probit P1 changes for every 1 standard deviation increase in the factor.

These values can be considered the unstandardized factor loadings and can be difficult to interpret given $y*$ for each item doesn't have a total variance of 1.0.

These values represent the unstandardized probit thresholds. The values can be interpreted as the expected probit that the observed response (y) equals zero when the factor is zero. That is, each value represents a predicted normal score when the total variance is *not* 1 as it is when interpreting z-scores.

A value of .430 is the expected probit that P1 equals zero, when the factor is zero (zero represents the mean level of the factor when the factor is standardized as it was here).

These values, similar to the unstandardized probit slope, can be difficult to interpret because the total variance of $y*$ for each item does not equal 1.

The factor variance was fixed to 1.0 in this particular example.

Figure 11.4 Selected output of one-factor model fit to dichotomous data using Mplus 6.0. *(continued)*

(continued)

These values represent the standardized relation between $y*$ and the factor.

A value of .752 is the correlation between the underlying latent response variable P1 and the factor. These values correspond to the parameters produced if the Delta scaling (i.e., total variance of $y* = 1$) were employed with a standardized factor. One can convert the unstandardized probit factor loadings above produced by the Theta scaling to the commonly reported standardized factor loadings produced by the Delta scaling using the conversion formulas from Table 11.4.

For P1, $\lambda_D = \dfrac{\lambda_T}{\sqrt{V(y*)}} = \dfrac{1.142}{\sqrt{2.304}} = .752$ $\quad V(y*) = 1.142^2 + 1$

These values represent the standardized probit threshold (z-scores).

The value .283 is the z-score that corresponds to the proportion (.611) who selected 0 for P1. These thresholds can be interpreted as the level of $y*$ needed to move from responding 0 to responding 1 for each item. A value of .283 indicates that respondents with $y*$ values less than .283 standard deviations above the mean of $y*$ for P1 would select 0, whereas respondents with $y*$ values greater than .283 for P1 would select 1. These values equal the threshold values produced if Delta scaling were employed. One can convert the unstandardized thresholds reported above from the Theta scaling to z-scores using the conversion formulas from Table 11.4.

For P1, $\tau_D = \dfrac{\tau_T}{\sqrt{V(y*)}} = \dfrac{.430}{\sqrt{2.304}} = .283$

These values represent IRT Discrimination parameters on a probit metric. These values match the unstandardized probit slopes presented above. This aligns with the Theta scaling conversion formula: $a = \lambda_T$.

The standardized factor loadings from the Delta scaling can also be used to produce IRT discrimination value.

$a = \dfrac{\lambda_D}{\sqrt{\theta_D}} = \dfrac{.752}{\sqrt{.434}} = 1.14$ $\quad \theta_D = 1 - \lambda_D^2 = 1 - .752^2$

One can multiply these probit IRT discrimination values by 1.7 to acquire IRT discrimination values on a logistic metric.

STANDARDIZED MODEL RESULTS
STDYX Standardization

	Estimate	S.E.	Est./S.E.	Two-Tailed P-Value
F1 BY				
P1	0.752	0.022	34.779	0.000
P2	0.679	0.026	26.398	0.000
P3	0.813	0.019	42.320	0.000
P4	0.747	0.021	35.520	0.000
P5	0.373	0.036	10.257	0.000
P6	0.767	0.020	38.403	0.000
P7	0.776	0.020	39.384	0.000
P8	0.774	0.020	38.902	0.000
P9	0.895	0.018	50.188	0.000
Thresholds				
P1$1	0.283	0.033	8.516	0.000
P2$1	-0.361	0.034	-10.751	0.000
P3$1	0.435	0.034	12.824	0.000
P4$1	-0.036	0.033	-1.099	0.272
P5$1	0.555	0.035	16.012	0.000
P6$1	-0.077	0.033	-2.353	0.019
P7$1	-0.036	0.033	-1.098	0.272
P8$1	-0.124	0.033	-3.765	0.000
P9$1	0.600	0.035	17.134	0.000
Variances				
F1	1.000	0.000	999.000	999.000

IRT PARAMETERIZATION IN TWO-PARAMETER PROBIT METRIC
WHERE THE PROBIT IS DISCRIMINATION*(THETA - DIFFICULTY)
Item Discriminations

F1 BY				
P1	1.142	0.076	15.091	0.000
P2	0.925	0.065	14.231	0.000
P3	1.398	0.098	14.327	0.000
P4	1.122	0.071	15.719	0.000
P5	0.402	0.046	8.829	0.000
P6	1.195	0.076	15.821	0.000
P7	1.229	0.078	15.690	0.000
P8	1.223	0.078	15.584	0.000
P9	2.003	0.200	10.012	0.000

Figure 11.4 (continued) Selected output of one-factor model fit to dichotomous data using Mplus 6.0.

Item Difficulties

P1$1	0.376	0.046	8.226	0.000
P2$1	-0.532	0.054	-9.783	0.000
P3$1	0.535	0.044	12.142	0.000
P4$1	-0.048	0.044	-1.097	0.273
P5$1	1.488	0.175	8.519	0.000
P6$1	-0.101	0.043	-2.343	0.019
P7$1	-0.046	0.042	-1.097	0.273
P8$1	-0.160	0.043	-3.735	0.000
P9$1	0.671	0.042	16.103	0.000

Variances

F1	1.000	0.000	0.000	1.000

R-SQUARE

Observed Variable	Estimate	S.E.	Est./S.E.	Two-Tailed P-Value	Scale Factors
P1	0.566	0.033	17.390	0.000	0.659
P2	0.461	0.035	13.199	0.000	0.734
P3	0.661	0.031	21.160	0.000	0.582
P4	0.557	0.031	17.760	0.000	0.665
P5	0.139	0.027	5.129	0.000	0.928
P6	0.588	0.031	19.202	0.000	0.642
P7	0.602	0.031	19.692	0.000	0.631
P8	0.599	0.031	19.451	0.000	0.633
P9	0.801	0.032	25.094	0.000	0.447

Residuals for Covariances/Correlations/Residual Correlations

	P1	P2	P3	P4	P5	P6	P7	P8	P9
P1	0.109								
P2	0.015	0.007							
P3	0.003	0.026	0.103						
P4	-0.014	-0.064	0.064	-0.032					
P5	-0.014	-0.042	-0.046	-0.031	-0.028				
P6	-0.049	-0.070	-0.066	-0.077	0.021	0.074			
P7	-0.040	0.038	-0.049	-0.033	-0.098	-0.010	0.058		
P8	-0.015	-0.086	-0.020	-0.034	0.094	0.024	0.026	0.029	

These values represent IRT difficulty parameters.

For P1, $b = \dfrac{\tau_T}{\lambda_T} = \dfrac{.430}{1.142} = .376$ $b = \dfrac{\tau_D}{\lambda_D} = \dfrac{.283}{.752} = .376$

When the factor (theta or latent trait) equals .376, there is a 50% likelihood of endorsing P1 (selecting 1 versus 0). Thus, a respondent who is .376 standard deviations above the mean of the latent factor would have a 50% chance of selecting 1 for P1. Given the latent factor is normally distributed and the majority of the area under the curve falls below .37 (65%), the majority of respondents would select 0 for P1.

The R-Square Estimate values are simply the standardized factor loadings squared: P1 = $.752^2 = .566$.

The Scale Factors can be used to produce the total variance of y^*.

$$V(y^*) = \frac{1}{\sqrt{Scale\ Factor}^2} = \frac{1}{\sqrt{.659}^2} = 2.30$$

These tetrachoric correlation residuals (requested by "RESIDUAL" on the "Output" command line) can be used to identify local areas of model-data misfit. The ability to easily produce these correlation residuals to diagnose local model-data misfit, in addition to the ability to assess global model-data fit (e.g., RMSEA, CFI), is one of the benefits of modeling ordered categorical data using an SEM approach.

Correlations among the residuals can be modeled by using the "with" option on the "model" command line (e.g., "P1 with P2;" would allow the residuals of P1 and P2 to correlate).

Figure 11.4 (continued)   Selected output of one-factor model fit to dichotomous data using Mplus 6.0.

the latent trait (i.e., assumption of local independence). Thus, the ability to relax the assumption of local independence and test for relationships among item residuals is not readily available in the IRT framework (Brown, 2006). Such complex items are often removed or combined for IRT analyses (Hill et al., 2007).

Moreover, complex items can be modeled as indicators of multiple factors. All SEM software packages can model the effects of multiple correlated or uncorrelated (e.g., bi-factor model) factors on an item. Specialized IRT software does exist to conduct multidimensional IRT (e.g., NOHARM, Fraser & McDonald, 1988; TESTFACT, Bock et al., 2004); however, most are limited to dichotomous items and one is additionally limited to a bi-factor model and exploratory models (TESTFACT).

Another advantage of the SEM approach is the ability to include co-variates believed to predict the latent variable (e.g., Muthén, 1988). Such covariates may be continuous or categorical (e.g., grouping variables) in nature. In addition to modeling the relation between the covariate(s) and factor, the covariate(s) can relate to individual items allowing for a test of uniform differential item functioning (DIF). More generally, there is simply greater flexibility in the SEM framework to model relations between the latent variable under study and both predictors and outcomes (full structural equation models). These structural relations provide validity evidence for the scores representing the construct of interest (Glöckner-Rist & Hoijtink, 2003).

There are, of course, limitations to conducting IRT in the SEM framework. As previously mentioned, current SEM software cannot accommodate the guessing parameter to estimate a 3PL IRT model. Moreover, researchers often employ IRT analyses to examine the degree to which items discriminate at all levels of the construct under study (Edelen & Reeve, 2007). Thus, researchers examine item characteristic curves and their inverse, item information curves, to determine which items function optimally for persons at varying "ability" levels (Rodebaugh et al., 2004). Although M*plus* offers an option to plot characteristic curves and information curves (Muthén & Muthén, 1998–2010) for individual items as well as the entire measure, this information is not utilized as often within the SEM framework.

SUGGESTIONS FOR HANDLING NONNORMALITY AND ORDERED CATEGORICAL DATA

Recommendations

Because so much information is associated with issues surrounding non-normal and ordered categorical data, Table 11.5 summarizes our recom-

TABLE 11.5 Recommendations for Dealing with Nonnormal and Ordered Categorical Data

Type of Data	Suggestions	Caveats/Notes
Continuous Data		
1. Approximately Normally Distributed	• Use ML estimation	• The assumptions of ML are met and estimates should be unbiased, efficient, and consistent.
2. Moderately Nonnormal (skew < 2, kurtosis < 7)	• Use ML estimation; fairly robust to these conditions • Use S–B scaling to correct χ^2 and standard errors for even slight nonnormality	• Given the availability of S–B scaling methods in the software packages, one could employ and report findings from both ML estimation and S–B scaling method.
3. Severely Nonnormal (skew > 2, kurtosis > 7)	• Use S–B scaling • Use bootstrapping	• S–B correction works well, currently much easier to implement than the bootstrap, and tends to be more sensitive to model misspecification than the bootstrap
Ordered Categorical Data		
1. Number of Ordered Categories is 6 or more	• Treat data as continuous and use S–B scaling methods with ML estimation • Treat data as categorical and use robust DWLS estimator	• Parameter estimates from S–B scaling equal ML-based estimates implying that they will be attenuated to some extent. • Robust DWLS estimators will adjust the parameter estimates, standard errors, and fit indices for the categorical nature of the data. • One could employ and report findings from both S–B scaling method and robust DWLS estimation.
2. Number of Ordered Categories is 5 or less	• Treat data as categorical and use robust DWLS estimator	• Robust DWLS estimators will adjust the parameter estimates, standard errors, and fit indices for the categorical nature of the data.

mendations when analyzing data of this type. This table is not meant to trivialize the complex issues surrounding this topic; instead, it may be treated as a supplement to the information presented in this chapter. Also, as explained throughout the chapter, the type of estimation method or technique employed is closely tied to the degree of nonnormality and/or the crudeness of the categorization. Therefore, researchers need to recognize the type of data they are modeling, in terms of both metric and distribution, before selecting a technique.

In brief, ML has been shown to be fairly robust if continuous data are only slightly nonnormal; therefore, we recommended its use in this situation (e.g., Chou et al., 1991; Green et al, 1997). If data are continuous and nonnormally distributed, we recommend the use of either the S–B scaling method in SEM software or bootstrapping.[11] Given the availability of the S–B scaling method, the ease of its use, and the empirical studies showing promising results, we can easily understand why this method is popular.

When modeling ordered categorical data, the research seems to indicate that if there are a large number of ordered categories, the data could be treated as continuous in nature. Specifically, if the variables have five categories or more, treating the data as continuous in nature and employing ML estimation with the S–B scaling methods appears to work well and has been recommended in this situation (see Rhemtulla et al., 2012). On the other hand, if ordered categorical variables have fewer than five categories, we recommend employing CVM to address both the metric and distribution. We specifically recommend employing a robust DWLS estimator because, unlike WLS, it avoids inverting a large asymptotic covariance matrix and has exhibited promising results.

When it is unclear which estimation method may be most appropriate to model ordered categorical data, we echo the suggestion by Wirth and Edwards (2007) to fit the model using multiple estimation methods. For example, if one is modeling items with five ordered categories, a comparison of results (global and local model fit, parameter estimates, standard errors) estimated via ML estimation with S–B adjustments and estimated via robust DWLS estimation would provide insight into the trustworthiness of the normal theory results. If the results are discrepant, one should report the robust DWLS results, whereas if the results align, one could report the results from ML estimation with S–B adjustments, footnoting that the robust DWLS estimator led to equivalent conclusions.

Strategies Not Recommended

An obvious omission from the recommendations table is the WLS estimator. Given the requirements of the WLS estimator (e.g., large sample size) and the lack of sensitivity to model misspecification (e.g., Olsson et al., 2000), we cannot recommend the use of this estimator as a method to analyze nonnormal or ordered categorical data. As noted throughout the chapter, other techniques outperform this estimator and should be employed.

A second technique that we cannot recommend, though commonly used to construct "more normally distributed" data (e.g., Marsh, Craven, & Debus, 1991), involves parceling items together (e.g., sum or average a subset of items). For example, parceling items with opposite skew has been conducted in order for the resulting parcel to have a better approximation of a normal distribution. Following the same logic, the parceling of ordered categorical items has been conducted to achieve a more continuous normal distribution allowing for the use of NT methods. Although it is true that the parcel may have properties that better approximate the assumptions underlying NT estimators, we cannot recommend the uncritical use of this technique as a strategy to deal with nonnormal or categorical data, because it results in ambiguous findings. As detailed at length in other sources (Bandalos & Finney, 2001), parceling can obscure the true relations among the variables leading to biased parameters estimates and fit indices. Further, increased ease of implementing alternative estimation techniques may reduce the desire to create parcels (Bandalos, 2008).

DIRECTIONS FOR FUTURE RESEARCH AND CONCLUSIONS

The purpose of this chapter was to review commonly used techniques to model nonnormal and categorical data and summarize research investigating their utility. Much of the previous research involving nonnormal and/or categorical data was concerned with comparing the performance of different estimation techniques (e.g., ML, WLS) under various conditions such as model size, sample size, and the observed variable distribution characteristics. An appropriate question at this point is where we go from here with respect to researching the effects of modeling categorical and nonnormal data.

ML Estimation with Latent Correlation Input

An area which may show promise for future research is analyzing ordered categorical data by applying robust ML procedures (i.e., Satorra–Bentler adjustments) to latent correlation input (e.g., polychoric correlations). As discussed previously, treating ordered categorical data as continuous and employing ML estimation with a covariance matrix can result in underestimated parameter estimates and biased standard errors and χ^2 values. Thus, one potential solution is to use the raw data to compute latent correlations (as discussed above) and apply ML estimation to the polychoric correlation matrix. Although the use of latent correlation input with ML estimation can provide accurate parameter estimates, the standard errors and χ^2 values will be biased (Holgado-Tello, Chácon-Moscoso, Barbero-García, & Vila-Abad, 2008; Jöreskog et al., 2000; Rigdon & Ferguson, 1991). That is, the weight matrix employed when using ML estimation with polychorics is simply the model-implied polychoric correlation matrix at each iteration, which doesn't include the distributional information housed in the asymptotic covariance matrix of the polychorics. Given that the S–B correction adjusts ML standard errors and χ^2 values for bias, the S–B correction paired with latent correlation input and ML estimation could recognize the categorical nature of the data and potentially provide unbiased standard errors and χ^2 values. This S–B adjustment simply involves estimating the asymptotic covariance matrix of the polychorics from the raw data and using it to adjust the ML–based standard errors and χ^2.

A few recent studies have examined the effectiveness of robust ML with latent correlation input for CFA models. Generally, parameter estimates and factor correlations have been found to be accurate (Holgado–Tello et al., 2008; Lei, 2009; Yang–Wallentin et al., 2010) even when data were binary or ordered categorical data were non–symmetric. Robust ML–based standard errors of parameter estimates showed relatively little bias when samples sizes were larger than 250; however, as with robust DWLS–based methods, negative bias in standard errors was observed if sample sizes were small ($N < 250$), higher levels of item non–normality were present, or model misspecification was present (Lei, 2009). In terms of model–data fit, the Satorra–Bentler ML χ^2 was comparable to the robust DWLS χ^2; it was sensitive to model misspecification (Yang–Wallentin et al., 2010) and was found to produce Type I error rates that were generally acceptable (Lei, 2009). Although findings seem to be optimistic, robust ML exhibited more convergence problems than robust DWLS when modeling latent correlations (Lei, 2009; Yang–Wallentin et al., 2010). In summary, robust ML with latent correlation input may be a promising technique for analyzing ordered categorical data; however, it has not yet received much attention. Further study is necessary as it would provide applied researchers with a greater understanding of conditions under which this procedure may be appropriate.[12]

Interpretation of Fit Indices

The increased use of robust DWLS in substantive research is to be expected, given its superior performance compared to WLS. However, there still remain questions regarding how to interpret the magnitude of robust DWLS–based fit indices. In general, the interpretation of various fit indices with respect to data-model fit has received a great deal of attention (e.g., Hu & Bentler, 1999; Marsh, Hau, & Wen, 2004). Moreover, the generalizability of guidelines for fit assessment established using ML estimation has received some attention. Specifically, Yu and Muthén (2002) suggested similar guidelines for robust DWLS-based fit indices and ML-based fit indices (CFI > .94 and RMSEA < .06 indicate good model-data fit). Unlike Yu and Muthén, Nye and Drasgow (2011) concluded that the fit index guidelines for ML-based fit indices were not applicable to robust DWLS-based indices. In fact, they noted that rules of thumb cannot be established for robust DWLS-based fit indices. Given the limited study of robust DWLS fit indices and the interpretation of their values, more work is needed in this area.

Full-Information Estimators

Beginning with Mplus 3.0, a full-information ML estimator (called ML with numerical integration) could be employed to analyze categorical variables.[13] This estimator uses information from the full multi-way frequency table of all categorical variables, which is why it is referred to as a "full information" technique. This differs from the estimators discussed throughout the chapter, which are "limited-information" techniques because they use bivariate information, or two-way frequency tables between pairs of variables. The full-information ML estimator is the current standard in IRT. The availability of the full-information ML estimation in Mplus provides opportunities for new research in the area of analyzing categorical data (e.g., feasibility of estimator with correlated multifactor models and complex items, comparability of WLSMV-based versus full-information ML-based parameter estimates and standard errors). A recent study compared the Mplus 6.0 limited-information WLSMV estimator and full-information ML estimator (MLR) integrating over a normal distribution when modeling non-normal latent factor distributions (DeMars, 2012). When fitting a multidimensional two-parameter model, the correlation between the two factors was estimated well by both estimators for both the platykurtic and negatively skewed latent factor distributions. The WLSMV difficulty and discrimination estimates were more biased than the MLR estimates when the latent factor distribution was negatively skewed, with the bias being greatest for well-discriminating easy or difficult items. Standard errors were relatively similar across the two estimators. The results suggest that full-information estimators, such as MLR, may be preferred when modeling dichotomous items coupled with the belief that the latent factor distribution is non-normally distributed.

In closing, given the presence of nonnormal and ordered categorical data in applied research, researchers need to not only recognize the properties of their data but to also utilize techniques that accommodate these properties. Simply using the default estimator from a computer package does not guarantee valid results. Understanding the issues surrounding various techniques, such as assumptions, robustness, and implementation in software programs, makes a researcher much more competent to handle issues that they might encounter.

APPENDIX A
WLS Syntax

The following syntax illustrates how to employ the WLS estimator with continuous and ordered categorical data. The model being estimated is a two-factor model with six indicators per factor. Each observed variable serves as an indicator to only one factor, all error covariances are fixed at zero and the factor correlation is freely estimated. The metric of the factor is set by constraining the factor variance to a value of 1.00.

Continuous Data

LISREL 8.80

```
Title First use PRELIS to obtain covariance matrix and asymptotic
covariance matrix
DA NI=12
LA
q1 q2 q3 q4 q5 q6 q7 q8 q9 q10 q11 q12
RA=example.dat
CO ALL
OU MA=CM SM=example.cov AC=example.acc BT XM
```

SIMPLIS Command Language Employed Using LISREL 8.80

```
!Second read matrices into SIMPLIS program
Title Illustrating WLS estimation with continuous data
Observed Variables q1 q2 q3 q4 q5 q6 q7 q8 q9 q10 q11 q12
Covariance matrix from file example.cov
Asymptotic matrix from file example.acc
Sample size 1000
Latent variables: fact1 fact2
Relationships
q1 q2 q3 q4 q5 q6 = fact1
q7 q8 q9 q10 q11 q12 = fact2
Options: WLS
Path Diagram
End of Problem
```

EQS 6.1

```
/TITLE
 Illustrating WLS estimation with continuous data
/SPECIFICATIONS
 VARIABLES= 12; CASES= 1000; DATAFILE = 'example.ess';
 MATRIX= raw; METHOD = agls;
/EQUATIONS
 V1 = *F1 + E1;
 V2 = *F1 + E2;
```

```
 V3  =  *F1 +  E3;
 V4  =  *F1 +  E4;
 V5  =  *F1 +  E5;
 V6  =  *F1 +  E6;
 V7  =  *F2 +  E7;
 V8  =  *F2 +  E8;
 V9  =  *F2 +  E9;
 V10 =  *F2 +  E10;
 V11 =  *F2 +  E11;
 V12 =  *F2 +  E12;
/VARIANCES
 F1 to F2 = 1;
 E1 to E12 = *;
/COVARIANCES
 F2, F1 = *;
/END
```

Mplus 6.0

```
TITLE: Illustrating WLS estimation with continuous data
Data: FILE IS example.dat;
VARIABLE: NAMES ARE q1 - q12;
ANALYSIS: ESTIMATOR = WLS;
MODEL:  f1 by q1-q6*;
        f2 by q7-q12*;
        f1 @ 1;
        f2 @ 2;
Output: STDYX; RES;
```

Ordered Categorical Data

LISREL 8.80

```
Title PRELIS Producing latent correlation matrix and asymptotic
covariance matrix.
DA NI=12
LA
q1 q2 q3 q4 q5 q6 q7 q8 q9 q10 q11 q12
RA=cat.dat
OR ALL
OU MA=KM SM=poly.cm AC=catex.acc BT XM
```

SIMPLIS Command Language Employed Using LISREL 8.80

```
Title Analyzing ordered categorical data with WLS estimation
Observed Variables q1 q2 q3 q4 q5 q6 q7 q8 q9 q10 q11 q12
Correlation matrix from file poly.cm
Asymptotic matrix from file catex.acc
Sample size 1000
Latent variables: fact1 fact2
```

```
Relationships
q1 q2 q3 q4 q5 q6 = fact1
q7 q8 q9 q10 q11 q12 = fact2
Options: WLS
Path diagram
End of problem
```

EQS 6.1

```
/TITLE
 Analyzing ordered categorical data with WLS estimation
/SPECIFICATIONS
 VARIABLES= 12; CASES= 1000; DATAFILE = 'example.ess';
 MATRIX= RAW; METHOD = AGLS; ANALYSIS=CORRELATION;
 CATEGORY=V1, V2, V3, V4, V5, V6, V7, V8, V9, V10, V11, V12;
/EQUATIONS
 V1 = *F1 + E1;
 V2 = *F1 + E2;
 V3 = *F1 + E3;
 V4 = *F1 + E4;
 V5 = *F1 + E5;
 V6 = *F1 + E6;
 V7 = *F2 + E7;
 V8 = *F2 + E8;
 V9 = *F2 + E9;
 V10 = *F2 + E10;
 V11 = *F2 + E11;
 V12 = *F2 + E12;
/VARIANCES
 F1 to F2 = 1;
 E1 to E12 = *;
/COVARIANCES
 F2, F1 = *;
/END
```

Mplus 6.0

```
TITLE: Analyzing ordered categorical data with WLS estimation
DATA: FILE IS cat.dat;
VARIABLE: NAMES ARE q1-q12;
          CATEGORICAL ARE q1-q12;
ANALYSIS: ESTIMATOR = WLS;
MODEL:    f1 BY q1-q6*;
          f2 BY q7-q12*;
          f1 @1;
          f2 @1;
Output: STDYX; RES;
```

APPENDIX B
S–B Scaling Syntax

The following syntax illustrates how to employ the S–B scaling methodology with continuous data and ordered categorical data treated as continuous. The model being estimated is a two-factor model with six indicators per factor. Each observed variable serves as an indicator to only one factor, all error covariances are fixed at zero and the factor correlation is freely estimated. The metric of the factor is set by constraining the factor variance to a value of 1.00.

Continuous and Ordered Categorical Data

SIMPLIS Command Language Employed Using LISREL 8.80

```
Title Illustrating SB chisq and standard errors
Observed variables q1 q2 q3 q4 q5 q6 q7 q8 q9 q10 q11 q12
Covariance matrix from file example.cov
Asymptotic matrix from file example.acc
Sample size 1000
Latent variables: fact1 fact2
Relationships
q1 q2 q3 q4 q5 q6 = fact1
q7 q8 q9 q10 q11 q12 = fact2
Options: ML
Path diagram
End of problem
```

EQS 6.1

```
/TITLE
 Illustrating SB chisq and standard errors
/SPECIFICATIONS
 VARIABLES= 12; CASES= 1000; DATAFILE = 'example.ess';
 MATRIX= raw; METHOD = ml, robust;
/EQUATIONS
 V1 = *F1 + E1;
 V2 = *F1 + E2;
 V3 = *F1 + E3;
 V4 = *F1 + E4;
 V5 = *F1 + E5;
 V6 = *F1 + E6;
 V7 = *F2 + E7;
 V8 = *F2 + E8;
 V9 = *F2 + E9;
 V10 = *F2 + E10;
 V11 = *F2 + E11;
 V12 = *F2 + E12;
```

```
/VARIANCES
 F1 to F2 = 1;
 E1 to E12 = *;
/COVARIANCES
 F2, F1 = *;
/END
```

Mplus 6.0

```
TITLE: Illustrating SB chisq and standard errors
Data: FILE IS example.dat;
VARIABLE: NAMES ARE q1 - q12;
ANALYSIS: ESTIMATOR = MLM;
MODEL:    f1 by q1* q2* q3* q4* q5* q6*;
          f2 by q7* q8* q9* q10* q11* q12*;
          f1 @ 1;
          f2 @ 1;
Output: STDYX; RES;
```

APPENDIX C
Robust DWLS Syntax

The following syntax illustrates how to employ the robust DWLS estimation with ordered categorical data. The model being estimated is a two-factor model with six indicators per factor and the factor correlation is freely estimated. The metric of the factor is set by constraining the factor variance to a value of 1.00. In LISREL, one specifies the latent correlation matrix (e.g., "Correlation matrix from File poly.cm") and the asymptotic covariance matrix of the latent correlations (e.g., "Asymptotic matrix from File catex.acc") produced in PRELIS, in addition to specifying "DWLS" on the options command line. In Mplus, there are two robust DLWS estimation options: WLSM or WLSMV. Simply type "WLSM" or "WLSMV" on the Mplus command line to specify the desired estimation method.

SIMPLIS command language employed using LISREL 8.80

```
Title Analyzing ordered categorical data with robust DWLS
estimation
Observed Variables q1 q2 q3 q4 q5 q6 q7 q8 q9 q10 q11 q12
Correlation matrix from file poly.cm
Asymptotic matrix from file catex.acc
Sample size 1000
Latent variables: fact1 fact2
Relationships
q1 q2 q3 q4 q5 q6 = fact1
q7 q8 q9 q10 q11 q12 = fact2
Options: dwls
Path diagram
End of problem
```

Mplus 6.0

```
TITLE:     Analyzing ordered categorical data with robust DWLS
              estimation
DATA:      FILE IS cat.dat;
VARIABLE:  NAMES ARE q1-q12;
           CATEGORICAL ARE q1-q12;
ANALYSIS:  ESTIMATOR=WLSMV;
MODEL:     f1 by q1- q6*;
           f2 by q7- q12*;
           f1 @ 1;
           f2 @ 1;
Output:  STDYX; RES;
```

ACKNOWLEDGEMENT

We thank Deborah Bandalos, Christine DeMars, Craig Enders, David Flora, Gregory Hancock, Cameron McIntosh, Ralph Mueller, and Dena Pastor for their helpful comments on this chapter.

NOTES

1. Bootstrapping has been used to accommodate non-normal continuous data. This method will not be discussed in this chapter; however, it is discussed in detail in the previous version of this chapter published in the first edition of Hancock and Mueller (2006), as well as in Hancock and Liu (2012).
2. SEM can be used to model both mean and covariance structures. However, for simplicity, we mainly focus our descriptions on the covariance structure.
3. Technically, this gives the reweighted least squares fit function, which is asymptotically equivalent to ML's well known fit function, $F = \log|\Sigma(\theta)| + tr[\Sigma(\theta)^{-1}\mathbf{S}] - \log|\mathbf{S}| - p$.
4. Enders (2010) provides a didactic treatment of the computation of robust standard errors, emphasizing the importance of both the first and second derivatives of the sample log-likelihood function in adjusting the standard errors for deviations of the data from normality (pp. 140–145).
5. LISREL 8.80 applies the Satorra–Bentler mean-adjustment (Satorra & Bentler, 1988, formula 41) to LISREL's normal theory weighted least squares (NTWLS) χ^2, not the ML minimum fit function χ^2 (Jöreskog et al., 2000, p. 196); thus, all approximate fit indices involving the χ^2 in their computations (e.g., CFI) are based off the mean-adjusted NTWLS χ^2. Moreover, Bryant and Satorra (2012) outlined potential problems associated with nested model χ^2 difference testing using the NTWLS mean-adjusted χ^2 values.
6. Prior to Mplus 6.0, employing the mean- and variance-adjustment to the chi-square (e.g., MLMV and WLSMV) involved estimating the optimal degrees of freedom. This resulted in the model degrees of freedom not matching the

usual degrees of freedom. However, Mplus 6.0 employs a new variance adjustment that instead uses a shift parameter, resulting in the usual degrees of
freedom being unchanged. The scale and shift parameter are chosen so the
expected value of the test statistic equals the degrees of freedom of the model
and the variance of the test statistic equals twice the degrees of freedom of the
model when H_0 is true. This new adjustment has been found to work as well
as the prior adjustment and is now the default when employing MLMV and
WLSMV (Asparouhov & Muthén, 2010). One still has the option of requesting
the mean- and variance-adjustment employed in Mplus prior to version 6.0.

7. The standard linear measurement model specifies that a person's score is a
    function of the relation (b) between the variable (y^*) and the factor (F) plus
    error (E): $y^* = bF + E$.

8. When there is an exogenous variable (x) influencing the factor (e.g., in
    MIMIC models), the sample statistics ($\mathbf{r}$) that are modeled are the probit
    thresholds and probit regression coefficients that result from predicting each
    y variable from the x predictor, in addition to the probit residual correlations
    between the y variables after controlling for x. The asymptotic covariance ma-
    trix of these sample statistics is constructed and estimates of model param-
    eters, standard errors, and model fit information are computed using a WLS
    estimator (Muthén & Muthén, 1998-2010). LISREL has a similar procedure
    when modeling predictors of a latent factor that is represented by ordered
    categorical indicators (Jöreskog, 2005).

9. For a single-group analysis, the choice of scaling of y^* (Delta or Theta) does
    not matter—values can be transformed from one scaling to the other using
    conversion formulas. However, in a multiple-group testing situation, the The-
    ta parameterization may be preferred (Millsap & Yun-Tein, 2004; Muthén &
    Asparouhov, 2002).

10. Currently, EQS (version 6.1) does not offer robust DWLS estimation with
    categorical variables. The EQS User's Manual (Bentler, 2004) recommends
    applying the Unweighted Least Squares (ULS) estimator with a robust cor-
    rection when ordered categorical data are present, which produces both the
    ULS mean-adjusted and ULS mean- and variance-adjusted χ^2 values in ad-
    dition to corrected standard errors. When implementing EQS, this option
    can be specified by including "LS, robust" as the METHOD. Although con-
    ceptually similar to robust DWLS (i.e., analyze latent correlations with a least
    squares estimator and apply corrections to standard errors and χ^2), EQS is
    employing a different estimator (ULS) than LISREL and Mplus (DWLS).
    Therefore, results can differ across software packages.

11. The following website provides the formula for computing the chi-square differ-
    ence test when employing MLM or WLSM estimation: http://www.statmodel
    .com/chidiff.shtml

12. In order to conduct robust ML estimation with latent correlational input in
    LISREL 8.80 (e.g., polychorics), item residual variances must be constrained
    to equal 1 minus the squared factor loading (Brown, 2006; Jöreskog et al.,
    2000; Yang-Wallentin et al., 2010). That is, the residual variance of y^* is not
    estimated, but is instead fixed to equal the reminder of $1 - \lambda^2 \text{Var}(F)$ for each
    item. The researcher must specify these constraints in the syntax when apply-

ing ML estimation to latent correlations (see Yang-Wallentin et al., 2010 for syntax). Because robust DWLS is not available in EQS 6.1, this approach is a possible option when analyzing ordered categorical data using EQS. Unlike LISREL, EQS automatically constrains the residual variance of $y*$ as described above and EQS provides both the mean- and the mean-and variance-adjusted χ^2 values. Mplus 6.0 does not allow one to employ ML estimation when variables are specified as categorical.

13. Full-information maximum likelihood for ordered categorical data is available in LISREL 8.80 for only exploratory factor analytic models.

REFERENCES

Asparouhov, T., & Muthén, B. (2006). *Robust chi square difference testing with mean and variance adjusted test statistics.* Retrieved from http://www.statmodel.com/download/webnotes/webnote10.pdf

Asparouhov, T., & Muthén, B. (2010). *Simple second order chi-square correction.* Retrieved from http://www.statmodel.com/download/WLSMV_new_chi21.pdf

Austin, J. T., & Calderón, R. F. (1996). Theoretical and technical contributions to structural equation modeling: An updated annotated bibliography. *Structural Equation Modeling: A Multidisciplinary Journal, 3,* 105–175.

Babakus, E., Ferguson, C., E., & Jöreskog, K. G. (1987). The sensitivity of confirmatory maximum likelihood factor analysis to violations of measurement scale and distributional assumptions. *Journal of Marketing Research, 24,* 222–228.

Bandalos, D. L. (in press). Relative performance of categorical diagonally weighted least squares and robust maximum likelihood estimation. *Structural Equation Modeling: A Multidisciplinary Journal.*

Bandalos, D. L. (2008). Is parceling really necessary? A comparison of results from item parceling and categorical variable methodology. *Structural Equation Modeling: A Multidisciplinary Journal, 15,* 211–240.

Bandalos, D. L., & Finney, S. J. (2001). Item parceling issues in structural equation modeling. In G. A. Marcoulides & R. E. Schumacker (Eds.), *New developments and techniques in structural equation modeling* (pp. 269–296). Mahwah, NJ: Lawrence Erlbaum Associates.

Bandalos, D. L., & Webb, M. (2005, April). *Efficacy of the WLSMV estimator for coarsely categorized and nonnormally distributed data.* Paper presented at the annual meeting of the American Educational Research Association, Montreal, Canada.

Beauducel, A., & Herzberg, P. Y. (2006). On the performance of maximum likelihood versus mean and variance adjusted weighted least squares estimation in CFA. *Structural Equation Modeling: A Multidisciplinary Journal, 13,* 186–203

Bentler, P. M. (1985–2010). *EQS for Windows* (Version 6.1) [Computer software]. Encinco, CA: Multivariate Software, Inc.

Bentler, P. M. (1990). Comparative fit indexes in structural equation models. *Psychological Bulletin, 107,* 238–246.

Bentler, P. M. (2004). *EQS 6 Structural Equations Program Manual.* Encinco, CA: Multivariate Software, Inc.

Bentler, P. M., & Dudgeon, P. (1996). Covariance structure analysis: Statistical practice, theory, and directions. *Annual Review of Psychology, 47*, 563–592.

Bock, R. D., Gibbons, R., Schilling, S.G., Muraki, E., Wilson, D. T., & Woods, R. (2004). *Testfact, version 4*, Scientific Software International.

Bollen, K. A. (1989). *Structural equation modeling with latent variables*. New York, NY: Wiley.

Bollen, K. A., & Barb, K. H. (1981). Pearson's R and coarsely categorized measures. *American Sociological Review, 46*, 232–239.

Brown, T. A. (2006). *Confirmatory factor analysis for applied research*. New York, NY: Guilford Press.

Browne, M. W. (1984). Asymptotic distribution-free methods in the analysis of covariance structures. *British Journal of Mathematical and Statistical Psychology, 37*, 62–83.

Bryant, F. B., & Satorra, S. (2012). Principles and practice of scaled difference chi-square testing. *Structural Equation Modeling: A Multidisciplinary Journal,19*, 372–398

Chou, C., & Bentler, P. M. (1995). Estimates and tests in structural equation modeling. In R. H. Hoyle (Ed.), *Structural equation modeling: Concepts, issues, and applications* (pp. 37–55). Thousand Oaks, CA: Sage.

Chou, C., Bentler, P. M., & Satorra, A. (1991). Scaled test statistics and robust standard errors for non-normal data in covariance structure analysis: A Monte Carlo study. *British Journal of Mathematical and Statistical Psychology, 44*, 347–357.

Curran, P. J., West, S. G., & Finch, J. F. (1996). The robustness of test statistics to nonnormality and specification error in confirmatory factor analysis. *Psychological Methods, 1*, 16–29.

DeMars, C. E. (2012). A comparison of limited-information and full-information methods in Mplus for estimating Item Response Theory parameters for non-normal populations. *Structural Equation Modeling: A Multidisciplinary Journal, 19*, 610–632.

DiStefano, C. (2002). The impact of categorization with confirmatory factor analysis. *Structural Equation Modeling: A Multidisciplinary Journal, 9*, 327–346.

DiStefano, C. (2003, April). *Considering the number of categories and item saturation levels with structural equation modeling*. Paper presented at the annual conference of the American Educational Research Association, New Orleans, Louisiana.

DiStefano, C., & Morgan, G. B. (in press). A comparison of diagonal weighted least squares robust estimation techniques for ordinal data. *Structural Equation Modeling: A Multidisciplinary Journal*.

Dolan, C. V. (1994). Factor analysis of variables with 2, 3, 5, and 7 response categories: A comparison of categorical variable estimators using simulated data. *British Journal of Mathematical and Statistical Psychology, 47*, 309–326.

Edelen, M. O., & Reeve, B. B. (2007). Applying item response theory (IRT) modeling to questionnaire development, evaluation, and refinement. *Quality of Life Research, 16*, 5–18.

Enders, C. K (2010). *Applied Missing Data*. New York, NY: Guilford.

Finch, J. F., West, S. G., & MacKinnon, D. P. (1997). Effects of sample size and nonnormality on the estimation of mediated effects in latent variable models. *Structural Equation Modeling: A Multidisciplinary Journal, 4*, 87–107.

Flora, D. B., & Curran, P. J. (2004). An empirical evaluation of alternative methods of estimation for confirmatory factor analysis with ordinal data. *Psychological Methods, 9,* 466 – 491.

Forero, C. G., Maydeu-Olivares, A., & Gallardo-Pujol, D. (2009). Factor analysis with ordinal indicators: A Monte Carlo study comparing DWLS and ULS estimation. *Structural Equation Modeling: A Multidisciplinary Journal, 16,* 625–641.

Foss, T., Jöreskog, K. G., & Olsson, U. H. (2011). Testing structural equation models: The effect of kurtosis. *Computational Statistics and Data Analysis, 55,* 2263–2275.

Fouladi, R. T. (2000). Performance of modified test statistics in covariance and correlation structure analysis under conditions of multivariate nonnormality. *Structural Equation Modeling: A Multidisciplinary Journal, 7,* 356–410.

Fraser, C., & McDonald, R.P. (1988). NOHARM: Least squares item factor analysis. *Multivariate Behavioral Research, 23,* 267–269.

Gierl, M. J., & Mulvenon, S. (1995, April). *Evaluating the application of fit indices to structural equation models in educational research: A review of the literature from 1990 through 1994.* Paper presented at the annual meeting of the American Educational Research Association, San Francisco, CA.

Glöckner-Rist, A., & Hoijtink, H. (2003). The best of both worlds: Factor analysis of dichotomous data using item response theory and structural equation modeling. *Structural Equation Modeling: A Multidisciplinary Journal, 10,* 544–565.

Green, S. B., Akey, T. M., Fleming, K. K., Hershberger, S. L., & Marquis, J. G. (1997). Effect of the number of scale points on chi-square fit indices in confirmatory factor analysis. *Structural Equation Modeling: A Multidisciplinary Journal, 4,* 108–120.

Hancock, G. R., & Liu, M. (2012). Bootstrapping standard errors and data-model fit statistics. In R. Hoyle (Ed.), *Handbook of structural equation modeling* (pp. 296–306). New York, NY: Guilford Press.

Herzog, W., Boomsma, A., & Reinecke, S. (2007). The model-size effect on traditional and modified tests of covariance structure. *Structural Equation Modeling: A Multidisciplinary Journal, 14,* 361–390.

Hill, C. D., Edwards, M. C., Thissen, D., Langer, M. M., Wirth, R. J., Burwinkle, T. M., & Varni, J. W. (2007). Practical issues in the application of item response theory: A demonstration using items from the Pediatric Quality of Life Inventory (PedsQL) 4.0 Generic Core Scales. *Medical Care, 45,* 39–47.

Holgado-Tello, F. P., Chácon-Moscoso, S., Barbero-García, I., & Vila-Abad, E. (2008). Polychoric correlations in exploratory and confirmatory factor analysis of ordinal variables. *Quality & Quantity, 44,* 153–166.

Hoogland, J. J., & Boomsma, A. (1998). Robustness studies in covariance structure modeling: An overview and a meta-analysis. *Sociological Methods & Research, 26,* 329–367.

Hu, L., & Bentler, P. M. (1998). Fit indices in covariance structure modeling: Sensitivity to underparameterized model misspecification. *Psychological Methods, 3,* 424–453.

Hu, L., & Bentler, P. M. (1999). Cutoff criteria for fit indexes in covariance structure analysis: Conventional criteria versus new alternatives. *Structural Equation Modeling: A Multidisciplinary Journal, 6,* 1–55.

Hu, L., Bentler, P. M., & Kano, Y. (1992). Can test statistics in covariance structure analysis be trusted? *Psychological Bulletin, 112*, 351–362.

Hutchinson, S. R., & Olmos, A. (1998). Behavior of descriptive fit indexes in confirmatory factor analysis using ordered categorical data. *Structural Equation Modeling: A Multidisciplinary Journal, 5*, 344–364.

Jöreskog, K. (1994). On the estimation of polychoric correlations and their asymptotic covariance matrix, *Psychometrika, 59*, 381–389,

Jöreskog, K. (2005). *Structural equation modeling with ordinal variables using LISREL.* Available at http://www.ssicentral.com/lisrel/techdocs/ordinal.pdf

Jöreskog, K., & Sörbom, D. (1996). *PRELIS 2: User's reference guide.* Chicago: Scientific Software International.

Jöreskog, K., & Sörbom, D. (2007). *LISREL 8.80 for Windows [Computer Software].* Lincolnwood, IL: Scientific Software International, Inc.

Jöreskog, K., Sörbom, D., du Toit, S., & du Toit, M. (2000). *LISREL 8: New statistical features.* Chicago: Scientific Software International.

Kamata, A., & Bauer, D. J. (2008). A note on the relation between factor analytic and item response theory models. *Structural Equation Modeling: A Multidisciplinary Journal, 15*, 136–153.

Kaplan, D. (2009). *Structural equation modeling: Foundations and extensions* (2nd ed.). Thousand Oaks, CA: Sage.

Lei, P. W. (2009). Evaluating estimation methods for ordinal data in structural equation modeling. *Quality & Quantity, 43*, 495–507.

MacCallum, R. C., & Austin, J. T. (2000). Applications of structural equation modeling in psychological research. *Annual Review of Psychology, 51*, 201–226.

Marsh, H. W., Craven, R. G., & Debus, R. (1991). Self-concepts of young children 5 to 8 years of age: Measurement and multidimensional structure. *Journal of Educational Psychology, 83*, 377–392.

Marsh, H. W., Hau, K. T., & Wen, Z. (2004). In search of golden rules: Comment on hypothesis-testing approaches to setting cutoff values for fit indexes and dangers in overgeneralizing Hu and Bentler's (1999) findings. *Structural Equation Modeling: A Multidisciplinary Journal, 11*, 320–341.

Micceri, T. (1989). The unicorn, the normal curve, and other improbable creatures. *Psychological Bulletin, 105*, 156–166.

Millsap, R. E., & Yun-Tein, J. (2004). Assessing factorial invariance in ordered-categorical measures. *Multivariate Behavioral Research, 39*, 479–515.

Muthén, B. O. (1984). A general structural equation model with dichotomous, ordered categorical, and continuous latent variable indicators. *Psychometrika, 49*, 115–132.

Muthén, B. O. (1988). Some uses of structural equation modeling in validity studies: Extending IRT to external variables. In H. Wainer & H.Braun (Eds.), *Test validity* (pp. 213–238). Hillsdale, NJ: Erlbaum

Muthén, B. O. (1993). Goodness of fit with categorical and other nonnormal variables. In K. A. Bollen & J. S. Long (Eds.), *Testing structural equation models* (pp. 205–243). Newbury Park, NJ: Sage.

Muthén, B. O. (1998–2004). *Mplus technical appendices.* Los Angeles, CA: Muthén & Muthén. Retrieved from http://www.statmodel.com/download/techappen.pdf

Muthén, B. O., & Asparouhov, T. (2002). *Latent variable analysis with categorical outcomes: Multiple-group and growth modeling in Mplus.* Retrieved from http://www.statmodel.com/download/webnotes/CatMGLong.pdf.

Muthén, B. O., du Toit, S., & Spisic, D. (1997). *Robust inference using weighted least squares and quadratic estimating equations in latent variable modeling with categorical and continuous outcomes.* Retrieved from http://pages.gseis.ucla.edu/faculty/muthen/articles/Article_075.pdf

Muthén, B. O., & Kaplan, D. (1985). A comparison of some methodologies for the factor analysis of nonnormal Likert variables. *British Journal of Mathematical and Statistical Psychology, 38,* 171–189.

Muthén, B. O., & Kaplan, D. (1992). A comparison of some methodologies for the factor analysis of non-normal Likert variables: A note on the size of the model. *British Journal of Mathematical and Statistical Psychology, 45,* 19–30.

Muthén, L. K., & Muthén, B. O. (1998–2010). *Mplus user's guide* (6th ed). Los Angeles, CA: Muthén & Muthén.

Nevitt, J., & Hancock, G. R. (2000). Improving the root mean square error of approximation for nonnormal conditions in structural equation modeling. *Journal of Experimental Education, 68,* 251–268.

Nevitt, J., & Hancock, G. R. (2004). Evaluating small sample approaches for model test statistics in structural equation modeling. *Multivariate Behavioral Research, 39,* 439–478

Nye, C. D., & Drasgow, F. (2011). Assessing goodness of fit: Simple rules of thumb simply do not work. *Organizational Research Methods, 14,* 548–570.

Olsson, U. H. (1979). Maximum likelihood estimation of the polychoric correlation coefficient. *Psychometrika, 44,* 443–460.

Olsson, U. H., Foss, T., & Troye, S. V. (2003). Does the ADF fit function decrease when kurtosis increases? *British Journal of Mathematical and Statistical Psychology, 56,* 289–303.

Olsson, U.H., Foss, T., Troye, S. V., & Howell, R. D. (2000).The performance of ML, GLS, and WLS estimation in structural equation modeling under conditions of misspecification and nonnormality. *Structural Equation Modeling: A Multidisciplinary Journal, 7,* 557–595.

Olsson, U. H., Troye, S. V., & Howell, R. D. (1999). Theoretical fit and empirical fit: The performance of maximum likelihood versus generalized least squares estimation in structural equation models. *Multivariate Behavioral Research, 34,* 31–58.

Oranje, A. (2003, April). *Comparison of estimation methods in factor analysis with categorized variables: Applications to NAEP data.* Paper presented at the annual conference of the American Educational Research Association, Chicago, Illinois.

Potthast, M. J. (1993). Confirmatory factor analysis of ordered categorical variables with large models. *British Journal of Mathematical and Statistical Psychology, 46,* 273–286.

Rhemtulla, M., Brosseau-Liard, P., & Savalei, V. (2012). When can categorical variables be treated as continuous? A comparison of robust continuous and categorical SEM estimation methods under suboptimal conditions. *Psychological Methods, 17,* 354–373.

Rodebaugh., T. L., Woods, C. M., Thissen, D. M., Heimberg, R. G., Chambless, D. L., & Rapee, R. M. (2004). More information from fewer questions: The factor structure and item properties of the original and brief fear of negative evaluation scale. *Psychological Assessment, 16,* 169–181.

Rosenberg, M. (1989). *Society and the adolescent self-image.* Revised edition. Middletown, CT: Wesleyan University Press.

Rigdon, E. E., & Ferguson, C. E. (1991). The performance of the polychoric correlation coefficient and selected fitting functions in confirmatory factor analysis with ordinal data. *Journal of Marketing Research, 28,* 491–497.

Satorra, A., & Bentler, P. M. (1988). Scaling corrections for chi-square statistics in covariance structure analysis. *Proceedings of the Business and Economic Statistics Section of the American Statistical Association,* (pp. 308–313).

Satorra, A., & Bentler, P. M. (1994). Corrections to test statistics and standard errors in covariance structure analysis. In A. von Eye & C. C. Clogg (Eds.), *Latent variables analysis: Applications for developmental research* (pp. 399–419). Thousand Oaks, CA: Sage.

Satorra, A., & Bentler, P. M. (2001). A scaled difference chi-square test statistic for moment structure analysis. *Psychometrika, 66,* 507–514.

Satorra, A., & Bentler, P.M. (2010). Ensuring positiveness of the scaled difference chi-square test statistic. *Psychometrika, 75,* 243–248.

Takane, Y., & de Leeuw, J. (1987). On the relationship between item response theory and factor analysis of discretized variables. *Psychometrika, 52,* 393–408.

Tremblay, P. F., & Gardner, R. C. (1996).On the growth of structural equation modeling in psychological journals. *Structural Equation Modeling: A Multidisciplinary Journal, 3,* 93–104.

Ullman, J. (2006). Structural equation modeling: Reviewing the basics and moving forward. *Journal of Personality Assessment, 87,* 35–50.

West, S. G., Finch, J. F., & Curran, P. J. (1995). Structural equation models with nonnormal variables: Problems and remedies. In R. H. Hoyle (Ed.) *Structural equation modeling: Concepts, issues, and applications* (pp. 56–75). Thousand Oaks, CA: Sage Publications.

Wirth, R. J., & Edwards, M. C. (2007). Item factor analysis: Current approaches and future directions. *Psychological Methods, 12,* 58–79.

Yang-Wallentin, F., Joreskog, K. G., & Luo, H. (2010). Confirmatory factor analysis with ordinal variables with misspecified models. *Structural Equation Modeling: A Multidisciplinary Journal, 17,* 392–423.

Yu, C., & Muthén, B. (2002, April). *Evaluation of model fit indices for latent variable models with categorical and continuous outcomes.* Paper presented at the annual meeting of the American Educational Research Association, New Orleans.

Yuan, K-H., Bentler, P. M., & Zhang, W. (2005). The effect of skewness and kurtosis on mean and covariance structure analysis: The univariate case and its multivariate implication. *Sociological Methods & Research, 34,* 240–258.

CHAPTER 12

ANALYZING STRUCTURAL EQUATION MODELS WITH MISSING DATA

Craig K. Enders

Researchers have traditionally dealt with missing data by discarding incomplete cases or by filling in missing scores with a single set of replacement values (i.e., single imputation). These ad hoc methods have fallen out of favor in the methodological literature because of their tendency to produce biased parameter estimates (Wilkinson & Task Force on Statistical Inference, 1999). In the 1970s, statisticians developed the underpinnings of two state-of-the art approaches, maximum likelihood missing data handling and multiple imputation (Beale & Little, 1975; Dempster, Laird, & Rubin, 1977; Finkbeiner, 1979; Rubin, 1978, 1987). However, as late as the mid-1990s, a lack of software options made it difficult to implement these techniques. Muthén, Kaplan, and Hollis (1987) had earlier proposed a multiple group approach that produced maximum likelihood estimates with incomplete data, but their model required complicated parameter constraints that became unwieldy with more than a few missing data patterns. Fortunately, the last decade has seen an explosion in missing data handling technology. All of the major structural equation modeling programs now implement maxi-

Structural Equation Modeling: A Second Course (2nd ed.), pages 493–519
Copyright © 2013 by Information Age Publishing
All rights of reproduction in any form reserved.

mum likelihood solutions, and a few also offer multiple imputation facilities. The methodology has grown to the point where most complete-data analyses now have a straightforward missing-data analog.

The primary purpose of this chapter is to provide an overview of *maximum likelihood estimation* (often referred to as *direct maximum likelihood* and *full information maximum likelihood*, or FIML) and *multiple imputation*. The chapter begins with a brief overview of Rubin's (1976) theoretical framework, as his missing data mechanisms provide an explanation for why these approaches are superior to older ad hoc techniques such as deletion and single imputation. The remainder of the chapter is largely devoted to describing and illustrating the use of maximum likelihood and multiple imputation in the context of a structural equation modeling (SEM) analysis. Throughout the chapter, I address a number of practical issues that arise in a missing data analysis, including the use of auxiliary variables and corrective procedures for nonnormal data. The chapter is by no means comprehensive, but it should serve as a springboard for accessing more detailed treatments of the topic (Enders, 2010; Little & Rubin, 2002; Schafer, 1997).

ARTIFICIAL DATA EXAMPLE

Throughout the chapter, I use an artificial data set that is loosely based on a study by Preti, Incani, Comboni, Petretto, and Masala (2006). These authors examined the mediating role of body dissatisfaction on the relation between childhood sexual abuse and eating disorder risk. Figure 12.1 shows a path diagram of the structural equation model that I use in the subsequent analysis examples. Briefly, the data contain the responses from 500 college-aged women, 90 of whom reported a history of sexual abuse prior to the age of 12. A set of seven questionnaire items defines the Body Dissatisfaction construct, and six questions serve as indicators of the Eating Disorder Risk construct. Finally, a dummy variable denotes unwanted sexual abuse prior to the age of 12 (i.e., 0 = no abuse, 1 = abuse). Although the variable does not appear in the analysis model, the data set also contains body mass index (BMI) scores.

Table 12.1 gives the complete-data descriptive statistics for the two groups of women. As seen in the table, participants with a previous history of sexual abuse have higher levels of body dissatisfaction and higher levels of eating disorder risk, on average. The questionnaire items are nonnormal, by definition, because they use a 7-point response format, but the score distributions are also somewhat asymmetric. I incorporated this nuance in order to illustrate the use of corrective procedures for nonnormal data. Table 12.1 also shows the overall missing data rate for each variable. Data from the questionnaire items are missing for a variety of reasons. For

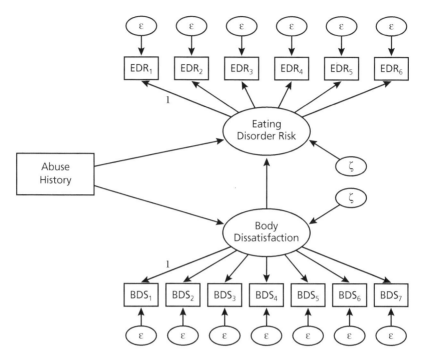

Figure 12.1 Structural equation model for the data analysis examples.

TABLE 12.1. Mean, Standard Deviation, Skewness, and Kurtosis Estimates by Abuse History

| Variable | No Abuse History (*n* = 410) | | | | Abuse History (*n* = 90) | | | | |
	M	*SD*	*S*	*K*	*M*	*SD*	*S*	*K*	Missing
BMI	22.423	2.803	−0.186	0.193	19.633	3.197	−0.161	0.364	8.0%
BDS_1	4.061	1.089	0.825	0.441	4.756	0.903	0.601	0.459	13.4%
BDS_2	4.027	1.009	0.821	0.722	4.722	1.006	0.655	0.127	12.2%
BDS_3	3.983	0.997	0.955	1.071	4.667	0.874	0.608	0.245	0.0%
BDS_4	4.034	1.101	0.750	0.272	4.733	0.934	0.731	0.464	12.4%
BDS_5	3.959	0.989	0.647	0.695	4.644	0.940	0.774	0.415	0.0%
BDS_6	3.946	1.010	0.724	0.723	4.689	1.013	0.530	0.572	0.0%
BDS_7	4.017	1.010	0.940	0.946	4.667	0.887	0.615	0.122	12.8%
EDR_1	4.012	0.932	0.760	0.913	4.778	1.014	0.661	0.054	0.0%
EDR_2	3.939	0.945	0.611	1.011	4.656	0.901	0.461	0.150	10.8%
EDR_3	4.017	0.983	0.883	0.617	4.644	1.053	0.465	0.116	12.0%
EDR_4	4.080	0.989	0.553	0.068	4.633	0.930	0.544	1.100	0.0%
EDR_5	3.968	0.965	0.704	0.804	4.678	0.981	0.545	−0.269	11.8%
EDR_6	4.056	0.958	0.423	0.435	4.656	0.889	0.349	−0.118	0.0%

some women, the propensity for missing data is related to higher body mass index values and previous sexual abuse history. For a small group of participants, the propensity for missing data is related to the would-be scores on the item that is missing (e.g., a woman that would have given a high score on a particular body dissatisfaction question has missing data on that item). Finally, participants with a past history of sexual abuse have a higher rate of missing data on the binary abuse variable, and BMI scores are missing from the upper tail of the BMI distribution. As seen in the table, the missing data rates for the incomplete variables ranged between 8% and 13.4%, and 10.4% of the abuse history scores were missing.

As an aside, researchers often ask whether there is a critical threshold for the missing data rate, below which it makes no difference which method you apply, and above which it is not appropriate to use maximum likelihood and multiple imputation. In truth, the "cause" of the missing data (i.e., the missing data mechanism) is more important than the amount of missing data. If the "missing at random" mechanism that I describe in the next section is tenable, then maximum likelihood and multiple imputation can yield consistent parameter estimates with missingness rates that are dramatically larger than those in the artificial data set. In contrast, if the missing at random mechanism does not hold, even a relatively small amount of missing data can introduce bias. Although it is impossible to generate useful rules of thumb, it seems safe to say that, relative to ad hoc missing data methods, the benefits of using maximum likelihood and multiple imputation (e.g., less bias, more power) increase as the missing data rate increases. Based on statistical theory and past research, there are virtually no situations where an ad hoc method would provide better performance.

THEORETICAL BACKGROUND

Rubin and colleagues (Little & Rubin, 2002; Rubin, 1976) developed a taxonomy of missing data mechanisms that explains why the choice of missing data procedure can impact structural equation model parameter estimates. In Rubin's theory, each participant has a latent probability of missing data on a variable Y, and this propensity for missingness might or might not be related to the variables in a particular analysis model. Depending on the nature of these relations, the researcher may need to estimate one or more parameters that describe the distribution of missingness, even though these parameters would not have been part of the analysis had the data been complete. From a technical perspective, Rubin's missing data mechanisms describe the conditions under which these nuisance parameters can be ignored. In practical terms, the mechanisms serve as assumptions for a missing data analysis.

To begin, the *missing completely at random* (MCAR) mechanism holds when the propensity for missing data on Y is unrelated to the observed scores of other variables and to the unobserved values of Y itself. Returning to the artificial data example, MCAR implies that a participant's likelihood of skipping one of the questionnaire items is unrelated to her scores on other variables (e.g., past history of sexual abuse) and to her would-be response to the item. MCAR is the most benign (and perhaps most unrealistic) of Rubin's mechanisms because it implies that the observed values in a data set are a random sample from the hypothetically complete data set. Consequently, basing an analysis on the complete cases (i.e., listwise or pairwise deletion) will produce unbiased parameter estimates because the observed data are no different from the missing data, on average.

The *missing at random* (MAR) condition is less restrictive in the sense that the propensity for missing data on Y can relate to the observed values of one or more other variables, but not to the would-be values of Y itself. Returning to the artificial data example, suppose that women with a past history of sexual abuse had a higher probability of skipping certain questionnaire items. This example would satisfy the MAR assumption, provided that the propensity for nonresponse is fully explained by abuse history (i.e., after controlling for abuse history, no residual relationship exists between missingness and the would-be scores on the questionnaire). The methodological literature often refers to the MAR mechanism as *ignorable missingness* because there is no need to explicitly model the propensity for missing data in the analysis. Rather, using an MAR-based missing data handling approach (i.e., maximum likelihood estimation or multiple imputation) to estimate the substantive model will produce consistent parameter estimates (i.e., the estimates get increasingly close to the population values as the sample size increases).

Finally, under the *not missing at random* (NMAR) mechanism, the propensity for missing data on Y can relate to the observed values of other variables as well as to the would-be values of Y itself. Returning to the example, suppose that women with a past history of abuse were more likely to skip the sexual history questions or participants with low body satisfaction were more likely to skip the items on that questionnaire. Both examples qualify as NMAR because the propensity for missingness on a variable is directly related to the would-be scores on that variable. An NMAR mechanism can also result from omitting a predictor of missingness from an analysis. For example, suppose that participants with high BMI values are more likely to skip certain questionnaire items. Technically, the analysis depicted in Figure 12.1 would qualify as NMAR because BMI is not included in the model. Unlike the first two examples, NMAR-by-omission is only capable of introducing bias when the partial correlation between the omitted variable (e.g., BMI) and the incomplete analysis variables (e.g., the questionnaire items) is rather strong (Collins, Schafer, & Kam, 2001).

Employing an inclusive analysis strategy that incorporates potential correlates of missingness (i.e., *auxiliary variables*) into the analysis can reduce or eliminate bias that results from NMAR-by-omission. However, the analysis becomes challenging when propensity for missing data on a variable is directly related to the would-be scores on that variable. In this situation, the only way to obtain unbiased estimates is to incorporate a submodel that explains the propensity for missing data (e.g., an additional logistic regression equation that uses one or more variables to predict a binary missing data indicator). Space limitations preclude a discussion of NMAR-based analysis procedures, but a number of resources are available to readers who are interested in learning about these models (Enders, 2010, 2011; Muthén, Asparouhov, Hunter, & Luechter, 2011).

Because NMAR models require assumptions that severely limit their utility (e.g., strict multivariate normality), some methodologists have argued that researchers should focus their efforts on MAR-based missing data handling techniques (Demirtas & Schafer, 2003; Enders & Gottschall, 2011b; Schafer & Graham, 2002). In line with this recommendation, maximum likelihood estimation and multiple imputation are the primary foci of this chapter. Although these methods work well when MAR holds, they can produce substantial bias under an NMAR mechanism. Unfortunately, there is no way to empirically test whether MAR is plausible for a given analysis because doing so would require knowledge of the missing scores (for the same reason, there is no way to verify that an NMAR mechanism is at play). Consequently, the accuracy of an MAR-based analysis ultimately relies on one or more untestable assumptions.

AD HOC MISSING DATA HANDLING TECHNIQUES

Researchers have traditionally relied on ad hoc techniques to deal with missing data. As I mentioned previously, these methods have long fallen out of favor in the methodological literature because of their tendency to produce biased parameter estimates. Because there is no longer any reason to use ad hoc missing data techniques, I provide a cursory discussion of these approaches. A number of resources are available for readers who want additional details (Enders, 2010; Little & Rubin, 2002; Schafer & Graham, 2002).

The most common ad hoc approach is to remove the incomplete cases, either completely (i.e., *listwise deletion*) or on an analysis-by-analysis basis (i.e., *pairwise deletion*). Although deletion methods are simple to implement, they require the stringent MCAR mechanism and can produce substantial bias when this assumption does not hold. Even if MCAR is plausible, eliminating data can dramatically reduce power. Because maximum likelihood estimation and multiple imputation retain the incomplete data records, they are generally superior with respect to power.

Single imputation techniques fill in missing scores with a single set of replacement values. These methods are diverse and include filling in the values with the arithmetic mean (e.g., mean imputation, averaging the available items), predicted scores from a regression equation (e.g., regression imputation, stochastic regression imputation), or the score values from another respondent (e.g., hot deck imputation, similar response pattern imputation). With few exceptions, single imputation methods tend to produce biased estimates, even under an MCAR mechanism. Bias aside, single imputation techniques are fundamentally flawed because they underestimate standard errors. At an intuitive level, imputing the data with guesses about the true score values should inflate standard errors. However, software programs treat the filled-in scores as real data and have no way of estimating this additional source of sampling error. As you will see, multiple imputation does not suffer from this problem because it provides a mechanism for estimating the missing-data sampling error.

MAXIMUM LIKELIHOOD ESTIMATION

Unlike conventional structural equation modeling analyses, maximum likelihood missing data handling uses a log likelihood function that requires raw data. Assuming a multivariate normal distribution in the population, the log likelihood for a single participant is

$$\log L_i = -\frac{k_i}{2}\log(2\pi) - \frac{1}{2}\log|\Sigma_i| - \frac{1}{2}(\mathbf{Y}_i - \mathbf{\mu}_i)^T \Sigma_i^{-1}(\mathbf{Y}_i - \mathbf{\mu}_i) \qquad (12.1)$$

where k_i is the number of observed variables for case i, $\mathbf{Y}_i$ is the individual's score vector, and are the population mean vector and covariance matrix, respectively (in a structural equation model such as that in Figure 12.1, $\mathbf{\mu}$ and Σ are the model-implied matrices). Equation 12.1 is the natural logarithm of the equation that defines the height of the multivariate normal distribution at a particular intersection of Y values. Consequently, the log likelihood value quantifies the relative probability of a participant's $\mathbf{Y}$ vector given a multivariate normal distribution with a particular model-implied mean vector and covariance matrix. The key component of the equation is the collection of terms known as Mahalanobis distance:

$$(\mathbf{Y}_i - \mathbf{\mu}_i)^T \Sigma_i^{-1}(\mathbf{Y}_i - \mathbf{\mu}_i). \qquad (12.2)$$

Mahalanobis distance is a squared z value that quantifies the standardized distance between a participant's data point and the model-implied means in μ, such that data points close to the center of the distribution (i.e., a small standardized distance) have a high log likelihood (i.e., a large relative probability).

With missing data, the sample log likelihood is the sum of the N individual log likelihood values from Equation 12.1. The goal of estimation is to identify the population parameter values that maximize the sample log likelihood (or equivalently, the population values that minimize the sum of the standardized distance values in Equation 12.2). Although it is not necessarily obvious, estimation does not require complete data. Returning to Equation 12.1, the i subscripts allow the size and the contents of the matrices to vary across participants with different missing data patterns. To illustrate, consider a simple analysis that estimates the mean vector and the covariance matrix for the first three body dissatisfaction items, BDS_1 to BDS_3. For participants with complete data, the log likelihood (i.e., relative probability) computation would utilize every parameter in and, as follows:

$$\log L_i = -\frac{3}{2}\log(2\pi) - \frac{1}{2}\log \begin{vmatrix} \hat{\sigma}^2_{BDS_1} & \hat{\sigma}_{BDS_{12}} & \hat{\sigma}_{BDS_{13}} \\ \hat{\sigma}_{BDS_{21}} & \hat{\sigma}^2_{BDS_2} & \hat{\sigma}_{BDS_{23}} \\ \hat{\sigma}_{BDS_{31}} & \hat{\sigma}_{BDS_{32}} & \hat{\sigma}^2_{BDS_3} \end{vmatrix}$$

$$-\frac{1}{2}\left(\begin{bmatrix} BDS_{1i} \\ BDS_{2i} \\ BDS_{3i} \end{bmatrix} - \begin{bmatrix} \hat{\mu}_{BDS_1} \\ \hat{\mu}_{BDS_2} \\ \hat{\mu}_{BDS_3} \end{bmatrix}\right)^T \begin{bmatrix} \hat{\sigma}^2_{BDS_1} & \hat{\sigma}_{BDS_{12}} & \hat{\sigma}_{BDS_{13}} \\ \hat{\sigma}_{BDS_{21}} & \hat{\sigma}^2_{BDS_2} & \hat{\sigma}_{BDS_{23}} \\ \hat{\sigma}_{BDS_{31}} & \hat{\sigma}_{BDS_{32}} & \hat{\sigma}^2_{BDS_3} \end{bmatrix}^{-1} \left(\begin{bmatrix} BDS_{1i} \\ BDS_{2i} \\ BDS_{3i} \end{bmatrix} - \begin{bmatrix} \hat{\mu}_{BDS_1} \\ \hat{\mu}_{BDS_2} \\ \hat{\mu}_{BDS_3} \end{bmatrix}\right).$$

For a participant who is missing BDS_3, the log likelihood computation would utilize only those parameter estimates that correspond to the observed data, as follows:

$$\log L_i = -\frac{2}{2}\log(2\pi) - \frac{1}{2}\log \begin{vmatrix} \hat{\sigma}^2_{BDS_1} & \hat{\sigma}_{BDS_{21}} \\ \hat{\sigma}_{BDS_{21}} & \hat{\sigma}^2_{BDS_2} \end{vmatrix}$$

$$-\frac{1}{2}\left(\begin{bmatrix} BDS_{1i} \\ BDS_{2i} \end{bmatrix} - \begin{bmatrix} \hat{\mu}_{BDS_1} \\ \hat{\mu}_{BDS_2} \end{bmatrix}\right)^T \begin{bmatrix} \hat{\sigma}^2_{BDS_1} & \hat{\sigma}_{BDS_{12}} \\ \hat{\sigma}_{BDS_{21}} & \hat{\sigma}^2_{BDS_2} \end{bmatrix}^{-1} \left(\begin{bmatrix} BDS_{1i} \\ BDS_{2i} \end{bmatrix} - \begin{bmatrix} \hat{\mu}_{BDS_1} \\ \hat{\mu}_{BDS_2} \end{bmatrix}\right).$$

The log likelihood computations would follow a similar logic for other configurations of missing data.

Although the estimation process does not impute the missing values, including the incomplete data vectors does imply probable values for the missing data. For example, in the previous equation, suppose that an individual had very low scores on the two body dissatisfaction items. Given the positive correlations among the questionnaire items, the multivariate normality assumption implies that the missing BDS_3 score should also fall in the lower tail of the BDS_3 distribution. Consequently, including the two observed scores in the estimation routine would implement a downward adjustment to the BDS_3 mean in order to account for the fact that the incomplete value is most likely a low score.

Standard Error Computations

Although maximum likelihood parameter estimates are consistent under an MAR mechanism, the corresponding standard errors may or may not be accurate. The second partial derivatives of the log likelihood function in Equation 12.1 largely determine maximum likelihood standard errors. Conceptually, these derivatives quantify the peakedness of the log likelihood function near its maximum, such that a large second derivative value (i.e., a steep log likelihood function) translates into a smaller standard error. With missing data, either the expected or the observed information matrix (i.e., −1 times the second derivative matrix) can generate standard errors; the former is termed "expected" because mathematical expectations replace certain quantities in the second derivative equations (e.g., a deviation score sum is replaced by its expected value of zero), whereas the latter is termed "observed" because the observed data solely determine the second derivative values. Kenward and Molenbergs (1998) showed that the expected information matrix yields standard errors that require the MCAR assumption, whereas the observed information matrix formulation assumes MAR. Under an MAR mechanism, the expected information matrix can produce standard errors that are much too small, even if the corresponding point estimates are accurate (Enders, 2010; Kenward & Molenberghs, 1998). Consequently, using the observed information to compute standard errors is almost always preferable. This is an important practical point because software programs offer different computational options. At the time of this writing, Mplus uses the observed information by default, whereas the user must specify this option in EQS and AMOS. LISREL computes standard errors based on the expected information.

Auxiliary Variables

I previously mentioned that the MAR assumption is defined relative to the variables in a particular analysis. For example, if the propensity for missing data on the questionnaire items is related to BMI scores, excluding BMI from the analysis would produce an NMAR mechanism. To improve the chances of satisfying MAR, the methodological literature generally recommends an inclusive analysis strategy that includes additional variables that (a) predict the propensity for missing data, or (b) correlate with the incomplete analysis variables (Collins et al., 2001; Graham, 2003; Rubin, 1996; Schafer & Graham, 2002). These so-called *auxiliary variables* can reduce bias (e.g., by making the MAR assumption more plausible) and can improve power (e.g., by allowing the incomplete analysis variable to borrow back lost information via its correlations with the auxiliary variables). Note that auxiliary variables can still benefit a maximum likelihood analysis, even if they are incomplete (Enders, 2008).

Graham (2003) proposed two methods for including auxiliary variables in a structural equation model analysis, the *extra dependent variable model* and the *saturated correlates model*. Because the saturated correlates model is implemented in some software packages, I limit the discussion to this approach. Graham outlined three formal rules for specifying the saturated correlates model: an auxiliary variable must correlate with (a) other auxiliary variables, (b) manifest predictor variables, and (c) residual terms of manifest endogenous variables (e.g., manifest dependent variables or manifest indicators of a latent construct). Importantly, these rules imply that the auxiliary variables should never correlate with latent variables or their disturbance terms.

To illustrate the saturated correlates model, suppose that a researcher formed scale scores from the body dissatisfaction and eating disorder risk questions and used these scale scores in a manifest variable path model. Figure 12.2 shows the saturated correlates model for this analysis, where

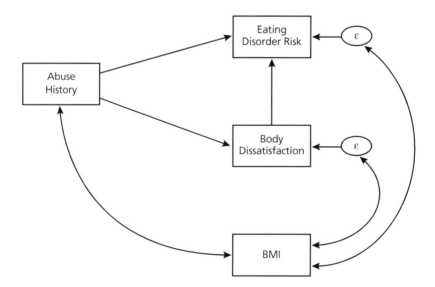

Figure 12.2 Saturated correlates model for a manifest variable regression. As an auxiliary variable, BMI correlates with the manifest exogenous variable and the residual terms of the manifest endogenous variables.

BMI is the sole auxiliary variable. Notice that the model specifies a correlation between BMI and sexual abuse history as well as correlations between BMI and the residual terms of the endogenous variables. Returning to the latent variable model in Figure 12.1, the saturated correlates model would introduce a correlation between BMI and abuse history and correlations between BMI and the 13 residual terms.

Including auxiliary variables in a structural equation analysis has no impact on the model fit statistic or its degrees of freedom because the saturated correlates model estimates all possible associations among the auxiliary variables and the analysis variables (i.e., the auxiliary variable portion of the model is saturated). By extension, the same conclusion holds for the RMSEA. However, the saturated correlates model does impact the SRMR and incremental fit indices such as the CFI. Auxiliary variables inappropriately reduce the magnitude of the SRMR because the standardized residuals for the auxiliary variable portion of the model are necessarily zero. Consequently, the SRMR calculation should exclude the residuals for all auxiliary variable parameters. In a similar vein, incremental fit indices require a special independence model. The appropriate null model should freely estimate the covariances among the auxiliary variables and the covariances between the auxiliary variables and other manifest variables because fixing these associations to zero would inappropriately penalize the null model, making the fit of the hypothesized model appear better than it should. At the time of this writing, Mplus and EQS implement auxiliary variable commands that automatically estimate the saturated correlates model, and both programs produce correct values for the SRMR and CFI. The program syntax in Appendix A demonstrates the Mplus auxiliary variable command, and Savalei and Bentler (2009, p. 485) described the corresponding EQS command.

Nonnormal Data

The maximum likelihood estimator that I described earlier assumes that endogenous manifest variables follow a multivariate normal distribution. A considerable body of literature has documented the impact of nonnormality on structural equation modeling analyses. This research suggests that parameter estimates are generally accurate, but standard error estimates and model fit statistics are distorted, with kurtosis largely determining the direction and magnitude of the bias (Yuan, Bentler, & Zhang, 2005). The prescription for dealing with normality violations is largely the same with or without missing data. Bootstrap and sandwich estimator (i.e., robust) standard errors (Arminger & Sobel, 1990; Yuan & Bentler, 2000) are available for missing data analyses, as are rescaled test statistics (Yuan & Bentler,

2000) and the bootstrap (i.e., Bollen-Stine) likelihood ratio test (Enders, 2002; Savalei & Yuan, 2009). The chapter by Finney and DiStefano in this volume gives additional details on these corrective procedures. At the time of this writing, both Mplus and EQS offer a variety of options for dealing with nonnormal missing data.

Incomplete Explanatory Variables

Maximum likelihood missing data handling is advantageous because it utilizes all of the available data—at least most of the time. Because the **Y** vector in Equation 12.1 does not necessarily include manifest variables, software packages might exclude cases with missing data on these variables, software packages generally exclude cases with missing data on these variables. In this situation, specifying the manifest predictor as a single-indicator latent variable is an easy fix because it effectively redefines the explanatory variable as an outcome without affecting its exogenous status in the model. This specification is the default in most SEM packages, but not in in Mplus. However, implementing this setup is quite easy in Mplus. Specifying the variance of the manifest predictor in the MODEL command (e.g., by listing the variable name) automatically recasts the variable as a single-indicator latent construct. However, when implementing this approach, it is very important to explicitly specify all possible covariances among the predictor variables. The program normally estimates these parameters by default, but defining a manifest predictor as a latent variable indicator eliminates these associations from the model. Failure to specify these parameters will produce incorrect parameter estimates. Enders (2010, pp. 116–118) gave additional details on this procedure, and Enders, Baraldi, and Cham (2012) outlined a similar approach for models with interaction terms.

Two-Stage Estimation

A brief discussion of the two-stage estimation strategy is warranted before illustrating maximum likelihood estimation (Enders, 2003; Savalei & Bentler, 2009; Yuan & Bentler, 2000). As its name implies, the two-stage approach estimates a model in two steps. The first step uses maximum likelihood missing data handling (i.e., the log likelihood in Equation 12.1) to estimate the mean vector and the covariance matrix (i.e., saturated model parameters). In the second stage, these summary statistics serve as input data for the main SEM analysis. The second step is straightforward because it uses the standard maximum likelihood fitting function for complete data.

With missing data, using a covariance matrix as input data is potentially problematic because no single value of N is applicable to the entire matrix; at an intuitive level, the elements with little or no missing data should have a larger N than the elements with missing data. The choice of N is vitally important, given its role in determining standard errors and model fit. Yuan and Bentler (2000) and Savalei and Bentler (2009) described a corrective procedure that uses the saturated model's information matrix to adjust the standard errors from the second stage. This correction is convenient because the standard error computations simply use the full N without regard to the missing data rate.

The advantage of the two-stage approach over direct maximum likelihood estimation primarily arises in models with incomplete explanatory variables or a large number of auxiliary variables. Because the first stage estimates a saturated covariance matrix, a variable's subsequent role in the structural equation model is irrelevant. Therefore, the two-stage approach eliminates the need to recast missing explanatory variables as a single-indicator latent variable. The two-stage estimator is also advantageous when working with auxiliary variables. Personal experience suggests that the saturated correlates model may fail to converge with even a modest number of auxiliary variables. In theory, the two-stage approach should be able to accommodate a much larger set of additional variables because the initial stage estimates a model with no restrictions, and the second stage uses only the covariance matrix elements that are relevant for estimating the hypothesized model.

Despite its potential advantages, the two-stage approach has limitations. For example, Savalei and Bentler's (2009) simulation results showed that the estimator was generally less powerful than direct maximum likelihood, particularly under an MAR mechanism. The two-stage approach did hold a slight advantage in small samples, but it is unclear whether this result generalizes to nonnormal data. Additional research is needed to clarify the advantages and disadvantages of the two-stage approach. Note that the estimator is currently an experimental feature in EQS 6.1 (see Savalei & Bentler, p. 485, for the necessary EQS commands).

Example Analysis

To illustrate maximum likelihood estimation, I used the artificial data set to estimate the model in Figure 12.1. In order to include the cases with missing abuse history scores, I recast this variable as a single-indicator latent variable. In addition, because the questionnaire items are not multivariate normally distributed, I used robust standard errors and rescaled fit statistics (e.g., the Satorra-Bentler chi-square test). As an alternative, I could have used bootstrap resampling to correct for the normality violations, but the bootstrap tends to yield very similar results as rescaling, particularly with larger samples such as this. Finally, I estimated the model with and without BMI as an auxiliary variable. The raw data for the analyses are available at http://www.appliedmissingdata.com/papers, and Appendix A gives the Mplus syntax for the auxiliary variable model. Note that the ESTIMATOR = MLR invokes robust standard errors and rescaled fit statistics, and the AUXILIARY command implements the saturated correlates model.

Table 12.2 gives the estimates and standard errors for selected parameters. To begin, notice that the saturated correlates model produced results that were similar to the Figure 12.1 model that omitted BMI. Because sampling variances (i.e., squared standard errors) are inversely related to the sample size, the ratio of the squared standard errors from the two analyses expresses the impact of the auxiliary variables as a proportion of the total sample size. In this example, the auxiliary variable reduced the standard errors for the abuse history regressions by an amount that was commensurate with a 2.3% increase in the total sample size. The modest improvement owes to the fact that BMI had rather weak correlations with the analysis model variables (i.e., rs of .10 to .20). In general, the benefit of including an auxiliary variable only becomes evident when these correlations exceed ±.40 (Collins et al., 2001). Although including a variable with low correlations is not problematic (Collins et al., 2001; Savalei & Bentler, 2009), identifying a small set of auxiliary variables that maximizes the shared variance with the analysis variables is a good strategy.

It is important to note that the substantive interpretations from a saturated correlates analysis are identical to those from a complete-data analysis

TABLE 12.2   Selected Maximum Likelihood and Multiple Imputation Parameter Estimates

Parameter	ML no Auxiliary Variable			ML with Auxiliary Variable			MI		
	Est.	SE	Std. Est.	Est.	SE	Std. Est.	Est.	SE	Std. Est.
Abuse → Dissatisfaction	0.710	0.103	0.356	0.705	0.102	0.354	0.687	0.102	0.355
Abuse → Risk	0.136	0.087	0.069	0.153	0.086	0.078	0.139	0.088	0.073
Dissatisfaction → Risk	0.702	0.076	0.712	0.698	0.076	0.709	0.698	0.074	0.709
Dissatisfaction Loading 2	1.028	0.094	0.692	1.026	0.094	0.692	1.030	0.089	0.694
Dissatisfaction Loading 4	1.182	0.099	0.750	1.184	0.099	0.751	1.179	0.097	0.749
Dissatisfaction Loading 7	1.169	0.101	0.784	1.169	0.101	0.785	1.172	0.095	0.787
Risk Loading 2	0.885	0.080	0.618	0.884	0.080	0.618	0.881	0.079	0.616
Risk Loading 3	1.005	0.082	0.670	1.007	0.082	0.671	1.003	0.080	0.668
Risk Loading 5	1.042	0.076	0.724	1.042	0.076	0.725	1.039	0.076	0.724

because the model uses correlations rather than regressions (i.e., two-headed arrows rather than single-headed arrows) to integrate the auxiliary variables. For example, the path connecting abuse history to body dissatisfaction represents the mean difference between the two groups, such that women who were sexually abused before the age of 12 scored significantly higher than women with no abuse history. In fact, the only noticeable difference with the missing data analysis is that the computer output will include estimates of the measurement intercepts and latent means, even when these parameters are not of substantive interest (the use of raw data in the Equation 12.1 log likelihood function necessitates a mean structure).

Turning to model fit, the rescaled test statistic from the saturated correlates analysis suggested adequate fit, (75) = 88.695, p = .133, as did other indices: RMSEA = .019, SRMR = .027, CFI = .994. It is important to reiterate that the inclusion of auxiliary variables should not materially affect the likelihood ratio test and the RMSEA because the saturated correlates model has the same degrees of freedom as the standard model. However, recall that auxiliary variables distort the SRMR and the CFI, making the fit the hypothesized model appear better than it should. Fortunately, using the AUXILIARY command to implement the saturated correlates model corrects this problem.

MULTIPLE IMPUTATION

Unlike maximum likelihood estimation, multiple imputation fills in the missing data with plausible replacement values. Multiple imputation is a three-step process that includes an *imputation phase*, an *analysis phase*, and a *pooling phase*. The imputation phase creates multiple copies of the data set, each of which contains different imputed values. Traditionally, methodologists recommended three to five data sets, but more recent research suggests a minimum of 20 (Graham, Olchowski & Gilreath, 2007). Having filled in the data, the researcher then estimates a structural equation model on each complete data set. Finally, the pooling phase combines the collection of estimates and standard errors from the analysis phase into a single set of values. Although this process sounds tedious, a number of software packages automate the analysis and pooling phases.

The imputation phase uses a regression-based procedure to fill in the missing values. Traditionally, researchers have employed imputation algorithms that assume a joint distribution for all variables. For example, Schafer's (1997) data augmentation algorithm is arguably the most common approach for multivariate normal data. Data augmentation is a two-step iterative algorithm that cycles between an imputation step (I-step) and a posterior step (P-step). The I-step uses an estimate of the mean vector and the covariance matrix to construct regression equations that predict the incomplete variables from the complete variables. To preserve the variability of the filled-in data, the algorithm augments the predicted values from these regression equations with normally distributed residual terms. The sum of a predicted score and random residual replaces each missing data point.

Generating unique imputations in each data set requires alternate estimates of the imputation regression coefficients (or equivalently, alternate estimates of the mean vector and covariance matrix that produce the regressions). The purpose of the P-step is to generate these new parameter values. The computational details of the P-step rely on Bayesian estimation principles, but the basic idea behind the P-step is conceptually simple: use the filled-in data from the I-step to update the mean vector and covariance matrix, then obtain new parameter values by adding a random residual term to each element in μ and Σ. More accurately, data augmentation uses Monte Carlo computer simulation to generate a new estimate of the mean vector and the covariance matrix from their respective posterior distributions. Conceptually, a posterior distribution is the Bayesian analog of a frequentist sampling distribution (e.g., the mean has a normally distributed posterior, the height of which describes the relative probability of different values for μ). Returning to the model in Figure 12.1, the P-step would "draw" a new set of means by generating 14 random numbers from a multivariate normal distribution, the center and spread of which depend on the filled-in data from the preceding I-step (e.g., the distribution is centered at the sample means from the filled-in data). The algorithm uses a similar procedure to sample a new covariance matrix from its posterior (i.e., sampling) distribution. At the next iteration, data augmentation uses the updated values of μ and Σ from the previous P-step to generate a new set of imputation regression equations, and the two-step process begins anew. The algorithm repeatedly cycles between the I-step and the P-step, often for thousands of iterations.

In the analysis phase, the researcher estimates a structural equation model on each filled-in data set. For example, in the subsequent analysis example, I fit the model in Figure 12.1 to each of the 100 data sets that I created in the imputation phase. This step produced 100 sets of parameter estimates and standard errors. Again, this step sounds tedious, but software packages that offer imputation facilities automate the model fitting process.

The final step of a multiple imputation analysis is to apply Rubin's (1987) rules for combining estimates and standard errors into a single set of values. For any parameter of interest, the pooled estimate is just the arithmetic average of the m estimates from the analysis phase. The pooling rules for standard errors are more complicated because they incorporate two sources of sampling variance. The within-imputation variance is the arithmetic average of the squared standard errors (i.e., the average sampling variance)

$$W = \frac{1}{m} \sum_{t=1}^{m} SE_t^2 \tag{12.3}$$

where the *t* subscript indexes the filled-in data sets. Because filled-in data sets generate the standard errors, the within-imputation variance estimates the sampling error that would have been obtained, had the data been complete.

The between-imputation variance quantifies the amount by which the missing data inflate standard errors. Computationally, the between-imputation variance uses the sample variance formula to compute the fluctuation of an estimate across the *m* imputed data sets

$$B = \frac{1}{m-1} \sum_{t=1}^{m} (\hat{\theta}_t - \bar{\theta})^2 \tag{12.4}$$

where is the estimate from data set *t* and is the pooled (i.e., average) estimate. The fact that the filled-in data sets contain different imputations is the sole reason why an estimate would exhibit non-zero between-imputation variation. Consequently, Equation 12.4 quantifies the additional uncertainty in an estimate that results from missing data.

Multiple imputation standard errors combine the within- and between-imputation variance into a single estimate of sampling error, as follows:

$$SE = \sqrt{W + B + B / m} \ . \tag{12.5}$$

Consistent with maximum likelihood estimation or ordinary least squares, this standard error serves as the denominator of a single-parameter test statistic

$$t = \frac{\bar{\theta} - \theta_0}{SE} \tag{12.6}$$

where $\bar{\theta}$ is the pooled estimate and θ_0 is some hypothesized value, typically zero. Rubin and colleagues (Barnard & Rubin, 1999; Rubin, 1987) derived complex degrees of freedom expressions for the *t* statistic, although some software packages (e.g., Mplus) use a standard normal distribution to generate a probability value.

Practical Issues

Several practical issues that warrant discussion arise during the imputation phase. For example, choosing an appropriate set of variables is a vital step in the process that dictates the accuracy and power of the subsequent analyses. At a minimum, the imputation phase should incorporate every variable in the structural equation model, but it should also include auxiliary variables. Unlike maximum likelihood estimation, the subsequent analysis need not include the auxiliary variables because the imputation process embeds the additional information into the imputed values. Although less obvious, the imputation process should also preserve the structure of the data. For example, if a researcher is interested in estimating a multiple group structural equation model, imputing the data separately for each subpopulation is the only way to adequately preserve group differences in the mean or the covariance structure (Enders & Gottschall, 2011a). On a related point, standard imputation algorithms are inappropriate for models with hidden subpopulations (e.g., latent class models, growth mixture models, factor mixture models) because there is no way to preserve class-specific parameters during the imputation process.

Recall from the previous section that the data augmentation algorithm assumes that the incomplete variables share a common joint distribution. Although this assumption is often sensible, it is not always the best choice. As an example, the model in Figure 12.1 uses a binary indicator of sexual abuse history. Had this variable been complete, it might have been reasonable to assume a multivariate normal distribution for the incomplete questionnaire items, despite the 7-point measurement scale. However, applying normality-based imputation to the incomplete binary variable is problematic because the algorithm would produce fractional imputations (e.g., a filled-in abuse history score of .673). Fortunately, methodologists have developed flexible imputation algorithms that can accommodate a variety of distribution shapes. For example, the *chained equations approach* (also known as *fully conditional specification* and *sequential regression imputation*) tailors the imputation model to each variable's measurement scale (Raghunathan, Lepkowski, Van Hoewyk, & Solenberger, 2001; Royston, 2005). As applied to the model in Figure 12.1, the chained equations algorithm would use logistic regression to impute the incomplete binary variable and would use linear regression to impute the questionnaire items. Mplus uses a comparable procedure, whereby incomplete binary and ordered categorical variables are imputed from a model that assumes an underlying continuous latent variable, such that discrete values arise from a series of thresholds in the latent variable distribution. The chapter by Finney and DiStefano in this volume gives additional details on this categorical variable model.

Finally, regardless of the algorithm employed, it is important to generate independent sets of imputed values. Markov Chain Monte Carlo algorithms (this includes data augmentation and the chained equations approach) introduce serial dependencies that can last for dozens and possibly hundreds of computational cycles. Sampling data sets at regular intervals in the data augmentation chain is one way to produce independent imputations. For instance, in the subsequent data analysis example, I generated 100 filled-in data sets. To do so, I allowed the MCMC algorithm to cycle for 60,000 iterations and saved a data set at every 600th imputation step (i.e., I specified 600 between-imputation or thinning iterations). In general, there is no harm in specifying an interval that is too large, but choosing an interval that is too small will attenuate the between-imputation component of the standard formula from Equation 12.5. Graphical and numerical diagnostic techniques can help guide the choice of between-imputation interval (e.g., see Enders, 2010; Schafer, 1997; Schafer & Olsen, 1998).

Example Analysis

At the time of this writing, Mplus 6 is the only structural equation modeling software package that implements all three phases of multiple imputation; LISREL and AMOS can generate multiply imputed data sets, but neither package automates the analysis and pooling phases, and EQS currently has no multiple imputation facilities. Consequently, I focus solely on Mplus for this example. The raw data for the analyses are available at http://www.appliedmissingdata.com/papers, and Appendices B and C give Mplus syntax files for the imputation and analysis phases, respectively.

Mplus offers a flexible set of imputation options. As I described in the previous section, the program can generate imputations for normally distributed variables as well as binary and ordered categorical variables. In addition, Mplus can use a saturated model for imputation (this is the standard approach that I described early in the chapter), or it can generate imputations that are consistent with a restricted model (e.g., it can impute the data with values that are consistent with the model in Figure 12.1). Although the latter option should produce slightly smaller standard errors, it carries the additional assumption that the hypothesized model is correct in the population. As a general rule, basing the imputations on a saturated model is usually a good strategy, particularly given that model-based imputation tends to produce negligible reductions in the standard errors. However, model-based imputation can be a useful strategy for reducing the complexity of a large imputation problem with many incomplete variables (e.g., a data set comprised of several multi-item questionnaires with item-level missingness).

I used a saturated imputation model to generate 100 filled-in data sets (a minimum of 20 is a rough rule of thumb; Graham et al., 2007) by saving a data set at every 600th imputation step in the MCMC algorithm. I used a numerical convergence diagnostic called the *potential scale reduction factor* (Gelman, Carlin, Stern, & Rubin, 1995) to determine the between-imputation interval, but a number of graphical diagnostic procedures are available for the same purpose (e.g., see Enders, 2010; Schafer, 1997; Schafer &

Olsen, 1998). Further, I used a categorical variable model to impute the binary abuse history variable and a multivariate normal model to impute the remaining variables, including the discrete questionnaire items. Although I could have used a threshold model to impute the questionnaire items, doing so dramatically increases the computational burden. Note that the multivariate normal model yields fractional imputations for discrete variables. Because recent research has recommended against rounding imputed values, I used the unrounded values for the subsequent analyses. Finally, to maintain consistency with the previous maximum likelihood analysis, I used BMI as an auxiliary variable during the imputation phase, and I implemented robust standard errors in the analysis phase.

Table 12.2 gives the estimates and standard errors for selected parameters. To begin, notice that maximum likelihood estimation and multiple imputation produced very similar results. This is not a surprise because the two procedures generally produce comparable estimates and standard errors, provided that the maximum likelihood analysis includes the same set of variables as the imputation phase of multiple imputation (Collins et al., 2001; Schafer, 2003). Consistent with maximum likelihood estimation, the substantive interpretations from the multiple imputation analysis are identical to those from a complete-data analysis. For example, the path connecting abuse history to body dissatisfaction represents the mean difference between the two groups, such that women who were sexually abused before the age of 12 scored statistically significantly higher than women with no abuse history.

Fit assessment is a key activity in most structural equation model analyses, yet this topic has received little attention in the multiple imputation literature. The basic logic is the same as that of the pooling phase: assess model fit separately in each data set and combine the fit statistics into a single value. Meng and Rubin (1992) outlined a procedure for pooling likelihood ratio tests that is applicable to the chi-square test of model fit in a structural equation analysis (some authors refer to this test as the D_3 statistic). In the original paper, Meng and Rubin referred this pooled test statistic to an F reference distribution with a complex degrees of freedom expression, but the Mplus variant of the test uses a chi-square distribution with the usual degrees of freedom.[1] Sampling distribution aside, the substantive interpretation of D_3 is identical to that of the usual chi-square test (i.e., $p < .05$ implies that the fit of the hypothesized model is statistically significantly worse than that of the saturated model). The pooled test statistic suggested adequate fit, (75) = 93.762, $p = .070$. As a comparison, recall that the maximum likelihood analysis produced a similar result, (75) = 88.695, $p = .133$. Until more research accumulates, it is a good idea to view the pooled likelihood ratio test with caution because methodologists have yet to examine its performance in the context of a structural equation model.

Unfortunately, there is no established procedure for pooling the fit indices that researchers routinely utilize in practice. Although purely ad hoc, I used the CFI, RMSEA, and SRMR values from the 100 analyses to form empirical sampling distributions. Because the distributions were approximately normal, the mean is a reasonable summary statistic: $M_{CFI} = .993$, $M_{RMSEA} = .021$, and $M_{SRMR} = .026$. Considered as a whole, the pooled fit statistics suggest adequate model fit. Again, it is important to underscore the ad hoc nature of this descriptive approach, but it is currently the only strategy available.

SUMMARY AND CONCLUSIONS

Methodologists regard maximum likelihood estimation and multiple imputation as state of the art missing data handling techniques (Schafer & Graham, 2002). In line with this view, the purpose of this chapter was to describe and illustrate the use of these methods in the context of a structural equation modeling analysis. The two methods are quite different in their approach; maximum likelihood uses all of the available to estimate the model parameters, whereas multiple imputation fills in the missing values prior to the analysis. Despite their differences, the two techniques produced indistinguishable estimates and standard errors in the data analysis examples. All things being equal, maximum likelihood and multiple imputation are asymptotically (i.e., in large samples) equivalent, so this result was not a surprise. Given their similarity, you may be asking, "Which approach should I use?" The answer is largely one of personal preference and software availability, but a number of practical issues may make one approach preferable to the other.

Maximum likelihood estimation has the advantage of being widely available in all major structural equation modeling software packages. The method is so simple to implement that it is usually transparent to the user (e.g., invoking maximum likelihood missing data handling requires only one or two keywords, and it is often the default setting). Despite its ease of use, there are a few limitations to maximum likelihood estimation. Although it is not true of every program, SEM packages that implement maximum likelihood typically require the user to adopt a common joint distribution for all exogenous variables (e.g., the multivariate normal). This is potentially problematic for analyses that include categorical outcome variables. The fact that software packages might exclude cases with incomplete manifest exogenous variables is a second potential limitation. However, recasting the incomplete predictors as single-indicator latent variables can solve this problem.

Turning to multiple imputation, flexibility is perhaps its most attractive feature. Because the imputation process is not model-specific (i.e., a saturated model typically generates the imputed values), a single set of imputed values can serve as input data for a variety of analyses. Furthermore, the role of a variable in the subsequent analysis phase is irrelevant. This makes dealing with incomplete explanatory variables simple because these variables simply serve as outcomes in the imputation regression model. Imputation is also a flexible when dealing with mixtures of categorical and continuous variables. In particular, the chained equations algorithm adapts the imputation model to each variable's measurement scale (e.g., a logistic model imputes incomplete binary variables; a linear model imputes continuous variables). Despite its considerable benefits, multiple imputation has its downsides. For one, the procedure arguably has a steeper learning curve than maximum likelihood. Researchers must pay attention to a number of subtle issues that are unique to imputation (e.g., the convergence of the imputation algorithm, the between-imputation interval, the preservation of special data structures). The subtleties can impact the accuracy of the analysis results, even when the MAR mechanism holds. Second, aside from the pooled likelihood ratio test, methodologists have yet to develop formal pooling rules for fit indices such as the CFI and RMSEA.

In summary, maximum likelihood estimation and multiple imputation are now viable missing data handling techniques for structural equation modeling analyses. Statistical theory and empirical computer simulation studies suggest that these methods are far superior to ad hoc approaches, both with respect to accuracy and power. Because the methods are asymptotically equivalent, the methods tend to produce comparable estimates and standard errors. Consequently, choosing between the approaches is usually a matter of personal preference and software availability.

APPENDIX A:
Mplus Syntax for Saturated Correlates Model

```
TITLE:
  Mplus saturated correlates analysis;
DATA:
  file = sem2ndcourse.dat;
VARIABLE:
  names = abuse bmi bds1—bds7 edr1—edr6;
  usevariables = abuse bds1—bds7 edr1—edr6;
  missing = all (-99);
  ! DEFINE BMI AS MISSING DATA AUXILIARY VARIABLE;
  auxiliary = (m) bmi;
ANALYSIS:
  ! ROBUST STANDARD ERRORS AND RESCALED TEST STATISTIC;
  estimator = mlr;
MODEL:
  ! DEFINE ABUSE AS SINGLE-INDICATOR LATENT VARIABLE;
  abuse;
  ! MEASUREMENT MODEL;
  bodydis by bds1—bds7;
  edrisk by edr1—edr6;
  ! STRUCTURAL MODEL;
  bodydis edrisk on abuse;
  edrisk on bodydis;
OUTPUT:
  sampstat standardized;
```

APPENDIX B:
Mplus Syntax for Imputation Phase of Multiple Imputation

```
TITLE:
  Mplus imputation phase program;
DATA:
  file = sem2ndcourse.dat;
VARIABLE:
  names = abuse bmi bds1—bds7 edr1—edr6;
  usevariables = abuse—edr6;
  missing = all (-99);
DATA IMPUTATION:
  ! VARIABLES TO BE IMPUTED;
  ! (C) DEFINES ABUSE AS CATEGORICAL VARIABLE;
  impute = abuse (c) bmi bds1 bds2 bds4 bds7 edr2 edr3 edr5;
  ! NUMBER OF IMPUTED DATA SETS;
  ndatasets = 100;
  ! PREFIX FOR IMPUTED DATA SET FILE NAMES;
  save = semimp*.dat;
  ! NUMBER OF BETWEEN-IMPUTATION ITERATIONS;
  thin = 600;
```

```
ANALYSIS:
  ! SATURATED IMPUTATION MODEL;
  type = basic;
  ! RANDOM NUMBER SEED FOR MCMC ALGORITHM;
  bseed = 48932;
  ! TECH8 GIVES THE POTENTIAL SCALE REDUCTION STATISTIC;
  tech8;
```

APPENDIX C:
Mplus Syntax for Analysis and Pooling Phase of Multiple Imputation

```
TITLE:
  Mplus analysis and pooling phase program;
DATA:
  ! MPLUS-GENERATED FILE THAT CONTAINS DATA SET NAMES;
  file = semimplist.dat;
  ! SPECIFY IMPUTATION DATA AS INPUT;
  type = imputation;
VARIABLE:
  names = abuse bmi bds1—bds7 edr1—edr6;
  usevariables = abuse bds1—bds7 edr1—edr6;
ANALYSIS:
  ! ESTIMATOR = ML GIVES POOLED LIKELIHOOD RATIO TEST;
  ! USE MLR TO POOL ROBUST STANDARD ERRORS;
  estimator = mlr;
MODEL:
  ! MEASUREMENT MODEL;
  bodydis by bds1—bds7;
  edrisk by edr1—edr6;
  ! STRUCTURAL MODEL;
  bodydis edrisk on abuse;
  edrisk on bodydis;
OUTPUT:
  standardized;
```

NOTE

1. Mplus produces the pooled likelihood ratio test with normal-theory maximum likelihood estimation (i.e., estimator = ML). Pooling sandwich estimator standard errors requires a separate analysis with the estimator set at MLR.

REFERENCES

Arminger, G., & Sobel, M. E. (1990). Pseudo-maximum likelihood estimation of mean and covariance structures with missing data. *Journal of the American Statistical Association, 85,* 195–203.

Beale, E. M. L., & Little, R. J. A. (1975). Missing values in multivariate analysis. *Journal of the Royal Statistical Society, Ser. B, 37,* 129–145.

Barnard, J., & Rubin, D. B. (1999). Small-sample degrees of freedom with multiple imputation. *Biometrika, 86,* 948–955.

Baraldi, A. N., Cham, H., & Enders, C. K. (2012). *Estimating moderated regression models with incomplete predictor variables.* Manuscript submitted for publication.

Collins, L. M., Schafer, J. L., & Kam, C-M. (2001). A comparison of inclusive and restrictive strategies in modern missing data procedures. *Psychological Methods, 6,* 330–351.

Demirtas, H., & Schafer, J. L. (2003). On the performance of random-coefficient pattern-mixture models for non-ignorable drop-out. *Statistics in Medicine, 22,* 2553–2575.

Dempster, A. P., Laird, N. M., & Rubin, D. B. (1977). Maximum likelihood from incomplete data via the EM algorithm. *Journal of the Royal Statistical Society, Ser. B, 39,* 1–38.

Enders, C. K. (2002). Applying the Bollen-Stine bootstrap for goodness-of-fit measures to structural equation models with missing data. *Multivariate Behavioral Research, 37,* 359–377.

Enders, C. K. (2003). Using the expectation maximization algorithm to estimate coefficient alpha for scales with item-level missing data. *Psychological Methods, 8,* 322–337.

Enders, C. K. (2008). A note on the use of missing auxiliary variables in full information maximum likelihood-based structural equation models. *Structural Equation Modeling: A Multidisciplinary Journal, 15,* 434–448.

Enders, C. K. (2010). *Applied missing data analysis.* New York, NY: Guilford.

Enders, C. K. (2011). Missing not at random models for latent growth curve analyses. *Psychological Methods, 16,* 1–16.

Enders, C. K., & Gottschall, A. (2011a). Multiple imputation strategies for multiple group structural equation models. *Structural Equation Modeling: A Multidisciplinary Journal, 18,* 35–54.

Enders, C. K., & Gottschall, A. C. (2011b). The impact of missing data on the ethical quality of a research study. In A. T. Panter & S. K. Sterba (Eds.), *The ethics of quantitative methodology* (pp. 357–382). New York, NY: Routledge.

Finkbeiner, C. (1979). Estimation for the multiple factor model when data are missing. *Psychometrika, 44,* 409–420.

Gelman, A., Carlin, J. B., Stern, H. S., & Rubin, D. B. (1995). *Bayesian data analysis.* Boca Raton, FL: Chapman & Hall.

Graham, J. W. (2003). Adding missing-data relevant variables to FIML-based structural equation models. *Structural Equation Modeling: A Multidisciplinary Journal, 10,* 80–100.

Graham, J. W., Olchowski, A. E., & Gilreath, T. D. (2007). How many imputations are really needed? Some practical clarifications of multiple imputation theory. *Prevention Science, 8,* 206–213.

Kenward, M. G., & Molenberghs, G. (1998). Likelihood based frequentist inference when data are missing at random. *Statistical Science, 13,* 236–247.

Little, R. J. A., & Rubin, D. B. (2002). *Statistical analysis with missing data* (2nd ed.). Hoboken, NJ: Wiley.

Muthén, B., Asparouhov, T., Hunter, A., & Leuchter, A. (2011). Growth modeling with non-ignorable dropout: Alternative analyses of the STAR*D antidepressant trial. *Psychological Methods, 16,* 17–33.

Muthén, B., Kaplan, D., & Hollis, M. (1987). On structural equation modeling with data that are not missing completely at random. *Psychometrika, 52,* 431–462.

Preti, A., Incani, E., Comboni, M. V., Petretto, D.R., & Masala, C. (2006). Sexual abuse and eating disorder symptoms: The mediator role of bodily dissatisfaction. *Comprehensive Psychiatry, 47,* 475–481.

Raghunathan, T. E., Lepkowski, J. M., Van Hoewyk, J., & Solenberger, P. (2001). A multivariate technique for multiply imputing missing values using a sequence of regression models. *Survey Methodology, 27,* 85–95.

Royston, P. (2005). Multiple imputation of missing values: Update. *The Stata Journal, 5,* 1–14.

Rubin, D. B. (1976). Inference and missing data. *Biometrika, 63,* 581–592.

Rubin, D. B. (1978). Multiple imputations in sample surveys—A phenomenological Bayesian approach to nonresponse. *Proceedings of the Survey Research Methods Section of the American Statistical Association,* 30–34.

Rubin, D. B. (1987). *Multiple imputation for nonresponse in surveys.* Hoboken, NJ: Wiley.

Rubin, D. B. (1996). Multiple imputation after 18+ years. *Journal of the American Statistical Association, 91,* 473–489.

Savalei, V., & Bentler, P. M. (2009). A two-stage ML approach to missing data: Theory and application to auxiliary variables. *Structural Equation Modeling: An Interdisciplinary Journal, 16,* 477–497

Savalei, V., & Yuan, K.-H. (2009). On the model-based bootstrap with missing data: obtaining a p-value for a test of exact fit. *Multivariate Behavioral Research, 44,* 741–763.

Schafer, J. L. (1997). *Analysis of incomplete multivariate data.* Boca Raton, FL: Chapman & Hall.

Schafer, J. L. (2003). Multiple imputation in multivariate problems when the imputation and analysis models differ. *Statistica Neerlandica, 57,* 19–35.

Schafer, J. L., & Graham, J. W. (2002). Missing data: Our view of the state of the art. *Psychological Methods, 7,* 147–177.

Schafer, J. L., & Olsen, M. K. (1998). Multiple imputation for multivariate missing-data problems: A data analyst's perspective. *Multivariate Behavioral Research, 33,* 545–571.

Wilkinson, L., & Task Force on Statistical Inference (1999). Statistical methods in psychology journals: Guidelines and explanations. *American Psychologist, 54,* 594–604.

Yuan, K-H., & Bentler, P. M. (2000). Three likelihood-based methods for mean and covariance structure analysis with nonnormal missing data. *Sociological Methodology, 30,* 165–200.

Yuan, K-H., Bentler, P. M., & Zhang, W. (2005). The effect of skewness and kurtosis on mean and covariance structure analysis: The univariate case and its multivariate implication. *Sociological Methods and Research, 24,* 240–258.

CHAPTER 13

MULTILEVEL STRUCTURAL EQUATION MODELING WITH COMPLEX SAMPLE DATA

Laura M. Stapleton

As researchers become aware of the wealth of large-scale datasets available at the national and international level (within the United States, for example, data are collected by the National Center for Education Statistics, the National Science Foundation, and the Center for Disease Control), questions of how to analyze such data are becoming more frequent. The choice of analysis depends first and foremost on the research question being asked. Once decided, the researcher typically needs to utilize special techniques in order to estimate model parameters and standard errors appropriately because the data were not collected using a simple random sampling procedure. In this chapter, I introduce the types of complex sampling designs that are used in national and international studies and illustrate the structural equation modeling (SEM) analysis tools that are available to researchers using data collected with such designs. In particular, because complex sampling designs often call for collecting data in multiple stages, for example, students nested within schools, the focus will be on analyzing multilevel models. But researchers should be aware that multilevel models

Structural Equation Modeling: A Second Course (2nd ed.), pages 521–562
Copyright © 2013 by Information Age Publishing
All rights of reproduction in any form reserved.

answer specific research questions and are not always necessary when data are nested. Multilevel models allow the researcher to examine relations among variables within a nested structure (such as patients within hospitals) as well as relations at a group level (in this case, hospitals). For other research questions, a single-level modeling approach may be appropriate and strategies for such modeling will be briefly discussed.

The first section will outline the characteristics of complex samples and typical sampling methods will be described. Consequences of naïvely analyzing complex sample data using traditional statistical methods that assume simple random sampling will then be discussed. The next section will introduce three different methods to model data from complex samples and the choice of method will depend on the research question(s) to be answered. Examples of research questions for each method will be provided and example program syntax for the multilevel analyses will be discussed in detail; the reader is cautioned that advances in software capabilities are happening at a rapid rate and thus the treatment provided here may not be state-of-the-art; thus, examining the version of the software being used is crucial. The last section of the chapter will point the reader toward outside resources for both applied examples of multilevel SEM analyses and methodological extensions of multilevel SEM.

REVIEW OF COMPLEX SAMPLING DESIGNS

National surveys and testing programs generally do not collect data based on simple random samples (SRS). Longford (1995) and Skinner, Holt, and Smith (1989) provided reviews of the special characteristics of such complex sample designs, including clustering, stratification, unequal selection probability, and non-response and post-stratification adjustment. In general, each of these elements of complex sampling designs requires attention during the analysis stage, but it is difficult to assess the overall impact of the sampling design on the statistical estimates from an analysis model when there are many sampling elements that require accommodation. Presented below is a treatment of selected complex sampling characteristics and their individual effects on model parameter estimation at the analysis stage.

Multi-Stage (Cluster) Sampling

An assumption in the use of traditional analysis techniques, such as single-level SEM, is that observations are independently sampled. However, most large data sets are collected using a multi-stage sampling technique—it is more efficient to collect data from twenty students in one school, for example, than to travel to twenty different schools to obtain data from one student at each. In multi-stage sampling, the first stage unit (or the first level of selection) is termed the *primary sampling unit* (PSU). Two-stage sampling examples include random selection of schools and then the selection of students from each school (here, schools are the PSUs). Another example is random selection of classrooms in a school and then the sampling of all students in the selected classrooms. As a result of this multi-stage sampling, data usually have some degree of dependence among observations (e.g., students who are in the same classroom tend to be more like each other than like students in another classroom). However, formulas for calculating standard errors that are incorporated into most statistical packages are based on an SRS design where independent observations are assumed (Lee, Forthofer, & Lorimor, 1989). Because these formulas assume that the correlation of the residuals across people is zero, a researcher might underestimate the sampling variance and thus standard errors, resulting in Type I error rates of unacceptably high levels; dramatic examples in an analysis of variance context are available in Kish and Frankel (1974) and Scariano and Davenport (1987). Likewise, in the SEM context, bias due to violation of the assumption of independent observations has been discussed in the estimation of standard errors of parameters (Hox, 2002; Kaplan & Elliot, 1997a) and χ^2 statistics (Asparouhov & Muthén, 2005, 2006; Muthén & Satorra, 1995; Stapleton, 2006). An analysis of nested, or clustered, data ignoring the dependencies in the data, therefore, may lead to the misidentification of statistically significant path coefficients where only random covariation exists and may lead to inappropriate rejection of hypothesized models.

Stratification

Stratification in the sampling design refers to the division of all elements in the population into mutually exclusive categories, where sampling is performed within each category. There are two reasons why survey researchers might choose a sample using stratification. First, a researcher might state, "I want to make sure I get a representative sample," and subsequently sample to ensure, for example, that the selected elements include some boys and some girls. Although the researcher might not realize it, the sample drawn was stratified by sex. This desire for representativeness is an informal expression of the second reason why stratified samples are taken. Stratification decreases the chances of getting a "bad" sample (the sample is guaranteed to be representative on the variable or variables chosen for stratification). Therefore, the statistical estimates from this sample will tend to be more precise and thus have smaller standard errors than if the sample were collected based on a completely random process. This advantage assumes, however, that the stratification variable is correlated with the response variable (Kalton, 1983). Recently, software for SEM has implemented the ability to take advantage of this strata information and estimates the more appropriate standard errors (Asparouhov, 2004). Alternatively, the strata can be modeled as fixed effects, either by incorporating dummy variables to represent strata or modeling the strata using a multisample approach (although this would be difficult as there are usually dozens of strata in large-scale datasets).

Any time sampling elements are stratified, a choice of whether to use *proportionate* or *disproportionate* sampling must be made. In proportionate sampling, the same selection probability is used in each stratum. In disproportionate sampling, the researcher can choose to use different selection probabilities in each stratum. If disproportionate sampling is used, sampling weights might be required to obtain unbiased parameter estimates in the analysis. When proportionate sampling is used, at least in a single-stage sampling design, inclusion of sampling weights into the analysis is not necessary. For example, suppose there are 1,000 students in a population (500 girls and 500 boys) and a sample of 100 is taken, making sure that 50 of the students are girls and 50 of the students are boys. A sampling rate of .1 (50/500) was used in both the girls' group and the boys' group and because this same sampling probability (π) was used within both strata, the use of sampling weights to adjust for differential selection probabilities is not required. In this case, the sampling technique is referred to as *proportionate stratified sampling*. However, if in the population there were 600 girls and 400 boys and 50 girls and 50 boys were each sampled, then $\pi = .08$ (50/600) for girls and $\pi = .125$ (50/400) for boys. This process would be described as disproportionate stratified sampling and strategies for analyzing these data are described in the next section.

Unequal Selection Probabilities

Elements in the population are associated with unequal selection probabilities when those elements are sampled at different rates across strata (also, elements can be selected with differential probabilities when using multi-stage sampling if clusters are of unequal sizes, although many sampling designs avoid this situation by using a technique called *probability-proportionate-to-size sampling*). Analyses ignoring these unequal inclusion probabilities can lead to biased parameter estimates when the probability of selection is correlated with the response variable. Kaplan and Ferguson (1999), Korn and Graubard (1995), and Lee et al. (1989) demonstrated, using SEM and regression techniques, that parameter estimate bias could result when unequal probability of selection is ignored in traditional, single-level analyses.

Often, the easiest way to incorporate unequal selection probabilities is to use sampling weights for observations. These weights can be built to reflect the original unequal sample inclusion probabilities, and help to compensate for differential response rates. Weights that can be found on national and international datasets represent the number of people that observation represents and therefore, when the weights on the dataset are summed, the sum will equal the population size. This raw weight, which typically ranges in number from 1 to something in the 1000s, must be carefully considered before included into an analysis. A user must be aware of how the software will treat the weight. The estimation of sampling variability (i.e., standard errors) when using weights is not invariant with respect to constant multiplicative factors of weights, so weights are typically transformed prior to analysis and the choice of weight transformation for the weights w_i can be important (Longford, 1995). To take a specific example, if an estimate of the mean is obtained, the standard error is usually calculated as the standard deviation of scores over the root sample size, but incorporating the weight would result in using the sum of the weights in the denominator instead of the sample size. If raw weights are used, the population size is in the denominator of the standard error estimate and the standard error estimate would be much too small. Thus, weights are typically transformed to be equal to the sample size (referred to as *weight normalization*). Other weight transformation choices are available. The choice of transformation becomes even more complicated in a multilevel SEM analysis, as weights may need to be provided at both the within- and between-portions of the model because clusters are selected with a given probability and then units within those clusters are selected at separate probabilities (Asparouhov, 2006; Rabe-Hesketh & Skrondal, 2006; Stapleton, 2002). Currently available SEM software performs the weight transformation process for the user and details on these defaults will be provided in a later section of this chapter.

An example dataset might help to clarify the sampling terms introduced above. The Early Childhood Longitudinal Study–Kindergarten (ECLS-K), in the base year, used a three-stage stratified sampling design (U.S. Department of Education, 2001). First, all schools in the target population were divided into sampling "groups" based on geographic area, size, and income (usually counties or sets of counties) and these groups were placed in 62 mutually-exclusive strata. These strata were defined by region of the nation, population size, race/ethnicity concentration, and per capita income. Twenty-four of these strata were "certainty" strata and contained just one geographic group (because there was only one group in the stratum, the group would automatically be selected into the sample, for example, New York City). In the remaining 38 strata, just two geographic groups were selected in each stratum. Once these 100 geographic groups were selected (as the PSUs), all the schools within each of the selected geographic groups were sorted by school characteristics and a subset of schools was selected using systematic sampling (a form of stratification called *implicit*). Selection of schools was therefore the second stage in the sampling design. Finally, in the third stage, approximately 24 children were sampled from each selected school. Prior to selection, students were placed into two different explicit strata: Asian/Pacific Islander students and all other students. Asian/Pacific Islander students were sampled at a rate about three times the sampling rate for the other students. Thus, there are two levels of clustering: schools within geographic groups and children within schools. There are also several different types of stratification: areas by geographic region, size and income; schools by school characteristics; and children by race/ethnicity. In addition, the children were sampled at disproportional sampling rates: Asian/Pacific Islander students were sampled at a rate higher than the other children, so they will be overrepresented in the sample. Given these issues, a statistical technique that assumes simple random sampling should not be used. Each of the characteristics of complex samples treated above—clustering, stratification, and disproportionate selection probabilities—can affect standard error and parameter estimates and how each is accommodated at the analysis stage is determined by the user's choice of analysis technique as discussed in the next section. National data collection agencies are not the only organizations collecting data using complex sampling methods. Within a university, for example, a thesis student might want to obtain a sample of undergraduates and therefore a sample is chosen of 50 freshmen, 50 sophomores, 50 juniors, and 50 seniors. This is a stratified sample and that fact would need to be addressed to obtain accurate standard errors for parameter estimates. If the number of students at each class level (freshman, sophomore, etc.) in the population is not equal, then the sample was chosen using disproportionate selection rates and sampling weights would need to be utilized to obtain accurate estimates of overall population characteristics; so, even what might appear to be a "simple" sampling process might make the analysis not so simple after all. In the next section, strategies to accommodate these aspects of complex sampling designs in SEM analyses will be discussed.

ANALYSIS OPTIONS

Three different analysis options will be discussed in this section, the choice of which hinges on the research question to be addressed: *design-based modeling, pooled within-group covariance matrix modeling*, and *multilevel modeling*. These options are shown using, as an example, the simple bivariate case in the graphs in Figures 13.1 and 13.2 where in each graph a variable repre-

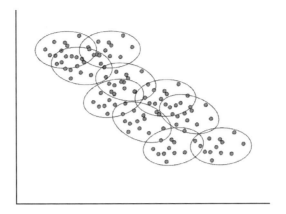

Figure 13.1 Graphical representation of clustered data.

senting "hours of TV watching" is shown on the *X* axis and a hypothetical latent factor, achievement, is shown on the *Y* axis.

Design-Based Modeling

Suppose a researcher has data from students who are clustered in schools (Figure 13.1) and is interested in overall population characteristics such as the strength of a hypothesized effect of the number of hours spent watching TV on a latent construct of achievement. In this case, the interest is not in disentangling the effects at the school level from the effects at the individual level (it is possible that, within schools, there is no relation between TV watching and achievement but across schools there is a relation). The interest, instead, is in examining the overall relation between TV watching and achievement, regardless of the level of the relation, and for this research question a typical single-level SEM model would be appropriate. Muthén and Satorra (1995) termed this type of approach with clustered data an *aggregate* analysis (shown in Figure 13.2a): a single regression line is estimated, but appropriate standard errors must be estimated. Given this example, the sampling design would need to be addressed: the multi-stage sampling of schools in geographic regions and students in schools, the disproportionate sampling of the students, and the stratification. As shown in Figure 13.2a, a naïve analysis that does not take account of the sampling design might result in standard error estimates represented by the small dotted lines, whereas a more appropriate, adjusted standard error is shown as the dashed line. Note that the estimated regression line could vary quite a bit from its current trajectory and still stay within the area demarked by the dashed line (although, importantly, note that a flat regression line would

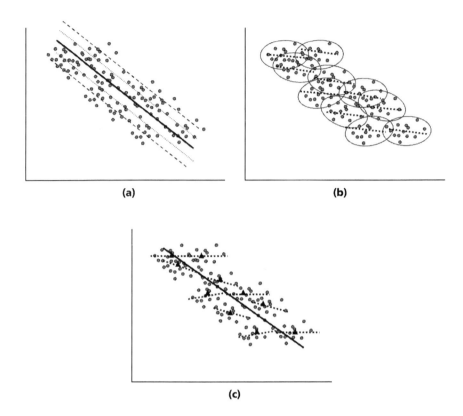

Figure 13.2 Graphical representation of modeling options with clustered data. (a) Aggregate modeling; (b) Pooled within-group modeling; (c) Multilevel modeling.

not fit within that area, suggesting the negative relation would still be statistically significantly different from zero). An analysis that adjusts estimates of standard errors given a sampling design but does not explicitly model the sampling design is referred to as a *design-based* analysis (Kalton, 1977).

Design-based SEM analyses can be conducted using some of the more recent versions of available software. Specifically, Mplus version 6.1, the *gllamm* package written for Stata version 11, and LISREL version 8.8 can fully accommodate a design-based analysis, including clustering and stratification at the PSU level with the inclusion of unconditional (overall) sampling weights. To undertake the analysis, the software uses either repeated replications or the Taylor Series approximation to obtain accurate estimates of standard errors (for simple descriptions of these techniques see Lee et al., 1989; Lohr, 1999; Stapleton, 2008) and uses estimates of standard error corrections to adjust the χ^2 data-model fit statistic.

Pooled Within-Group Covariance Matrix Modeling

If the researcher's interest is in testing a model that explains the relations *within* a school between hours of TV watching and a latent construct of achievement and ignores between-school relations, then data could be modeled as suggested by Hox (2002), using a pooled within-cluster covariance matrix. This type of analysis is shown in Figure 13.2b; note that a common slope is estimated in each of the schools (and in this example, there appears to be a small negative relation, on average, between the variable TV watching and the latent achievement construct within schools). Calculation of the pooled within-cluster covariance matrix would take place outside of the SEM software and the matrix would be analyzed as if it was a single-level covariance matrix and interpretation would be focused on relative relations within schools. The pooled within-cluster covariance matrix can be calculated as:

$$\mathbf{S}_{PW} = (N - G)^{-1} \sum_{g=1}^{G} \sum_{i=1}^{n_g} w_{gi} (\mathbf{y}_{gi} - \overline{\mathbf{y}}_g) (\mathbf{y}_{gi} - \overline{\mathbf{y}}_g)'$$

where G is the total number of clusters, N is the total sample size, $\mathbf{y}_{gi}$ is the vector of responses for person i within cluster g, $\overline{\mathbf{y}}_g$ is the vector of cluster means, and w_{gi} is the transformed unconditional (overall) sampling weight for person i within cluster g. When modeling the covariance matrix, the analyst should indicate to the software that the appropriate number of cases is $(N - G)$, not N. Note that for the ECLS-K example, weights (with an appropriate transformation or scaling) would still need to be used in the calculation of the matrix because children were selected at disproportionate rates within schools. Also, because there is stratification of schools in the sampling design that is not being addressed, estimates of standard errors using this procedure likely will be somewhat overestimated because the calculations do not recognize that the sample was drawn specifically to be more representative than a simple random sample would typically be.

Multilevel Modeling

A third choice in analyzing data with a nested structure is to use multilevel SEM. This type of analysis allows the researcher to examine within-cluster relations (e.g., if a student watches more TV than other students in his school, how does this relate to his relative achievement) and between-cluster relations (e.g., what is the relation between the average hours of TV watched at a school and the average achievement in the school). The first question is similar to the question answered by the previously discussed analysis technique of modeling the pooled within-cluster covariance matrix. However, multilevel modeling provides three advantages: (1) the researcher obtains an estimate of the amount of variance that exists in the measured variables between clusters and can model that variance; (2) at the within-cluster level the analysis will be more powerful than an analysis using only the within-cluster covariance matrix because the between-cluster covariance matrix provides additional information about the within-cluster relations; and (3) variability across clusters in the relations among variables within clusters can be assessed. Figure 13.2c shows conceptually the modeling that is done. In this analysis, within-school relations are estimated and relations among the school means (indicated by the triangles) are also examined; in this example, there is little relation between TV watching and achievement within schools, on average, but there is a negative relation between the average TV watching and average achievement across the schools. Additionally, note that the relation within schools varies; in some schools there is a negative relation of relative TV hours and relative achievement and in other schools there is none. A multilevel analysis would allow one to evaluate school-level predictors of this variability in slope.

In general, when elements of the sampling design are included in the analytic model the strategy is referred to as a *model-based* analysis (Kalton, 1977). If a multi-level analysis is being undertaken to address the fact that multi-stage sampling was used to select students from within schools, then the analytic approach is model-based. However, just the use of a multilevel analysis does not address the many possible other components of the sampling design that depart from a simple random sample, such as stratification and disproportionate selection rates. Accommodating these other components of the sampling design is covered in Asparouhov (2006) and Rabe-Hesketh and Skrondal (2006). For purposes of this chapter, only the more simple case of a multi-stage sample with random sampling at each level will be considered. The next section of this chapter develops the logic of multilevel SEM and presents example analyses.

THE MULTILEVEL STRUCTURAL
EQUATION MODELING PROCESS

For simplicity, the example in this section will consider multi-stage sampling only, setting aside other complex sampling issues. To supply data for the example, a dataset representing a population of over 700,000 students was created that reflects a hierarchy (first-grade students within 1,000 elementary schools), and the relation between hours of TV watching (a measured variable) and achievement (a latent variable indicated by three test scores) was varied both within and between schools. To gather data for the example, a two-stage sample was drawn from this population; first, schools were randomly selected and then students within those schools were randomly selected with equal probability. The sample consists of 100 schools selected at random, with subsequent random selection of 2% of the students within each selected school. Note that the probability of a school being selected was $\pi_g = .1$ and the probability of a student being selected given the school's selection was $\pi_{i|g} = .02$, resulting in an overall (unconditional) probability of selection for each student of $\pi_g \pi_{i|g} = .002$ (where the inverse of this probability, 500, is the raw weight for each student). Because neither clusters nor students were selected with differential selection probabilities, all weights are equal and there is no reason to include them in the analysis. This dataset is used for the example analyses and is available from the author. The dataset contains nine variables:

SCHLID	the ID number for the school
STUDID	the ID number for the student
TV	number of hours of television typically watched
TEST1	score on a standardized exam (for example, letters)
TEST2	score on a standardized exam (for example, shapes)
TEST3	score on a standardized exam (for example, sounds)
FEMALE	sex of the student (0 = male, 1 = female)
MOS	measure of size for the PSU (school total enrollment)
AVETV	the school mean hours of reported TV watching

Note that two variables are ID variables, five variables are student-level variables, and the last two are cluster-level variables. Further note that the last variable, AVETV, is an aggregated variable, based on student-level responses. This variable will be used to demonstrate a difference between multilevel SEM and hierarchical linear modeling in a later section of the chapter. The number of students per school in the sample ranged from 4 to 26 and descriptive statistics for the student-level data are shown in Table 13.1.

TABLE 13.1 Selected Descriptive Statistics for the Example Data Set (N = 1,410)

	Mean	Standard Deviation	Minimum	Maximum
TV	6.00	1.13	3	9
TEST1	49.76	8.99	24	82
TEST2	49.97	9.19	23	80
TEST3	50.08	9.39	16	86
GENDER	0.49	0.50	0	1

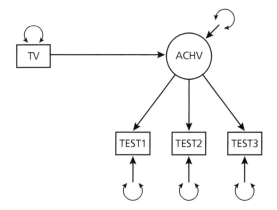

Figure 13.3   Theoretical model of interest.

Suppose the researcher's interest is to test the model shown in Figure 13.3. TV represents "hours of TV " (which is assumed in this model to be measured without error). TEST1, TEST2, and TEST3 are all test scores on various knowledge tests and the latent construct of achievement (ACH) is assumed to be indicated by these three observed variables. Prior to modeling, a question that might arise with these clustered data is, "How much of the variance in the four manifest variables—TV watching and the three test scores—might be attributed to school differences as opposed to purely individual differences?" The *intraclass correlation* (ICC),

$$\rho = \frac{\sigma_B^2}{\sigma_B^2 + \sigma_W^2},$$

is defined for a population as a ratio of the univariate between-cluster variance to the total variance and can be estimated from the components of an ANOVA for a given sample:

$$\hat{\rho} = \frac{MS_B - MS_W}{MS_B + (n. - 1)MS_W}$$

where $n.$ is the sample size per cluster if balanced. If clusters do not contain equal numbers of students, then $n.$ is calculated such that

$$n. = \frac{N^2 - \sum_{g=1}^{G} n_g^2}{N(G-1)}$$

where G is the number of clusters, N is the total sample size, and n_g is the number of elements within cluster g (Kenny & Judd, 1986).

TABLE 13.2   ANOVA Components and Estimates of ICCs for the Example Data Set

	MS_W	MS_B	ICC
TV	1.05	4.18	.17
TEST1	65.20	287.38	.19
TEST2	70.52	269.68	.17
TEST3	70.79	318.59	.20

In this example, because there are varying numbers of students per school, an estimate for $n.$ is needed. The total number of students in the sample is 1,410 and the number of schools is 100. The sum of the squared cluster sizes is 21,972, thus an $n.$ of 14.08 is obtained, quite close to the average cluster size of 14.10. Components from an ANOVA using SCHLID as the factoring variable and the resulting estimates of the ICC for each of the four variables are shown in Table 13.2.

The ICCs contained in Table 13.2 indicate that about 20 percent of the variance in each of the variables is a function of differences in cluster means and 80 percent is variability among individuals within clusters. Previous research indicates that with geographically-determined clusters, the ICC tends to be relatively low on demographic variables (such as age and gender) and higher for socioeconomic variables and attitudes (Kalton, 1977). In educational studies involving classrooms, ICC values have been found, in some cases, to be rather high, such as .3 and .4 for mathematics achievement (Muthén, 1997). However, in their recent extensive study of existing data sets, Hedges and Hedberg (2007) found that, on average, across grades kindergarten through 12 the unconditional ICC for mathematics and reading achievement averaged around .22. The value of the ICC should be reviewed to determine the appropriateness of multilevel modeling to address a researcher's hypothesis. If the ICC values are close to 0, obtaining convergence of multilevel structural equation models is unlikely.

Considering a Single-Level Analysis

Before continuing with the multilevel SEM example, consider how the example data could be used without running a multilevel model. First, a researcher could naïvely run the model as if the data were not clustered. If the model shown in Figure 13.3 is imposed on the data for the 1,410 students, a model χ^2 value of 1.943 with 2 degrees of freedom is obtained with a maximum likelihood (ML) estimate of the regression path from TV watching to the latent ability construct of -2.590 and a standard error of .172. Given these results, a conclusion would be drawn that for each hour of

TV watched, the value of the latent ability construct is expected to decrease by 2.590 points; the standardized path coefficient is –.493. If the same model were run but with acknowledgement to the SEM software of the dependency of persons nested within schools (as explained in Stapleton, 2006), adjusted χ^2 fit statistics and standard errors would be obtained. The χ^2 value in this example is adjusted downward, at 1.277, still with 2 degrees of freedom and the regression parameter estimate is still –2.590 but the standard error is now estimated to be .214. This increase in the standard error reflects the imprecision in the estimate of the regression parameter due to having dependent observations. Assuming the causal model was correct, the conclusion would be that there is a negative effect of TV watching on ability. Note that, in this design-based analysis, no reference was made to schools and therefore the school context is not hypothesized to play a role in the causal relation between TV watching and ability.

An alternative approach that might be investigated is a pooled within-cluster analysis. The pooled within-school covariance matrix was obtained using SAS software with the PROC CANDISC command and the model in Figure 13.3 was applied to that covariance matrix (shown in Table 13.3) specifying that $N = 1,310$ observations. From this analysis, a model χ^2 value of 1.562 with 2 degrees of freedom was obtained with an ML estimate of the regression path of the latent ability construct on TV watching of –2.568 with a standard error of .180. For each hour of TV watched, the value of the latent ability construct relative to one's schoolmates is expected to decrease by 2.568 points; the standardized path coefficient is –.476. In this interpretation, the relation within schools is investigated—children who watch more television as compared to their school peers have lower abilities as compared to their peers. Given that the relation within school is nearly the same as the relation overall from the design-based analysis, it is likely that there is little school contextual effect to be found in a multilevel analysis.

Considering a Multilevel Analysis

In the initial formulation of multilevel SEM available in software to applied researchers, the focus had been to parse out the within-cluster and

TABLE 13.3   Pooled Within-Cluster Covariance Matrix for the Example

	TV	TEST1	TEST2	TEST3
TV	1.052			
TEST1	–2.657	65.197		
TEST2	–2.721	32.319	70.517	
TEST3	–2.983	31.968	33.420	70.785

between-cluster variance for each variable and then model within-cluster relations on the within-cluster variance components while simultaneously modeling between-cluster relations on the between-cluster variance components (Muthén & Satorra, 1995). A requirement in this type of model is that the within-cluster covariance matrices are assumed to be equal across all clusters (i.e., $\Sigma_{W1} = \Sigma_{W2} = \ldots = \Sigma_{Wg}$), allowing for a single pooled within-cluster matrix (Σ_W) to be used in estimation. This restriction can be thought of as random effects modeling; an assumption is made that the relations between variables is the same across all clusters. The statistical theory behind this assumption is available in McDonald and Goldstein (1989). More complex multilevel models with Σ_{Wg} differing across schools are now supported in recent versions of SEM software and is addressed in a later section of this chapter. For simplicity, the initial, more restricted, multilevel SEM modeling process is discussed next. As an example of this multilevel variance parsing, imagine it is hypothesized that the variance of the TV variable is comprised of two parts. There is variance due to differences in schools: perhaps some schools are in neighborhoods where children are not closely observed and are free to watch television as much as they desire and other schools are in neighborhoods where children tend to be scheduled with extracurricular activities and do not watch much television on average. Also, there is variance due to differences in individuals: students in the various schools are different from their own classmates due to factors such as intelligence, personality, and parenting styles. So, the variance in the TV variable, σ^2_{TV}, is comprised of some function of σ^2_B (between-school variance) and some function of σ^2_W (within-school variance), as shown in Figure 13.4, with the between and within processes denoted as factors (shown with dotted lines to differentiate them from traditional "construct" factors).

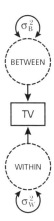

Figure 13.4 Parsing variance of observed variables into between and within components.

Seemingly, this model is underidentified with only one variable (at least two variances will need to be estimated). However, with the use of both the sample within- and between-school covariance matrices, the model becomes identified. Muthén (1994) and Muthén and Satorra (1995) discussed the theoretical background of the estimation and proposed to use multiple group modeling to model sample within- and between-cluster matrices simultaneously. Notice that this is not a true multiple group situation—there are not two separate groups. The estimates of the between- and pooled within-school covariance matrices are calculated such that:

$$\mathbf{S}_B = (G-1)^{-1} \sum_{g=1}^{G} w_g n_g (\overline{\mathbf{y}}_g - \overline{\mathbf{y}})(\overline{\mathbf{y}}_g - \overline{\mathbf{y}})'$$

$$\mathbf{S}_{PW} = (N-G)^{-1} \sum_{g=1}^{G} \sum_{i=1}^{n_g} w_{i|g} (\mathbf{y}_{gi} - \overline{\mathbf{y}}_g)(\mathbf{y}_{gi} - \overline{\mathbf{y}}_g)'$$

where N is the total sample size, G is the number of clusters in the sample, $\overline{\mathbf{y}}_g$ is the vector of cluster mean responses, $\overline{\mathbf{y}}$ is the vector of grand mean responses, y_{gi} is a vector of responses of individual i in cluster g, n_g is the number of observations within cluster g, w_g is the sampling weight of cluster g, and $w_{i|g}$ is the conditional sampling weight for person i given the selection of cluster g. The information contained in $\mathbf{S}_{PW}$ includes unbiased estimates of within-cluster variance and covariance ($E[\mathbf{S}_{PW}] = \Sigma_W$) but the information in $\mathbf{S}_B$ contains both within- and between-cluster variance and covariance information ($E[\mathbf{S}_B] = \Sigma_W + n.\Sigma_B$). Muthén (1994) showed that multilevel SEM parameters can be estimated in the balanced case (equal number of students in each school) by minimizing the following fit function:

$$F_{ML} = G\left\{\log\left|\hat{\Sigma}_W + n.\hat{\Sigma}_B\right| + tr[\hat{\Sigma}_W + n.\hat{\Sigma}_B]^{-1} - \log|\mathbf{S}_B| - p\right\}$$
$$+ (N-G)\left\{\log\left|\hat{\Sigma}_W\right| + tr[\Sigma_W^{-1}] - \log|\mathbf{S}_{PW}| - p\right\}$$

where $n.$ is the common group size, where $\hat{\Sigma}_W$ and $\hat{\Sigma}_B$ represent the model-implied within- and between-cluster covariance matrices, and where Σ_W is assumed equivalent across clusters in the population. This estimator has been termed MUML by some researchers (e.g., Muthén & Satorra, 1995); however, it has also been referred to as *limited information maximum likelihood* and *pseudo-balanced maximum likelihood* (Hox & Maas, 2001). This multilevel SEM estimation can be used in the more typical unbalanced case and can be undertaken, in either the balanced or unbalanced case, using conventional SEM software that allows for multigroup modeling. In order to estimate the fit function, the two components of $\mathbf{S}_B$ and $\mathbf{S}_{PW}$ must be modeled simultaneously with specific constraints and the G, $N - G$, and n. values provided in the syntax. This approach to modeling was demonstrated in the first edition of this text. More recent versions of software alleviate the need to do this manual multigroup set-up. Readers may come across analyses that have used this method of simultaneously modeling with the overall between and within-cluster covariance matrices.

A more general type of estimation for multilevel structural equation models that can accommodate both random coefficients and missing (at random) data is full information ML estimation (Mehta & Neale, 2005). This estimation has been implemented in the most recent versions of software and uses the EM algorithm as described in Raudenbush and Bryk (2002). Both Mplus (Muthén & Asparouhov, 2008) and LISREL (du Toit & du Toit, 2008) use a version of a Fisher scoring algorithm to obtain faster convergence. In this estimation, the likelihood of the obtained data given a specific model is a function of the individual data vector likelihoods as well as the cluster-level likelihoods.

When estimating multilevel structural equation models, it is suggested that a step-wise approach be undertaken that allows the researcher to evaluate the plausibility of the model at each level. This section proposes a set of steps to take when undertaking a two-level SEM analysis. Hox (2002) proposed five steps for undertaking multilevel SEM analyses, starting with the evaluation of the pooled within covariance matrix. While his steps are helpful, they are difficult to undertake unless the researcher uses the manual calculation of pooled within matrices outside the SEM software. Therefore, here I provide an alternate set of steps that can be accomplished using current versions of software. The model-building nature of these steps allows the researcher to evaluate the plausibility of the hypothesized model at each level of the model. If a researcher's first step is to evaluate the full hypothesized model at both the within and between levels simultaneously, misfit is difficult to diagnose. Specifically, due to differential sample sizes at the within-cluster level and the between-cluster level, current indices of fit are dominated by the within-cluster portion of the model (Ryu & West, 2009). The listing of the steps below is followed by a full description and example of conducting the analysis for each step using the Mplus software; the use of other SEM software is discussed in a subsequent section.

Step 1. Evaluate Descriptive Information of All Variables
Step 2. Run Baseline Models for Both the Within- and Between-Cluster Levels
Step 3. Run a Theoretical Model at the Within Level, Saturated at the Between Level
Step 4. Run a Theoretical Model at the Between Level, Saturated at the Within Level

Step 5.   Run a Model with Theory Imposed at Both Levels
Step 6.   Evaluate Random Coefficients at the Within Level

These steps may not be applicable to every theoretical model of interest, but can aid the multilevel modeler in evaluating the plausibility of a hypothesized model.

Step 1. Evaluate Descriptive Information of All Variables

With any analysis, an important step is to inspect each of the analysis variables, including the distribution (e.g., minimum, maximum, standard deviations, skew/kurtosis, central tendency), as well as bivariate correlations with other variables in the analysis. This inspection may be related to assumptions of the modeling technique or related to general appropriateness of the data values (outlier/data entry error inspection). Considering two-level analyses, an important consideration is the ICC of the variables, or how much homogeneity exists within clusters. If the value of the ICCs of all variables is close to zero, multilevel SEM will be very difficult as there will be minimal between-cluster variance to model and estimation convergence may be a problem. In order to obtain estimates of the ICCs for variables, the best option is to run a multilevel "null" or "unconditional" model for each variable separately. For Mplus, this syntax is shown in Figure 13.5 and reflects the model graphically shown in Figure 13.4. Note that although the data set contains nine variables, the user should indicate which variables will be used in the analysis with the *USEVARIABLES* statement. Also, the variable that in-

```
TITLE: MULTILEVEL NULL MODEL FOR TV VARIABLE
DATA:
   FILE IS "C:\TEMP\TS_SRS.dat";
VARIABLE:
   NAMES ARE SCHLID STUDID GENDER TV TEST1
        TEST2 TEST3 MOS AVETV;
   USEVARIABLES ARE SCHLID TV;
   CLUSTER IS SCHLID;
ANALYSIS:
   TYPE IS TWOLEVEL;
MODEL:
 %WITHIN%
    TV;
 %BETWEEN%
    TV;
OUTPUT: STANDARDIZED SAMPSTAT;
```

Figure 13.5   Mplus Syntax for Step 1 in Multilevel SEM.

dicates the "level 2" clustering, or school in this case, must be identified with the *CLUSTER IS* option in the *VARIABLE:* section. In the *ANALYSIS:* section, the *TYPE IS TWOLEVEL* option indicates that this will be a multilevel analysis. With the *TYPE IS TWOLEVEL* option, the researcher models the relations that are hypothesized to exist within clusters in the *%WITHIN%* section of the *MODEL:* statement and the relations that are hypothesized to exist among clusters in the *%BETWEEN%* section of the *MODEL:* statement. Because there is only one variable being modeled, just the variance is estimated by the *TV;* statement at each level and the intercept is estimated at the between level. At the within side of the model, cluster-mean centering is used and thus the intercept within clusters will be 0 by definition.

From the output, the estimated ICC is .167 and matches what was hand-calculated from ANOVA components and presented in Table 13.2.[1] Also, the output shows that the estimated within-cluster variance (the lower part of Figure 13.4) is 1.051 and the estimated between-cluster variance (the upper part of Figure 13.4) is .210. Note that the ratio of the between-cluster variance to the total variance, .210/(.210 + 1.051), is .167. This value of the ICC suggests that there is a reasonable amount of variability to be modeled at the cluster level in developing a cluster-level model. This process of ICC evaluation should be undertaken for each variable in the model. If some variables have limited cluster-level variability (e.g., ICC = .01) and the analysis encounters convergence problems, the user may opt to define that variable as a *WITHIN* variable only and not attempt including it at the cluster level. Note that evaluating the size of the ICC with respect to the need for running a multilevel analysis should not be done without also considering the number of observations within each cluster. With extremely large clusters, even an ICC as low as .01 can adversely affect estimates of the standard error in a single-level analysis. Specifically, the underestimation of the standard error can be approximated by calculating the *design effect*: $1 + (n - 1)\text{ICC}$. With a design effect of 1, the standard error of the estimate of the mean is not underestimated. When the design effect becomes larger than 1, the standard error is underestimated by a factor of the square root of the design effect (Kish, 1965). As is evident from the formula above, with an ICC as low as .01 but with clusters of size 100, the design effect would be approximately 2, which would result in standard errors needing to be multiplied by 1.4 to approximate the true sampling variability of the estimate. This adjustment formula can be applied to standard errors of the mean, but will likely result in over-correction for standard errors of regression coefficients (Kish, 1995; Kish & Frankel, 1974).

From this same model, the estimate of the between-cluster intercept is 5.991. This estimate differs slightly from the overall mean of the TV variable, 6.00, shown in Table 13.1. This intercept estimate is the average of the school averages; because the schools in our sample have varying numbers of students, the overall average and the average of the school averages may not be the same values.

Step 2. Run Baseline Models for Both the Within- and Between-Cluster Levels

Before attempting to analyze the theoretical model of interest, a review of model testing strategies is needed. Evaluating model fit in multilevel structural models has not been the subject of extensive research. The model fit indices that are provided with SEM software can confound information about the between-level and within-level relations. For example, the comparative fit index (CFI) is an oft-used index that considers the degree of misfit of the model as compared to the possible degree of misfit of a baseline model (assuming a model wherein all variables are unrelated). In single-level models, values of .95 or .90 have been suggested for use as a cut-off criterion for determining a close fitting model, however dependence on these cut-offs is cautioned against (Kline, 2011). For the partially-saturated models that will be analyzed in steps 3 and 4 of the model testing process (where fit of the within-cluster model and the between-cluster model is tested respectively), appropriate baseline models are not those where the variables are unrelated at *both* the within- and between-cluster levels. Therefore, in this step of model testing, baseline models should be run from which χ^2 test information will be used to test the plausibility of the imposed theoretical model in subsequent steps.

Specifically, Step 2 involves running two models: 1) one that allows all variables to covary at the between-cluster level and restricts all variable covariances to values of 0 at the within-cluster level, and 2) one that restricts all variable covariances to values of 0 at the between-cluster level and allows all variables to covary at the within-cluster level.[2] The syntax for these two analytic models is shown in Figure 13.6. From these analyses, the baseline (also called independence) χ^2 value for the within-cluster model is 1048.370 with 6 degrees of freedom, and the baseline (independence) χ^2 value for the between-cluster model is 33.335 with 6 degrees of freedom. Part of the explanation for the very large differences in these values is that the former is based on a null model for the 1410 observations while the latter is based on a null model for only 100 observations; the χ^2 test statistic is a function of the sample size. Note that these two fit indices suggest that the within-cluster independence model does not fit the data well and, similarly, the between-cluster independence model does not satisfy a test of exact fit ($p<.001$ for both tests). These two test values will be used in steps 3 and 4 of the multi-level model testing process.

This first syntax is used to obtain a baseline test statistic for the within-cluster model

```
TITLE:  BASELINE MODEL
  DATA:
    FILE IS "C:\TEMP\TS_SRS.dat";
  VARIABLE:
    NAMES ARE SCHLID STUDID GENDER TV TEST1 TEST2 TEST3 MOS AVETV;
    USEVARIABLES ARE SCHLID TV TEST1 TEST2 TEST3;
    CLUSTER IS SCHLID;
  ANALYSIS:
    TYPE IS TWOLEVEL;
    ESTIMATOR IS ML;
  MODEL:
   %WITHIN%
    TV WITH TEST1@0 TEST2@0 TEST3@0;
    TEST1 WITH TEST2@0 TEST3@0;
    TEST2 WITH TEST3@0;
   %BETWEEN%
    TV WITH TEST1 TEST2 TEST3;
    TEST1 WITH TEST2 TEST3;
    TEST2 WITH TEST3;

  OUTPUT:   STANDARDIZED SAMPSTAT;
```

This second syntax is used to obtain a baseline test statistic for the between-cluster model (only the relevant model statements are shown—all else remains as above).

```
  MODEL:
   %WITHIN%
    TV WITH TEST1 TEST2 TEST3;
    TEST1 WITH TEST2 TEST3;
    TEST2 WITH TEST3;
   %BETWEEN%
    TV WITH TEST1@0 TEST2@0 TEST3@0;
    TEST1 WITH TEST2@0 TEST3@0;
    TEST2 WITH TEST3@0;

  OUTPUT:   STANDARDIZED SAMPSTAT;
```

Figure 13.6   Mplus syntax for Step 2 in multilevel SEM—the baseline model.

Step 3. Run a Theoretical Model at the Within Level, Saturated at the Between Level

The next step entails testing the plausibility of the within-level hypothesized relations. In this analysis, all variables of theoretic interest are included in the model but the variables at the between-cluster level are allowed to freely covary with each other. The model is therefore saturated at the between level and should fit perfectly. The within-cluster model of interest should be imposed on the within portion of the model. Poor fit of this model suggests that the theory imposed at the within-cluster level is not plausible.

The syntax to run this model is just a slight alteration to that shown in the upper portion of Figure 13.6 as in Step 2 but with the following statements in the *%WITHIN%* section: *ACH_W BY TEST1 TEST2 TEST3; ACH_W ON TV.* Selected output from this model is shown in Figure 13.7.

Note that although the between model remains saturated as it was in the first model of Step 2, the between-level covariance estimates do not completely replicate the estimated cluster-level covariance matrix that can be obtained from the *SAMPLE STATISTICS* section in the Mplus output. The discrepancy is due to model estimation on the within side of the model; because the model on the within side did not account perfectly for the within-level covariance matrix, some of that misfit is absorbed at the between-cluster level. The χ^2 test of model fit for this model is 1.681 with 2 df, suggesting good fit at the within-cluster level given that the test value is not significantly different from zero. Mplus provides standardized root mean square residual (SRMR) values for the within and between covariance matrices, and for this example these values are .007 and .002, respectively. Based on recommendations of Ryu and West (2009), a level-specific model CFI can be calculated as:

$$1 - \frac{\max[\chi^2_{PSw} - df_{PSw}, 0]}{\max[\chi^2_{PSiw} - df_{PSiw}, 0]}$$

where χ^2_{PSw} and df_{PSw} represent the test value and degrees of freedom respectively from the partially saturated within-cluster model in Step 3, and where χ^2_{PSiw} and df_{PSiw} represent the test value and degrees of freedom respectively from the baseline (independence) model partially saturated within-cluster model in Step 2. In this example, the numerator of the second component would be solved as 0, so the level-specific CFI would have a value of 1.0.

If the fit were poor of this model at Step 3, as determined by a significant exact test value or based on guidelines provided by such authors as Hu and Bentler (1999) or Fan and Sivo (2005) for the SRMR or level-specific CFI fit indices, some evaluation of the within-level theory should be conducted. If fit is good at this stage, then the next step would be to impose the hypothesized relations at the between-cluster level.

```
MODEL RESULTS
                                                    Two-Tailed
                   Estimate     S.E.   Est./S.E.   P-Value

Within Level

  ACH_W    BY
     TEST1           1.000      0.000    999.000    999.000
     TEST2           1.038      0.057     18.048      0.000
     TEST3           1.056      0.059     17.812      0.000

  ACH_W      ON
     TV             -2.565      0.179    -14.324      0.000

  Variances
     TV              1.051      0.041     25.629      0.000

  Residual Variances
     TEST1          34.522      1.970     17.521      0.000
     TEST2          37.404      2.129     17.573      0.000
     TEST3          36.662      2.156     17.006      0.000
     ACH_W          23.669      2.085     11.352      0.000

Between Level

  TV       WITH
     TEST1          -0.473      0.248     -1.905      0.057
     TEST2          -0.647      0.239     -2.705      0.007
     TEST3          -0.762      0.278     -2.738      0.006

  TEST1    WITH
     TEST2           3.869      1.981      1.953      0.051
     TEST3           8.423      2.432      3.463      0.001

  TEST2    WITH
     TEST3           5.705      2.200      2.593      0.010

  Means
     TV              5.991      0.054    111.006      0.000
     TEST1          49.761      0.450    110.654      0.000
     TEST2          49.946      0.428    116.571      0.000
     TEST3          50.151      0.484    103.707      0.000

  Variances
     TV              0.210      0.041      5.182      0.000
     TEST1          15.190      2.820      5.386      0.000
     TEST2          12.949      2.501      5.177      0.000
     TEST3          17.899      3.327      5.380      0.000
```

Figure 13.7 Selected Mplus results of modeling in Step 3—the *theoretical within* model.

Step 4. Run a Theoretical Model at the Between Level, Saturated at the Within Level

At this point, the researcher should run a saturated model at the within-level and impose the hypothesized model at the between level. Tests of data-model fit then evaluate the plausibility of the between-level model. For this example, the latter model syntax specified for Step 2 is used but with the following statements in the *%BETWEEN%* model: *ACH_B BY TEST1 TEST2 TEST3; ACH_B ON TV;* Note that there is no requirement that the between-level model be a mirror image of the within-level model. Here, it is hypothesized that there is a latent construct at the school level, perhaps "achievement orientation," that is reflected by average performance on the three assessments, and this achievement orientation of the school is a function of the average amount of TV watched by children in a school. The nature of latent constructs at the cluster level can be perplexing; the notion of a within-cluster latent factor, such as a particular student's level of self-efficacy or achievement, can be easily understood as representing a construct that differentiates students within a cluster on their attitudes or abilities. But what does a school-level latent factor based on those same variables represent? In this example, it is hypothesized that the average test scores are indicators of the emphasis on achievement at the school. Schools with high emphasis on achievement would therefore be expected to have a relatively high value of the average TEST1 score, a relatively high value of the average TEST2 score, and so on. Using a similar example of math achievement scores, but nested within classrooms, Muthén (1991) hypothesized that the classroom-level construct was a measure of selection effects due to tracking and differences in curricula while the within-classroom construct was a measure of a general math achievement trait. The importance of considering the interpretation of construct meaning at each level was first brought forward by Cronbach (1976). Marsh and colleagues (2012) provide an excellent review of considerations for the multilevel modeler and highlight the differences between contextual constructs and climate constructs at the cluster level.

While cluster-level constructs can be based on cluster-level scores (such as years of experience of the principal or principal ratings of school characteristics), they can also be based on aggregates of individual-level scores. Contextual constructs are cluster aggregates of individual characteristics (e.g., average achievement of the students, average SES of the students, average age of the students). It is not expected that students respond in the same way to these measures and these measures are not interchangeable across students. Climate measures, on the other hand, are measures that refer to the cluster itself although they are assessed at the individual level. For example, students may be asked to rate the safety of their school using multiple items. What is intended to be measured is the safety of the *school* and therefore it is expected that students in the same school should respond in similar ways and the measures should be interchangeable across students in the same school. Given the differences in these two types of constructs, a researcher should expect differences in the magnitudes of the ICCs (with climate measures having higher values). For climate measures, the inter-pretation of the between-cluster and within-cluster constructs will be very different. Extending the safety example, at the school level, the construct represents the safety of the school; at the within-school level, a construct (if even modeled) would represent discrepancy of the student's perception of safety from the norm of the other students in the school. Careful consid-eration of the potential meaning of each modeled construct, at each level, should be undertaken prior to positing a specific theoretical model.

Turning back to the example, the between-level model (with a saturated within-cluster model) has a χ^2 value of 2.673 with 2 df, suggesting good model fit, given an exact test. The level-specific CFI, using the between-level independence model χ^2 of 33.335 from Step 2, is .975, slightly lower than the CFI of .999 that is reported in the Mplus output. The SRMR at the between level is .049 and at the within level is .001. Model estimates are shown in Figure 13.8. Note that although the model is saturated at the within-cluster level, within-level estimates may not be perfect replications of within-cluster covariances given that the estimates adjusted for slight misfit at the between-cluster level. Given reasonable fit at the between-cluster lev-el, one can proceed to simultaneous evaluation of both levels of the model.

Step 5. Run a Model with Theory Imposed at Both Levels

Combining the *%WITHIN%* model statements from Step 3 and the *%BETWEEN%* model statements from Step 4, both hypothesized models are imposed simultaneously. The estimates from each level of the model should not have changed dramatically from Steps 3 and 4; if they have, the model should be considered unstable. This imposed model is shown in Figure 13.9. Although the same observed variables may be referenced at each level of the model in the Mplus syntax (such as TEST1), any factors modeled must have unique references across the levels (e.g., ACH_W and ACH_B to refer to the within and between achievement factors). Selected output for this analysis is shown in Figure 13.10.

For this model, the χ^2 test statistic is 4.214 with 4 df ($p = .38$) and the SRMR values are .007 and .049 for the within and between levels, respec-tively. In reviewing the change in the path coefficient from TV to ACH at both the within and between levels from Steps 3 and 4 to Step 5, both estimates changed by just .002, from −2.565 to −2.567 and from −2.671 to

```
MODEL RESULTS
                                                      Two-Tailed
                        Estimate     S.E.   Est./S.E.  P-Value
Within Level

 TV       WITH
    TEST1               -2.648      0.240   -11.056     0.000
    TEST2               -2.720      0.249   -10.941     0.000
    TEST3               -2.983      0.252   -11.835     0.000

 TEST1    WITH
    TEST2               32.226      2.065    15.609     0.000
    TEST3               31.989      2.072    15.438     0.000

 TEST2    WITH
    TEST3               33.321      2.152    15.487     0.000

 Variances
    TV                   1.051      0.041    25.622     0.000
    TEST1               65.087      2.538    25.641     0.000
    TEST2               70.315      2.740    25.666     0.000
    TEST3               70.789      2.766    25.597     0.000

Between Level

 ACH_B    BY
    TEST1                1.000      0.000   999.000   999.000
    TEST2                0.761      0.267     2.851     0.004
    TEST3                1.395      0.396     3.526     0.000

 ACH_B    ON
    TV                  -2.671      0.931    -2.868     0.004

 Means
    TV                   5.990      0.054   111.105     0.000

 Intercepts
    TEST1               65.761      5.597    11.750     0.000
    TEST2               62.123      5.410    11.484     0.000
    TEST3               72.470      6.573    11.025     0.000

 Variances
    TV                   0.210      0.041     5.172     0.000

 Residual Variances
    TEST1                9.536      2.208     4.318     0.000
    TEST2                9.682      2.110     4.589     0.000
    TEST3                6.726      3.198     2.104     0.035
    ACH_B                4.255      2.057     2.068     0.039
```

Figure 13.8 Selected Mplus results of modeling in Step 4—the *theoretical between* model.

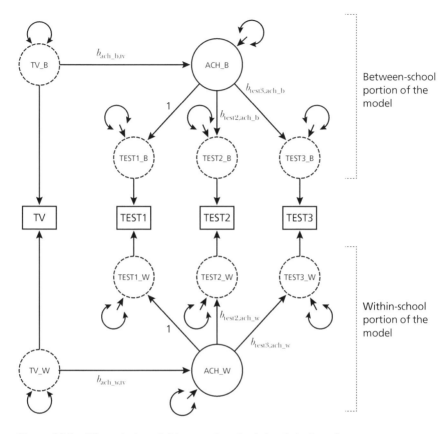

Figure 13.9   *Theoretical* model imposed on both levels in Step 5.

−2.673. Note that all model estimates should be compared in this way and not just one (as was done here for didactic purposes). Given the stability of the estimates, proceeding to interpret the coefficients is now reasonable.

Model Interpretation. First, from the estimates shown in Figure 13.10, the within-school loadings from ACH_W to TEST2 and TEST3 are statistically significantly different from zero ($p < .001$ for both). To obtain a statistical test of the loading path from ACH_W to TEST1, a model with this loading free and another indicator chosen as the referent with a loading fixed to 1.0 would need to be run. Next, within schools, the path from TV to ACH is estimated to be −2.567 and is significantly different from zero ($p < .001$). This finding suggests that, given this model, for each additional hour of TV that a student watches (relative to other students in the school), that student's latent achievement is expected to decrease by about 2.6 units (where achievement is scaled in the metric of Test 1).

```
MODEL RESULTS
                                                    Two-Tailed
                       Estimate     S.E.   Est./S.E.  P-Value
Within Level

 ACH_W    BY
    TEST1               1.000      0.000    999.000   999.000
    TEST2               1.038      0.058     18.036     0.000
    TEST3               1.057      0.059     17.827     0.000

 ACH_W       ON
    TV                 -2.567      0.179    -14.355     0.000

 Residual Variances
    TEST1              34.536      1.970     17.531     0.000
    TEST2              37.385      2.127     17.573     0.000
    TEST3              36.668      2.156     17.011     0.000
    ACH_W              23.626      2.082     11.346     0.000

Between Level

 ACH_B    BY
    TEST1               1.000      0.000    999.000   999.000
    TEST2               0.764      0.268      2.855     0.004
    TEST3               1.402      0.399      3.513     0.000

 ACH_B       ON
    TV                 -2.673      0.929     -2.879     0.004

 Intercepts
    TEST1              65.774      5.582     11.784     0.000
    TEST2              62.175      5.400     11.514     0.000
    TEST3              72.606      6.578     11.038     0.000

 Residual Variances
    TEST1               9.570      2.206      4.338     0.000
    TEST2               9.686      2.110      4.591     0.000
    TEST3               6.697      3.205      2.089     0.037
    ACH_B               4.207      2.045      2.057     0.040
```

Figure 13.10 Selected Mplus results of modeling in Step 5—the *theoretical* model.

Turning to the between-school model, the loading paths from ACH_B to TEST2 and TEST3 are statistically significantly different from zero ($p = .013$ and $p < .001$, respectively). Note that there is less information at the between-school level—data from just 100 schools—so tests of these between-level estimates generally will be less powerful than at the within-school level, resulting in smaller test statistics. Between schools, the path from TV to ACH_B is estimated to be −2.673 and statistically significantly different from zero ($p =. 001$). This finding suggests that for each additional hour of TV watched on average by students at a school, the school's average achievement is expected to decrease by about 2.7 units. Note, however, that this estimate is confounding both between- and within-school relations. If there is a within-school relation, it will be reflected in this between-school relation because group-mean centering of TV was used at the within-school level. To obtain an unconfounded estimate of the between-school level relation (referred to as a *contextual* or *compositional* effect), the path estimate for the within-group relation is subtracted from the between-group relation ($\beta_b - \beta_w$) (Raudenbush & Bryk, 2002, pp. 139–140). In this example, taking the relation between TV and ACH, (-2.673) – (-2.567), a value of -0.106 is obtained; there is thus a very small negative contextual effect of average school TV watching and average school achievement. The contextual effect can be interpreted as the expected difference in the outcome (ACH) between two individuals who watched the same amount of TV, but who attend schools that differ by one hour in average TV watching.

Note that using this method of subtracting to obtain the estimate of the compositional effect does not provide a formal test of the effect. To obtain such a test, the model could be respecified to include a constraint that the two paths equal each other; if this constraint results in a statistically significant change in model fit, then a contextual effect can be presumed to be present in the population. Alternately, the Mplus software includes a simple way of using syntax to obtain a test of the difference in the two paths. In this process, each of the estimates is named and a request is made for a test of whether the difference in the two is statistically significantly different from zero. The model statement syntax to accomplish this test is shown in Figure 13.11. In running this syntax, a contextual effect parameter estimate value of $-.106$ is obtained; its standard error estimate is .957 resulting in a nonsignificant test statistic of $-.111$ ($p = .912$). Therefore, it can be inferred

```
MODEL:
 %WITHIN%
   ACH_W BY TEST1 TEST2 TEST3;
   ACH_W ON TV (gamma10);
 %BETWEEN%
   ACH_B BY TEST1 TEST2 TEST3;
   ACH_B ON TV (gamma01);
MODEL CONSTRAINT:
  NEW(betac);
  betac = gamma01 - gamma10;
```

Figure 13.11 Syntax to test a contextual effect with Mplus.

that, beyond the within-school relation of TV hours and achievement, there appears to be no effect of school-average TV watching on achievement.

Turning to the standardized solution, shown in Figure 13.12, using these or the unstandardized estimates above, the construct reliability of the achievement factor could be evaluated at both the within-school and between-school levels (Muthén, 1991). Additionally, Raykov and Penev (2009)

```
STANDARDIZED MODEL RESULTS

STDYX Standardization
                                                        Two-Tailed
                        Estimate     S.E.   Est./S.E.    P-Value
Within Level

 ACH_W     BY
    TEST1                0.685      0.022     31.480      0.000
    TEST2                0.684      0.022     31.430      0.000
    TEST3                0.694      0.022     31.932      0.000

 ACH_W     ON
    TV                  -0.476      0.026    -18.146      0.000

 Residual Variances
    TEST1                0.531      0.030     17.791      0.000
    TEST2                0.532      0.030     17.841      0.000
    TEST3                0.518      0.030     17.158      0.000
    ACH_W                0.773      0.025     30.942      0.000

Between Level

 ACH_B     BY
    TEST1                0.611      0.111      5.531      0.000
    TEST2                0.506      0.130      3.886      0.000
    TEST3                0.791      0.116      6.849      0.000

 ACH_B     ON
    TV                  -0.513      0.132     -3.875      0.000

 Intercepts
    TEST1               16.828      1.505     11.180      0.000
    TEST2               17.234      1.666     10.345      0.000
    TEST3               17.152      1.529     11.221      0.000

 Residual Variances
    TEST1                0.626      0.135      4.638      0.000
    TEST2                0.744      0.132      5.653      0.000
    TEST3                0.374      0.183      2.044      0.041
    ACH_B                0.737      0.136      5.433      0.000
```

Figure 13.12 Mplus standardized output from Step 5—the *theoretical* model.

proposed a method to evaluate maximal reliability for a construct, assuming no contextual effect.

Looking at the standardized structural relations, given this model, a one standard deviation increase in TV watching (relative to the other students within a school) is expected to result in a .476 standard deviation drop in relative achievement. As modeled, about 23% of the within-school variance in achievement has been explained (at the end of the Mplus output, R-squared values are provided). Similarly, given this model, a one standard deviation increase in *average* TV watching for a school (relative to the averages for other schools) is expected to result in a .513 standard deviation drop in average achievement. As modeled, over 26 percent of the between-school variance in average achievement is explained. Recall, however, that part of this relation is just a function of the within-group relation. There is a small negative relation in the sample between a school's average hours of TV watched and the school's average achievement, above and beyond the relation at the within-school level.

Step 6. Evaluate Random Coefficients at the Within Level

Up to this point, some of what appears in Figure 13.2c has been modeled, but a single common regression coefficient has been estimated in each of the schools (i.e., −2.567). Is it possible that this relation differs across the schools? To answer this question, a model would be run allowing the path from TV to ACH_W to differ across schools. For this model, two changes must be made to the syntax: In the *ANALYSIS* statement, the *TYPE IS TWOLEVEL RANDOM;* command is used, and in the *MODEL %WITHIN%* statement, *s | ACH_W ON TV;* is included. Note that the results from this analysis will not provide fit statistics, only deviance values to undertake information-based model comparison. The following values were obtained: Log-likelihood = −16,700.395, AIC = 33,436.790, BIC = 33,531.314, and the sample size adjusted BIC = 33,474.134. The change in the model from Step 5 to Step 6 can be evaluated by examining the change in these deviance values. For the fixed coefficient model in Step 5, the following values were obtained: Log-likelihood = −16,700.708, AIC = 33,435.416, BIC = 3,524.689, sample size adjusted BIC = 33,470.686. Because a parameter has been added to the model (the estimation of a slope latent variable), the model is expected to have better data fit, as evidenced by the better log-likelihood value for the model in Step 6; however, note that the AIC, BIC, and sample size adjusted BIC all increased from Step 5 to Step 6, suggesting that the more parsimonious fixed coefficient model be retained. With this random coefficient model, the estimate of the slope variance across schools is .003, with a standard error of .073, suggesting a nonsignificant amount of variance.

In other words, the relation between TV watching and relative achievement within schools does not statistically differ across the schools. If the relation did differ across schools, a researcher might try to explain the variability by adding school-level variables to explain why the relation between TV and ACH might be stronger or weaker.

Other Software

LISREL version 8.8 (Jöreskog & Sörbom, 2006) and EQS 6.1 (Bentler & Wu, 2003) both offer estimation for multilevel models. In this section only the final theoretical model from Step 5 will be shown using the LISREL and EQS software. I hope that this section, while limited, will help the researcher who is just getting started with multilevel SEM. All three software packages considered in this chapter provide multilevel modeling capabilities that are fairly robust and users should use the software with which they are most comfortable.

LISREL. When using SIMPLIS syntax in LISREL, the only estimation option for multilevel analyses is full-information ML estimation as discussed above (du Toit & du Toit, 2001). The syntax for running the example model is shown in Figure 13.13. The analyst needs to specify the relations that exist at the between- and within-school levels (in this case, by using the *GROUP* statements to separate the two sections of the model). When SIMPLIS encounters the *$CLUSTER* statement (as seen on the third line of the syntax), a multilevel model is run. Note that LISREL (unlike Mplus and EQS) requires variables that are presumed to be measured without error still need to be modeled as latent variables. Thus, in the syntax, the factor FTV is indicated by the manifest variable TV (note that the loading is set to 1 and the residual variance for TV is set to zero). The parameter estimates from LISREL are shown in Table 13.4 and are very similar to the estimates from Mplus. Note that LISREL first calculates and provides the unrestricted covariance matrix for both within-school and between-school relations and therefore the elements on the diagonal of the matrices can be used to calculate ICCs for variables of interest, as shown earlier.

EQS. To run a multilevel model in EQS, as shown in Figure 13.14, in the */SPECIFICATIONS* statement, the *MULTILEVEL* and *CLUSTER* options should be included. Note that the *CLUSTER=* option should be followed by the EQS variable name for the clustering variable. In the example, SCHLID was the first variable on the file so is referred to as "V1" in the syntax. In the first */EQUATIONS, /VARIANCES,* and */COVARIANCES* statements, the within-group model should be indicated. There is a second section with */EQUATIONS, /VARIANCES,* and */COVARIANCES* statements and these should contain the between-group model. Estimation in EQS is done

```
GROUP BETWEEN SCHOOLS
  RAW DATA FROM FILE TS_SRS.PSF
  $CLUSTER SCHLID
  LATENT VARIABLES
    FTV ACH
  RELATIONSHIPS
    TV=1*FTV
    TEST1=1*ACH
    TEST2 TEST3 = ACH
    ACH = FTV

  SET THE ERROR VARIANCE OF TV TO 0
  SET THE VARIANCE OF FTV FREE
  SET THE ERROR VARIANCE OF ACH FREE
  SET THE ERROR VARIANCE OF TEST1 FREE
  SET THE ERROR VARIANCE OF TEST2 FREE
  SET THE ERROR VARIANCE OF TEST3 FREE

GROUP WITHIN SCHOOLS
  RAW DATA FROM FILE TS_SRS.PSF
  RELATIONSHIPS
    TV=1*FTV
    TEST1=1*ACH
    TEST2 TEST3 = ACH
    ACH = FTV

  SET THE ERROR VARIANCE OF TV TO 0
  SET THE VARIANCE OF FTV FREE
  SET THE ERROR VARIANCE OF ACH FREE
  SET THE ERROR VARIANCE OF TEST1 FREE
  SET THE ERROR VARIANCE OF TEST2 FREE
  SET THE ERROR VARIANCE OF TEST3 FREE

  LISREL OUTPUT: ND=3 SC
```

Figure 13.13 SIMPLIS syntax for Step 5—the *theoretical* model.

slightly differently as compared to LISREL and Mplus with an ML method described in Bentler and Liang (2003). Nonetheless, the results from this run correspond closely to the estimates from LISREL and Mplus and are shown in Table 13.4.

ADDITIONAL ISSUES IN MULTILEVEL SEM

Given this brief introduction to multilevel models in an SEM framework, the reader likely has questions regarding specific issues in multilevel modeling. In addition, the dataset used in the examples contained clustering only;

TABLE 13.4   A Comparison of Parameter (and Standard Error) Estimates Across Software for Multilevel SEM

	Mplus 6.1		LISREL 8.8		EQS 6.1		Mplus 6.1 Manifest covariate	
Between–Group Estimates								
$b_{test2,ach}$	0.764	(.268)	0.763	(.259)	0.764	(.252)	0.822	(.255)
$b_{test3,ach}$	1.402	(.399)	1.404	(.445)	1.404	(.435)	1.373	(.329)
$b_{ach,tv}$	−2.673	(.929)	−2.676	(.970)	−2.682	(.950)	−2.583	(.699)
c_{test1}	9.570	(2.206)	9.569	(2.339)	9.572	(2.273)	9.573	(2.178)
c_{test2}	9.686	(2.110)	9.677	(2.099)	9.675	(2.045)	9.593	(2.144)
c_{test3}	6.697	(3.205)	6.681	(3.270)	6.687	(3.195)	7.164	(2.866)
c_{ach}	4.207	(2.045)	4.200	(2.048)	4.196	(1.994)	4.111	(1.949)
c_{tv}	.210	—	.210	(.041)	$.210^2$	(.040)	.287	—
Within–Group Estimates								
$b_{test2,ach}$	1.038	(.058)	1.038	(.058)	1.038	(.056)	1.036	(.057)
$b_{test3,ach}$	1.057	(.059)	1.057	(.059)	1.057	(.057)	1.055	(.059)
$b_{ach,tv}$	−2.567	(.179)	−2.567	(.180)	$−2.570^a$	(.170)	−2.571	(.179)
c_{test1}	34.536	(1.970)	34.536	(1.974)	34.536	(1.902)	34.508	(1.970)
c_{test2}	37.385	(2.127)	37.386	(2.132)	37.387	(2.055)	37.400	(2.127)
c_{test3}	36.668	(2.156)	36.670	(2.147)	36.670	(2.070)	36.678	(2.155)
c_{ach}	23.626	(2.082)	23.627	(2.077)	23.628	(2.002)	23.675	(2.084)
c_{tv}	1.051	—	1.051	(.041)	1.051^a	(.040)	.978	—
χ^2 value	4.214		4.214		4.215		4.193	

[a] The path estimate and the TV variance estimate and their respective standard error estimates are rescaled; EQS was not able to run with the original scaling of the TV variable and a new variable (TV*10) was created to allow for model estimation. The resultant estimates therefore needed to be rescaled by multiplying by 10 and dividing by 100, respectively, to obtain comparable values to the other software programs.

usually the accommodation of other aspects of data from complex sampling designs would be needed. Therefore, the following topics will be briefly addressed in this section: incorporating sampling weights (from disproportionate stratified sampling), accommodating more complex (e.g., higher order) sampling designs, modeling with between-level variables, modeling with missing data, and comparisons with hierarchical linear regression. For more detailed explanations of analyses and extensions, suggested readings are provided at the end of the chapter.

Incorporating Sampling Weights. Sampling weights can be accommodated with each of the methods of multilevel modeling addressed above. When using Mplus, the option *WEIGHT IS [variable name]* can be used in the *VARIABLE:* statement. Note that no matter what scale has been applied to the weight (i.e., raw, normalized, effective sample size), Mplus automatically scales the weight so that the sum of the weights is equivalent to the sample size. Also, the *WEIGHT IS* option will include a weight at the within-level portion of the model only and the modeling will assume that the clusters were selected with equal probability, which often is not the case. A cluster-level weight can be included by use of the *BWEIGHT IS* command. Although LISREL has the ability to incorporate sampling weights for a design-based analysis, at this time LISREL cannot accommodate design weights in two-level models (S. du Toit, personal communication, December 20, 2011). EQS allows the inclusion of only a single weight variable, and thus if schools are selected with unequal probability, results from EQS estimation may not be appropriate. To include a weight variable, within the */SPECIFICATIONS* statement, include *WT=[variable]*.

Higher Order Sampling Designs. National probability samples sometimes include more than two stages of sampling; it is possible, for example, that first school districts are sampled, then schools within those districts, then children within those schools. Or, perhaps schools are first selected, then classrooms within those schools, then children within those classrooms. In these cases, modeling data as if the clusters are independent of each other would yield standard errors at the cluster level that are negatively biased. An additional issue that has not been addressed in the example syntax for this chapter is stratification. If the schools in the example data set had been stratified by urbanicity (e.g., urban, suburban, rural), then the standard error estimates might have been positively biased; the estimates may be more precise than suggested, given that the sample was designed to include different types of schools. Of the three programs considered in this chapter, only Mplus currently can accommodate the extra sampling design information represented by additional stages of sampling and stratification. With the *ANALYSIS=TWOLEVEL COMPLEX* command and the inclusion of PSU and stratum information, appropriate standard errors and adjusted χ^2 statistics are provided (Asparouhov & Muthén, 2006).

Between-Level Variables. The inclusion of between-level variables also can be accomplished with all of the approaches to multilevel SEM that have been addressed in this chapter. For example, a researcher may want to theorize that school size (MOS) has an effect on average school achievement. When modeling with a school-level variable using the Mplus *TYPE IS TWOLEVEL* option, the analyst must declare in the syntax that school size is a cluster-level variable by including the option *BETWEEN IS MOS* in the *VARIABLE:* statement. Once identified as a school-level variable, the school-level variable can then be modeled in the equations found in the *%BETWEEN%* section of the *MODEL:* statement. To include a cluster-level variable in LISREL, the multilevel model will need to be set up using LISREL programming tools and not the SIMPLIS syntax. In EQS, the between-level variable is simply added into the second */EQUATIONS, /VARIANCES, /COVARIANCES* section of the syntax.

Missing Data. Because the default estimation in each of the software packages is full information ML, missing data are accommodated under the assumption that data are missing at random (i.e., conditioned on the variables in the model). For more information on how missing data are accommodated with full information ML estimation, see the chapter by Enders in this volume.

Comparisons With Hierarchical Linear Regression. One last issue to consider is the difference between running multilevel models in SEM software and with hierarchical linear modeling software. Aside from the ability to easily model latent constructs and between-level causal models in multilevel SEM, a basic difference of estimation currently exists between software packages. Suppose a simple two-level regression was run with the example data of TEST2 on TV in both Mplus and PROC MIXED in SAS with group mean centering and average TV viewing as a predictor at level 2. Different estimates of the between-level regression path would be obtained. With Mplus, the estimate would be −2.574 at the within-school level and −3.079 at the between-school level. With PROC MIXED, the estimate would be −2.586 at the within-school level, very close to the estimate for Mplus, but it would be −2.813 at the between-school level. SEM software packages estimate multilevel *latent covariate* models, where the covariate (in this example, average TV hours) is assumed to be a latent variable (Lüdtke, Marsh, Robitzsch, Trautwein, Asparouhov, & Muthén, 2008). Because it is assumed that only a sample of observations are obtained at the within-cluster level, the between-cluster level measure based on those within-cluster data is measured with some unreliability. Furthermore, some measures at the within-level are clearly intended to measure a between-level construct. For example, a question about "classroom climate" may be posed to students in the classroom and any variability in those measures within classrooms (represented as an ICC < 1.0) suggests unreliability in the measurement of the classroom variable. This unreliability of measurement is modeled in multilevel SEM and thus regression paths at the between level of the model will be disattenuated as compared to how the data are modeled in software for hierarchical linear modeling where it is assumed that group means are measurement error free. All three SEM programs considered here use the latent covariate approach. Sometimes, however, this default approach is not appropriate. For example, if the sample is actually a census of all children in a classroom, then the measure does not suffer from sampling error. Furthermore, if the cluster-level measure of interest is a formative aggregation, that is, a variable representing group characteristics such as proportion female or average income, then variability (ICC < 1.0) in the measures at the within-cluster level should not be of concern as it does not reflect unreliability of measurement at the cluster level. In these cases the latent covariate approach would not be appropriate and the multilevel SEM user will need to model with a between-level variable that has been averaged per group outside of the SEM software.

Additional Resources

I have attempted to provide some guidance to researchers interested in using multilevel modeling in an SEM framework, but this limited treatment has only touched on the details of estimation, conceptualization, and interpretation. There are many resources for those who are interested in learning more about multilevel SEM and multilevel linear regression. Researchers should, of course, turn to the software user's manuals and technical addenda for their software of choice. Chapters on multilevel SEM methods can be found in texts on multilevel and advanced SEM analysis by Hox (2010) and Heck and Thomas (2009); the multilevel SEM chapter in Heck and Thomas (2009) provides several examples of path analysis and structural models and has a helpful section on centering and compositional effects. More advanced topics can be found in recent edited volumes by Hox and Roberts (2011) and de Leeuw and Meijer (2008). Helpful examples of the research questions that can be addressed with multilevel SEM are provided by Muthén and Asparouhov (2011) in their chapter in the Hox and Roberts volume. Two somewhat older but very nice examples of multilevel path analysis can be found in articles by Kaplan and Elliott (1997a and 1997b), and an early example of multilevel confirmatory factor analysis is provided in Harnqvist, Gustafsson, Muthén, and Nelson (1994). For readers who are interested in learning about multilevel regression modeling, texts by Snijders and Bosker (2012), Heck, Thomas, and Tabata (2010), and Raudenbush and Bryk (2002) are excellent places to start the learning process. An additional focus of this chapter has been to introduce readers to the issues of complex sampling and the analysis of such data. The text by Lee et al. (1989) is an excellent brief introduction to the issues in sampling and the problems that these issues present in any analysis.

SUMMARY

In this chapter I reviewed several concepts regarding data that are collected via complex sampling procedures and introduced three methods of modeling such data in an SEM framework: *design-based modeling*, *pooled within-group covariance matrix modeling*, and *multilevel modeling*. Readers should remember that traditional, single-level statistical procedures rely on an assumption that observations are independent. Usually, with cluster samples, that independence has been compromised. Diagnostic tools, such as the calculation of ICCs, are available to assess the level of dependency and if observations are found not to be independent the researcher must then choose an appropriate course of action. However, a sample that was collected using cluster sampling does not necessarily require a multilevel analysis. The research question specifies the level of analysis required and therefore should dictate the analysis tool.

If one chooses multilevel SEM as the appropriate tool with which to answer a specific research question, a variety of software is available. Currently, Mplus version 6.1 offers the most flexibility and choice of estimation techniques; however, new versions of software are being released at a rapid pace and users should determine whether new versions of SEM software have increased capabilities in these types of analyses. The steps outlined in this chapter and those originally suggested by Hox (2002) allow the researcher to test, in a hierarchical fashion, the plausibility of hypothesized models at each level of the nested data structure. These steps offer a logical progression in multilevel model testing and researchers are encouraged to address each of the questions regarding the level at which (co)variation exists whenever multilevel analyses are undertaken.

As software becomes more advanced, methodological exploration of new estimation approaches and the robustness of the resultant statistics will be required. I hope that this chapter has introduced a new set of researchers to various issues in analyzing complex sample data with multilevel SEM methods and that these researchers can extend the use of multilevel SEM in applied settings and further methodological study of the technique.

NOTES

1. Note that Mplus includes a *TYPE=TWOLEVEL BASIC* option that will provide ICCs of all variables provided in the *USEVARIABLES* statement. The user should be careful in using this option; if the list of variables includes any aggregated variables (i.e., cluster-level means of individual-level variables, such as AVETV) then estimates of ICCs for the individual-level variables will be affected.
2. Alternately, a different set of covariances might be set to 0 to specify a more plausible baseline model. See Widaman and Thompson (2003) for discussion of this strategy.

REFERENCES

Asparouhov, T. (2004). *Stratification in multivariate modeling. Web Notes: No. 9.* Retrieved from: http://www.statmodel.com/mplus/examples/webnotes/Mplus Note921.pdf

Asparouhov, T. (2006). General multi-level modeling with sampling weights. *Communications in Statistics: Theory and Methods, 35,* 439–460.

Asparouhov, T., & Muthén, B. (2005). Multivariate statistical modeling with survey data. *Proceedings of the Federal Committee on Statistical Methodology Research Conference.* Retrieved from: http://www.fcsm.gov/05papers/Asparouhov_Muthen_IIA.pdf, retrieved September 26, 2006.

Asparouhov, T., & Muthén, B. (2006). Multilevel modeling of complex survey data. *Proceedings of the Joint Statistical Meeting in Seattle, August 2006. ASA section on Survey Research Methods,* 2718–2726.

Bentler, P. M., & Liang, J. (2003). Two-level mean and covariance structures: Maximum likelihood via an EM algorithm. In S. Reise & N. Duncan (Eds.) *Multilevel modeling: Methodological advances, issues, and applications* (pp. 53–70). Mahwah, NJ: Lawrence Erlbaum Associates.

Bentler. P. M., & Wu, E. J. (2003). *EQS 6.1 for Windows: User's guide.* Encino, CA: Multivariate Software.

Cronbach, L. J. (1976). *Research on classrooms and schools: Formulations of questions, design and analysis.* Stanford, CA: Stanford Evaluation Consortium.

de Leeuw, J., & Meijer, E. (2008). *Handbook of multilevel analysis.* New York, NY: Springer.

du Toit, M., & du Toit, S. (2001). *Interactive LISREL: User's guide.* Lincolnwood, IL: Scientific Software International.

du Toit, M., & du Toit, S. (2008). Multilevel structural equation modeling. In J. de Leeuw & M. Meijer (Eds.), *Handbook of multilevel analysis,* (pp. 435–478). New York, NY: Springer.

Fan, X., & Sivo, S. A. (2005). Sensitivity of fit indexes to misspecified structural or measurement model components: Rationale of the two-index strategy revisited. *Structural Equation Modeling: A Multidisciplinary Journal, 12,* 343–367.

Harnqvist, K., Gustafsson, J. E., Muthén, B., & Nelson, G. (1994). Hierarchical models of ability at class and individual levels. *Intelligence, 18,* 165–187.

Heck, R. H., & Thomas, S. L. (2009). *An introduction to multilevel modeling techniques* (2nd ed.). New York, NY: Routledge.

Heck, R. H., Thomas, S. L., & Tabata, L. (2010). *Multilevel and longitudinal modeling with PASW/SPSS.* New York, NY: Routledge/Taylor & Francis.

Hedges, L. V., & Hedberg, E. C. (2007). Intraclass correlation values for planning group-randomized trials in education. *Educational Evaluation and Policy Analysis, 29,* 60–87.

Hox, J. (2002). *Multilevel analysis: Techniques and applications.* Mahwah, NJ: Lawrence Erlbaum Associates.

Hox, J. (2010). *Multilevel analysis: Techniques and applications* (2nd ed.). New York, NY: Routledge.

Hox, J. J., & Maas, C. (2001). The accuracy of multilevel structural equation modeling with pseudobalanced groups and small samples. *Structural Equation Modeling: A Multidisciplinary Journal, 8*, 157–174.

Hox, J., & Roberts, J. K. (2011). *Handbook of advanced multilevel analysis.* New York, NY: Routledge.

Hu, L., & Bentler, P. M. (1999). Cutoff criteria for fit indexes in covariance structure analysis: Conventional criteria versus new alternatives. *Structural Equation Modeling: A Multidisciplinary Journal, 6*, 1–55.

Jöreskog, K. G., & Sörbom, D. (2006). *LISREL 8.8 for Windows* [Computer software]. Skokie, IL: Scientific Software International, Inc.

Kalton, G. (1977). Practical methods for estimating survey sampling errors. *Bulletin of the International Statistical Institute, 47*, 495–514.

Kalton, G. (1983). Models in the practice of survey sampling. *International Statistical Review, 51*, 175–188.

Kaplan, D., & Elliott, P. R. (1997a). A didactic example of multilevel structural equation modeling applicable to the study of organizations. *Structural Equation Modeling: A Multidisciplinary Journal, 4*, 1–24.

Kaplan, D., & Elliott, P. R. (1997b). A model-based approach to validating education indicators using multilevel structural equation modeling. *Journal of Educational and Behavioral Statistics, 22*, 323–347.

Kaplan, D., & Ferguson, A. J. (1999). On the utilization of sample weights in latent variable models. *Structural Equation Modeling: A Multidisciplinary Journal, 6*, 305–321.

Kenny, D. A., & Judd, C. M. (1986). Consequences of violating the independence assumption in analysis of variance. *Psychological Bulletin, 99*, 422–431.

Kish, L. (1965). *Survey sampling.* New York, NY: John Wiley & Sons, Inc.

Kish, L. (1995). Methods for design effects. *Journal of Official Statistics, 11*, 55–77.

Kish, L., & Frankel, M. R. (1974). Inference from complex samples. *Journal of the Royal Statistical Society, 36* (Series B), 1–37.

Kline, R. B. (2011). *Principles and practice of structural equation modeling* (3rd ed.). New York, NY: The Guilford Press.

Korn, E. L., & Graubard, B. I. (1995). Examples of differing weighted and unweighted estimates from a sample survey. *The American Statistician, 49*, 291–295.

Lee, E. S., Forthofer, R. N., & Lorimor, R. J. (1989). *Analyzing complex survey data.* Newbury Park, CA: Sage Publications, Inc.

Lohr, S. L. (1999). *Sampling: Design and analysis.* Pacific Grove, CA: Duxbury Press.

Longford, N. T. (1995). *Model-based methods for analysis of data from 1990 NAEP trial state assessment.* Washington, DC: National Center for Education Statistics: 95–696.

Lüdtke, O., Marsh, H. W., Robitzsch, A., Trautwein, U., Asparouhov, T., & Muthén, B. (2008). The multilevel latent covariate model: A new, more reliable approach to group-level effects in contextual studies. *Psychological Methods, 13*, 203–229.

Marsh, H. W., Ludtke, O., Nagengast, B., Trautwein, U., Morin, A. J. S., Abduljabbar, A. S., & Koller, O. (2012). Classroom climate and contextual effects: Conceptual and methodological issues in the evaluation of group-level effects. *Educational Psychologist, 47*, 106–124.

McDonald, R. P., & Goldstein, H. (1989). Balanced versus unbalanced designs for linear structural relations in two-level data. *British Journal of Mathematical and Statistical Psychology, 42,* 215–232.

Mehta, P. D., & Neale, M. C. (2005). People are variables too: Multilevel structural equations modeling. *Psychological Methods, 10,* 259–284.

Muthén, B. O. (1991). Multilevel factor analysis of class and student achievement components. *Journal of Educational Measurement, 28,* 338–354.

Muthén, B. O. (1994). Multilevel covariance structure analysis. *Sociological Methods & Research, 22,* 376–398.

Muthén, B. O. (1997). Latent variable modeling of longitudinal and multilevel data. In A. E. Raftery (Ed.), *Sociological Methodology 1997* (pp. 453–480). Cambridge, MA: Blackwell.

Muthén, B., & Asparouhov, T. (2008). Growth mixture modeling: Analysis with non-Gaussian random effects. In G. Fitzmaurice, M. Davidian, G. Verbeke, & G. Molenberghs (Eds.), *Longitudinal data analysis* (pp. 143–165). Boca Raton, FL: Chapman & Hall/CRC Press.

Muthén, B., & Asparouhov, T. (2011). Beyond multilevel regression modeling: Multilevel analysis in a general latent variable framework. In J. Hox & J. K. Roberts (Eds.), *Handbook of advanced multilevel analysis* (pp. 15–40). New York, NY: Taylor & Francis.

Muthén, B. O., & Satorra, A. (1995). Complex sample data in structural equation modeling. In P. V. Marsden (Ed.), *Sociological methodology* (pp. 267–316). Washington, D.C.: American Sociological Association.

Rabe-Hesketh, S., & Skrondal, A. (2006). Multilevel modeling of complex survey data. *Journal of the Royal Statistical Society, Series B, 60,* 23–56.

Raudenbush, S. W., & Bryk, A. S. (2002). *Hierarchical linear models: Applications and data analysis methods.* Newbury Park, CA: Sage Publications, Inc.

Raykov, T., & Penev, S. (2009). Estimation of maximal reliability for multiple-component instruments in multilevel designs. *British Journal of Mathematical and Statistical Psychology, 62,* 129–142.

Ryu, E., & West, S. G. (2009). Level-specific evaluation of model fit in multilevel structural equation modeling. *Structural Equation Modeling: A Multidisciplinary Journal, 16,* 583–601.

Scariano, S. M., & Davenport, J. M. (1987). The effects of violations of independence assumptions in the one-way ANOVA. *The American Statistician, 41,* 123–129.

Skinner, C. J., Holt, D., & Smith, T. M. F. (1989). *Analysis of complex surveys.* Chichester, England: John Wiley & Sons.

Snijders, T. A. B., & Bosker, R. J. (2012). *Multilevel Analysis: An introduction to basic and advanced multilevel modeling* (2nd ed.). London: Sage Publications.

Stapleton, L. M. (2002). The incorporation of sample weights into multilevel structural equation models. *Structural Equation Modeling: A Multidisciplinary Journal, 9,* 475–502.

Stapleton, L. M. (2006). An assessment of practical solutions for structural equation modeling with complex sample data. *Structural Equation Modeling: A Multidisciplinary Journal, 13,* 28–58.

Stapleton, L. M. (2008). Variance estimation using replication methods in structural equation modeling with complex sample data. *Structural Equation Modeling: A Multidisciplinary Journal, 15,* 183–210.

U. S. Department of Education (2001). *ECLS-K, Base Year Public-Use Data File, Kindergarten Class of 1998–99: Data Files and Electronic Code Book: (Child, Teacher, School Files).* Washington, D.C.: National Center for Education Statistics.

Widaman, K. F., & Thompson, J. S. (2003). On specifying the null model for incremental fit indexes in structural equation modeling. *Psychological Methods, 8,* 16–37.

CHAPTER 14

BAYESIAN STRUCTURAL EQUATION MODELING

Roy Levy and Jaehwa Choi

Bayesian approaches are receiving an increasing amount of attention in structural equation modeling (SEM) as viable, if not preferable, alternatives to more traditional modeling approaches. With recent theoretical developments, practical applications, and production of software serving as a guide, we anticipate that Bayesian concepts and practices are likely to play an increasing role in the future of SEM and related modeling techniques. This chapter describes Bayesian approaches to SEM by presenting concepts associated with Bayesian modeling, illustrating procedures for conducting SEM analyses in this framework, and discussing points of departure from frequentist approaches. Conceptual and practical strengths and weaknesses of the current state of the art will be discussed with an eye toward areas for future development.

We do not assume that readers are familiar with principles of Bayesian statistical modeling. Thus, before discussing their application to SEM, we review foundational principles of Bayesian inference. In addition, we discuss key concepts associated with Markov chain Monte Carlo (MCMC) estimation, the emergence of which has precipitated the explosion in applications of Bayesian inference to SEM and other complex modeling

Structural Equation Modeling: A Second Course (2nd ed.), pages 563–623
Copyright © 2013 by Information Age Publishing

techniques. Readers interested in more comprehensive introductory treatments of Bayesian data analysis, modeling, and estimation might find such treatments in Gelman, Carlin, Stern, and Rubin (2003), Jackman (2009), Gill (2007), Lynch (2007), and Press (1989). For classic presentations of normal-theory or linear models such as ANOVA and regression that form the basis for SEM, see Box and Tiao (1973) and Lindley and Smith (1972); see also Gelman and Hill (2007) for a more modern presentation. Barnett (1999) and Jaynes (2003) provided thorough philosophical and conceptual accounts contextualized in comparison to other inferential paradigms.

INTRODUCTION TO BAYESIAN INFERENCE

Review of Frequentist Inference

To facilitate the exposition of Bayesian approaches to inference in SEM, we briefly review the frequentist approach that has historically been the dominant paradigm in SEM, highlighting maximum likelihood as it will serve as a launching point for Bayesian inference. Letting $\mathbf{X}$ and θ denote the observed data and full collection of model parameters, respectively, a model is typically constructed by specifying the conditional distribution of the data given the model parameters, denoted $P(\mathbf{X}|\theta)$. In the frequentist paradigm, most model parameters are treated as fixed (i.e., constant) and unknown, and model fitting and parameter estimation comes to finding point estimates for the parameters. Assuming multivariate normality of the observed variables, common parameter estimation routines within this tradition include maximum likelihood (ML; Bollen, 1989), normal theory generalized least squares (GLS; Browne, 1974), and weighted least squares (WLS; Browne, 1984). These estimation routines can provide consistent, efficient, and unbiased parameter estimates, asymptotic standard errors, and an omnibus test of data-model fit if sample size is adequate and if the model is properly specified (see Bollen, 1989, for more details of these estimation routines in SEM).

Because closed-form solutions are generally not available in SEM, parameter estimates in all of these traditional estimation procedures are typically obtained via iterative methods, in which initial values are evaluated and subsequent candidate values are obtained via a numerical search algorithm using the first and/or second derivatives of the target function for optimization (Jöreskog, 1969; Lee, 2007; Mulaik, 2009). These estimation methods are known as *gradient-based* or *Newtonian* search algorithms: using the first and/or second derivatives of a target function, the algorithm iteratively searches for the parameter estimates by optimizing the target function (for more details of the general gradient-based search algorithms, see Süli & Mayers, 2003; for more specific details in SEM contexts, see Lee, 2007; Mulaik, 2009).

For example, at a conceptual level, ML estimation seeks to answer the question, "What are the values of the parameters that yield the highest probability of observing the values of the data that were in fact observed?" That is, ML estimation seeks the single best choice for the parameters, where the notion of "best" is operationalized as maximizing the likelihood function. To pursue this, ML proceeds as follows. Once values for **X** are observed, they can be entered into the conditional probability expression to induce a likelihood function. That is, the likelihood function, denoted $L(\theta|\mathbf{X})$, is the same expression as the conditional probability expression $P(\mathbf{X}|\theta)$, where the difference in notation reflects that when viewed as a likelihood function, the values of the data are known and the expression is viewed as varying over different possible values of θ. ML estimation then comes to finding the values of θ that maximize $L(\theta|\mathbf{X})$. Figure 14.1 illustrates a hypothetical likelihood function for a single parameter, where the maximum occurs where the parameter value is C; thus C is the ML estimate of the parameter.

Metaphorically speaking, the goal of ML is to find the highest peak among all of the mountains in a particular range. Determining parameter estimates using gradient-based estimation is akin to climbing to the top of a mountain blindfolded, where the climber is unable to broadly survey the terrain, and must instead incrementally move in ways that seem optimal in light of the limited knowledge possessed by the feel of the mountain's slope at her feet. The biggest challenge is that there might be multiple peaks that one could aim for (i.e., multiple *local maxima* exist), and that the peak that one ends up climbing depends on where one starts the climb, as some peaks are hidden from one's view at a particular location (i.e., the *global maximum* might not be reached). For example, if one starts climbing from point B in Figure 14.1, one will reach the desired *global* maximum C (the overall highest peak). If one starts climbing from another point, say A, one might wind up only at a *local* maximum, D (a tall peak, but not the overall highest one). Unfortunately, there is no guarantee that one will obtain the

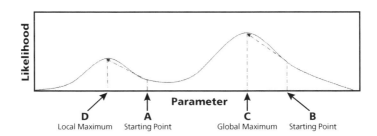

Figure 14.1   Visualization of the ML estimation procedure.

global maximum as the stopping point of the iterative search algorithm will depend on the actual starting value. Difficulties with local maxima and other threats to model convergence are usually exacerbated in complex models and small samples, which might pose problems for SEM analysts working in these contexts.

In frequentist approaches, uncertainty in estimation is captured by standard errors of the estimates, computed using the matrix of second-order partial derivatives (the Hessian matrix), which can be employed to construct confidence intervals. See Bollen (1989, ch. 4), Lee (2007, ch. 3), and Mulaik (2009, ch. 7) for descriptions of ML and other approaches to estimation of SEM in this paradigm, and see complementary chapters in these sources on the use of these estimates in constructing test statistics for evaluating data-model fit. Importantly, these estimation routines rely on asymptotic arguments to justify the calculation of parameter estimates, standard errors, or the assumed sampling distributions of the parameter estimates and associated test statistics (Bollen, 1989).

Moreover, the interpretations of point estimates, standard errors, and confidence intervals derive from the frequentist perspective in which the parameters are treated as *fixed* (constant) and it is inappropriate to discuss fixed parameters probabilistically. Thus, in frequentist inference, distributional notions of uncertainty and variability concern the parameter estimates, and are rooted in the distributional properties of the data. Specifically, the standard error is a measure of the variability of the parameter estimator, which is the variability of the parameter *estimates* upon repeated sampling of the data from the population. Likewise, the probabilistic interpretation of a confidence interval rests on the sampling distribution of the parameter estimates upon repeated sampling of data, and applies to the process of interval estimator construction. These notions refer to the variability and likely values *of a parameter estimator* (be it a point or interval estimator), that is, the distribution of *parameter estimates* upon repeated sampling, not the variability and likely values *of the parameter itself*. As discussed in detail in the next section, the Bayesian approach does not model the parameters as fixed values. Importantly, this conceptual distinction has implications for parameter estimation, quantifying uncertainty in estimation, and interpreting the results from fitting models to data.

Finally, in most applications of SEM certain unknowns are viewed as varying (random) rather than fixed. Specifically, the values of the latent variables (i.e., factor scores) are viewed in this way. However, invoking the distributional assumption (e.g., normality assumption of latent variables) allows for the construction of the likelihood in a way that does not involve the values of latent variables and, hence, they do not enter into the estimation; see Equation 14.18 and the surrounding discussion below. The same holds for frequentist SEM estimation strategies such as GLS and WLS (Bollen, 1989).

Bayesian Inference

Bayes' Theorem and the Mechanics of Bayesian Inference

Bayesian approaches to inference and statistical modeling share several features with frequentist approaches reviewed above. Both approaches treat the data as random and accordingly assign the data a distribution. As discussed above, when this distribution is structured as conditional on model parameters, and values of the data are observed, it may be viewed as a likelihood function. As such, the likelihood function induced by observing values for the data plays a key role in Bayesian inference just as in ML inference.

However, a Bayesian approach differs from frequentist approaches in several key ways. As briefly reviewed above, frequentist approaches treat some of the model parameters as fixed. As a consequence, distributional notions of uncertainty and variability (e.g., confidence intervals, standard errors) are attached to the parameter estimators, and are rooted in the sampling variability of the data.

In contrast, a Bayesian approach is characterized by treating the model parameters—indeed, all entities—as *random* and models them via distributions.[1] In a Bayesian analysis, the model parameters are assigned a *prior distribution*, a distribution chosen by the analyst to possibly reflect substantive, a priori knowledge, beliefs, or assumptions about the parameters. Bayesian inference is then conducted by combining the prior distribution with the likelihood to yield the *posterior distribution*, which in a Bayesian analysis constitutes the solution that results from fitting the model to the data. Bayes' theorem states that the posterior distribution for the model parameters θ given the data $\mathbf{X}$ is

$$P(\theta|\mathbf{X}) = \frac{P(\mathbf{X}|\theta)P(\theta)}{P(\mathbf{X})} = \frac{P(\mathbf{X}|\theta)P(\theta)}{\int_{\theta} P(\mathbf{X}|\theta)P(\theta)\,d\theta} \propto P(\mathbf{X}|\theta)P(\theta), \quad (14.1)$$

where $P(\theta|\mathbf{X})$ is the posterior distribution, $P(\theta)$ is the prior distribution, $P(\mathbf{X}|\theta)$ is the likelihood expressed as a conditional distribution of $\mathbf{X}$ given θ, and $P(\mathbf{X})$ is the marginal distribution of the data, which, as seen in the middle of Equation 14.1, may be obtained by integrating[2] the conditional distribution of the data over the prior distribution of the parameters. The marginal distribution of the data in the denominator serves as a normalizing constant to ensure that the resulting posterior density integrates to 1. Dropping the denominator on the right side of Equation 14.1 reveals that the posterior distribution is proportional to the product of the likelihood and the prior.

It is instructive to consider the conceptual interpretations of the right side of Equation 14.1. The posterior distribution can be interpreted as the

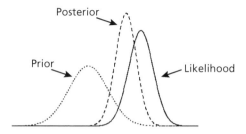

Figure 14.2 Visualization of a prior (dotted line), likelihood (solid line), and posterior (dashed line).

result of updating the prior distribution in light of the information in the data (contained in the likelihood). Alternatively, it may be interpreted as taking what the data imply about the likely values of the parameters (contained in the likelihood) and tempering that in light of what was believed, assumed, or known a priori about the parameters before observing the data (contained in the prior). Graphically, these concepts are illustrated in Figure 14.2, where the posterior distribution is seen to be a synthesis or compromise between the information in the prior and the information in the likelihood.

The posterior distribution constitutes the solution obtained from fitting the model to the data. The posterior can be summarized using familiar tools for characterizing distributions. In SEM, posterior means and standard deviations are common choices for point estimates and measures of variability, respectively. Percentiles may be reported or used to construct intervals. For example, a 95% central *credibility interval* is formed by calculating the 2.5th and 97.5th percentiles of the posterior. Unlike frequentist confidence intervals, a credibility interval constitutes a direct probabilistic statement about the unknown parameter (i.e., the parameter falls in this interval with probability .95). Similarly, probabilistic statements can be asserted for ranges of parameters or for the relations among parameters (e.g., the probability that one parameter exceeds a selected value, or the probability that one parameter exceeds another).

On the Use of Distributions for Parameters in Modeling

We expect that readers steeped in frequentist traditions might not be immediately comfortable with probability statements about parameters such as those in the previous paragraph. How can we talk about a parameter probabilistically? And when we say that a parameter falls within an interval with probability .95, is there a single unknown true value for the parameter? A distribution? Both? As Bayesian approaches differ from frequentist approaches that have dominated SEM, it is worth discussing the key difference between them, namely the treatment of parameters as random and the accompanying use of probabilistic expressions for parameters, both in the form of prior distributions in specifying models and the resulting posterior distribution as a solution.

It would be misleading to suggest there is a single, monolithic Bayesian philosophy or perspective on the related issues of (a) treating the parameters as random, (b) how to select prior distributions, (c) interpretations of prior and posterior distributions, (d) the definition of probability and meaning of probabilistic statements more generally, (e) the nature of statistical modeling, and (f) associated ontological commitments. Rather, there is considerable debate within the Bayesian community, as Bayesian analyses have been motivated (and criticized) from a number of perspectives on these issues. Covering these issues and the competing philosophical perspectives in any depth is beyond the scope and intent of this chapter. The reader is referred to Barnett (1999), Jaynes (2003), Williamson (2010), Gelman (2011), Senn (2011), and Gelman and Shalizi (2013) for such discussions. Briefly, we put forth two perspectives that motivate Bayesian analyses and support interpretations of parameters probabilistically. One perspective draws on the interpretation of probability as the language of uncertainty (and certainty). As such we assign prior distributions to unknown parameters to reflect our uncertainty about them, regardless of whether the analyst conceives of a parameter as having one true value or not (i.e., as fixed or random in the conventional sense).

A slightly different approach draws from the perspective that parameters and distributions are just some of the building blocks in a model that is necessarily wrong, though hopefully useful (Box, 1976). A model represents a simplification of the world, where salient features are represented while others are suppressed, constructed by the analyst to facilitate the desired inferences. A prior distribution is just one of the many assumptions involved in modeling, on par with other assumptions such as the distribution of the data or likelihood (e.g., linearity or normality), assumptions regarding sampling (e.g., independence of subjects), parameters (e.g., a variance is greater than 0), and relations among parameters (e.g., equality). Like specifying other parts of a model with these assumptions—e.g., in SEM, it is common to specify a likelihood that reflects assumptions of normality and independence of subjects—specifying a prior distribution is then just an instance of building in assumptions or substantive a priori beliefs to make the model consistent with what is known about the real-world situation (Gelman et al., 2003). We then engage in probability-based reasoning through the machinery of Bayes' theorem. The result of this process is the posterior distribution, which represents our uncertainty regarding the parameter (whether it is conceived as a fixed, random, or simply a convenient fiction) conditional on the data and the assumed model expressed in terms of the prior distribution and likelihood.

Modeling parameters distributionally is a flexible approach to specifying models that may be viewed as a generalization of conventional approaches from frequentist traditions. For an example from SEM, consider the case of loadings for observables. In a frequentist approach, if an observable is specified as loading on a latent variable, the loading will be estimated using the information in the likelihood. Specifying an observable as not loading on a latent variable may be viewed as specifying the loading for that observable on that latent variable to be 0. Each of these may be modeled in a Bayesian analysis that specifies parameters distributionally. The use of a uniform prior over the real line yields a posterior that is driven only by the likelihood, and is akin to the frequentist approach where the parameter is included. The situation where the loading is constrained to 0 may be modeled in a Bayesian approach by specifying a prior for this parameter with all of its mass at 0. Thus, at one end of the spectrum, the use of a prior distribution with all of its mass at one point (in this example, 0) can be viewed as corresponding to the conventional approach in which the parameter is fixed to that value. Viewing a prior distribution as a model assumption, this suggests that fixing a parameter by concentrating all the mass of the prior at that value represents a particularly strong assumption or belief about the parameter. At the other end of the spectrum, the use of uniform distribution reflects a very weak assumption or belief about the parameter. Importantly, in contrast to a frequentist approach, a Bayesian approach affords for room in between. By choosing a prior distribution of a particular form (shape, central tendency, variability, etc.) we can encode assumptions of varying strengths or beliefs of various levels of certainty. This is particularly natural if a prior can be constructed where one of its parameters captures this strength. Continuing with the example, using a normal prior for a loading allows for the specification of the certainty via the prior variance. A prior variance of 0 indicates the parameter is constrained to be equal to the value of the prior mean (e.g., 0); increasing the prior variance represents a weakening of this assumption. For more examples from SEM, see Muthén and Asparouhov (2012), who advocated the use of small-variance priors for parameters over the traditional approach of fixing parameters at certain values. This flexibility in using distributional representations to encode a continuum of possibilities has been exploited to great advantage in other modeling contexts. For example, in multilevel modeling this same machinery allows for partial pooling among groups, which lies between the extremes of complete pooling or no pooling among groups (Gelman & Hill, 2007).

Choosing Prior Distributions

As Equation 14.1 and the preceding discussion highlights, a key aspect of Bayesian inference is the incorporation of prior distributions into an analysis. How then is an analyst to specify prior distributions? There are several issues to consider when specifying prior distributions. First, prior distributions should be consistent with, and embody, substantive knowledge about the parameters. This knowledge may be based on the meaning of the parameter and the role that it plays. For example, a distribution with support over the entire real number line (e.g., a normal distribution) might be an appropriate prior for a loading or path coefficient in SEM, but it would be inappropriate for variances, which cannot be negative. In addition to these theoretical considerations, an important source for choosing prior distributions can come from prior research. For example, if prior research (say, from analyses of an already existing dataset, or from published research) has yielded or informed beliefs for the parameters, the prior distribution can be based on such beliefs. Indeed, the posterior distribution from one study can serve as a prior distribution for another study. In this way, the use of a prior distribution allows for the formal inclusion of past research into the current analysis. Having flexible prior distributions allows for more nuanced specifications and sensitivity analyses (e.g., using a beta distribution for an unknown proportion allows for a variety of shapes over the unit interval, or using a *t* distribution rather than a normal distribution allows for thicker-tailed distributions).

The choice of prior distributions also impacts the necessary computation needed to obtain or estimate posterior distributions. In particular, *conjugate* prior distributions are those that, when combined with the likelihood, yield a posterior distribution that is a member of the same family as the prior. For example, the use of a beta prior for the unknown parameter of a binomial distribution, combined with the binomial likelihood induced by observing the data, yields a posterior that is also a beta distribution (e.g., Gelman et al., 2003). Historically, the use of conjugate priors was important to facilitate analytical computation of posterior distributions. In light of modern simulation-based estimation methods (described in a following section), this issue is less critical than before. However, the choice of conjugate or semi-conjugate (Gelman et al., 2003) priors can ease the computational burden and greatly speed up the necessary computing time. Nevertheless, modern estimation techniques have, in a sense, freed Bayesian analysts from a dependence on conjugate priors. To that extent, we encourage consideration of substantive knowledge as the primary criterion. In the vast majority of situations, we should gladly pay the price of computational complexity for accuracy of modeling one's belief structure. If the prior is flexible, computationally efficient, and easily interpretable, all the better.

A few final points regarding the choice of priors are worth making, as researchers not steeped in the Bayesian tradition might feel apprehensive about the use of prior distributions. First, sensitivity analyses in which solutions from models using different priors can be compared to reveal the robustness of the inferences to the priors or the unanticipated effects of the priors (for examples in SEM, see Dunson, Palomo, & Bollen, 2005; Muthén & Asparouhov, 2012). Second, the posterior represents a balancing of the contributions of the data, in terms of the likelihood and the prior. As the contribution of the data increases, the posterior becomes less dependent on the prior and more closely resembles the likelihood in shape, under most circumstances.[3] Thus, in the case of unimodal and symmetric posteriors, the posterior mean (or median and mode) and the posterior standard deviation are asymptotically equivalent to the maximum likelihood estimate and standard error, respectively. Importantly, even when posterior point estimates and standard deviations are *numerically similar* to their frequentist counterparts, they carry *conceptually different* interpretations; the former afford probabilistic reasoning about the parameters, the latter do not. In general, there are two main ways to decrease the dependence of the posterior on the prior: (1) increase the sample size of the data, and (2) choose a diffuse prior (e.g., a uniform prior, a normal prior with large variance).

Controversy about Bayesian Inference

There has been considerable debate over the legitimacy of Bayesian statistical inference, and we shall not attempt to cover all the relevant issues here. Briefly, we summarize the key elements of the debate and our position. Importantly, there is no debate over the mechanics of Bayes' theorem, as it follows from the law of total probability. The arguments stem from whether it should be applied to treat certain problems of inference. Though some applications have enjoyed relatively widespread acceptance (e.g., the use of Bayes' theorem to yield posterior probabilities of group membership in latent class and mixture models), frequentist approaches in statistical inference have dominated much of statistical modeling in the last 100 years. This includes the development of SEM, as reflected in textbooks (e.g., Bollen, 1989) and quantitative training (Aiken, West, & Millsap, 2008). However, there are two main classes of arguments in favor of adopting a Bayesian approach. The first is philosophical in nature, and consists of arguments of why it is appropriate to treat parameters as random, a few of which have been briefly presented in the section "On the Use of Distributions for Parameters in Modeling." These arguments have countervailing positions, and the debate between them often spins into debates over different interpretations or groundings of probability. These sorts of arguments have received considerable attention in the literature, and we will not venture into them here (see Barnett, 1999, for relevant background, discussion, criticism from a variety of perspectives, and additional references), other than to say that we believe that Bayesianism as a philosophical position is defensible, if not preferable. For those who either do not share this philosophical position or do not think philosophical positions have any bearing, the second class of arguments are more practical in nature and advocate adopting a Bayesian approach to the extent that it is useful. We support these arguments, which advance the notion that adopting a Bayesian approach allows one to fit models and reach conclusions about research questions more readily than can be done using frequentist approaches, and allows for the construction and use of complex models that are more aligned with features of the real world and the data. In the concluding section of this chapter, we review a number of areas and applications where a Bayesian approach outperforms a frequentist approach, or facilitates modeling and inference in contexts in which frequentist approaches are challenging or intractable.

BAYESIAN SEM

In the following sections we develop Bayesian approaches to SEM. To facilitate the exposition of the unique features of a Bayesian approach, we first present SEM as it is viewed in conventional approaches, via systems of structural equations and common assumptions. In subsequent subsections, we develop several Bayesian approaches by building on this foundation.

Traditional Formulation of SEM

Beginning with a structural model for latent variables, let Ξ be an $(N \times K)$ matrix containing the values of the latent variables (factor scores) for N subjects along K exogenous latent variables. At the individual level, $\xi_i = (\xi_{i1}, \dots, \xi_{iK})'$ is the $(K \times 1)$ vector of K exogenous latent variables for subject i (i.e., the transpose of the ith row of Ξ), with mean vector κ and covariance matrix Φ across subjects. Further, let $\mathbf{H}$ be an $(N \times L)$ matrix containing the values of the latent variables (factor scores) for N subjects along L endogenous latent variables. At the individual level, $\eta_i = (\eta_{i1}, \dots, \eta_{iL})'$ is the $(L \times 1)$ vector of L endogenous latent variables for subject i (i.e., the transpose of the ith row of $\mathbf{H}$). The L endogenous latent variables contained in η_i are related to each other and to ξ_i through the latent structural regression model

$$\eta_i = \alpha + \mathbf{B}\eta_i + \Gamma\xi_i + \zeta_i, \tag{14.2}$$

where α is an $(L \times 1)$ vector of intercepts, $\mathbf{B}$ is an $(L \times L)$ matrix of path coefficients relating the η_i to each other, Γ is an $(L \times K)$ matrix of path coefficients relating η_i to ξ_i, and ζ_i is an $(L \times 1)$ vector of disturbance variables, where it is typically assumed that $\zeta_i \sim N(\mathbf{0}, \mathbf{\Psi})$. We assume that $\mathbf{B}$ has zeros on the diagonal and η_i can be rearranged such that $\mathbf{B}$ is lower triangular. The notational distinction between exogenous and endogenous latent variables aids in the exposition of how the model is constructed. To combine them, let $M = K + L$ be the number of latent variables, and let $\mathbf{F}_i = (\xi_i, \eta_i) = (\xi_{i1}, \ldots, \xi_{iK}, \eta_{i1}, \ldots, \eta_{iL})'$ denote the $(M \times 1)$ vector of all the latent variables for subject i.

Turning to the observable variables, let $\mathbf{X}$ be an $(N \times J)$ matrix containing the potentially observable values from N subjects to J observable variables. These observables are related to a set of latent variables via a factor analytic measurement model

$$\mathbf{X}_i = \mathbf{\tau} + \mathbf{\Lambda} \mathbf{F}_i + \mathbf{\delta}_i, \qquad (14.3)$$

where $\mathbf{X}_i = (X_{i1}, \ldots, X_{iJ})'$ is the $(J \times 1)$ vector of observed values from subject i (i.e., the transpose of the ith row of $\mathbf{X}$), $\mathbf{\tau} = (\tau_1, \ldots, \tau_J)'$ is a $(J \times 1)$ vector of intercepts, $\mathbf{F}_i$ is the $(M \times 1)$ vector of latent variables for subject i, $\mathbf{\Lambda}$ is a $(J \times M)$ matrix of loadings, and $\mathbf{\delta}_i$ is a $(J \times 1)$ vector of errors, where it is typically assumed that $\mathbf{\delta}_i \sim N(\mathbf{0}, \mathbf{\Theta})$. For simplicity we further assume that $\mathbf{\Theta}$ is diagonal; this assumption may easily be relaxed in ways discussed later. The reader should note that despite the similarity in symbols, $\mathbf{\Theta}$ and its elements (e.g., the jth diagonal element θ_{jj}) refer to error (co)variances and should be distinguished from $\mathbf{\theta}$ which we have used to refer to the full collection of parameters.

Path diagrammatic graphical representations are common in SEM for specifying models. Figure 14.3 presents such a representation of a model with: two exogenous, potentially correlated latent variables; one endogenous latent variable with direct effects from the two exogenous latent variables; a second endogenous latent variable with a direct effect from the other endogenous variable; where each latent variable has three observed variables as indicators.

This section was intended as a typical presentation of SEM, where systems of equations (i.e., Equations 14.2 and 14.3) play the central role. It should be noted that common path diagrammatic representations in SEM (Figure 14.3) convey the same information as expressed in the equations. In contrast, a Bayesian perspective views the model more in terms of probability distributions. The familiar equations from SEM will of course play a key role, but as will be seen the equations manifest themselves in terms of specifying parameters for certain distributions. In the next few sections, we develop several Bayesian approaches to SEM, where we shift our emphasis from thinking *equationally*, as we have done so far, to thinking *distributionally*.

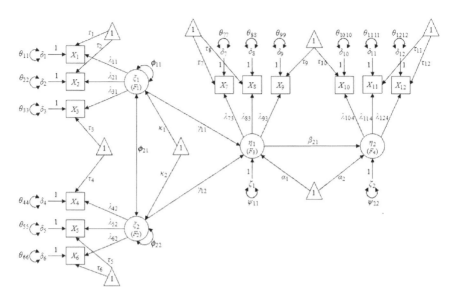

Figure 14.3 Path diagram for a structural equation model. The parenthetical labeling of the latent variables in the ovals reflects the notation combining the exogenous and endogenous latent variables.

Bayesian SEM with Subject-Level Data

In a Bayesian analysis, fitting a model involves obtaining the posterior distribution for all unknowns, given the observed data, as in Equation 14.1. In SEM, the collection of unknowns are: values of the exogenous and endogenous latent variables (Ξ and $\mathbf{H}$, respectively); means and covariances for the exogenous latent factors (κ and Φ, respectively); parameters for the structural equations relating the latent variables (α, $\mathbf{B}$, Γ, and Ψ); and parameters for the measurement model relating the observables to the latent variables (τ, Λ, and Θ). In Bayesian SEM, once values for the data are observed, the posterior distribution following Bayes' theorem is given by

$$P(\Xi,\mathbf{H},\kappa,\Phi,\alpha,\mathbf{B},\Gamma,\Psi,\tau,\Lambda,\Theta|\mathbf{X}) \propto P(\mathbf{X}|\Xi,\mathbf{H},\kappa,\Phi,\alpha,\mathbf{B},\Gamma,\Psi,\tau,\Lambda,\Theta) \quad (14.4)$$

$$\times P(\Xi,\mathbf{H},\kappa,\Phi,\alpha,\mathbf{B},\Gamma,\Psi,\tau,\Lambda,\Theta).$$

In order to write the normalized version of the posterior distribution (i.e., the middle representation in Equation 14.1), the marginal distribution of the data would be needed (i.e., the denominator in Equation 14.1). To obtain this value, we would need to integrate the right side of Equation 14.4 over the entire parameter space, which is computationally intractable in complex applications, including many in SEM. As discussed in a later section, modern Bayesian estimation allows for the estimation of posterior distributions without needing to analytically evaluate the marginal distribution of the data.

Specifying the model involves specifying the terms on the right side of Equation 14.4, which constitute specifying the full probability model for all the entities in the model (Gelman et al., 2003). We will treat each in turn.

The Conditional Probability of the Data

The first term on the right side of Equation 14.4 is the conditional probability of the data. When values of the data are known and the parameters are unknown, this is viewed as the likelihood function. Note that it is written here using all the parameters in the model on the right side of the conditioning bar. Conditional independence assumptions inherent in the model allow for the following simplification:

$$P(\mathbf{X}|\Xi,\mathbf{H},\kappa,\Phi,\alpha,\mathbf{B},\Gamma,\Psi,\tau,\Lambda,\Theta) = P(\mathbf{X}|\Xi,\mathbf{H},\tau,\Lambda,\Theta). \qquad (14.5)$$

The term on the right side of Equation 14.5 is the conditional probability of the data given the latent variables and the factor analytic measurement model parameters. In words, Equation 14.5 expresses that, given the latent variables Ξ and $\mathbf{H}$, the data are conditionally independent of the parameters that govern the distribution of the latent variables κ, Φ, α, $\mathbf{B}$, Γ, and Ψ.

Assumptions of independence of subjects and conditional independence assumptions inherent in the model allow for the following factorization of the conditional probability of the data (likelihood function):

$$P(\mathbf{X}|\Xi,\mathbf{H},\tau,\Lambda,\Theta) = \prod_{i=1}^{N} P(\mathbf{X}_i|\xi_i,\eta_i,\tau,\Lambda,\Theta) = \prod_{i=1}^{N}\prod_{j=1}^{J} P(X_{ij}|\xi_i,\eta_i,\tau_j,\lambda_j,\theta_{jj}). \quad (14.6)$$

More specifically, the assumption of independence of subjects allows for the factorization of the conditional probability of the observables into the product over subjects (the middle of Equation 14.6). Recalling that $\mathbf{F}_i = (\xi_i, \eta_i)$ denotes the full collection of latent variables for subject i, then following the assumption of normality of errors the conditional distribution of the observable values from subject i is

$$P(\mathbf{X}_i|\mathbf{F}_i,\tau,\Lambda,\Theta) \sim N(\tau + \Lambda\mathbf{F}_i,\Theta). \qquad (14.7)$$

Importantly, though the expressions in Equations 14.3 and 14.7 differ on the surface, they represent the same model. Combining the deterministic expression in Equation 14.3 with the assumption that $\delta_i \sim N(\mathbf{0}, \Theta)$ implies the probabilistic expression in Equation 14.7. Formulating the model in this latter way is an important step towards our aim of thinking distributionally. We emphasize this approach throughout the development of the Bayesian approach. Doing so is aligned with the machinery of probability-based reasoning via Bayes' theorem, as well as modern approaches to building Bayesian models, as described in the following section "Reconceiving Bayesian SEM."

The assumption that Θ is diagonal facilitates the further factorization into the product over observables (the right side of Equation 14.6), where the assumption of normality of errors implies that the conditional distribution of the observable value from subject i to observable variable j is

$$P(X_{ij} \mid \mathbf{F}_i, \tau_j, \lambda_j, \theta_{jj}) \sim N(\tau_j + \mathbf{F}_i \lambda_j', \theta_{jj}). \tag{14.8}$$

Equations 14.6 through 14.8 define the likelihood function implied by the traditional model described above.

The Prior Distribution—Begun

In a Bayesian analysis, all unknown entities are treated as random variables and assigned a prior distribution, which is the second term on the right side in Equation 14.4. The conditional independence assumptions in the model allow for the factorization of the prior distribution as

$$P(\Xi, \mathbf{H}, \kappa, \Phi, \alpha, \mathbf{B}, \Gamma, \Psi, \tau, \Lambda, \Theta) = P(\Xi \mid \kappa, \Phi) \times P(\mathbf{H} \mid \Xi, \alpha, \mathbf{B}, \Gamma, \Psi) \tag{14.9}$$
$$\times P(\kappa, \Phi, \alpha, \mathbf{B}, \Gamma, \Psi, \tau, \Lambda, \Theta).$$

To construct the prior distribution, we will draw heavily on assumptions of *exchangeability* (e.g., Lindley & Novick, 1981). A full treatment of exchangeability is beyond the scope of this chapter. The reader is referred to Jackman (2009) for a textbook introduction and Barnett (1999) for historical origins and connections to familiar assumptions of independence. For our purposes it is sufficient to say that, conceptually, two entities are exchangeable if we have the same beliefs about them. The first term on the right side of Equation 14.9 is the prior distribution for exogenous latent variables. An exchangeability assumption amounts to meaning that we have the same beliefs about each subject prior to observing the data. This implies that the joint distribution of the subjects' exogenous latent variables may be factored into the product of common prior distributions for each ξ_i (Lindley & Smith, 1972),

$$P(\Xi \mid \kappa, \Phi) = \prod_{i=1}^{N} P(\xi_i \mid \kappa, \Phi). \tag{14.10}$$

Assuming normality, the prior for the exogenous latent variables for each subject is

$$P(\xi_i | \kappa, \Phi) \sim N(\kappa, \Phi). \qquad (14.11)$$

Subject independence and exchangeability assumptions imply that the joint distribution of the subjects' endogenous latent variables, the second term on the right side of Equation 14.9, may be factored into the product of common prior distributions for each η_i

$$P(\mathbf{H} | \Xi, \alpha, \mathbf{B}, \Gamma, \Psi) = \prod_{i=1}^{N} P(\eta_i | \xi_i, \alpha, \mathbf{B}, \Gamma, \Psi). \qquad (14.12)$$

The structural model in Equation 14.2 and the assumption of normality of disturbances imply that

$$P(\eta_i | \xi_i, \alpha, \mathbf{B}, \Gamma, \Psi) \sim N(\alpha + \mathbf{B}\eta_i + \Gamma\xi_i, \Psi). \qquad (14.13)$$

On the surface it is unusual to see the term on the left side of the conditioning bar (here, η_i) appear in the right side of the expression. However, recalling that η_i is ordered such that $\mathbf{B}$ is lower triangular, we might write this model in a hierarchical manner where any element of η_i depends on the previous elements of η_i. If Ψ is additionally assumed to be diagonal, we can write the model in terms of the value of each endogenous latent variable

$$P(\eta_{il} | \xi_i, \eta_{i(<l)}, \alpha_l, \beta_l, \gamma_l, \psi_{ll}) \sim N(\alpha_l + \beta_l\eta_{i(<l)} + \gamma_l\xi_i, \psi_{ll}), \qquad (14.14)$$

where $\eta_{i(<l)}$ refers to the values for endogenous latent variables that preceded the lth endogenous latent variable, β_l and γ_l are the lth rows of $\mathbf{B}$ and Γ, and ψ_{ll} is the lth diagonal element of Ψ.

Interlude—What Bayes and ML Share

Before continuing with the specification of the prior distribution, it is worth noting that the preceding developments underlie ML as well as Bayesian analyses. This is true despite what appear to be differences; for example, typical frequentist formulations of SEM do not express the distribution of observables as conditional on the latent variables as in Equations 14.7 and 14.8. Rather, they are expressed in terms of other parameters, in this case latent variable means (i.e., κ for the exogenous latent variables and the model-implied values for the endogenous latent variables; see e.g., Equations 8.44 and 8.45 of Bollen, 1989). These differences can be reconciled by recognizing what is being conditioned on, and what is being marginalized over. For simplicity of illustration, consider a model with only exogenous factors, $\mathbf{F}_i = \xi_i$. Equation 14.7 implies that

$$E(\mathbf{X}_i | \xi_i, \tau, \Lambda, \Theta) = \tau + \Lambda \xi_i \qquad (14.15)$$

with an explicit conditioning on ξ_i. Marginalizing over the distribution of latent variables (see Equation 14.11),

$$E(\mathbf{X}_i | \kappa, \tau, \Lambda, \Theta) = \tau + \Lambda \kappa. \qquad (14.16)$$

This latter expression is akin to typical frequentist presentations (e.g., Equation 8.44 of Bollen, 1989). A similar analysis yields an analogous connection between the current formulation of the model for endogenous latent variables in Equation 14.13 and that of a typical frequentist formulation (e.g., Equation 8.43 of Bollen, 1989).

Such frequentist formulations may be viewed as expressing the model only with reference to the parameters that a frequentist approach treats as fixed, and without reference to the parameters that a frequentist approach treats as random (i.e., the latent variables/factor scores). Indeed, doing so is how the usual ML estimation is derived in SEM. In typical ML estimation in SEM, the following steps are taken. First, we construct the joint distribution of the observable and latent variables:

$$P(\mathbf{X}, \Xi, \mathbf{H} | \kappa, \Phi, \alpha, \mathbf{B}, \Gamma, \Psi, \tau, \Lambda, \Theta) = P(\mathbf{X} | \Xi, \mathbf{H}, \tau, \Lambda, \Theta) \times P(\Xi | \kappa, \Phi) \qquad (14.17)$$

$$\times P(\mathbf{H} | \Xi, \alpha, \mathbf{B}, \Gamma, \Psi).$$

The first term on the right side of Equation 14.17 is the conditional distribution of the data, defined in Equations 14.6 through 14.8. The second and third terms on the right side of Equation 14.17 are the distributions of the exogenous and endogenous latent variables for subjects, respectively, defined in Equations 14.10 through 14.14. A marginal likelihood is then obtained by integrating Equation 14.17 over the distributions of the exogenous and endogenous latent variables

$$P(\mathbf{X} | \kappa, \Phi, \alpha, \mathbf{B}, \Gamma, \Psi, \tau, \Lambda, \Theta) = \iint_{\Xi \mathbf{H}} P(\mathbf{X} | \Xi, \mathbf{H}, \tau, \Lambda, \Theta) \qquad (14.18)$$

$$\times P(\Xi | \kappa, \Phi) \times P(\mathbf{H} | \Xi, \alpha, \mathbf{B}, \Gamma, \Psi) d\Xi d\mathbf{H}.$$

Given the linear relations and normality assumptions in the model as formulated above, the resulting joint distribution in Equation 14.17 is normal, and so the distribution in Equation 14.18 is also normal. In ML, when values for the observables are known, the result of Equation 14.18 is viewed as a (marginal) likelihood, which is maximized.

The Prior Distribution—Resumed

Returning to the Bayesian analysis, the final term on the right side of Equation 14.9 is the prior distribution for the remaining parameters. In what follows, we adopt the common practice of assuming a priori independence between types of parameters and employ conjugate or semi-conjugate priors (Gelman et al., 2003; Rowe, 2003; see Lee, 2007, for approaches using recursive conditional distributions).

Following an exchangeability assumption regarding the observables, the intercepts are assigned a common prior distribution; likewise for the loadings and error variances. Intercepts and loadings are typically assigned normal prior distributions

$$\tau_j \sim N(\mu_\tau, \sigma_\tau^2), \tag{14.19}$$

$$\lambda_{jm} \sim N(\mu_\lambda, \sigma_\lambda^2), \tag{14.20}$$

and error variances are frequently assigned inverse-gamma distributions

$$\theta_{jj} \sim G^{-1}(\alpha_\theta, \beta_\theta), \tag{14.21}$$

where $G^{-1}(\ )$ denotes the inverse gamma distribution, and μ_τ, σ_τ^2, μ_λ, σ_λ^2, α_θ, and β_θ are parameters that govern these prior distributions that are chosen by the analyst.

Turning to the distributions for the latent variables (factor scores), an assumption of exchangeability regarding the exogenous latent variables allows for the specification of a common prior for all elements of κ; normal distributions are typically employed for the elements of κ:

$$\kappa_k \sim N(\mu_\kappa, \sigma_\kappa^2), \tag{14.22}$$

where μ_κ and σ_κ^2 are parameters that govern these prior distributions that are chosen by the analysts. An inverse-Wishart distribution, denoted by $W^{-1}(\ )$, is typically used for Φ (Arminger & Muthén, 1998; Lee, 2007; Rowe, 2003),

$$\Phi \sim W^{-1}(d\Phi_0, d), \tag{14.23}$$

where, in the parameterization of Spiegelhalter, Thomas, Best, and Lunn (2007), Φ_0 is a matrix of values reflecting prior expectation for Φ and $d \geq K$ is a chosen weight, with smaller values for d yielding a more diffuse prior distribution.

The remaining parameters concern the latent structural model. As was the case in specifying the parameters of the measurement model, exchangeability assumptions imply the use of a common prior for the same types of parameters and the normality assumptions motivate choosing normal priors for the intercepts and path coefficients

$$\alpha_l \sim N(\mu_\alpha, \sigma_\alpha^2), \tag{14.24}$$

$$\beta_{ll'} \sim N(\mu_\beta, \sigma_\beta^2), \tag{14.25}$$

$$\gamma_{lk} \sim N(\mu_\gamma, \sigma_\gamma^2), \tag{14.26}$$

where $\beta_{ll'}$ is the element in row l and column l' of **B** capturing the (partial) path coefficient from $\eta_{l'}$ to η_l, γ_{lk} is the element in row l and column k of **Γ** capturing the (partial) path coefficient from ξ_k to η_l, and μ_α, σ_α^2, μ_β, σ_β^2, μ_γ, and σ_γ^2 are parameters chosen by the analyst to define the prior distribution. Similarly, exchangeability and normality assumptions motivate the use of inverse-gamma distributions for all disturbances

$$\psi_{ll} \sim G^{-1}(\alpha_\psi, \beta_\psi), \tag{14.27}$$

where α_ψ and β_ψ are parameters that govern these prior distributions that are chosen by the analyst.

The above presentation has characterized the model in a fairly "open" manner, without imposing many of the restrictions seen in applications. As in traditional modeling strategies, identification of the location and scale of the latent variables may be imposed by fixing values of parameters (e.g., constraining loadings, factor means). Likewise, restrictions on parameters based on theory may be imposed by constraining parameters as appropriate (e.g., constraining loadings so that each measured variable reflects only one latent variable). On the other hand, many of the assumptions implied in the above presentation can be relaxed. Several examples are worth noting. The assumption that all loadings follow the same normal prior distribution (implicit in Equation 14.20) may be relaxed by specifying different prior means and variances for different loadings. Likewise, the assumptions of common priors inherent in Equations 14.19, 14.21, 14.22, and 14.24 through 14.27 may be relaxed. If covariances among the errors are desired, instead of modeling each error variance univariately, a multivariate approach may be taken by modeling **Θ** (or some of its submatrices) via an inverse-Wishart distribution as was done for **Φ**; the same holds for **Ψ** if covariances among the disturbances are desired. Conversely, assumptions of independence between exogenous latent variables may dictate univariate priors for each variance, rather than the matrix approach described above for **Φ**. Moreover, analysts need not make the same choices for the forms of the prior distributions. For examples, see Gelman (2006) and Gelman, Jakulin, Pittau, and Su (2008) on the use of alternative priors for variances and path coefficients. Finally, as discussed in the final section, Bayesian approaches can easily handle extensions of the model presented here (e.g., supporting of multiple observed or latent groups, clustered observations or other multilevel contexts).

Reconceiving Bayesian SEM

The above description of a Bayesian approach to SEM first presents the model in traditional terms that readers familiar with SEM should recognize, and then overlays a Bayesian perspective by modeling unknown parameters with prior distributions. Given the relative recency of the emergence of Bayesian approaches to SEM, we suspect most readers will find this approach most accessible for learning Bayesian SEM.

It is worth developing the model from a different perspective to highlight key principles of modern Bayesian modeling, specifically one that lays out the model by starting with the data and "building out" the model by specifying distributions in a *hierarchical* manner following assumptions of exchangeability. This perspective has come to dominate Bayesian modeling for complex models (Jackman, 2009), and applies to SEM as well as other modeling approaches.

We begin with the joint distribution of the data, $\mathbf{X}$. According to the adopted model, the distribution of $\mathbf{X}$ is as described in Equations 14.5 and 14.6, which introduces certain unknown parameters. That is, to model the joint distribution of the data, we have introduced the parameters τ, Λ, Θ, and $\mathbf{F} = (\Xi, \mathbf{H})$. As unknown entities, they require prior distributions. Starting with the first of these, we tackle the challenge of specifying a J-variate prior distribution for τ by assuming exchangeability, which allows for the specification of a common univariate prior for all elements of τ, given in Equation 14.19. The parameters that govern this (common) prior distribution are sometimes referred to as *hyperparameters* and in the current case are values chosen by the analyst. Similarly, an assumption of exchangeability allows for the specification of a common univariate prior for the elements of Λ with its hyperparameters in Equation 14.20. Likewise for Θ in Equation 14.21.

To model the prior distribution of the latent variables, $\mathbf{F}$ is partitioned into the exogenous and endogenous latent variables. Beginning with the exogenous latent variables Ξ, an assumption of exchangeability with respect to subjects allows us to specify a common prior for all subjects' exogenous latent variables as in Equation 14.11. This common prior involves other parameters, κ and Φ. As parameters that govern a prior distribution, these may also be regarded as hyperparameters. However, as unknown entities, these also require prior distributions. An assumption of exchangeability with respect to the latent exogenous variables allows for common univariate prior to be specified for the elements of κ in Equation 14.22 with its own hyperparameters. A prior distribution for Φ, Equation 14.23, with its own hyperparameters, then completes the prior for Ξ.

To recap this development for the exogenous latent variables, Ξ is a $N \times K$ matrix of unknowns, and in a Bayesian analysis we are therefore tasked with specifying a $N \times K$-variate prior distribution. To achieve this, we recursively make use of exchangeability assumptions to construct the prior in a hierarchical fashion. Equation 14.11 represents the first level of this hierarchy, and introduces (hyper)parameters κ and Φ, for which (hyper)prior distributions are needed. These are specified in the next level of the hierarchy, in Equations 14.22 and 14.23. To specify the former, again an assumption of exchangeability is invoked to allow for a common univariate prior for the elements of κ, in Equation 14.22. In short, by invoking exchangeability assumptions and approaching the problem in a hierarchical manner, we have solved the problem of specifying a $N \times K$-variate prior distribution with the following (collecting Equations 14.11, 14.22, and 14.23)

$$P(\xi_i \mid \kappa, \Phi) \sim N(\kappa, \Phi) \text{ for all } i;$$

$$\kappa_k \sim N(\mu_k, \sigma_k^2) \text{ for all } k;$$

$$\Phi \sim W^{-1}(d\Phi_0, d).$$

To complete the model, we turn to the endogenous latent variables **H**. These are expressed as distributions in Equations 14.12 and 14.13; to specify these distributions we introduce the (hyper)parameters α, **B**, Γ, and Ψ, which themselves require (hyper)prior distributions. A similar approach of invoking exchangeability allows for their hierarchical specification as in Equations 14.24 through 14.27.

This line of reasoning develops the model in a slightly different manner than SEM has typically been conceived. It is distinctly Bayesian in nature in that it (a) focuses on the specifications of (conditional) distributions, and (b) for every parameter that is introduced, a (prior) distribution—itself possibly conditional on other parameters—is specified. Combining this approach with assumptions of exchangeability allows for the efficient construction of prior distributions. The process concludes when there are no more unknown parameters in need of distributional specification; in the current case, all the terms on the right sides of Equations 14.19 through 14.27 are known values chosen by the analyst.

As seen, assumptions of exchangeability greatly facilitate the efficient construction of joint prior distributions via specifying a common distribution for its elements. This is not a requirement of Bayesian modeling. If assumptions of exchangeability are not warranted, the model can still be formulated. For example, if we cannot assume exchangeability between the measurement quality of two observables that measure a latent variable, we need not assign their loadings the same prior distribution as was done in Equation 14.20. Rather, each loading can have a different prior.

What this reflects is that if exchangeability cannot be assumed, we might construct the model to allow for an assumption of *conditional exchangeability* (Jackman, 2009; Lindley & Novick, 1981), which means elements are exchangeable conditional on some other relevant entity. For example, in multiple-group modeling, subjects from *different* groups might not be assumed to be exchangeable, but subjects may be assumed to be exchangeable *within* groups, that is, *conditional* on the grouping variable. In that case, a conditionally exchangeable prior for Ξ is given by

$$P(\xi_{ig} \mid \kappa_g, \Phi_g) \sim N(\kappa_g, \Phi_g) \text{ for all } i \text{ in group } g;$$

$$\kappa_{kg} \sim N(\mu_{kg}, \sigma_{kg}^2) \text{ for all } k \text{ in group } g;$$

$$\Phi_g \sim W^{-1}(d_g \Phi_{0g}, d_g),$$

where the additional subscripting by g indicates group-specific parameters.

Bayesian SEM with Summary-Level (or Moment-Level) Data

An alternative approach to model construction works with the summary level data. We present this in terms of an analysis of the covariance structure; extensions to models for mean structures follow similarly. The procedure employs the second-order moments contained in the covariance matrix of the observables, denoted by $\mathbf{S}$, as the data. Assuming multivariate normality of $\mathbf{X}$, this approach models the (summary) data using the Wishart probability density

$$\mathbf{S} \sim W(\Sigma(\theta)), \tag{14.28}$$

where $\Sigma(\theta)$ is the model-implied covariance matrix based on the model parameters θ. Following this approach, the likelihood function is given by $P(\mathbf{S} \mid \theta)$, where here θ contains the model parameters $\Phi, \mathbf{B}, \Gamma, \Psi, \Lambda,$ and Θ. The posterior distribution is then given as proportional to the product of this likelihood and the prior

$$P(\Phi, \mathbf{B}, \Gamma, \Psi, \Lambda, \Theta \mid \mathbf{S}) \propto P(\mathbf{S} \mid \Phi, \mathbf{B}, \Gamma, \Psi, \Lambda, \Theta) P(\Phi, \mathbf{B}, \Gamma, \Psi, \Lambda, \Theta). \tag{14.29}$$

The prior distribution may be specified in the same manner as above. Note that, in contrast to the individual-level approach, the specification of the distribution of the data does not involve the subjects' values of the latent variables (factor scores).

Comparing the Individual-Level and Summary-Level Approaches

There are a number of important features of the individual- and summary-level approaches with implications for researchers wishing to employ Bayesian SEM. First, both parameterizations utilize distributions rooted in normality assumptions that are well known (indeed, the assumed Wishart distribution for the covariance matrix in Equation 14.28 lies at the root of the ML fit function; see Bollen, 1989). Similarly, both approaches are easily represented and implemented using matrix algebra, in terms of matrix-based regression structures for the individual-level data or the model-implied covariance matrix for the summary-level data.

A critical aspect of the modern methods of estimating posterior distributions (discussed in a following section) is that these routines involve many iterations and may be computationally demanding. The time needed to conduct the estimation increases with each additional parameter to be estimated. Importantly, because the summary-level approach takes the covariance matrix (i.e., the second-order moments) as data, the number of parameters—and hence, the numerical cost—does not increase as sample size increases. In contrast, in the individual-level approach, each additional subject brings with it M latent variables in need of estimation. Thus, all else being equal, the summary-level approach is likely to be faster, especially in situations with large samples and many latent variables (Choi, Levy, & Hancock, 2006). The estimation-time advantage of the summary-level approach is likely to be mitigated as technological advances increase the computational power available to researchers. Recent implementations in AMOS and Mplus (see Appendices A, B, and D) represent a considerable step in this direction. A related point is that it is common in the SEM literature to report results in terms of moments rather than the raw data; hence, the summary-level approach may be more easily implemented for secondary analysis.

However, the use of the Wishart probability density function in modeling **S** in the summary-level approach follows from an assumption that the observed variables are normally distributed (Bollen, 1989). Although ML parameter estimates under such assumptions are often robust to violations of normality, the estimates of the standard errors and the model χ^2 can be severely biased (e.g., Yuan & Bentler, 1998). An extension of this approach to non-normally distributed data (including discrete data) is not straightforward. Although WLS (Browne, 1984) has been developed and been widely used to address this issue within the traditional gradient-based estimation paradigm, corresponding developments in MCMC estimation in SEM have not yet been realized for analyzing moment-level data. In contrast, the individual-level approach allows for more flexibility in terms of choosing non-normal distributions when modeling the (individual-level) data (e.g., Zhang, Lai, Lu, & Tong, 2013).

Graphical Model Representation

Bayesian models, including Bayesian structural equation models, may be represented as a particular type of graphical model (Lunn, Spiegelhalter, Thomas, & Best, 2009). These representations are similar in many ways to the path diagrammatic representations that have long been popular in SEM; importantly, there are several key differences. We briefly describe the graphical modeling approach to Bayesian modeling, and draw connections to and distinctions from conventional SEM path diagrams.

A graphical model representation of a structural equation model for individual-level data is depicted in Figure 14.4 and consists of the following elements. All entities in the model are represented in the graph as *nodes*. Following SEM path modeling conventions, we use rectangles to represent observable entities and ovals to represent latent or unknown entities. As in typical path modeling in SEM, the entities represented as nodes include the observable variables and latent variables (i.e., the rectangle with X_{ij}, and the ovals with ξ_i and η_{il}). In contrast to traditional SEM path diagrams, the graphical model here also represents the remaining parameters (i.e., loadings, structural paths, intercepts, means, variances) as nodes in the graph. A set of directed edges (one-headed arrows) indicate stochastic dependence between the elements they connect. Reflecting the history of graphical models in the field of genealogy, a node at the destination of an edge is referred to as a *child* of the node at the source of the edge, its *parent*. For example, ξ_i is a child of Φ and κ_k, and is a parent of η_{il} and X_{ij}.

An important class of graphical models with connections to Bayesian modeling are acyclic, directed graphs (frequently referred to as *directed acyclic graphs*, DAGs, of which Figure 14.4 is an example). A directed graph is one in which all the edges are directed (i.e., unidirectional) so that there is a "flow" of dependence. An acyclic graph is one in which, when moving along paths from any node, it is impossible to return to that node. The graph also contains a number of *plates* associated with indexes, used to efficiently represent many nodes. For example, the node for X_{ij} resides within a plate for i and a plate for j, indicating that there is a node X_{ij} for $i = 1, \ldots, N$ and $j = 1, \ldots, J$ (i.e., there is a node for X from every subject for every observ-

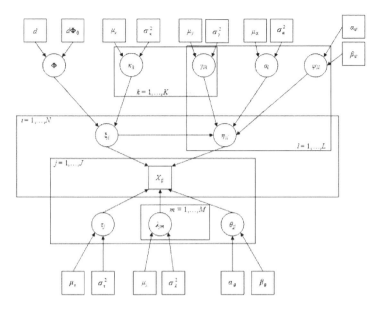

Figure 14.4 Example DAG for SEM for individual-level data.

able variable). Importantly, DAGs differ from path diagrams in SEM in that the former do not have nondirectional (two-headed) arrows. To specify correlational structures in a DAG akin to those typically conveyed by nondirectional arrows, a multivariate approach to specifying the distribution of the variables may be adopted. This approach is employed in Figure 14.4 where the exogenous latent variables, instead of being specified separately, are specified as a group following a multivariate distribution with covariance structure Φ, allowing for nondirectional associations among the exogenous latent variables.

The structure of the graph conveys how the model structures the joint distribution. Letting $\mathbf{Z}$ denote the full collection of entities (in SEM: observable and latent variables, loadings, intercepts, structural paths, variances, covariances, etc.), the joint distribution $P(\mathbf{Z})$ may be factored according to the structure of the graph as

$$P(\mathbf{Z}) = \prod_{\mathbf{z} \in \mathbf{Z}} P\left[\mathbf{z} \mid pa(\mathbf{z})\right], \tag{14.30}$$

where $pa(\mathbf{z})$ stands for the parents of $\mathbf{z}$; if $\mathbf{z}$ has no parents, $P[\mathbf{z} \mid pa(\mathbf{z})]$ is taken as the unconditional (marginal) distribution of $\mathbf{z}$. Thus, the graph reflects the dependence and (conditional) independence relationships in the model (see Pearl, 2009).

Returning to the example, Figure 14.4 is the graphical model for a structural equation model in which (a) a set of observable variables depend on exogenous and endogenous latent variables, (b) all errors and disturbances are uncorrelated with all other entities, and (c) the endogenous latent variables have direct effects from exogenous latent variables only (i.e., $\mathbf{B} = \mathbf{0}$). Note that at periphery of the graph, the nodes are all rectangles, reflecting that entities at the highest level of the hierarchical specification of the model involve known entities (i.e., the analyst chooses values for the parameters that govern the prior distributions at the highest level of the model). If there is an unknown entity in the model, it is assigned a prior distribution with parameters (sometimes referred to as hyperparameters). If these (hyper)parameters are unknown, they are assigned a prior distribution with their own set of (hyper)parameters; this continues until the (hyper)parameters are known (chosen) values.

Mimicking the presentation of the individual-level model in the equations above, the DAG in Figure 14.4 is fairly "open" in that it does not contain the restrictions seen in typical applications. If a loading is fixed to a value (say, to identify the model), its node in the DAG would be a rectangle rather than an oval, it would not be specified as a child of any other node, and would in most cases not be specified inside a plate (unless multiple loadings are being fixed to that value); this is the case in the example presented in the "Illustration of Bayesian SEM" section. DAGs are highly flexible ways to represent models, and can be used to convey additional information, for example, if a loading is fixed to a known value (e.g., for identification purposes). Thus, similar to the case of path diagrams, the use of a graphical model is a convenient visual representation for constructing and communicating the model. Furthermore, the structure of the graph aids in structuring the computations in the estimation, a point we return to in the following section on estimation.

However, the use of DAGs to represent Bayesian structural equation models masks certain features that are made more explicit in path diagrams. For example, path diagrams might be more efficient for deriving or communicating model-implied moments or partitioning direct, indirect, and non-directional effects. Ultimately, the choice of graphical representation depends on the purpose at hand. A DAG representation of a model aids in thinking in Bayesian terms, namely, in terms of viewing model parameters as having the same status as observable and latent person variables, and in terms of the specification of a model via the specification of conditional probability distributions.

MARKOV CHAIN MONTE CARLO ESTIMATION

Bayesian model fitting involves estimating the posterior distribution. In simple models, analytical solutions might be available. However, for SEM with latent variables—particularly highly complex models—evaluating the possibly high-dimensional integrals in the denominator in Equation 14.1 quickly becomes intractable. In SEM, certain choices for distributional forms (e.g., assumptions of multivariate normality of the data and associated conjugate priors) yield posterior distributions of known form, in which case an empirical approximation may be obtained by simulating values using Monte Carlo procedures (Lee, 2007).

However, in models that do not employ conjugate priors and/or have complicated features (e.g., latent interactions, latent mixtures, multilevel structures), the posterior distribution is often not of known form. A highly flexible framework for estimating distributions is provided by MCMC estimation, in which a series of possibly dependent draws are simulated from a distribution of interest (Brooks, 1998; Gelfand & Smith, 1990; Gilks, Richardson, & Spiegelhalter, 1996; Smith & Roberts, 1993; Spiegelhalter et al., 2007; Tierney, 1994). To begin, initial values for all parameters are specified. Subsequent values for the parameters are repeatedly drawn, creating a sequence that constitutes the *chain*. Given that certain general conditions hold (see, e.g., Jackman, 2009; Roberts, 1996; Tierney, 1994), a properly constructed chain is guaranteed to converge to its stationary distribution. This is leveraged in Bayesian analyses in that we construct a Markov chain that has the posterior distribution of interest as its stationary distribution. As such, MCMC estimation in Bayesian analyses consists of drawing from a series of distributions that is, in the limit, equal to drawing from the posterior distribution.

Though both the gradient-based methods used in the frequentist paradigm and MCMC procedures used in Bayesian paradigm are iterative and rely on a sequence of values, there are two key differences between frequentist and Bayesian estimation—and their respective gradient- and MCMC-based instantiations—worth noting. The first concerns the function being evaluated. In frequentist (e.g., gradient-based) estimation, the main interest is parameter estimation, which comes to maximizing or minimizing a target function (e.g., based on the likelihood), whereas in Bayesian (e.g., MCMC) estimation the goal is estimating a distribution of the parameters of interest. Second, frequentist (e.g., gradient-based) estimation seeks to optimize a function (e.g., maximize the likelihood function) and is akin to finding the highest point in a mountain range (Figure 14.1). Bayesian (e.g., MCMC) estimation is not used to optimize (i.e., reach a single point), but seeks to represent the whole distribution. Continuing with the mountain range metaphor, instead of seeking the highest peak, Bayesian (e.g., MCMC) estimation seeks to map the entire terrain of peaks, valleys, and plateaus. Thus, whereas the use of a traditional estimation routine in a frequentist paradigm can be viewed as seeking the highest peak of a mountain range, the use of MCMC in a Bayesian paradigm can be viewed as mapping the entire terrain of a different mountain range.[4]

The mountain ranges may be very similar; as noted above, in situations with highly diffuse priors and/or large sample sizes, the posterior distribution strongly resembles the likelihood. Trivially, in the case of completely diffuse priors (i.e., a uniform prior over the entire support), the posterior is effectively a normalized likelihood. In this case, MCMC estimation provides the whole map of the normalized likelihood, and the highest peak in such case is theoretically equivalent to the point estimate of traditional ML estimation methods. Similarly, as sample size increases, the likelihood comes to dominate the posterior so that, asymptotically, posterior modes are equivalent to ML estimates. The similarity between ML and Bayes modal estimates asymptotically, as sample size increases and/or the prior becomes more diffuse, provides a connection between frequentist and Bayesian estimation. MCMC may therefore provide an attractive option for exploring, maximizing, or finding multimodality in likelihoods.

In the following subsections we describe the key ideas of the most popular MCMC routines used for Bayesian SEM.

Gibbs Sampling

Let $\theta = (\theta_1, \theta_2, \ldots, \theta_R)'$ denote the R parameters in the model; in SEM, these include loadings, structural paths, error and disturbance variances, covariances, intercepts, means, and, in the case of the individual-level specification, the latent variables (factor scores) for each subject. Let $P(\theta_r | \mathbf{X}, \theta_{(r)})$ denote the conditional distribution of the rth model parameter given the data ($\mathbf{X}$) and all other model parameters $\theta_{(r)}$. This is known as the *full conditional distribution* for the rth model parameter. It can be shown that a joint distribution may be defined by the complete set of such full conditional distributions (Besag, 1974; Gelfand & Smith, 1990). Thus, in a Bayesian analysis, the joint posterior distribution of model parameters may be defined as the complete set of full conditional posterior distributions. That is, the joint posterior $P(\theta | \mathbf{X})$ may be defined by $P(\theta_1 | \mathbf{X}, \theta_{(1)})$, $P(\theta_2 | \mathbf{X}, \theta_{(2)})$, ..., $P(\theta_R | \mathbf{X}, \theta_{(R)})$. Sampling from the joint posterior then involves iteratively sampling from these full conditional distributions.

Let $\theta_r^{(t)}$ denote the value of model parameter r at iteration t. Gibbs sampling (Gelfand & Smith, 1990; Geman & Geman, 1984; see also Brooks, 1998; Casella & George, 1992; Gilks et al., 1996) consists of proceeding with the following steps:

1. Assign initial values for the parameters yielding the collection $\theta_1^{(0)}$, $\theta_2^{(0)},\ldots,\theta_R^{(0)}$ where the superscript of $t = 0$ conveys that these are initial values.

2. For $r = 1,\ldots,R$, draw values for parameter θ_r from its full conditional distribution given the observed data and the current values of all other model parameters. In other words, for each parameter θ_r, we obtain the value of the chain at iteration $(t+1)$ by drawing from its full conditional distribution $P(\theta_r|X,\theta_1^{(t+1)},\ldots,\theta_{r-1}^{(t+1)},\theta_{r+1}^{(t)},\ldots,\theta_R^{(t)})$. One cycle is given by sequentially drawing values from

$$\theta_1^{(t+1)} \sim P(\theta_1|\mathbf{X},\theta_2^{(t)},\ldots,\theta_R^{(t)})$$

$$\theta_2^{(t+1)} \sim P(\theta_2|\mathbf{X},\theta_1^{(t+1)},\theta_3^{(t)},\ldots,\theta_R^{(t)})$$

$$\vdots$$

$$\theta_R^{(t+1)} \sim P(\theta_R|\mathbf{X},\theta_1^{(t+1)},\ldots,\theta_{R-1}^{(t+1)}).$$

3. Repeat step 2 for some large number T iterations.

Importantly, the independence assumptions inherent in the model great-ly reduce the set of parameters that need to be conditioned on in each of distributions in step 2. For example, in drawing a value for a subject's latent variable(s), the assumption of independence between subjects im-plies that the values of the latent variable(s) for the remaining subjects can be ignored. It is here that the use of DAGs in conceptualizing the model can greatly aid in computation, as it can be shown that the full conditional distribution for any entity depends on its parents, its chil-dren, and the parents of its children; all other parameters can be ignored (Lunn et al., 2009).

If these full conditional distributions are of familiar form, sampling from them may proceed using Monte Carlo procedures. However, in complex models, it might be the case that full conditional distributions are not of known form. In these cases, more complex sampling schemes are required. The following sections describe two such schemes.

Metropolis-Hastings Sampling

To simplify notation, let $\pi(\theta) = P(\theta|\mathbf{X})$ denote the target distribution (i.e., the posterior distribution of interest). In Metropolis-Hastings sam-pling (Hastings, 1970; see also Brooks, 1998; Chib & Greenberg, 1995; Gilks et al., 1996) the basic idea is that in lieu of drawing a value from the diffi-cult-to-sample-from target distribution, we draw a value from a distribution we *can* easily sample from, and then decide to accept or reject that value in such a way that, in the limit, the resulting series of values from this process converges to the target distribution. The procedure consists of conducting the following steps:

1. Initialize the parameters by assigning a value for $\theta^{(t)}$ for $t = 0$.
2. Draw a *candidate value* $\mathbf{y} \sim q(\mathbf{y} \mid \theta^{(t)})$ from a *proposal* distribution q that we can easily sample from (e.g., a normal distribution).
3. Accept $\mathbf{y}$ as the $(t + 1)$th iteration for θ with probability

$$\alpha(\mathbf{y} \mid \theta^{(t)}) = \min\left[1, \frac{\pi(\mathbf{y})q(\theta^{(t)} \mid \mathbf{y})}{\pi(\theta^{(t)})q(\mathbf{y} \mid \theta^{(t)})}\right].$$

Retain the current value of $\theta^{(t)}$ for $\theta^{(t+1)}$ with probability $1 - \alpha(\mathbf{y} \mid \theta^{(t)})$.
4. Repeat steps 2 and 3 for some large number T iterations.

The term $\alpha(\mathbf{y} \mid \theta^{(t)})$ in step 3 is known as the *acceptance probability* and involves evaluating the posterior distribution π and the proposal distribution q at both the current and candidate values. The proposal distribution q may be any distribution that is defined over the support of the stationary (posterior) distribution π. As such, the Metropolis-Hastings algorithm is an extremely flexible approach to estimating posterior distributions. The formulation here of the proposal distribution q as possibly conditional on the current value of the chain in step 2 is intentional, and facilitates the exposition of the Metropolis sampler below. More generally, q need not be conditional on the current value of the chain.

Metropolis Sampling

In Metropolis sampling (Metropolis, Rosenbluth, Rosenbluth, Teller, & Teller, 1953; see also Brooks, 1998; Gilks et al., 1996), q is chosen so that it is symmetric with respect to its arguments, $q(\mathbf{y} \mid \theta^{(t)}) = q(\theta^{(t)} \mid \mathbf{y})$. The acceptance probability in step 3 of the Metropolis-Hastings sampler then simplifies to

$$\alpha(\mathbf{y} \mid \theta^{(t)}) = \min\left[1, \frac{\pi(\mathbf{y})}{\pi(\theta^{(t)})}\right].$$

A popular choice for q is the normal distribution centered at the current value of the chain.

Single-Component-Metropolis or Metropolis-Within-Gibbs

The Gibbs sampler is attractive because it decomposes the joint posterior set of parameters into more manageable univariate components. However, it is limited in that it requires the full conditionals to be of known form to facilitate sampling. The Metropolis (and Metropolis-Hastings) sampler is attractive because it can sample from distributions even when they are not of known form. The *single-component-Metropolis*, also termed the Metropolis-within-Gibbs sampler, combines the component decomposition approach of the Gibbs sampler with the flexibility of Metropolis (or, if asymmetric proposal distributions are used, with Metropolis-Hastings). Specifically, the full conditionals need to be constructed as in Gibbs sampling. When they are of familiar form, they can be sampled from directly. When the full conditional for a parameter is not of known form, a Metropolis (or Metropolis-Hastings) step may be taken where a candidate value is drawn from a proposal distribution q and accepted with probability α as the next value in the chain for that parameter.

How MCMC Overcomes the Obstacles of Bayesian Modeling

As stated above, in Bayesian analyses of complex models, the posterior distribution is usually only known up until a constant of proportionality (see Equations 14.1 and 14.4). Inspection of the Metropolis-Hastings and Metropolis samplers reveals that the stationary distribution π (i.e., the posterior distribution) appears in both the numerator and denominator of the acceptance probability α and therefore only needs to be known up to a constant of proportionality. Thus, MCMC alleviates the need to perform the integration over the parameter space in the denominator of Equation 14.1 to obtain the posterior distribution. This is the key feature of MCMC estimation that permits the estimation of complex Bayesian models. Prior to the advent of MCMC, the application of Bayesian modeling was limited because of the difficulty in analytically evaluating or empirically approximating the marginal distribution of the data (in the denominator of Equation 14.1). Practically speaking, MCMC sidesteps the prohibitively complex integration in Equation 14.1 by repeatedly sampling from the posterior distribution. The resulting drawn values from the posterior distribution (i.e., the chain's stationary distribution) constitute an empirical approximation to the posterior.

Practical Issues with MCMC

Conducting MCMC estimation to yield a series of draws brings with it two immediate questions. First, although there are a variety of MCMC estimation methods that guarantee the chain will converge to its stationary distribution (i.e., the posterior distribution in the Bayesian framework) (see, for example, ch. 4 of Jackman, 2009; Roberts, 1996; Tierney, 1994, for technical details of conditions that ensure convergence), there is no guarantee *when* that will occur—so how are we to determine when the chain has converged? Second, once the chain has converged, how do we employ the iterations to facilitate inference? We begin with the first question, as it is the more difficult issue. To start, it is important to recognize that, whereas in a frequentist paradigm we seek *convergence to a point*, in the Bayesian paradigm we seek *convergence to a distribution* (i.e., a collection of points that occur with a specific relative frequency). Returning to the mountain metaphor, whereas convergence to the proper solution in frequentist approaches means we have reached the highest peak, convergence in the Bayesian approach means we have mapped the elevation of the entire terrain accurately. A number of approaches to diagnosing convergence have been proposed, and many can be viewed as comparing summaries of one subset of the draws to another. Examples include viewing time-series plots of draws from one or multiple chains as a graphical check of when the draws stabilize, tests invoking normal statistical theory for whether the mean of the draws from a later part of the chain differs significantly from that from an earlier part of the chain (Geweke, 1992), ANOVA-like analyses of multiple chains run in parallel examining the ratio of total-variability (i.e., from all chains) to within-chain variability (Brooks & Gelman, 1998; Gelman & Rubin, 1992). The reader is referred to Cowles and Carlin (1996), Gill (2007), Jackman (2009), and Sinharay (2004) for thorough treatments of MCMC convergence assessment.

Turning to the issue of how to summarize the results from MCMC, inferences are based on the draws obtained after the chain has converged. The draws prior to this point are referred to as *burn-in iterations* and are discarded. The draws obtained after convergence form an empirical approximation to the posterior, and inferences are made using familiar methods for summarizing distributions. Histograms or smoothed density plots for each parameter graphically convey the marginal posterior distributions. Numerical summaries of the posterior distribution are estimated via summaries of the series of draws; posterior means, standard deviations, percentiles, and credibility intervals are estimated by those quantities computed on the draws from the chain. Similarly, scatterplots, contour plots, and correlations between draws for parameters provide evidence about their dependence. Commonly, point estimates of posterior central tendency and variability (e.g., posterior means and standard deviations) and credibility intervals (e.g., a 95% central credibility interval) are reported in documentation.

A practical issue that influences both convergence and summarizing results in MCMC is the serial dependence among draws. A chain that moves well throughout the support of the distribution is said to be *mixing* well. A chain that that seems to get stuck at a certain region, or one that moves slowly throughout the support, is said to be mixing poorly. Slower mixing chains usually require more iterations to converge. The dependence between draws may be captured by the autocorrelation at various lags where values close to zero indicate approximate independence between draws. One approach to managing high autocorrelations involves reparameterizing the model in some way (Dunson et al., 2005). Alternatively, one can employ MCMC with block samplers in which highly correlated parameters are sampled jointly. When it comes to summarizing results, the serial dependence in a set of stored draws (i.e., those obtained after convergence) may be mitigated by thinning the chain, in which only some of the draws are stored. For example, if the autocorrelation for a chain is approximately 0 at a lag of 10, then thinning the chain by 10 (i.e., saving every 10th draw) would produce a set of draws that are approximately independent. Note that thinning does not produce a better representation of the posterior, as the dropped draws constitute a loss of information. Rather, if one has a limited number of draws that can be saved, say, due to computational limits, it may be preferable to store approximately independent draws accomplished via thinning. The serial dependence also motivates the use of multiple chains run in parallel from different start values; though draws are dependent *within* a chain, they are independent *between* chains.

There are a number of other issues in implementation of MCMC that are of importance for those wishing to program their own MCMC algorithms. Popular software packages for Bayesian SEM to some degree hide or address these via defaults; see the appendices for more details.

BAYESIAN APPROACHES TO ADDITIONAL ANALYSES

Adopting a Bayesian approach to modeling facilitates not only the fitting of models, but also a number of supporting analyses related to SEM. Here, we describe two types of such analyses.

Data-Model Fit Assessment

A number of data-model fit procedures for Bayesian analyses have been proposed. Here, we briefly introduce several of the more popular methods;

the reader is referred to Levy (2011) for a more detailed presentation of these and other methods in the context of SEM.

We follow the Bayesian literature in referring to functions that summarize the (lack of) fit between the data and the model as *discrepancy measures*, which in the SEM literature are often referred to as *test statistics* or *fit indices*. Familiar examples from SEM include residual and standardized residual covariances, functions based on the likelihood ratio (*LR*), standardized root mean square residual (*SRMR*), and the root mean square error of approximation (*RMSEA*). Computations of these quantities involve two ingredients: the data and the model parameters. We generically denote a discrepancy measure (e.g., *LR*, *SRMR*, *RMSEA*) as $D(\mathbf{X}; \theta)$ to convey that it is a function with the data and the model parameters as arguments.

In frequentist approaches, the discrepancy measure is calculated using the observed data and point estimate for the model parameters, such as the ML estimate. An analogous approach may be utilized in a Bayesian analysis by employing a point estimate for the posterior distribution for the model parameters, such as the posterior mean. Lee (2007) provided a clear illustration of this procedure in the context of SEM by using the posterior mean of θ as a point estimate in order to facilitate the computation of residuals.

Alternatively, the analysis might employ the full posterior distribution of θ in calculating the discrepancy measure, yielding the posterior distribution of the discrepancy measure. In practice, this distribution is estimated in much the same way that MCMC estimates posterior distributions. Specifically, each of the draws from the posterior distribution are used to calculate the value of the discrepancy measure. Let $\theta^{(1)} | \mathbf{X}, \ldots, \theta^{(T)} | \mathbf{X}$ denote the *T* draws from the posterior distribution. Using each of these in the calculation of discrepancy measure yields the collection $D(\mathbf{X}; \theta^{(1)} | \mathbf{X}), \ldots, D(\mathbf{X}; \theta^{(T)} | \mathbf{X})$, which is an empirical approximation to the posterior distribution of the discrepancy measure. Such an approach allows for supporting probabilistic statements regarding the discrepancy measures (e.g., the 95% central probability interval for *SRMR*, or the probability that *SRMR* is less than .10).

Perhaps the most widely employed Bayesian approach to data-model fit assessment is posterior predictive model checking (PPMC; Gelman, Meng, & Stern, 1996; Meng, 1994; Rubin, 1984; see also Levy, 2011). In this framework, the posterior distribution for a discrepancy measure is referred to as the *realized values* of the discrepancy measure, where the name reflects that these values come from computations with the actual (realized) data. PPMC employs the posterior predictive distribution, which is the distribution of future, or *predicted* data, conditional on the observed data and the model:

$$P(\mathbf{X}^{postpred} | \mathbf{X}) = \int_{\theta} P(\mathbf{X} | \theta) P(\theta | \mathbf{X}) d\theta. \tag{14.31}$$

Conceptually, the first term on the right hand side of (14.31) may be viewed as analogous to the classical notion of a sampling distribution (i.e., the conditional distribution of the data, given the values of the model parameters). The second term on the right hand side reflects the Bayesian aspect, namely, by viewing the values of the model parameters in terms of a distribution rather than in terms of a point (or a point estimate). In practice, the posterior predictive distribution is empirically approximated via simulation by taking each of the T draws from the posterior and using them to generate a *posterior predicted dataset* via the model. For draw t, the posterior predicted dataset is $\mathbf{X}^{postpred(t)} \sim P(\mathbf{X} \mid \theta = \theta^{(t)})$. The discrepancy measure is then evaluated using the draws for the model parameters and the posterior predicted datasets to yield $D(\mathbf{X}^{postpred(1)}; \theta^{(1)} \mid \mathbf{X}^{obs}), \ldots, D(\mathbf{X}^{postpred(T)}; \theta^{(T)} \mid \mathbf{X}^{obs})$ that form an empirical approximation to the posterior predictive distribution of the discrepancy measure.

The logic of PPMC is as follows. It is an open question whether or not the data and the model are consistent with one another. PPMC operates by comparing (a) the discrepancy between the model and the observed data to (b) the discrepancy between the model and the posterior predicted data, which are known to be consistent with the model. The extent that they are comparable (disparate) constitutes evidence of data-model fit (misfit). This can be done via a graphical representation such as a scatterplot, or via a posterior predictive p-value (Gelman, 2003; Meng, 1994):

$$p_{\text{post}} = P(D(\mathbf{X}^{postpred}; \theta \mid \mathbf{X}) \geq D(\mathbf{X}; \theta \mid \mathbf{X})). \tag{14.32}$$

In a simulation environment, p_{post} is estimated as the proportion of the T draws in which the posterior predicted discrepancy $D(\mathbf{X}^{postpred(t)}; \theta^{(t)} \mid \mathbf{X})$ exceeds the corresponding realized discrepancy $D(\mathbf{X}; \theta^{(t)} \mid \mathbf{X})$.

Model Comparisons

A number of options are available for comparing SEM models in a Bayesian framework. For any two models M_1 and M_2 the ratio of posterior probabilities (i.e., the posterior odds) of the latter to the former is

$$\frac{P(M_2 \mid \mathbf{X})}{P(M_1 \mid \mathbf{X})} = \frac{P(M_2)}{P(M_1)} \times \frac{\displaystyle\int_{\theta_{(2)}} P(\mathbf{X} \mid \theta_{(2)}) P(\theta_{(2)} \mid \mathbf{X}) \, d\theta_{(2)}}{\displaystyle\int_{\theta_{(1)}} P(\mathbf{X} \mid \theta_{(1)}) P(\theta_{(1)} \mid \mathbf{X}) \, d\theta_{(1)}}, \tag{14.33}$$

where $\theta_{(1)}$ is the collection of parameters for M_1 and $P(M_1)$ is the prior probability for M_1; analogous definitions hold for $\theta_{(2)}$ and M_2. The first term on the right side of Equation 14.33 is the ratio of prior probabilities (i.e., the prior odds) and the last term is the ratio of the marginal likelihoods under each model, also referred to as the *Bayes factor* (e.g., Kass & Raftery, 1995; Raftery, 1993). Note that the Bayes factor has the form of a likelihood ratio (and in certain cases specializes to the familiar likelihood ratio; Kass & Raftery, 1995). Equation 14.33 reveals that, functionally, the Bayes factor effectively serves to modify the prior odds based on the likelihoods of the data and, as such is sometimes interpreted as the weight of evidence in favor of one model over the other (Gill, 2007). Importantly, the Bayes factor is not limited to applications in which the models are hierarchically related (nested). See Kass and Raftery (1995) for recommendations on computing and interpreting Bayes factors as well as Gelman et al. (2003) and Gill (2007) for critical discussions and additional references.

In situations where computing the Bayes factor may be difficult, Kass and Raftery (1995) advocated approximating it based on the Bayesian information criterion (*BIC*; Schwarz, 1978):

$$BIC = -2\ell(\hat{\theta}\,|\,\mathbf{X}) + p + N, \tag{14.34}$$

where $\ell(\hat{\theta}\,|\,\mathbf{X})$ is the maximized log-likelihood and p is the number of parameters. Letting S be half the difference in two models' *BIC* values, the Bayes factor may be approximated by $e^{(-2S)}$. More common uses of *BIC* and related information criteria (Burnham & Anderson, 2002) conduct model selection via examining the values for each model and selecting the model with the smallest value. Building off this tradition of comparing values of information criteria, Spiegelhalter, Best, Carlin, and van der Linde (2002) introduced the deviance information criterion (*DIC*) as

$$DIC = \overline{D(\theta)} + p_D = 2\overline{D(\theta)} - D(\overline{\theta}) + 2p_D \tag{14.35}$$

where $\overline{D(\theta)}$ is the posterior mean of the deviance (negative of twice the log-likelihood function), p_D is a complexity measure defined as the difference between the posterior mean of the deviance and the deviance evaluated at the posterior mean, $D(\overline{\theta})$. Similar to the ways other information criteria are employed, model selection here involves choosing the model with the smallest *DIC* value.

Raftery (1993), Song, Lee, and Zhu (2001), Song and Lee (2002a), and Lee and Song (2003b) discussed the use of the Bayes factors and *BIC* in the context of SEM. Applications of *DIC* to comparing structural equation models can be found in Lee, Song, and Tang (2007), Muthén (2010), and Lee, Song, and Cai (2010). A number of other methods are also available for model comparison in SEM. See Gill (2007) for an overview and critical discussion of several methods, including those discussed here. Song, Xia,

Pan, and Lee (2011) recently proposed a new measure, evaluated for each model via PPMC, to be used for model comparison in SEM. In addition, Raftery discussed the Bayesian approach of model averaging, in which no one single model is selected, but the results of different models are pooled. To date, little research has been done on the relative merits and limitations of these approaches to model comparison in Bayesian SEM.

ILLUSTRATION OF BAYESIAN SEM

To illustrate an application of Bayesian SEM, we conduct analyses of widely available data associated with example 5.11 in the Mplus User's Manual (Muthén & Muthén, 1998–2010). The model is depicted in traditional path diagram form in Figure 14.3. The model is identified by fixing the means of the exogenous latent variables (κ_1 and κ_2) and the intercepts for the endogenous latent variables (α_1 and α_2) to 0, and fixing the loadings for the first observable indicator for each latent variable (λ_{11}, λ_{42}, λ_{73}, and $\lambda_{10\,4}$) to 1. The DAG for the model is given in Figure 14.5. The DAG appears overly complex, in large part because the identification constraints imply that the parameters associated with a number of observables are fixed, and therefore these need to reside outside of the plates. The benefit of this complexity, however, is the explicit representation of every entity in the model in terms of the other entities that it depends on.

The DAG has been split into two portions to avoid additional crossing of arrows. Figure 14.5a represents the portion of the model associated with the first six observables and the exogenous latent variables. Figure 14.5b represents the portion of the model associated with the latter six observables and the endogenous latent variables. The bottom portions of Figures 14.5a and 14.5b are similar, as they both represent the factor analytic measurement model. The differences in the top portions reflect the differences in the specifications for the exogenous and endogenous latent variables. The exogenous latent variables (depicted as nodes with grey shading) are the link between Figures 14.5a and 14.5b.

The appendices contain inputs for analyses of this example using AMOS, WinBUGS, and Mplus. In the present discussion, we focus on the results from the latter. The model was fit using the following priors:

$$\tau_j \sim N(0,\ 10)\ j = 1,\dots,12; \tag{14.36}$$

$$\lambda_{j1} \sim N(0,\ 10)\ j = 2,3;\ \lambda_{j2} \sim N(0,\ 10)\ j = 5,6;$$
$$\lambda_{j3} \sim N(0,\ 10)\ j = 8,9;\ \lambda_{j4} \sim N(0,\ 10)\ j = 11,12; \tag{14.37}$$

$$\theta_{jj} \sim G^{-1}(1,\ 1)\ j = 1,\dots,12; \tag{14.38}$$

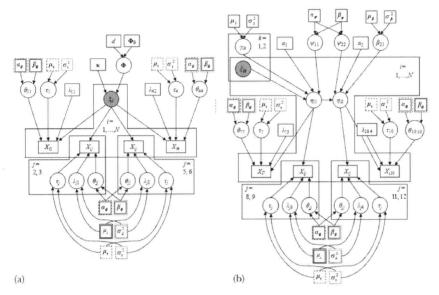

(a)   (b)

Figure 14.5 DAG for the illustration. For ease of presentation, the DAG in two fragments: (a) the model for the exogenous latent variables; (b) the model for the endogenous latent variables. The DAG fragments are linked via the nodes for the exogenous latent variables (ξ) which are highlighted with a grey background. For spacing purposes, certain nodes are presented multiple times. Nodes with the same outlining shape (other than the single solid line) represent the same nodes (e.g., the double-lined boxes for μ_λ toward the bottom of both fragments reflects that these values are the same).

$$\Phi \sim W^{-1}(\mathbf{I}^{(2)}, 2) \text{ where } \mathbf{I}^{(2)} \text{ is the } 2 \times 2 \text{ identity matrix;} \quad (14.39)$$

$$\gamma_{1k} \sim N(0, 10) \ k=1,2; \quad (14.40)$$

$$\beta_{21} \sim N(0, 10); \quad (14.41)$$

$$\psi_{ll} \sim G^{-1}(1, 1) \ l = 1,2. \quad (14.42)$$

Three chains were run in parallel in Mplus without specifying any particular number of iterations. In this case, Mplus runs the chains and determines the convergence by monitoring the potential scale reduction factor of Gelman and Rubin (1992) for each parameter, discarding the first half of the draws from the chains. A trace plot of the full history of the chains from this analysis for λ_{21} is given in Figure 14.6. The vertical line at iteration 150 indicates that the chains reached Mplus's convergence criteria by 150 iterations. In addition, it is seen that the chains are mixing quite well through-

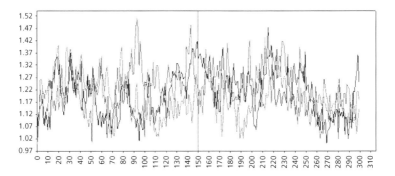

Figure 14.6   Trace plot of the history of values for λ_{21}.

out the space from the outset; they move throughout the support so well that it is difficult to tell the chains apart. The trace plots for all the other parameters yielded similar patterns, which constitutes *prima facie* evidence of fast convergence and adequate mixing throughout the posterior distribution. By default, Mplus reports summaries of the posterior distribution based on the same number of iterations after convergence as was needed to achieve convergence (i.e., 150 in the current case).

To obtain more draws, the model was fit using three chains with 1000 iterations per chain; Mplus's automatic discarding of the first half of the iterations for each chain results in 1500 iterations (500 from each chain) used to summarize the posterior distribution.

The overall data-model fit is evaluated using the model *LR* discrepancy measure (referred to as chi-square by Mplus) in a PPMC framework. Figure 14.7 contains a scatterplot, produced by Mplus, of the realized (observed) and posterior predicted (replicated) values with the unit line added

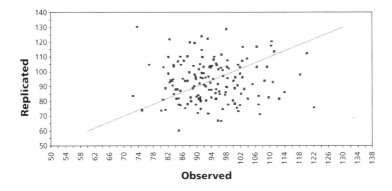

Figure 14.7   Scatterplot of realized (observed) and posterior predicted (replicated) values of the likelihood ratio statistic, with the unit line added as a reference.

as a reference, where it is seen that the realized values are not systematically different from their posterior predicted counterparts. This is summarized via the estimated value of p_{post}, the proportion of points above (or to the left of) the line, which is .503. These results suggest adequate data-model fit as captured by the *LR*.

The posterior distribution may be represented graphically. To illustrate, Figure 14.8 contains a plot, produced by Mplus, of the marginal posterior distribution of the γ_{11} in the standardized metric as a smoothed density of the draws from the chains. This representation highlights how, in contrast to frequentist estimation, the solution in a Bayesian analysis is a distribution, not a point. This affords summary of the results in terms of characteristics of a distribution, namely the posterior is fairly unimodal and symmetric around .45, which is the mean, median, and mode (to two decimals). Further, it facilitates probabilistic interpretations of the parameter. The probability that γ_{11} (in the standardized metric) is larger than 0 is 1.0, and there is a .95 probability that the parameter is between .35 and .54. Note that these expressions refer to the probability for the parameter; in a Bayesian analysis we can employ probability-based reasoning and probabilistic interpretations for the parameters directly, rather than for estimators as in frequentist analysis.

We advocate the consideration of plots of marginal distributions for all parameters of interest when interpreting the results; however, plots akin to Figure 14.8 for the remaining parameters are not presented due to space considerations. The posterior distribution may also be summarized numerically. Table 14.1 contains summaries of the marginal posterior distributions, including posterior, means, standard deviations, and 95% central credibility intervals for the parameters of the model in the standardized metric (except subjects' latent variables/factor scores, which are not included due

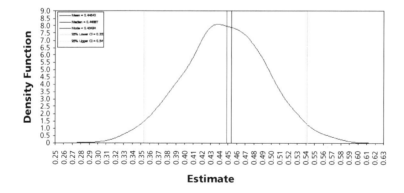

Figure 14.8 Smoothed density for the marginal posterior for γ_{11} in the standardized metric from Mplus. The horizontal axis would be better interpreted as referring to the parameter, as the distribution is of a parameter, not of estimates.

TABLE 14.1 Summaries of the Posterior Distribution of the Standardized Solution

Structural Model

Structural Paths					Covariance and Variance Terms				
Parameter	Mean	SD	2.5%	97.5%	Parameter	Mean	SD	2.5%	97.5%
γ_{31}	0.45	0.05	0.35	0.54	ϕ_{21}	−0.03	0.06	−0.15	0.10
γ_{32}	0.63	0.04	0.54	0.72	ψ_{33}	0.41	0.06	0.31	0.52
β_{43}	0.59	0.05	0.49	0.68	ψ_{44}	0.65	0.06	0.54	0.76

Measurement Model

Exogenous Latent Variables					Endogenous Latent Variables				
Parameter	Mean	SD	2.5%	97.5%	Parameter	Mean	SD	2.5%	97.5%
λ_{11}	0.68	0.03	0.61	0.74	λ_{73}	0.77	0.03	0.71	0.82
λ_{21}	0.78	0.03	0.71	0.84	λ_{83}	0.72	0.03	0.66	0.78
λ_{31}	0.63	0.04	0.56	0.70	λ_{93}	0.73	0.03	0.67	0.79
λ_{42}	0.64	0.04	0.57	0.72	$\lambda_{10\,4}$	0.65	0.04	0.57	0.72
λ_{52}	0.64	0.04	0.56	0.71	$\lambda_{11\,4}$	0.63	0.04	0.53	0.70
λ_{62}	0.63	0.04	0.55	0.71	$\lambda_{12\,4}$	0.52	0.05	0.42	0.61
τ_1	0.01	0.04	−0.07	0.10	τ_7	0.04	0.05	−0.06	0.13
τ_2	0.04	0.04	−0.05	0.12	τ_8	0.05	0.05	−0.04	0.14
τ_3	0.01	0.04	−0.08	0.09	τ_9	0.06	0.05	−0.03	0.15
τ_4	0.07	0.04	−0.01	0.16	τ_{10}	−0.01	0.05	−0.09	0.08
τ_5	0.06	0.05	−0.02	0.15	τ_{11}	0.03	0.05	−0.06	0.12
τ_6	0.06	0.05	−0.03	0.15	τ_{12}	0.03	0.04	−0.06	0.11
θ_{11}	0.53	0.05	0.45	0.63	θ_{77}	0.41	0.04	0.33	0.50
θ_{22}	0.40	0.05	0.29	0.49	θ_{88}	0.48	0.04	0.40	0.56
θ_{33}	0.60	0.05	0.51	0.69	θ_{99}	0.46	0.04	0.38	0.55
θ_{44}	0.59	0.05	0.49	0.68	$\theta_{10\,10}$	0.58	0.05	0.48	0.67
θ_{55}	0.59	0.05	0.49	0.68	$\theta_{11\,11}$	0.61	0.06	0.51	0.72
θ_{66}	0.60	0.05	0.50	0.70	$\theta_{12\,12}$	0.73	0.05	0.63	0.82

to space considerations). Consideration of the plots and the summaries in Table 14.1 support the following interpretations for the model. The measurement quality appears strong. The posterior means for the standardized loadings for the observables ranged from a low of .52 to a high of .78. Viewing values of .4 or higher for standardized loadings as indicative of solid measurement, for each indicator there is a probability greater than or equal to .975 that the standardized loading exceeds .4. Similar probabilistic interpretations may be constructed for other interpretable values if desired.

At the structural level, the exogenous latent variables do not appear to be strongly correlated; the posterior mean and standard deviation for ϕ_{21} in the standardized metric are −.03 and .06, respectively, and the 95% central credibility interval is (−.15, .10). Both exogenous latent variables have moderately strong direct effects on the first endogenous latent variable—for γ_{11}: posterior mean .45, standard deviation .05, and 95% central credibility interval of (.35, .54); for γ_{12}: posterior mean .63, standard deviation .04, and 95% central credibility interval of (.54, .72). Similarly the first endogenous variable has a moderately strong effect on the second endogenous latent variable—for β_{21}: posterior mean .59, standard deviation .05, and 95% central credibility interval (.49, .68).

ADVANTAGES OF, CHALLENGES WITH, AND OPPORTUNITIES FOR BAYESIAN SEM

The above description and illustration of Bayesian SEM treats foundational models, characterized by the analysis of a single group using linear relations and normal distributions. These models have had a long and productive association with frequentist approaches to modeling and estimation (Bollen, 1989). Nevertheless, a Bayesian approach to SEM offers a number of potential advantages over traditional frequentist approaches in these contexts. The phrasing in describing the results of the illustration in the preceding section reveals perhaps the chief advantage, wherein probabilistic statements are formulated for the parameters, not parameter estimates. Likewise, in a Bayesian paradigm we can make probabilistic statements about multiple parameters (e.g., the probability that one path coefficient exceeds another, one group's factor mean exceeds another) or models. Thus, a Bayesian approach allows us to use the language of probability to *directly* discuss what is of inferential interest—parameters, hypotheses, models, etc.—rather than indirectly as in frequentist approaches. This allows the analyst to "say what they mean and mean what they say" (Jackman, 2009, p. xxviii), and overcome the limitations and avoid the pitfalls of frequentist probabilistic machinery (e.g., Goodman, 2008; Wagenmakers, 2007).

A Bayesian approach also allows for straightforward estimation for subjects' latent variables (factor scores). This supports the examination of subjects' values, which might be of direct interest, such as when ordering subjects in terms of their latent variable values or assessing the change in a subject's latent variable over time. These can also be used in service of related inquiries, such as identifying outlying subjects in terms of the latent variable itself, examining if the results for structural coefficients are especially influenced by particular subjects, or examining whether such relations hold over the range of the latent variable distribution akin to what is done in observed variable regression diagnostics (Dunson et al., 2005).

On a more technical level, using a Bayesian approach allows for multimodality of the sort that is depicted in Figure 14.1—often undetectable by traditional approaches to estimation—to be seen in posterior distributions (Scheines, Hoijtink, & Boomsma, 1999). Unlike a frequentist approach, a Bayesian approach to modeling does not rely on asymptotic arguments or assumptions to justify estimation routines or data-model fit procedures (Scheines et al., 1999; Levy, 2011), and early research has provided evidence of Bayesian estimation's ability to provide better results for estimation and data-model fit than traditional procedures in small samples in SEM (Ansari & Jedidi, 2000; Lee & Song, 2004b; Muthén & Asparouhov, 2012; Price, 2012).

A Bayesian approach allows for the incorporation of prior distributions into the analysis so that the model reflects the substantive beliefs about the situation (Gelman et al., 2003). For example, if expert beliefs or previous studies have indicated that the standardized path coefficient between two variables is between .1 and .4, then a prior distribution can be used to reflect that the value in the current study probably is most likely in that area, or at least not −.9.

Prior distributions can also serve as a source of information for parameters that are underidentified in the frequentist sense (Scheines et al., 1999). Muthén and Asparouhov (2012) lamented the conventional practice of fixing parameters to particular values (or equality) in deference to requirements for identification in frequentist approaches. Instead, they advocated the use of a Bayesian approach with small-variance priors for parameters affording the analyst opportunities to fit models that are more closely aligned with substantive theory but underidentified in the frequentist sense.

If the incorporation of prior information is not desired, the influence of the prior can generally be reduced by using diffuse priors. The construction of priors is an area of active research in Bayesian analysis. Examples include comparisons of allegedly diffuse priors for variance components (Gelman, 2006), or strategies for forming weakly informative priors that encode minimal prior knowledge (Gelman et al., 2008). It should be noted that eliciting prior distributions from subject matter experts unfamiliar with the language of probability can prove difficult. For strategies for eliciting and codifying prior information, see Choy, O'Leary, and Mengersen (2009), Press (1989), von Winterfeldt and Edwards (1986), Kadane and Wolfson (1998), O'Hagan (1998), and the accompanying discussion papers in a special issue of *The Statistician*.

Moreover, a Bayesian approach and MCMC estimation can serve as a unifying framework for modeling as it naturally allows for the incorporation of aspects that are prevalent in applications of SEM. For example, Bayesian procedures to model ignorable and nonignorable missingness in data are well developed (Enders, 2010; Lee & Song, 2004a; Lee & Tang, 2006; Song & Lee, 2002b), and a Bayesian perspective offers a unified approach to latent variables and missing data (Jackman, 2000). Similarly, the inclusion of multilevel structures in SEM to model the hierarchical organization of data is easily accomplished in a Bayesian approach (Ansari & Jedidi, 2000; Ansari, Jedidi, & Dube, 2002; Levy, Mislevy, & Behrens, 2011); indeed, viewing parameters either as deterministic or probabilistically dependent on covariates in frequentist multilevel modeling overlays with traditional Bayesian conceptions of specifying prior distributions for parameters (Lindley & Smith, 1972).

Moreover, it is straightforward to impose ordinal constraints (Scheines et al., 1999) as well as define and make inferences about additional parameters. For example, a Bayesian analysis of models with latent quadratic effects simply defines these terms in the model and proceeds as planned with essentially no difference to construction or estimation (Arminger & Muthén, 1998; Lee, Song, & Tang, 2007), which stands in contrast to approaches within the frequentist paradigm (see, e.g., Marsh, Wen, Hau, & Nagengast, this volume). Similarly, Bayesian approaches to mediation analysis simply define indirect effects as new parameters with no difference in model construction or estimation (Chen, Choi, Weiss, & Stapleton, in press; Yuan & MacKinnon, 2009). Importantly, nonlinearities of this sort pose challenges for frequentist approaches to estimation of parameters and standard errors, as products of normal distributions are not normal.

Indeed, a key advantage of a Bayesian approach to SEM lies in the power of MCMC to estimate nonstandard models that pose considerable challenges for ML and least-squares estimation (Lee, 2007; Levy, 2009; Levy et al., 2011; Muthén & Asparouhov, 2012). In addition to those mentioned above, examples of such applications include models with quadratic, interaction, and other nonlinear relationships among latent and observable variables in cross-sectional and longitudinal designs (Arminger & Muthén, 1998; Harring, Weiss, & Hsu, 2012; Lee & Song, 2004a; Lee et al., 2007; Lee & Tang, 2006; Lee & Zhu, 2000; Song & Lee, 2002a, 2004; Song, Lee, & Hser, 2009), covariates (Lee et al., 2007), ordered dichotomous and polytomous data (Lee & Song, 2003a, 2004a; Lee & Zhu, 2000; Shi & Lee, 1998; Song & Lee, 2004), unordered dichotomous and polytomous data (Lee et al., 2010; Song, Xia, & Lee, 2009) multiple groups (Song & Lee, 2001, 2002a), latent mixtures (Lee & Song, 2003b; Zhu & Lee, 2001), latent means and growth curve models (Zhang, Hamagami, Wang, & Nesselroade, 2007), multilevel models (Song & Lee, 2004), models that employ alternative non-normal distributions of the data (Lee, 2007; Zhang et al., 2013) or non-normal distributions of the latent variables (Lee, Lu, & Song, 2008; Song et al., 2009; Yang & Dunson, 2010) as well as models that integrate many of these features (Lee & Song, 2008).

Research directly comparing Bayesian and frequentist methods in SEM is fairly limited. However, work conducted to date has suggested that Bayesian approaches perform as well or better, particularly in situations where frequentist theory breaks down, such as with small samples or nonlinear relationships (Chen et al., in press; Harring et al., 2012; Lee & Song, 2004b; Lee, Song, & Poon, 2004; Muthén & Asparouhov, 2012; Price, 2012). Furthermore, adopting a Bayesian approach and MCMC estimation can aid in overcoming misconceptions associated with traditional modeling and estimation. For example, see Wirth and Edwards (2007) and Levy (2009) for a discussion of how traditional estimation strategies yield countervailing misconceptions about models for discrete data from SEM and item response modeling perspectives, and how a Bayesian approach to modeling and MCMC estimation can aid in freeing researchers from possible misconceptions.

Conducting Bayesian analyses in SEM is not without its difficulties. A conceptual challenge to the emergence of Bayesian SEM lies in that a Bayesian perspective is rooted in probability-based reasoning and emphasizes that we think distributionally—including using probability distributions to model and reason about uncertainty, specifying models via (conditional) distributions, and making inferences by summarizing distributions via probabilistic statements about unknown entities—which is a different mindset than that associated with model-fitting, parameter estimation, and hypothesis testing in the frequentist paradigm that dominates current practice and training.

Similarly, MCMC estimation is quite different than traditional estimation procedures, and has been criticized as being difficult to implement. However, some researchers have argued that, for complex models, it might actually be *easier* to set up an MCMC estimation routine than it is to proceed through the necessary steps in frequentist estimation routines (e.g., Ansari & Jedidi, 2000; Ansari et al., 2002; Muthén & Asparouhov, 2012; Segawa, Emery, & Curry, 2008; Yuan & MacKinnon, 2009; Zhang et al., 2007). MCMC allows analysts to estimate models without requiring mathematical and/or numerical procedure necessary to obtain (a) derivatives in frequentist approaches to estimation, or (b) the marginal distribution of the data that serves as the normalizing constant in a Bayesian approach (i.e., in the denominator of Equation 14.1). It is this flexibility that has supported the adoption of Bayesian approaches in the complex modeling scenarios listed above. Advances in tools for conducting traditional estimation and broad perspectives on modeling have supported the development of frequentist estimation routines and software for complicated and nuanced models (e.g., Muthén, 2002; Muthén & Muthén, 1998–2010; Rabe-Hesketh, Skrondal, & Pickles, 2004a; Rabe-Hesketh, Skrondal, & Pickles, 2004b). Nevertheless, when researchers want to push the boundaries of even these complex modeling paradigms, a Bayesian approach to modeling powered by MCMC may be a beneficial or in some cases the only feasible option (Levy, 2009; Lee & Song, 2008; Muthén & Asparouhov, 2012; Segawa et al., 2008). Moreover, recent developments in software means that the user is released from some of the burdens of setting up MCMC estimation.

Practically, to employ MCMC the user must be capable of (a) accessing software that affords MCMC estimation, (b) making relevant choices for successful practice, and (c) understanding the results. Research on best practices for choices about types and specifics of MCMC algorithms, the number of chains, start values, methods for determining convergence, and summarizing inferences from MCMC is needed in SEM (see Sinharay, 2004, for examples of this type of research in other latent variable models).

A related challenge is that software for conducting MCMC estimation is not as widespread and fully integrated into SEM practice. Until recently, analysts wishing to conduct Bayesian estimation of SEM would have either needed to code their own MCMC algorithms (e.g., Choi et al., 2006) or turn to general purpose software such as WinBUGS (Spiegelhalter et al., 2007) or one of its variants. As a flexible package for conducting MCMC estimation, WinBUGS has often been the tool of choice for analysts looking to specify complex Bayesian statistical models; to date, the majority of applications of Bayesian SEM have used WinBUGS. Recently, MCMC estimation has been implemented in AMOS (Arbuckle, 2007) and Mplus (Muthén & Muthén, 1998–2010), which are likely to be more accessible for users steeped in SEM traditions but new to Bayesian modeling. Importantly, these packages automate more of the processes involved in conducting MCMC estimation than WinBUGS, but at the expense of flexibility (see Appendix D). We suspect expansions that include more detailed MCMC estimation guidelines and more automated analyses are likely, and that SEM analysts looking to employ Bayesian methods will increasingly turn to these more familiar SEM programs. Appendices A–C provide example inputs for Mplus, AMOS, and WinBUGS; Appendix D provides more details on the capabilities of the various software programs. Regardless of the software, MCMC estimation is almost always more time-consuming than other approaches because of Monte Carlo simulation nature of the method. However, the speed differential between MCMC and frequentist approaches has diminished with efficient implementations in mainstream SEM software such as AMOS and Mplus, and is likely to further shrink for all software as computing power increases.

APPENDIX A

This appendix contains input for running the example in AMOS. In AMOS, the model is specified via constructing a path diagram. A screenshot of the path diagram in AMOS for the analysis of the example model is given in Figure 14.A. Additional information on AMOS's capabilities is available in Appendix D. A data file is available in a web appendix available at https://sites.google.com/a/asu.edu/roylevy/papers-software/FilesAccompanying-BayesianSEM.zip. In this diagram, we adopt the convention of "spelling out" Greek letters in all capitals.

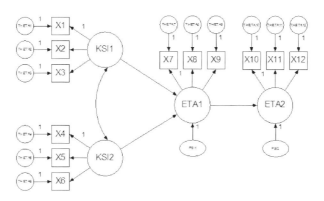

Figure 14.A Screenshot of SEM diagram specification in AMOS.

APPENDIX B

This appendix contains input code for running the example in Mplus. Specifically, this code runs the second analysis described in the chapter, in which two chains are run for 1000 iterations, and posterior means are computed. In addition, the code requests ten imputed data sets to illustrate a way to save draws for subjects' latent variables (factor scores). More details on this code and additional information on Mplus's other capabilities is available in Appendix D. This code, the data file, output files, are available in a web appendix available at https://sites.google.com/a/asu.edu/roylevy/papers-software/FilesAccompanyingBayesianSEM.zip. In this code, we adopt the convention of "spelling out" Greek letters in all capitals. By default, Mplus uses the identifying constraints adopted in example in this chapter.

```
TITLE:          Input file to conduct final analysis
DATA:           FILE IS ex5.11.dat;
VARIABLE:       NAMES ARE X1-X12;
ANALYSIS:       ESTIMATOR = BAYES;
                CHAINS = 3;
                FBITERATIONS = 1000;
                POINT = MEAN;
MODEL:          KSI1 BY X1-X3 (LAMBDA1-LAMBDA3);
                KSI2 BY X4-X6 (LAMBDA4-LAMBDA6);
                ETA1 BY X7-X9 (LAMBDA7-LAMBDA9);
                ETA2 BY X10-X12 (LAMBDA10-LAMBDA12);
                ETA2 ON ETA1 (BETA21);
                ETA1 ON KSI1 KSI2 (GAMMA11-GAMMA12);
                KSI1 WITH KSI2 (PHI21);
                KSI1 (PHI11);
                KSI2 (PHI22);
                X1-X12 (THETA1-THETA12);
                ETA1-ETA2 (PSI1-PSI2);
                [X1-X12] (TAU1-TAU12);
MODEL PRIORS:
                LAMBDA2-LAMBDA3~N(0, 10);
                LAMBDA5-LAMBDA6~N(0, 10);
                LAMBDA8-LAMBDA9~N(0, 10);
                LAMBDA11-LAMBDA12~N(0, 10);
                BETA21~N(0, 10);
                GAMMA11~N(0, 10);
                GAMMA12~N(0, 10);
                PHI11~IW(1,2);
                PHI22~IW(1,2);
                PHI21~IW(0,2);
                THETA1-THETA12~IG(1,1);
                PSI1-PSI2~IG(1,1);
                TAU1-TAU12~N(0,10);
```

```
DATA IMPUTATION:
                IMPUTE = ALL (c);
                PLAUSIBLE = latent.out;
                SAVE = impute.*.out;
                NDATASETS = 10;
OUTPUT:         TECH1;
                TECH8;
                STANDARDIZED;
PLOT:           TYPE = PLOT2;
```

APPENDIX C

This appendix contains the code that specifies the model for the illustration suitable for use in WinBUGS. Note that, unlike the other software, all the entities that are assigned distributions are specified explicitly. Additional information on WinBUGS's capabilities is available in Appendix D, along with pointers to resources for running WinBUGS. A web appendix available at https://sites.google.com/a/asu.edu/roylevy/papers-software/Files AccompanyingBayesianSEM.zip contains an annotated version of this code, a file with data formatted for WinBUGS, initial values for three chains, and an output file containing the trace plots from an analysis with three chains run for 1000 iterations, and summary statistics from the analysis, discarding the first 500 iterations from each chain as burn-in. In this code, we adopt the convention of "spelling out" Greek letters in all capitals.

```
model{
PHI.0[1,1] <- 1;
PHI.0[1,2] <- 0;
PHI.0[2,1] <- 0;
PHI.0[2,2] <- 1;
d <- 2;

for (k in 1:K){
      for (kk in 1:K){
            dxPHI.0[k,kk] <- d*PHI.0[k,kk];
      }
}

inv.PHI[1:K,1:K] ~ dwish(dxPHI.0[ , ], d);
PHI[1:K,1:K] <- inverse(inv.PHI[ , ]);

for(k in 1:K){
      KAPPA[k] <-0;
}

for (i in 1:N){
      XI[i, 1:K] ~ dmnorm(KAPPA[], inv.PHI[,]);
}
```

```
for(l in 1:L){
      ALPHA[l] <- 0;
      inv.PSI[l,l] ~ dgamma(1, 1);
      PSI[l,l] <- 1/inv.PSI[l,l];
}

BETA[2,1] ~ dnorm(0, .1);
for(k in 1:K){
      GAMMA[1,k] ~ dnorm(0, .1);
}

for (i in 1:N){
      MU.ETA[i,1] <- ALPHA[1] + GAMMA[1,1]*XI[i,1] + GAMMA[1,2]*
        XI[i,2]
      MU.ETA[i,2] <- ALPHA[2] + BETA[2,1]*ETA[i,1]

      for (l in 1:L){
            ETA[i,l] ~ dnorm(MU.ETA[i,l], inv.PSI[l,l]);
      }
}

for(j in 1:J){
      TAU[j] ~ dnorm(0, .1);
      inv.THETA[j,j] ~ dgamma(1, 1);
      THETA[j,j] <- 1/inv.THETA[j,j];
}

LAMBDA[1, 1] <- 1.0;
LAMBDA[4, 2] <- 1.0;
LAMBDA[7, 3] <- 1.0;
LAMBDA[10, 4] <- 1.0;

for (j in 2:3){
      LAMBDA[j,1] ~ dnorm(0, .1);
}
for (j in 5:6){
      LAMBDA[j,2] ~ dnorm(0, .1);
}
for (j in 8:9){
      LAMBDA[j,3] ~ dnorm(0, .1);
}
for (j in 11:12){
      LAMBDA[j,4] ~ dnorm(0, .1);
}
for (i in 1:N){
      for (j in 1:3){
            MU.X[i,j] <- TAU[j] + LAMBDA[j,1]*XI[i,1];
      }
      for (j in 4:6){
            MU.X[i,j] <- TAU[j] + LAMBDA[j,2]*XI[i,2];
      }
```

```
for (j in 7:9){
        MU.X[i,j] <- TAU[j] + LAMBDA[j,3]*ETA[i,1];
}
for (j in 10:12){
        MU.X[i,j] <- TAU[j] + LAMBDA[j,4]*ETA[i,2];
}
for (j in 1:J){
        X[i,j] ~ dnorm(MU.X[i,j], inv.THETA[j,j]);
}
}
}
```

APPENDIX D

In this appendix we describe the use of Mplus, AMOS, and WinBUGS software to conducting Bayesian SEM analyses, and discuss some of their key features as of April 2012. We fully expect future developments to expand upon the current capabilities.

Mplus

Beginning with version 6, Mplus implemented MCMC estimation routines applicable for a number of models (Muthén & Muthén, 1998-2010). A number of papers about Mplus's implementation and capabilities are available for download at http://www.statmodel.com/papers.shtml, including descriptions of types of data supported, defaults, and options for prior distributions. We briefly survey some of the key features likely to be routinely used by most users. Bayesian estimation may be requested by using the 'ESTIMATOR = BAYES' option in the ANALYSIS command (see the accompanying code in Appendix B). By default, Mplus will run two chains in parallel, and chooses the number of iterations to discard as burn-in based on the criterion that the potential scale reduction factor of Gelman and Rubin (1992) is sufficiently close to 1 for all parameters. The convergence criteria are calculated on the second half of the chains, a window that expands as more iterations are run. Once the convergence criteria are reached, the chains are stopped, the first half of each is discarded, and the second half of each are then used in the computations of the summaries of the posterior distribution that is printed in the usual output file.

A number of options are available to the user, including the number of chains, iterations (per chain), the stringency of the convergence criterion, and the point estimate to be used in output. Appendix B contains the code for the example in the paper in which three chains are run for 1,000 iterations each and the posterior means based on the latter halves of each chain are reported in the output. This input file, and the data file, output file, and the .gh5 file (discussed below) are included in a web appendix available at https://sites.google.com/a/asu.edu/roylevy/papers-software/Files AccompanyingBayesianSEM.zip.

In the output file, Mplus reports for each parameter an estimate summarizing the posterior distribution (the default is the posterior median), the posterior standard deviation, a one-tailed p-value representing the posterior probability the parameter exceeds 0, and the 95% central credibility interval. The output file also includes a summary of the results for PPMC analyses of the LR discrepancy measure (see, e.g., Levy, 2011), which Mplus refers to as chi-square, including the p_{post} value and the 95% posterior interval for the difference between realized (observed) and posterior predicted (replicated) values, and the computation of the DIC (Spiegelhalter, Best, Carlin, & van der Linde, 2002).

The code also illustrates how plots can be requested using the 'TYPE = PLOT2' option in the 'PLOT' command. For each parameter, the available plots include histograms and smoothed densities of the posterior distributions, trace plots of the history of the draws, and plots of the within-chain autocorrelations. In addition, scatterplots and histograms are available for PPMC analyses of the LR (chi-square) discrepancy measure; additional discrepancy measures are available for models with discrete data (Asparouhov & Muthén, 2010). The estimated p_{post} is printed on the plot. Importantly, requesting the plots produces a .gh5 file (newer versions of Mplus) or a .gph file (older versions of Mplus), which is used by Mplus to produce the plots. This file contains some technical output related to the estimation, including all the drawn values for all the iterations of the model parameters excluding the latent variables. Thus, if the user wished to analyze more iterations other than is done by Mplus, they could be extracted from the .gh5 (or .gph) file and analyzed separately. Examples of such situations include conducting additional convergence diagnostics or data-model fit analyses outside of Mplus, or if the analyst desires a certain number of iterations but does not need to discard the entire first half of a chain. (Mplus allows for the customization of the convergence criterion, but does not allow for the direct choice of the number of iterations to burn-in.)

It should be noted that the .gh5 (.gph) file does not include the draws for the values for the latent variables ('factor scores' in Mplus terminology). The draws for these values are not available; summaries of these draws are available if the 'SAVE = FSCORES' option in the SAVEDATA command is used. To obtain draws from the posterior distribution for the subjects' latent variables, imputed datasets may be requested via the 'DATA IMPUTATION' command, as illustrated in the code in Appendix B for 10 draws. Importantly, though these may be viewed as draws from the posterior distribution, these are not the same draws used in MCMC estimation of the remaining parameters. Finally, the example code requests additional output to be printed via selections for the 'OUTPUT' command: 'TECH1' and 'TECH8' request the printing of some details regarding the model, and 'STANDARDIZED' requests the printing of various standardized solutions, including that summarized in Table 14.1.

AMOS

AMOS 16 (Arbuckle, 2007) facilitates model construction via a graphical approach and allows for testing and refining of the model via drag-and-drop drawing tools. As such, the model is specified graphically (see Appendix A). By default, completely diffuse uniform prior distributions are used; users wishing to specify different priors may select from a few parametric forms for certain parameters or use a graphical drawing tool for specifying a variety of shapes of priors.

Among other options, the user is allowed to specify the number of iterations, the number of burn-in iterations to discard, and the convergence criterion via a dialog box (Figure 14.D). In connection with the latter, AMOS produces a convergence diagnostic based on Gelman, et al. (2003). Options for AMOS output include a variety of summary statistics, including the posterior mean, median, standard deviation, minimum, maximum, skewness, kurtosis, and central credibility interval for each parameter. AMOS also produces plots of the history (trace) of the draws, histograms, frequency polygons, as well as plots of autocorrelations for each parameter, as well as bivariate distributions and contour plots for examining joint (marginal) posterior distributions for pairs of parameters. AMOS calculates estimates of p_{post} for LR for data-model fit, and the DIC for model comparisons. See Arbuckle (2007) for additional details on AMOS's capabilities and mechanisms for requesting options.

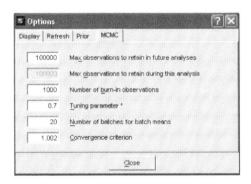

Figure 14.D Screenshot of AMOS' dialog box for specifying MCMC options.

WinBUGS

WinBUGS (Spiegelhalter, Thomas, Best, & Lunn, 2007) and its variants are highly flexible software packages for conducting MCMC estimation for a wide variety of statistical models. Indeed, WinBUGS is perhaps the most flexible software for fitting statistical models, and as such has been the software package of choice for fitting SEM and related models. As a tradeoff for this considerable flexibility, there are no defaults for model specification; the onus falls to the user to specify the model in full in terms of specifying equations and distributions. Models may be defined via writing code or specifying a DAG. Code for the example used in the chapter is given in Appendix C (for additional examples of WinBUGS code for SEM, see, e.g., Lee, 2007). A web appendix available at https://sites.google.com/a/asu.edu/roylevy/papers-software/FilesAccompanyingBayesianSEM.zip contains an annotated version of this code, a file with data formatted for WinBUGS, initial values for three chains, and an output file containing the trace plots from an analysis with three chains run for 1000 iterations, and summary statistics from the analysis, discarding the first 500 iterations from each chain as burn-in.

WinBUGS allows for user to control a number of features. Most users will be concerned with specifying the number of chains, the number of iterations, and the number of iterations to discard as burn-in. WinBUGS contains a number of key features for users to investigate when analyzing results, including summary statistics, plots of smoothed densities of the posterior distributions, bivariate scatterplots of draws, trace plots of the history of the draws, plots of the within-chain autocorrelations, and running plots of the Brooks-Gelman-Rubin diagnostic (Brooks & Gelman, 1998; Gelman & Rubin, 1992) for diagnosing convergence.

WinBUGS also computes the *DIC* used for model comparisons. And though there are not explicit calculations for data-model fit, the flexibility of the code allows for conducing residual or PPMC analyses by implementing the computations within the body of the code, or by reading in the iterations from WinBUGS into another statistical package and programming the computations in that environment (Levy, 2011).

This latter approach relies on WinBUGS' capabilities to save the draws to files that can be then read into other software; see http://www.mrc-bsu.cam.ac.uk/bugs/winbugs/remote14.shtml for a list statistical software packages that interface with WinBUGS. For example, a number of packages for interfacing WinBUGS with R are available, including packages to prepare the data to be suitable for WinBUGS, call WinBUGS, and read in the draws for additional analyses such as convergence diagnostics and summarizing results (Plummer, Best, Cowles, & Vines, Smith, 2009; Sturtz, Ligges, & Gelman, 2005). Researchers new to running WinBUGS either as stand-alone or via another environment would be well served to start by viewing a brief tutorial on steps to run analyses in WinBUGS available at http://www.mrc-bsu.cam.ac.uk/bugs/winbugs/winbugsthemovie.html.

NOTES

1. In some approaches to Bayesian modeling, certain unknown entities (i.e., those at the highest level of a hierarchical specification) are not assigned prior distributions, and instead are estimated using frequentist strategies (see, e.g., Gill, 2007, p. 425).
2. If the parameters are discrete, the integral is replaced by a summation over the discrete states of the parameters.
3. Exceptions include situations where the support of the prior does not include the values suggested by the data, for example, a prior that restricts a mean to be positive when the data suggests it is negative.
4. If the goal is to find the modal value of the posterior distribution, gradient-based estimation routines may be applied to a posterior distribution (see Mislevy, 1986, for details in the context of item response models).

REFERENCES

Aiken, L. S., West, S. G., & Millsap, R. E. (2008). Doctoral training in statistics, measurement and methodology in psychology: Replication and extension of Aiken, West, Sechrest, and Reno's (1990) survey of PhD programs in North America. *American Psychologist, 63,* 32–50.

Ansari, A., & Jedidi, K. (2000). Bayesian factor analysis for multilevel binary observations. *Psychometrika, 65,* 475–498.

Ansari, A., Jedidi, K., & Dube, L. (2002). Heterogeneous factor analysis models: A Bayesian approach. *Psychometrika, 67,* 49–78.

Arbuckle, J. L. (2007). *AMOS 16.0 user's guide.* Chicago, IL: SPSS Inc.

Arminger, G., & Muthén, B. O. (1998). A Bayesian approach to nonlinear latent variable models using the Gibbs sampler and the Metropolis-Hastings algorithm. *Psychometrika, 63,* 271–300.

Asparouhov, T., & Muthén, B. (2010). Bayesian analysis using Mplus: Technical implementation. Technical Report. Version 3. Available online at http://www.statmodel.com/papers.shtml

Barnett, V. (1999). *Comparative statistical inference* (3rd ed.). West Sussex: John Wiley & Sons Ltd.

Besag, J. (1974). Spatial interaction and the statistical analysis of lattice systems (with discussions). *Journal of the Royal Statistical Society, Series B, 36,* 192–236.

Bollen, K. A. (1989). *Structural equations with latent variables.* New York, NY: Wiley.

Box, G. E. (1976). Science and statistics. *Journal of the American Statistical Association, 71,* 791–799.

Box, G. E., & Tiao, G. C. (1973). *Bayesian inference in statistical analysis.* Reading: Addison-Wesley.

Brooks, S. P. (1998). Markov chain Monte Carlo method and its application. *The Statistician, 47,* 69–100.

Brooks, S. P., & Gelman, A. (1998). Convergence assessment techniques for Markov chain Monte Carlo. *Statistics and Computing, 8,* 319–335.

Browne, M. W. (1974). Generalised Least Squares Estimators in the analysis of covariance structures. *South African Statistical Journal, 8,* 1–24.

Browne, M.W. (1984). Asymptotically distribution-free methods for the analysis of covariance structures. *British Journal of Mathematical and Statistical Psychology, 37,* 1–21.

Burnham, K. P., & Anderson, D. R. (2002). *Model selection and multimodel inference: A practical information-theoretic approach.* New York, NY: Springer-Verlag.

Casella, G., & George, E. I. (1992). Explaining the Gibbs sampler. *The American Statistician, 46,* 167–174.

Chen, J., Choi, J., Weiss, B. A., & Stapleton, L. (in press). An empirical evaluation of mediation effect analysis using Markov chain Monte Carlo and alternative estimation methods. *Structural Equation Modeling: A Multidisciplinary Journal.*

Chib, S., & Greenberg, E. (1995). Understanding the Metropolis-Hastings Algorithm. *The American Statistician, 49,* 327–335.

Choi, J., Levy, R., & Hancock G. R. (April, 2006). *Markov chain Monte Carlo estimation method with covariance data for structural equation modeling.* Paper presented at the annual meeting of the American Educational Research Association, San Francisco, CA.

Choy, S. L., O'Leary, R., & Mengersen, K. (2009). Elicitation by design in ecology: Using expert option to inform priors for Bayesian statistical models. *Ecology, 90,* 265–277.

Cowles, M. K., & Carlin, B. P. (1996). Markov chain Monte Carlo convergence diagnostics: A comparative review. *Journal of the American Statistical Association, 91,* 883–904.

Dunson, D. B., Palomo, J., & Bollen, K. (2005). *Bayesian structural equation modeling.* Technical Report #2005-5. Statistical and Applied Mathematical Sciences Institute.

Enders, C. K. (2010). *Applied missing data analysis.* New York, NY: Guilford.

Gelfand, A. E., & Smith, A. F. M. (1990). Sampling-based approaches to calculating marginal densities. *Journal of the American Statistical Association, 85,* 398–409.

Gelman, A. (2003). A Bayesian formulation of exploratory data analysis and goodness-of-fit testing. *International Statistical Review, 71,* 369–382.

Gelman. A. (2006). Prior distributions for variance parameters in hierarchical models. *Bayesian Analysis, 1,* 515–533.

Gelman, A. (2011). Induction and deduction in Bayesian data analysis. *Rationality, Markets and Morals, 2,* 67–78.

Gelman, A., Carlin, J. B., Stern, H. S., & Rubin, D. B. (2003). *Bayesian data analysis* (2nd ed.). Boca Raton: Chapman and Hall/CRC.

Gelman, A., & Hill, J. (2007). *Data analysis using regression and multilevel hierarchical models.* New York, NY: Cambridge University Press.

Gelman, A., Jakulin, A., Pittau, M. G., & Su, Y. S. (2008). A weakly informative default prior distribution for logistic and other regression models. *The Annals of Applied Statistics, 2,* 1360–1383.

Gelman, A., Meng, X. L., & Stern, H. (1996). Posterior predictive assessment of model fitness via realized discrepancies. *Statistica Sinica, 6,* 733–807.

Gelman, A., & Rubin, D. B. (1992). Inference from iterative simulation using multiple sequences. *Statistical Science, 7,* 457–472.

Gelman, A., & Shalizi, C. R. (2013). Philosophy and the practices of Bayesian statistics. *British Journal of Mathematical and Statistical Psychology, 66,* 8–38.

Geman, S., & Geman, D. (1984). Stochastic relaxation, Gibbs distributions, and the Bayesian restoration of images. *IEEE Transactions on Pattern Analysis and Machine Intelligence, 6,* 721–741.

Geweke, J. (1992). Evaluating the accuracy of sampling-based approaches to the calculation of posterior moments. In J. M. Bernardo, A. F. M. Smith, A. P. Dawid, & J. O. Berger (Eds.), *Bayesian Statistics 4* (pp. 169–193). Oxford: Oxford University Press.

Gilks, W. R., Richardson, S., & Spiegelhalter, D. J. (Eds.). (1996). *Markov chain Monte Carlo in practice.* London: Chapman & Hall.

Gill, J. (2007). *Bayesian methods: A social and behavioral sciences approach* (2nd ed.). New York, NY: Chapman and Hall/CRC.

Goodman, S. (2008). A dirty dozen: Twelve *p*-value misconceptions. *Seminars in Hematology, 45,* 135–140.

Harring, J. R., Weiss, B. A., & Hsu, J. C. (2012). A comparison of methods for estimating quadratic effects in structural equation models. *Psychological Methods, 17,* 193–214.

Hastings, W. K. (1970). Monte Carlo sampling methods using Markov chains and their applications. *Biometrika, 57,* 97–109.

Jackman, S. (2009). *Bayesian analysis for the social sciences.* West Sussex, UK: John Wiley & Sons, Ltd.

Jackman, S. (2000). Estimation and inference are missing data problems: Unifying social science statistics via Bayesian simulation. *Political Analysis, 8,* 307–332.

Jaynes, E. T. (2003). *Probability theory: The logic of science.* Cambridge, UK: Cambridge University Press.

Jöreskog, K. G. (1969). A general approach to confimatory maximum likelihood factor analysis. *Psychometrika, 34,* 183–202.

Kadane, J. B., & Wolfson, L. J. (1998). Experiences in elicitation. *The Statistician, 47,* 3–19.

Kass, R. E., & Raftery, A. E. (1995). Bayes factors. *Journal of the American Statistical Association, 90,* 773–795.

Lee, S. Y. (2007). *Structural equation modeling: A Bayesian approach.* West Sussex: John Wiley & Sons, Ltd.

Lee, S. Y., Lu, B., & Song, X. Y. (2008). Semiparametric Bayesian analysis of structural equation models with fixed covariates. *Statistics in Medicine, 27,* 2341–2360.

Lee, S. Y., & Song, X.-Y. (2003a). Bayesian analysis of structural equation models with dichotomous variables. *Statistics in Medicine, 22,* 3073–3088.

Lee, S. Y., & Song, X. Y. (2003b). Bayesian model selection for mixtures of structural equation models with an unknown number of components. *British Journal of Mathematical and Statistical Psychology, 56,* 145–165.

Lee, S. Y., & Song, X. Y. (2004a). Bayesian model comparison of nonlinear structural equation models with missing continuous and ordinal categorical data. *British Journal of Mathematical and Statistical Psychology, 57,* 131–150.

Lee, S. Y., & Song, X. Y. (2004b). Evaluation of Bayesian and maximum likelihood approaches in analyzing structural equation models with small samples. *Multivariate Behavioral Research, 39,* 653–686.

Lee, S. Y., & Song, X. Y. (2008). On Bayesian estimation and model comparison on an integrated structural equation model. *Computational Statistics and Data Analysis,* 4814–4827.

Lee, S. Y., Song, X. Y., & Cai., J. H. (2010). A Bayesian approach for nonlinear structural equation models with dichotomous variables using logit and probit links. *Structural Equation Modeling: A Multidisciplinary Journal, 17,* 280–302.

Lee, S. Y., Song, X. Y., & Poon, W. Y. (2004). Comparison of approaches in estimating interaction and quadratic effects of latent variables. *Multivariate Behavioral Research, 39,* 37–67.

Lee, S. Y., Song, X. Y., & Tang, N. S. (2007). Bayesian methods for analyzing structural equation models with covariates, interaction, and quadratic latent variables. *Structural Equation Modeling: A Multidisciplinary Journal, 14,* 404–434.

Lee, S. Y., & Tang, N. S. (2006). Bayesian analysis of nonlinear structural equation models with nonignorable missing data. *Psychometrika, 71,* 541–564.

Lee, S.-Y., & Zhu, H. T. (2000). Statistical analysis of nonlinear structural equation models with continuous and polytomous data. *British Journal of Mathematical and Statistical Psychology, 53,* 209–232.

Levy, R. (2009). The rise of Markov chain Monte Carlo estimation for psychometric modeling. *Journal of Probability and Statistics, vol. 2009,* Article ID 537139, 18 pages. doi:10.1155/2009/537139

Levy, R. (2011). Bayesian data-model fit assessment for structural equation modeling. *Structural Equation Modeling: A Multidisciplinary Journal, 18,* 663–685

Levy, R., Mislevy, R. J., & Behrens, J. T. (2011). MCMC in educational research. In S. Brooks, A. Gelman, G. L. Jones, & X. L. Meng (Eds.), *Handbook of Markov chain Monte Carlo: Methods and applications* (pp. 531–545). London: Chapman & Hall/CRC.

Lindley, D. V., & Novick, M. R. (1981). The role of exchangeability in inference. *The Annals of Statistics, 9,* 45–58.

Lindley, D. V., & Smith, A. F. M. (1972). Bayes estimates for the linear model. *Journal of the Royal Statistical Society, Series B, 34,* 1–41.

Lunn, D., Spiegelhalter, D., Thomas, A., & Best, N. (2009). The BUGS project: Evolution, critique and future directions. *Statistics in Medicine, 28,* 3049–3067.

Lynch, S. M. (2007). *Introduction to applied Bayesian statistics and estimation for social scientists.* New York, NY: Springer.

Meng, X. L. (1994). Posterior predictive *p*-values. *The Annals of Statistics, 22,* 1142–1160.

Metropolis, N., Rosenbluth, A. W., Rosenbluth, M. N., Teller, A. H., & Teller, E. (1953). Equations of states calculations for fast computing machines. *Journal of Chemical Physics, 21,* 1087–1091.

Mislevy, R. J. (1986). Bayes modal estimation in item response models. *Psychometrika, 51,* 177–196.

Mulaik, S. A. (2009). *Linear causal modeling with structural equations.* Boca Raton, FL: CRC Press.

Muthén, B. O. (2002). Beyond SEM: General latent variable modeling. *Behaviormetrika, 29,* 81–117.

Muthén, B. O. (2010). *Bayesian analysis in Mplus: A brief introduction.* Technical Report. Version 3. Retrieved April 17, 2012 from http://www.statmodel.com/download/IntroBayesVersion%203.pdf

Muthén, B. O., & Asparouhov, T. (2012). Bayesian SEM: A more flexible representation of substantive theory. *Psychological Methods, 17,* 313–335.

Muthén, L. K., & Muthén, B. O. (1998–2010). *Mplus user's guide* (6th edition). Los Angeles, CA: Muthén & Muthén.O'Hagan, A. (1998). Eliciting expert beliefs in substantial practical applications. *The Statistician, 47,* 21–35.

O'Hagan, A. (1998). Eliciting expert beliefs in substantive practical applications. *The Statistician, 47,* 21–35.

Pearl, J. (2009). *Causality: Models, reasoning, and inference* (2nd ed). Cambridge: Cambridge University Press.

Plummer, M., Best, N., Cowles, K., & Vines, K. (2009). coda: Output analysis and diagnostics for MCMC. R package version 0.13-4. http://cran.r-project.org/web/packages/coda/index.html.

Press, S. J. (1989). *Bayesian statistics: Principles, models, and applications.* New York, NY: John Wiley & Sons Ltd.

Price, L. R. (2012). Small sample properties of Bayesian multivariate autoregressive time series models. *Structural Equation Modeling: A Multidisciplinary Journal, 19,* 51–64.

Rabe-Hesketh, S., Skrondal, A., & Pickles, A. (2004a). Generalized multilevel structural equation modelling. *Psychometrika 69,* 183–206.

Rabe-Hesketh, S., Skrondal, A., & Pickles, A. (2004b). *GLLAMM Manual* (Second Edition). U.C. Berkeley Division of Biostatistics Working Paper Series. Working Paper 160. http://www.bepress.com/ucbbiostat/paper160.

Raftery, A. E. (1993). Bayesian model selection in structural equation modeling. In K.A. Bollen & J.S. Long (Eds.), *Testing structural equation modeling* (pp. 163–180). Beverly Hills: Sage.

Roberts, G. O. (1996). Markov chain concepts related to sampling algorithms. In W. R. Gilks, S. Richardson, & D. J. Spiegelhalter (Eds.), *Markov chain Monte Carlo in practice* (pp. 45–57). London: Chapman & Hall.

Rowe, D. B. (2003). *Multivariate Bayesian statistics: Models for source separation and signal unmixing.* Boca Raton, FL: CRC Press.

Rubin, D. B. (1984). Bayesianly justifiable and relevant frequency calculations for the applied statistician. *Annals of Statistics, 12,* 1151–1172.

Scheines, R., Hoijtink, H., & Boomsma, A. (1999). Bayesian estimation and testing of structural equation models. *Psychometrika, 64,* 37–52.

Schwarz, G. (1978). Estimating the dimension of a model. *The Annals of Statistics, 6,* 461–464.

Segawa, E., Emery, S., & Curry, S. J. (2008). Extended generalized linear latent and mixed model. *Journal of Educational and Behavioral Statistics, 33,* 464–484.

Senn, S. (2011). You may believe you are a Bayesian but you are probably wrong. *Rationality, Markets and Morals, 2,* 48–66.

Shi, J. Q., & Lee, S. Y. (1998). Bayesian sampling-based approach for factor analysis model with continuous and polytomous data. *British Journal of Mathematical and Statistical Psychology, 51,* 233–252.

Sinharay, S. (2004). Experiences with Markov chain Monte Carlo convergence assessment in two psychometric examples. *Journal of Educational and Behavioral Statistics, 29,* 461–488.

Smith, A. F. M., & Roberts, G. O. (1993). Bayesian computation via the Gibbs sampler and related Markov chain Monte Carlo methods. *Journal of the Royal Statistical Society, Series B, 55,* 3–23.

Song, X. Y., & Lee, S. Y. (2001). Bayesian estimation and test for factor analysis model with continuous and polytomous data in several populations. *British Journal of Mathematical and Statistical Psychology, 54,* 237–263.

Song, X. Y., & Lee, S. Y. (2002). A Bayesian approach for multigroup nonlinear factor analysis. *Structural Equation Modeling: A Multidisciplinary Journal, 9,* 523–553.

Song, X. Y., & Lee, S. Y. (2002b). Analysis of structural equation model with ignorable missing continuous and polytomous data. *Psychometrika, 67,* 261–288.

Song, X. Y., & Lee, S. Y. (2004). Bayesian analysis of two-level nonlinear structural equation models with continuous and polytomous data. *British Journal of Mathematical and Statistical Psychology, 57,* 29–52.

Song, X. Y., Lee, S. Y., & Hser, Y. I. (2009). Bayesian analysis of multivariate latent curve models with nonlinear longitudinal latent effects. *Structural Equation Modeling: A Multidisciplinary Journal, 8,* 378–396.

Song, X. Y., Lee, S. Y., & Zhu, H. T. (2001). Model selection in structural equation models with continuous and polytomous data. *Structural Equation Modeling: A Multidisciplinary Journal, 16,* 245–266.

Song X. Y., Xia Y. M., & Lee, S. Y. (2009). Bayesian semiparametric analysis of structural equation models with mixed continuous and unordered categorical variables. *Statistics in Medicine, 28,* 2253–2276.

Song X. Y., Xia Y. M., Pan, J. H., & Lee, S. Y. (2011). Model comparison of Bayesian semiparametric and parametric structural equation models. *Structural Equation Modeling: A Multidisciplinary Journal, 18,* 55–72.

Spiegelhalter, D. J., Best, N. G., Carlin, B. P., & van der Linde, A. (2002). Bayesian measures of model complexity and fit. *Journal of the Royal Statistical Society, B, 64,* 583–639.

Spiegelhalter, D. J., Thomas, A., Best, N. G., Lunn, D. (2007). *WinBUGS user manual: version 1.4.3.* Cambridge: MRC Biostatistics Unit. Online at http://www.mrc-bsu.cam.ac.uk/bugs/winbugs/contents.shtml

Sturtz, S., Ligges, U., & Gelman, A. (2005). R2WinBUGS: A package for running WinBUGS from R. *Journal of Statistical Software, 12,* 1–16.

Süli, E., & Mayers, D. (2003). *An introduction to numerical analysis.* Cambridge: Cambridge University Press.

Tierney, L. (1994). Markov chains for exploring posterior distributions. *Annals of Statistics, 22,* 1701–1728.

Yang, M., & Dunson, D. B. (2010). Bayesian semiparametric structural equation models with latent variables. *Psychometrika, 75,* 675–693.

Yuan, K-H., & Bentler, P. M. (1998). Structural equation modeling with robust covariances. *Sociological Methodology, 28,* 363–396.

Yuan, Y., & MacKinnon, D. P. (2009). Bayesian mediation analysis. *Psychological Methods, 14,* 301–322.

von Winterfeldt, D., & Edwards, W. (1986). *Decision analysis and behavioral research.* New York, NY: Cambridge University Press.

Wagenmakers, E. J. (2007). A practical solution to the pervasive problems of *p* values. *Psychonomic Bulletin & Review, 14,* 779–804.

Williamson, J. (2010). *In defence of objective Bayesianism.* New York, NY: Oxford University Press.

Wirth, R. J., & Edwards, M. C. (2007). Item factor analysis: Current approaches and future directions. *Psychological Methods, 12,* 58–79.

Zhang, Z., Hamagami, F., Wang, L., & Nesselroade, J. R. (2007). Bayesian analysis of longitudinal data using growth curve models. *International Journal of Behavioral Development, 31,* 374–383.

Zhang, Z., Lai, K., Lu, Z., & Tong X. (2013). Bayesian inference and application of robust growth curve models using Student's *t* distribution. *Structural Equation Modeling: A Multidisciplinary Journal, 20,* 47–78.

Zhu, H. T., & Lee, S. Y. (2001). A Bayesian analysis of finite mixtures in the LISREL model. *Psychometrika, 66,* 133–152.

CHAPTER 15

USE OF MONTE CARLO STUDIES IN STRUCTURAL EQUATION MODELING RESEARCH

Deborah L. Bandalos and Walter Leite

Users of quantitative methods often encounter situations in which the data they have gathered violate one or more of the assumptions underlying the particular statistic(s) they wish to use. For example, data might be nonnormally distributed when the statistic a researcher plans to use requires normality of the variables' distributions. In cases such as these, the focus turns to the *robustness* of the statistic; that is, will use of the statistical procedure still yield relatively unbiased values of parameter estimates, standard errors, and/or test statistics if certain assumptions, such as normality, are violated? In structural equation modeling (SEM), the inclusion of many observed variables, the concomitant estimation of numerous model parameters, the need for large sample sizes, the inclusion of coarsely categorized and/or nonnormally distributed observed variables, and the possible misspecification of the model will almost guarantee assumption violations in the majority of published studies. And even in the rare situation in which data per-

Structural Equation Modeling: A Second Course (2nd ed.), pages 625–666
Copyright © 2013 by Information Age Publishing

fectly meet the requirements of the statistical estimator, the lack of known sampling distributions for most of the commonly used goodness-of-fit statistics in SEM has resulted in the need for Monte Carlo studies to determine values at which one should infer a model to be incorrect in the population. Other violations include the presence of nonlinear relations or of nonindependent data. Finally, although the maximum likelihood (ML) and generalized least squares (GLS) estimators commonly used in SEM studies do have known sampling distributions, these distributions are known only asymptotically. This means that, although the properties of these estimators can be shown to hold as sample size approaches infinity, it is not always clear how large a sample is required for them to hold in practice.

In order to investigate the questions raised by problems such as these, simulation or Monte Carlo studies are used. According to Harwell, Stone, Hsu, and Kirisci (1996), the term "Monte Carlo" was first applied to simulations by Metropolis and Ulam (1949). Such investigations were used during World War II in studies of atomic energy. We distinguish Monte Carlo studies from simulation studies by noting that Monte Carlo studies are statistical sampling investigations in which sample data are typically generated in order to obtain empirical sampling distributions. We use the term simulation study to refer to studies that involve data generation of some type but do not necessarily involve the generation of sample data. For example, a simulation study might entail the generation of population data or might simply generate a set of data to demonstrate a statistical analysis. In this chapter, we will focus on Monte Carlo studies in which a model of some particular type (e.g., measurement, structural, growth model) is specified and many samples of data are then generated from that model structure. The data in these samples are typically generated in such a way that they violate one or more of the assumptions of the estimator being used. The samples are then analyzed using SEM software and the results are aggregated across samples to determine what effect the assumption violations had on the parameter estimates, standard errors, goodness-of-fit statistics, and/or other outcomes of interest. Serlin (2000) described the utility of Monte Carlo studies as "provid[ing] the information needed to help researchers select the appropriate analytical procedures under design conditions in which the underlying assumptions of the procedures are not met" (p. 231). In SEM studies, however, the lack of viable nonparametric alternatives may often result in estimators being used even when assumptions are known to have been violated. This introduces an additional situation in which Monte Carlo studies can assist applied researchers: determining the degree to which the obtained parameter estimates, standard errors and fit statistics might be compromised if assumptions are violated.

Alternatives to Monte Carlo Studies

In some cases, Monte Carlo studies are not the best way to address the types of problems raised above. One of these situations is when an analytical solution is available for the problem. As noted by Harwell et al. (1996), "A Monte Carlo study should be considered only if a problem cannot be solved analytically, and should be performed only after a compelling rationale for using these techniques has been offered" (p. 103). For example, if it were possible to derive the sampling distribution of a fit index such as the Comparative Fit Index (CFI) analytically, it would be preferable to obtain regions of model rejection in that manner rather than by conducting a Monte Carlo study to determine the critical value at which the CFI produces the desired rejection rates. However, in many cases an analytical solution is not possible. One reason for this is that statistical tests are typically developed using statistical theory based on the properties of known mathematical distributions such as the normal distribution. If the data do not conform to such a distribution then the sampling distribution no longer holds. In SEM studies the sampling distributions of test statistics also depend on the correct specification of the model. If correct model specification (or the degree of misspecification) of the model cannot be assumed then exact sampling distributions cannot be derived. In some situations, Monte Carlo studies are used to provide insight into the behavior of a statistic even when mathematical proofs of the problem being studied are available. One reason for this is the fact that theoretical properties of estimators do not always hold under real data conditions. Nevertheless, results of Monte Carlo studies should always be related back to exact statistical theory as much as possible.

Another situation in which Monte Carlo studies might not be appropriate is when a population study is sufficient. In a population study, a model (or models) is specified in order to study the effect of certain conditions *in the population*. This type of study is useful for situations in which the degree of sampling variability and/or the stability of results such as parameter values is not of interest. For example, Kaplan (1988) was interested in the differential impact of population model misspecification on maximum likelihood (ML) and two-stage least squares (2SLS) estimation. Because he was interested in the effects in the population it was not necessary to generate sample data.

Advantages and Disadvantages of Monte Carlo Studies

As noted previously, Monte Carlo studies can be illuminating in situations such as the following:

- Violations of assumptions such as:
  - Normality
  - Independence of observations
  - Model misspecification
  - Linearity of effects
- Investigation of issues such as:
  - Small sample behavior of asymptotically-based statistics
  - Continuousness of observed variable distributions
  - Effects of estimating large numbers of model parameters
  - Effects of analyzing correlation matrices rather than covariance matrices
  - Properties of goodness-of-fit statistics for which no mathematical distribution exists

Perhaps most importantly, Monte Carlo studies make it possible to study the potential interactions of assumption violations with other problems.

On the other hand, some disadvantages of Monte Carlo studies should be noted. First, Monte Carlo studies are essentially experimental studies in which the variables of interest are manipulated. As such, they share the limitations of all experimental studies. More specifically, Monte Carlo studies are dependent on the representativeness of the conditions modeled. If these conditions are not similar to those found in real data, the usefulness of the study would be severely limited. This problem is exacerbated by the extreme complexity of most real world data. In many cases, real data sets include such complications as complex sampling structures, data that are missing in a nonrandom fashion, and the existence of heterogeneous subgroups (see the respective chapters by Stapleton, Enders, and Pastor and Gagné, all in this volume). Unfortunately, the inclusion of all factors known to affect the behavior of the statistic under study could render the Monte Carlo study impossibly complex. Thus, as in any study, Monte Carlo studies must balance practicality with fidelity to the real world phenomena one is attempting to model.

Below we will discuss planning Monte Carlo studies, including the study design and choice of dependent and independent variables. In the following sections we will present information relevant to data generation, automation of Monte Carlo studies using SAS/IML and R software, analysis of data from Monte Carlo studies, and issues of data management. These sections will be illustrated with an example of a Monte Carlo study designed to investigate the behavior of the DWLS (mean- and variance-adjusted Diagonally Weighted Least Squares) estimation method available in the Mplus program. We will conclude by offering suggestions for those engaged in or contemplating Monte Carlo studies. Readers interested in Monte Carlo studies should also review work by Paxton, Curran, Bollen, Kirby, and Chen

(2001), Fan, Felsővályi, Sivo, and Keenan (2001), and by Muthén and Muthén (2002). The Paxton et al. article provided an illustration of the design and implementation of a Monte Carlo study while the Muthén and Muthén paper demonstrated how to conduct a Monte Carlo study to determine the sample size needed to obtain unbiased parameter estimates and standard errors under nonnormality and missing data conditions. The book by Fan et al. provided detailed information on conducting Monte Carlo studies using SAS for a wide variety of statistical procedures.

PLANNING MONTE CARLO STUDIES

In this section we focus on the decisions to be made in planning a Monte Carlo study. First, factors involved in determining a research question will be presented. Next, the selection of the independent variables in the study and appropriate levels for these will be discussed, followed by suggestions for the determination of appropriate dependent variables and their formulations. The actual data generation process will be discussed and illustrated in the next section.

Determining the Research Question

Essentially, the determination of a research question in Monte Carlo research is no different from that in other areas. Good research questions should address gaps in the literature and be designed to advance understanding of a particular topic. In Monte Carlo research, additional stipulations are that no analytical solution is available to answer the research question of interest and that sampling stability or variability is of interest so that a population study would not be sufficient to answer the question. Also, as noted previously, the research questions should be informed as much as possible by statistical theory.

Research questions in Monte Carlo studies are often developed to answer problems that occur in either applied or methodological research. For example, in our own work, we frequently encounter ordered categorical data such as those arising from the use of Likert scales. Because popular SEM estimators such as ML are based on the assumption that data are continuous, the question of proper estimation for ordered categorical data arises. The diagonally weighted least squares (DWLS) estimator with corrections to the model χ^2 and standard errors was designed to yield more accurate estimates of standard errors and model χ^2 values than normal theory (NT) based estimators for such data. As explained by Arnold-Berkovits (2002), the corrected (DWLS) estimator, should result in more stable estimation than earlier WLS estimators because estimation is based on a diagonal weight matrix rather than the full weight matrix required by previous WLS estimators. This is due to the fact that the full WLS weight matrix can become extremely large with even a moderate number of variables, resulting in unstable estimation unless sample size is quite large (2000 or greater). Use of a diagonal weight matrix in DWLS in lieu of the full weight matrix should result in more stable estimation with smaller samples. Although based on strong statistical theory, DWLS estimation has not been studied extensively in practice. Thus, while DWLS should provide more stable results in applications involving categorized data, there is still some uncertainty regarding the conditions under which this advantage will be manifested. Studies of the DWLS estimator by Muthén, du Toit, and Spisic (1997), Yu and Muthén (2002), Arnold-Berkovits (2002), Flora and Curran (2004), Beaudecel and Herzberg (2006), and Lei (2009) have indicated that fit indices, standard errors, and parameter estimates based on this estimator are biased when small sample sizes are combined with non-normality of the data. The study by Muthén et al. also revealed that analysis of a confirmatory factor analysis (CFA) model resulted in less standard error bias than did analysis of a full latent variable path model. However, because the CFA model included fewer variables, the effect may have been due to model size rather than to the type of model. One research question of interest in our example study is therefore whether DWLS displays a sensitivity to model size and/or model type.

Related to the issue of the research question is the specification of research hypotheses. In most studies, research hypotheses flow naturally from the review of the literature. Essentially, the point of one's literature review is to demonstrate why the study is needed. In the case of Monte Carlo research, this should involve a discussion of why analytical or population studies would be unsatisfactory and how the research questions are related to statistical theory. However, it has not been typical for specific research hypotheses to be stated in Monte Carlo research. While these hypotheses might be implicit in the design of the study, our belief is that these should be stated explicitly. Doing so makes it much easier for readers to judge whether the hypotheses are reasonable, both in terms of being answerable by the study design and of being related to statistical theory and previous research. It also allows for a more direct link with the study results and discussion. In the context of our example study, one such hypothesis would be that the DWLS estimation will display less sensitivity to model size than earlier categorical variable methodology (CVM) estimators. This is because the method of estimation for these estimators is not as complex as that used for the earlier CVM estimators, and should therefore result in more stable estimation, even in the presence of larger models.

CHOICE OF INDEPENDENT VARIABLES[1]
AND THEIR LEVELS

Choice of a Model and Model Characteristics

The first decision Monte Carlo researchers must make is the determination of what type of model to use. Four aspects of model choice are typically of interest: model type, model size, model complexity, and model parameters. The most basic choice with regard to model type is that of the basic type(s) of models to be simulated. At the time of the first edition of this book, the choice was basically one between path, CFA, or full structural (i.e., latent variable path) model. More recently, however, interest in latent growth, mixture, multilevel, and other more complex models has burgeoned, with a concomitant increase in Monte Carlo studies based on such models. Researchers might be specifically interested in only one of these, or might want to investigate different types of models, thus making the type of model one of the independent variables in the study. Model size is usually operationalized as the number of parameters to be estimated in the model, or as model degrees of freedom. A closely related determination is that of model complexity. Within any of the model types, complexity can be increased through the addition of such features as cross-loading indicators, reciprocal paths, correlated disturbances, or nonlinear effects, if these are of interest. Finally, values must be chosen for all parameters to be estimated. These values represent the population values of interest, although when sample data are generated from these parameters their statistical values will, of course, vary to some extent because of sampling error. These four aspects of model choice are discussed in more depth in the following sections.

Model type. The research question(s) might govern the choice of a model. For example, investigation of the effects of cross-loading indicators could be of most interest in CFA models. However, researchers should keep in mind that the full information nature of commonly used estimators in SEM means that facets of the model influencing estimates of one parameter may be propagated to other, possibly unexpected, parts of the model (Kaplan, 1988). In most situations it is probably best to include more than one type of model in Monte Carlo studies because this allows for greater generalization to applied settings. However, the nature of the research question should govern this choice. As in any study, there is a tradeoff between the inclusion of many independent variables with few levels of each, or fewer independent variables with more levels of each. For areas in which little previous research has been conducted, it may be more informative to choose one representative model but to include a wider variety of other independent variables, and/or more levels of these variables. For areas in which a particular model has been studied extensively under many different conditions, a researcher might choose to broaden the research base by studying other models under some of the same conditions. For example, historically the bulk of the Monte Carlo studies in SEM have been conducted with CFA models; in a review of 62 Monte Carlo studies in SEM, Hoogland and Boomsma (1998) found that 89% used CFA models. This suggests that future Monte Carlo studies should incorporate different types of models. Once the model(s) to be studied have been decided upon, the next choice is to determine the exact structure of the generating, or population model(s).

There are two basic ways in which this can be done. One is to review the applied and methodological literature to determine which configurations of the model are most commonly encountered in practice. A population model could then be constructed as a sort of composite of the model structures found in these studies. The advantage of this method is that the model and its characteristics can be manipulated experimentally in order to investigate the conditions of interest. The disadvantage is that the constructed model, as a composite of features from many different models, may not reflect real world conditions. The second and probably less common way of obtaining a population model is to base it on an actual data set. In this method, the researcher would treat an existing data set as the "population," and define a model fit to those data as the population model. The population parameters would be defined as the estimated model parameters from that data set. The advantage of using this method is that it is more likely to reflect real world conditions and therefore to produce results that are generalizable to these conditions. The disadvantage is that the researcher may not be able to manipulate all model characteristics that may be of interest. This approach has been endorsed by MacCallum (2003) because it does not make the unrealistic assumption that a model fits exactly in the population. MacCallum further argued that Monte Carlo studies based on perfectly fitting population models "are of only limited value to users of the methods" (p. 135) because they do not address the question of how methods perform when the model of interest is incorrect in the population. In his article, MacCallum suggested other methods of incorporating model error into a population model, such as including several minor nuisance or method factors in a CFA model.

In the context of our example study, we were interested in both CFA and full structural models. Because only one previous study (Muthén et al., 1997) has investigated both CFA and structural equation models, we wanted to expand on that study by including both types of models.

Model size. Another important consideration is the size of the model. This is typically determined by the number of observed variables. In their review of the literature, Hoogland and Boomsma (1998) found that the number of observed variables in SEM Monte Carlo studies ranged from 4 to 33, with a mean of 11.6 variables. Another way of conceptualizing model size is as the number of parameters to be estimated relative to the number of observed variables, usually quantified as the degrees of freedom for a model. This was the method used in a meta-analysis of Monte Carlo studies on the robustness of the χ^2 statistics in SEM by Powell and Schafer (2001), who found that model degrees of freedom ranged from 3 to 104. In this meta-analysis, model degrees of freedom were found to have large effects on the maximum likelihood and asymptotically distribution-free (ADF) χ^2 statistics. Model size is therefore an important model characteristic in SEM Monte Carlo studies, and should be chosen carefully. In most cases, the best approach would be to vary model size as one of the independent variables in the study design. Failure to take model size into account in this way can result in misleading conclusions. For example, an early study of ADF estimators with four observed variables, Muthén and Kaplan (1985) concluded that ADF-based χ^2 estimates showed little bias. However, in a later study that included models with up to 15 variables (Muthén & Kaplan, 1992), these authors found that the sensitivity of ADF χ^2 tests increased with sample size.

Because of these findings, we were interested in including models with large numbers of variables in the example study. We were also interested in studying the issue of whether model size or model type (CFA or structural model) have differential impacts on model fit. We included four levels of number of variables, ranging from 8 to 24 variables. As noted previously, within each of the four levels of number of variables we also included two models: one CFA and one full structural model. This resulted in a total of eight different models. Figure 15.1 shows two illustrative models: an eight variable, two-factor CFA model and a full SEM model with two exogenous and two endogenous factors, each measured by six variables. For estimation with binary categorical data, the smallest model has 17 parameters to be estimated (6 factor loadings + 1 factor covariance + 8 threshold values + 2 factor variances). However, note that this is only if the variables have two categories; with more categories, more thresholds would need to be estimated.

Model complexity. Related to model size is model complexity. Model complexity has often been operationalized as the number of free parameters in a model, with more free parameters indicating greater complexity. Model complexity can be introduced by adding such things as cross-loading indicators, reciprocal paths, correlated disturbances or measurement error terms, or nonlinear effects such as interactions or quadratic terms. More recently, Preacher (2006) has argued that, while the number of free parameters represents one aspect of model complexity, other features of a model also contribute to its complexity. Preacher defined model complexity as the ability of a model to fit different data patterns, regardless of whether the model is "true" in the sense of representing the data-generating process. Thus, a more complex model can be thought of as one with a greater abil-

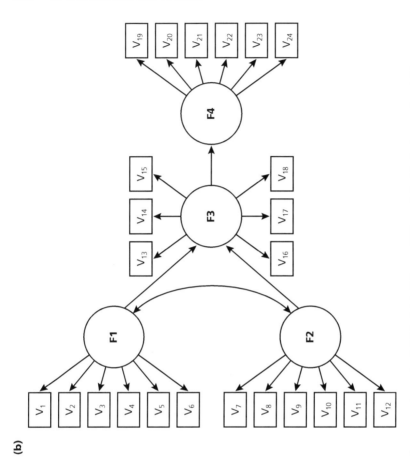

Figure 15.1 Path diagrams of two models from the example study: (a) Model 1: 2-factor CFA model with 4 observed variables per factor; (b) Model 2: 4-factor structural model with 6 observed variables per factor.

ity to adapt to differing data patterns. For example, a one-factor model is more complex than a simplex model because, while the simplex model fits well only to band-diagonal correlational patterns, the one-factor model can provide a good fit to a larger variety of correlational patterns. Thus, consideration of model complexity is important because it can play an important role in the fit of a model. When different models are included in a Monte Carlo study, researchers should therefore consider the degree to which the models differ in complexity. Unfortunately, this is more easily said than done. Currently, procedures for assessing model complexity as described by Preacher involve fitting the models to large sets of randomly generated data. Until more practical methods of assessing model complexity are developed, we recommend that researchers pay close attention to two model features noted by Preacher as contributing to model complexity. These are the number of zero covariances implied by the model and the imposition of equality or other constraints on model parameters. As an example of a model with implied zeros, consider a two-factor CFA model with uncorrelated factors. The lack of correlation between the two factors implies zero covariances for variables measuring different factors. Such a model has a lower complexity than a model that allows the factors to correlate. Similarly, when parameters are constrained to equality, the model will have lower complexity than the same model without equality constraints. This is because the model with equality constraints will not provide a good fit to as many different data structures as the unconstrained model.

Readers might wonder why a researcher would want to complicate her/ his model by including such parameters, as these may confound the effects of other independent variables. One reason is that many of the data sets analyzed in practice include such effects, so models that include these may be more generalizable to conditions encountered by applied researchers. A related reason is that, because ignoring such effects may bias the estimation of other model parameters, it might be of interest to determine the amount of bias that could be introduced by such omissions. Finally, as noted previously, complexity affects data-model fit. In the current example study we did not include complex models because the behavior of the DWLS estimator has not been studied extensively even in the context of simple models. We therefore felt that the inclusion of complicating factors was premature at this stage of investigation.

Model parameter values. Choice of parameter values is an important consideration because these are related to the reliability of factor indicators as well as the strength of relationships among both observed and latent variables. These in turn affect the levels of power of tests of individual parameter estimates. In choosing the size of factor loadings, the researcher should keep in mind that indicator reliability may be defined as:

$$\frac{\lambda_i^2}{\lambda_i^2 + \theta_{ii}} \, , \qquad\qquad (15.1)$$

where λ_i and θ_{ii} are the unstandardized factor loading and error variance of indicator i, respectively. Thus, indicator reliability depends not just on the value of the loading, but also on the ratio of loading size to the size of the error variance. This is also equivalent to the square of the standardized loading.

Hoogland and Boomsma (1998) reported standardized factor loading sizes ranging (in absolute value) from .3 to 1.0 and factor correlations from 0 to .5 in their review of Monte Carlo research in SEM. These authors did not provide values for structural parameters, but common standardized values range from .3 to .7. The values chosen should represent what is typically seen in applied research in the field of study to which one hopes to offer guidance. A review of applied studies in the literature of that field should be done to determine if one's choices are reasonable from this point of view. One issue regarding factor loading values is whether all loadings for a particular factor should be set to the same value or whether loadings should have mixed values. If mixed loadings are chosen it is still possible to vary loading magnitude by choosing sets of loadings whose average values differ within high, low, or medium ranges. Of course, researchers could choose to include sets of both varied and equal loadings.

Because most previous studies of the DWLS estimator have included only fairly high loading conditions (i.e., loadings of .7 to .8), we decided to include both low (.4 to .5) and high (.7 to .8) loading conditions in the example study. As discussed by Enders and Finney (2003), loadings in the range of .4 to .5 are not uncommon in applied studies. These researchers found, not surprisingly, that lower loadings resulted in less power to detect model misspecifications for NT based estimators.

Other Independent Variables

As with the choice of a model, the choice of other independent variables to include in the design of a Monte Carlo study should be based on a review of the literature. Aside from choices about aspects of the model(s) to be investigated, other common independent variables in SEM Monte Carlo studies include:

- Sample size
- Level of nonnormality
- Degree of model misspecification
- Degree of categorization of observed variables
- Method of estimation

Here again, Monte Carlo research is not essentially different from other quantitative research. Determining the appropriate independent variables and their levels should be based on a review of the applied and/or methodological research.

Sample size. Powell and Schafer (2001) found that sample size affected values of some types of estimators but not of others, so the choice of whether to include sample size as an independent variable might depend on the type(s) of estimator to be studied. It would also be important to determine if there was any statistical theory to suggest how sample size would affect the estimators of interest so that appropriate research hypotheses could be posed. Sample size would also be an important factor to consider if sampling stability was of interest, because stability will obviously be better with larger samples. The interest in this case may be in determining the smallest sample size at which reasonably stable estimates can be obtained. Researchers should keep in mind that fit statistics, parameter values, and standard errors will typically have different thresholds for stability. Powell and Schafer (2001) found that sample size varied from 25 to 9600 in the Monte Carlo studies they reviewed. A typical range of values is from 100 to 1000, unless the researcher is interested in focusing on extremely large or small sample sizes.

Previous studies of the DWLS estimator incorporated sample sizes ranging from 100 to 1000. The results of these studies indicated poor performance for smaller sample sizes if the data were nonnormally distributed and/or had small numbers of categories. DWLS performance also deteriorated with larger models at the smaller sample sizes. Because the number of variables in the models for the example study varied quite a bit, we chose different sample size levels for the different models. For the smallest model with 8 variables we varied sample size at 150 and 300, while for the largest model with 24 variables we used sample sizes of 450 and 950. This maintained the same ratio of number of subjects to number of estimated parameters.

Type of distribution. Distributional type was included as an independent variable in 76% of the studies in Hoogland and Boomsma's (1998) review of robustness studies in SEM. Given the assumptions of normality underlying common estimators such as ML and GLS, this is not surprising. Powell and Schafer (2001) found that researchers used several methods of generating nonnormal data. One of these was to generate multivariate normal data and then categorize certain variables in such a way as to yield the desired degrees of skew and kurtosis. Over half of the studies investigated by these authors used this approach. This is also the method used in Mplus MONTECARLO procedures and in the example study. A less commonly used variation on this approach was to censor variables' data from above or below to obtain the chosen degree of nonnormality. One issue to consider if nonnormality is induced using these methods is the form of the underlying distribution. While most studies begin with normally distributed data and categorize the data in such a way as to yield nonnormal distributions, another approach is to assume that the underlying distribution itself is nonnormal and categorize the data so that it matches that distribution. This might be a more realistic approach in some situations (see Bandalos & Enders, 1996, for a discussion).

Another common approach to generating nonnormal data involves creating a covariance matrix with a specified structure from which sample data can be generated, and inducing the desired levels of univariate skew and kurtosis for the observed variables using procedures such as those developed by Vale and Maurelli (1983). Data from a multivariate normal distribution can also be transformed into a nonnormal multivariate χ^2 or t distribution (for examples, see Hu, Bentler, & Kano, 1992). Mattson (1997) proposed a method in which the distributions of the latent common and unique factors are transformed in such a way as to yield the desired levels of nonnormality in the observed variables. This method was reviewed by Reinartz, Echambadi, and Chinn (2002) who found it to work well when implemented in three commonly used software packages (SAS, Mathematica, and LISREL/PRELIS). Finally, Muthén and Muthén (2002) proposed a method in which nonnormally distributed data are obtained as a mixture of two normally distributed subpopulations having different means and variances for the observed variables. Because the majority of these methods control the univariate distributions, and not the multivariate distributions, the level of multivariate normality should be computed and reported after the data generation process has been completed. A related strategy is to vary the multivariate skew and kurtosis rather than the separate univariate values. This method is preferable because it is the multivariate and not the univariate parameters that are assumed to be within normal limits for SEM NT estimators. However, there currently exist no published guidelines regarding the levels of multivariate skew and kurtosis to which these estimators are robust (although see the chapter by Finney and DiStefano in this volume).

One problem in generating data from populations with prespecified levels of skew and kurtosis is that these characteristics have high levels of sampling variability. Thus, although the desired levels might be obtained for the population, for any given sample they might deviate considerably from these values. This is especially true for kurtosis, and with higher levels of both skew and kurtosis. A final point has to do with the choice of skewness and kurtosis values. This choice is made difficult by the fact that levels of skewness and kurtosis vary widely in the applied literature. As noted by Micceri (1989), distributions "exhibited almost every conceivable type of contamination" (p. 162), making it difficult to generalize about common levels of skew and kurtosis. Given this, the best course of action in choosing distributional levels may be to consult both the applied and Monte Carlo literature most relevant to the applied area to which the study will be targeted. Researchers might also want to explore the limits of NT estimators by increasing the level of nonnormality systematically in order to determine the degree of nonnormality that can be tolerated without severe consequences to parameter estimates, standard errors, or overall fit.

Level of nonnormality was of interest in our example study. Previous studies of the DWLS estimator included levels of absolute skewness (sk) of 2.3 or less, and kurtosis (k) of 5.3 or less. In our study we wanted to extend the research base by including more severe levels of nonnormality and designated the following target levels for "moderate" and "severe" nonnormality, respectively: sk = 2.2, k = 3.0; and sk = 3.0, k = 7.0. Clearly, these designations are arbitrary, to some extent, and ours are based on relative comparisons to those previously studied. We included a condition in which data were normally distributed for comparison (for two-category data, a uniform distribution was created as a normal distribution is not possible).

Model misspecification. MacCallum (2003) has argued that Monte Carlo researchers should focus more on models that contain some level of population misspecification. Population lack of fit has been referred to as "error of approximation" and is arguably a more realistic conceptualization for most phenomena in the behavioral sciences (Browne & Cudeck, 1992; Cudeck & Henly, 1991). From this point of view, consideration of models with some degree of error of approximation is likely to provide more useful information to applied researchers who are attempting to model such phenomena. Estimation methods have also been found to be differentially sensitive to model misspecification (Olsson, Foss, Troye, & Howell, 2000; Yuan & Chan, 2005). For example, Olsson et al. found that fit indexes obtained from ML estimation were more sensitive to misspecification than those obtained from GLS or WLS. Model misspecification is therefore an important independent variable in many Monte Carlo studies. One way in which lack of fit is studied is through generating sample data from a population model with a perfect fit, misspecifying the model in some way, and then fitting the generated sample data to the misspecified model. Models misspecifications are often categorized as errors of omission and errors of inclusion. As the names imply, the first type involves omitting the estimation of model parameters that were present in the population model, while the second refers to estimating parameters that were not present in the population. Although some researchers argue that errors of inclusion do not really constitute errors in a strict sense, because they do not result in lack of fit as measured by the χ^2 statistic, they are commonly included in discussions of misspecification because they do affect model degrees of freedom and can result in inflated standard errors. Commonly omitted parameters include secondary factor loadings (i.e., cross-loadings), correlations between residuals, structural paths, and even entire factors. Incorrectly included parameters are typically modeled as extra structural parameters. Errors of omission have been found to be most serious and are therefore the object of most study.

Another method of obtaining misspecified data is to use an actual large data set as the "population" and to generate samples from it. Because no model will fit perfectly in even a large sample, some degree of model misspecification will be present in both the "population" and sample data. Samples can be generated by either taking repeated samples of a given size with replacement or by obtaining a covariance matrix based on the "population" data and generating random samples from that parent correlational structure. This method was recommended by MacCallum (2003) who also suggested introducing lack of fit into a simulated population model by including such things as extra minor factors or method effects.

One problem with studies involving misspecification is that the degree of misspecification is often not quantified, making it difficult to model the effect of misspecification on values of fit indices, parameters, and standard errors. In some cases model misspecifications that are assumed to be serious might not have as large an effect on these values as do presumably less serious misspecifications. This is due to the fact that statistical tests of model parameters have differential levels of power, depending on their correlations with other model parameters (Kaplan & Wenger, 1993). While many Monte Carlo researchers simply categorize misspecifications as small, medium, or large, a more informative method of quantifying misspecification due to errors of omission is to estimate the power associated with the test of the omitted parameter, using methods such as those proposed by MacCallum, Browne, and Sugawara (1996). Alternatively, because the root mean square error of approximation (RMSEA) is a measure of model lack of fit per degree of freedom, this index could be used to quantify the degree of model misspecification. Procedures such as these not only assure that misspecifications correspond to the desired levels but allow for more accurate modeling of the effects of model misspecification.

The degree to which the model χ^2 and other fit indices based on a particular estimator are able to detect model misspecification is one of the most important criteria for evaluating that estimator. Given this, we gave serious consideration to including misspecified models in the example study. However, because it appeared to us that the behavior of this estimator had not been studied sufficiently even in the context of correctly specified models, we decided not to include model misspecification as a design facet in the study. This decision was also motivated by concerns about the size of the study, which at this point consists of 8 (models) × 2 (loading sizes) × 2 (sample sizes) × 3 (levels of nonnormality), or 96 cells, and there was still one more condition we wanted to vary. We therefore decided to retain the issue of model misspecification for a future study.

Categorization. Because many of the measures used in the social sciences are not at the interval level, many applied researchers are interested in the degree to which the inclusion of categorical or ordinal level data will affect the results of SEM analyses. Categorization of generated data is usually accomplished through generating continuous data and categorizing it through the imposition of cutpoints at different places along the continuum. This method can be used to obtain nonnormal univariate distributions and is the method that was used to create the categorical, nonnormally distributed data in the example study, as noted previously. In this method, the level of nonnormality is controlled through the choice of cutpoints. For example, to obtain data with two categories, we used a single cutpoint. Use of a cutpoint of 1.39, as shown in the Mplus program in a later section, resulted in skewness and kurtosis levels of approximately 3.0 and 7.0, respectively. In this context, it should be noted that categorized data do not allow for the generation of perfectly normal distributions, especially when the number of categories is small. Because SEM results are affected most severely by distributions with smaller numbers of categories, Monte Carlo research in this area commonly focuses on variables with 2, 3, or 4 categories. With 5 or more categories, the effects of categorization on parameter estimates and standard errors are generally negligible (Dolan, 1994). For more discussion of the analysis of categorical data, see Finney and DiStefano (this volume).

Because the DWLS estimator of interest in our example study has been developed for use with categorical data, we were particularly interested in including this as a design facet. In general, studies of categorical variables have found that performance does not improve substantially when the number of categories is increased beyond five. We therefore decided to use data with two, three, and five categories.

Estimation method. As mentioned previously, the various estimators used in SEM are not equally sensitive to the effects of the other possible independent variables that have been discussed. Some estimators, such as the ADF estimator, have been formulated specifically to be less sensitive to the effects of nonnormality. Other estimators commonly studied include those incorporating adjustments to the chi-square statistic and standard errors in order to adjust for the biasing effects of data nonnormality or categorization. For details on these estimators, the reader is referred to the chapter by Finney and DiStefano (this volume). As noted previously, estimators have also been found to be differentially sensitive to model misspecification. Clearly the choice of which estimators to study would be based on the type of data and model(s) included in the study. Often, however, the ML estimator is included as a basis of comparison even when it is not the estimator of interest. In the current study we followed this convention and included the ML estimator along with the DWLS estimator.

Interaction effects. It should be clear from reading the previous sections that the various independent variables that might be included in an SEM Monte Carlo study do not operate in isolation but interact with each other. For example, estimation methods are differentially sensitive to such things as sample size, model size, level of misspecification, and nonnormality. Therefore possible interactive effects should be anticipated when designing Monte Carlo studies, and the relevant variables should be included. Of course, as in any study, the researcher must make compromises between the inclusion of multiple design characteristics and resources available. For Monte Carlo studies, time and computer capacity are two important resources that usually limit the scope of a study. This means that the researcher must choose the independent variables and their levels carefully. Researchers should also keep in mind that the interpretation of high level interactions among multiple variables with several levels each may be very difficult, both for researchers to explain and for readers to comprehend. Focusing on fewer carefully chosen variables may be more informative in the long run.

To sum up the research design for our example study thus far, there are eight models, two magnitudes of factor loadings, two sample sizes, three levels of nonnormality, and three levels of categorization, resulting in $8 \times 2 \times 2 \times 3 \times 3$, or 288 cells in the design. Data from each cell of the design will be analyzed using both ML and DWLS estimation methods and results will be saved. In the next section we discuss choices for dependent variables.

CHOICE OF DEPENDENT VARIABLES

The primary focus in many Monte Carlo studies is on the effects of the independent variables on parameter estimates, standard errors, and/or fit indices. For each of these outcomes, bias and efficiency are typically of most interest. Bias refers to a systematic difference between a sample estimate and the corresponding population value. Efficiency has to do with the sampling variability of a statistic, with statistics that vary least across samples being preferred.

Parameter Estimates

Bias. Parameter estimate bias is, in general terms, the average deviation of a parameter estimate from its true value. Bias can be calculated in several different ways. The most basic formulation for bias, sometimes termed *raw bias*, is

$$Bias(\hat{\theta}_i) = \sum_{j=1}^{n_r} \frac{(\hat{\theta}_{ij} - \theta_i)}{n_r} \qquad (15.2)$$

where $\hat{\theta}_{ij}$ is the jth sample estimate of the ith true parameter value θ_i, and n_r is the number of replications within the cell. Because raw bias values can be difficult to interpret, another form of bias known as *relative bias* can be calculated as the average deviation of the sample estimate from its true value, relative to the true value, or:

$$ rBias(\hat{\theta}_i) = \sum_{j=1}^{n_r} \left(\frac{\hat{\theta}_{ij} - \theta_i}{\theta_i} \right) / n_r. \qquad (15.3) $$

Both raw and relative bias values can be multiplied by 100 to obtain percentage bias values. If certain parameter values, such as factor loadings or error variances, are affected in the same way by the study conditions, bias across these parameters is sometimes averaged as well. If these parameter estimates are affected differentially, however, bias should be reported separately for the individual parameters of interest. Not surprisingly, different researchers have suggested differing criteria for what constitutes "serious" bias. Muthén, Kaplan, and Hollis (1987) have suggested that values of relative bias less than .10 to .15 might be considered negligible. Hoogland and Boomsma (1998) have offered the more stringent criterion that the value of relative bias be less than .05 for a parameter estimate to be considered unbiased.

Efficiency. The efficiency of parameter estimates is often measured by the mean squared error (MSE), calculated as the squared difference between the sample estimates and their population value, or:

$$ \sum_{j=1}^{n_r} \frac{(\hat{\theta}_{ij} - \theta_i)^2}{n_r}. \qquad (15.4) $$

In some applications, n_r in the denominator is replaced by $n_r - 1$, but with large numbers of replications the difference will be negligible. When parameter estimate bias is present, the quantity in Equation 15.4 is a combination of squared bias and the variance of the parameter estimate, and represents the overall accuracy of parameter estimation. If parameter estimates are unbiased, this quantity is a measure of the sampling variance of an estimate and its square, known as the root mean square error (RMSE), is sometimes called the empirical standard error of the parameter estimates. If parameter estimates are biased, the population value θ_i can be replaced by the mean of the parameter estimates across replications to obtain a measure that does not reflect bias.

Coverage. The percentage of times the population value is contained within the confidence interval of a parameter estimate is also sometimes used as an indicator of bias in standard errors. As an example, a 95% confidence interval around the parameter estimate should contain the population value in 95% of replications. If empirical coverage is only 90%, this suggests that the confidence interval is too wide (e.g., Enders, 2001). However, note that parameter estimate bias can also result in low coverage values. Collins, Schafer, and Kam (2001) suggested that coverage values lower than 90% are problematic.

Standard Errors. Relative standard error bias is assessed in a manner similar to relative parameter estimate bias, that is, as the deviation of each sample standard error from its population value relative to the population value, averaged across replications. However, as Hoogland and Boomsma (1998) pointed out, there are different methods of estimating the population values. These are:

- Obtain estimates as the square root of the diagonal of the asymptotic covariance matrix of the parameter estimates ($\hat{\theta}$) using the population parameter values for the parameters.
- Calculate the empirical standard errors as the standard deviation of parameter estimates over a large number of replications as in Equation 15.3.

The first method is only accurate if the assumptions of the estimation method are met. If these assumptions are violated, as when nonnormally distributed data are used for ML or GLS estimation methods, the obtained standard errors will be incorrect. The accuracy of the second method is not dependent on any distributional assumptions.

After the population value is decided upon, the relative standard error bias can be calculated as:

$$rBias\left(\widehat{SE}(\hat{\theta}_i)\right) = \sum_{j=1}^{n_r} \left(\frac{\widehat{SE}(\hat{\theta}_i)_j - SE(\hat{\theta}_i)}{SE(\hat{\theta}_i)} \right) / n_r \qquad (15.5)$$

where $SE(\hat{\theta}_i)$ is an estimate of the population standard error of $\hat{\theta}_i$, and $\widehat{SE}(\hat{\theta}_i)_j$ is the estimated standard error of $\hat{\theta}_i$ for the jth replication. As with parameter estimate bias, standard error bias can be calculated either within or across cells of the design. Hoogland and Boomsma (1998) suggested that "acceptable" levels of bias not exceed 10% of the absolute value of Equation 15.5.

Fit Indices

Values of the χ^2 statistic have been studied most often as outcome variables in Monte Carlo studies. Effects on actual χ^2 values as well as rejection rates are typically studied. Rejection rates are obtained in Monte Carlo studies by calculating the number of replications for which the χ^2 value for a model would be larger than the critical value at a given nominal level for Type I error. Relative bias of actual χ^2 values can be calculated as the average difference between the sample values and the population value, expressed as a proportion of the population value. Because the population value of the model χ^2 statistic is equal to its degrees of freedom, this can be expressed as:

$$Bias(\hat{\chi}^2) = \sum_{j=1}^{n_r} \frac{\hat{\chi}_j^2 - df}{df} / n_r. \qquad (15.6)$$

Because most fit indices other than χ^2 have unknown sampling distributions, values of these indices are often described by comparing their average values across replications to the cut-off criteria suggested in the literature (e.g., Hu & Bentler, 1999).

Equation 15.6 provides a way of quantifying average values of the model χ^2 statistic. However, high average χ^2 values do not necessarily mean that there will be a large number of model rejections, because the high mean values could result from only a few cases. Researchers are therefore also interested in the "tail behavior," or percentage of sample χ^2 values that fall into the region of rejection. This is often quantified as the proportion of Type I errors that are made. Bradley (1978) suggested as "stringent" and "liberal" criteria that the empirical Type I error rate π lie in the range $\alpha \pm 0.1\alpha$ or $\alpha \pm .5\alpha$, respectively. Robey and Barcikowski (1992) have outlined more complicated procedures that can be used to determine critical values to test the hypothesis that $\pi = \alpha$. Most recently, Serlin (2000) has amended these procedures so that the range for α would depend on its value, with the distance between α and π becoming more liberal with smaller values of α.

Other Dependent Variables

Other choices for dependent variables in SEM Monte Carlo studies are as varied as are the studies themselves. Other commonly used variables include power levels and Type II error rates, proportions of convergent and/or admissible solutions, and values or patterns of modification indices.

For our example, we were interested in bias in parameter estimates and standard errors as well as bias and Type I error rejection rates for the DWLS estimator.

GENERATING DATA FOR MONTE CARLO STUDIES

Once the research questions have been developed and the study design has been mapped out, the researcher is ready to begin simulating the data to be used in the study. This process typically begins with the creation of a population covariance matrix. If more than one model is to be used in the study, a population covariance matrix is created for each, and data are generated from each of these. The sample data are then analyzed and the results from these are saved and evaluated. In some software packages, such as Mplus, both the generation and analysis of the sample data can be accomplished within the same program. In others such as LISREL, different programs must be used to generate the sample data and conduct the analyses. For most programs, the data generation and analysis procedures can be automated using software such as SAS/IML or R. However, before samples can be generated, two important decisions must be made: the choice of a random seed and the choice of the number of replications. These issues will be addressed in the following sections.

Generating a Population Matrix

A common method of creating population covariance matrices is to fix all parameter values to the desired population values in an SEM software package. The covariance matrix implied by these values can then be saved and used as a basis for generating the sample data. This method is used in LISREL and other programs. In Mplus and EQS the population and sample matrices are generated and analyzed using the same program. If population covariance matrices are created separately they should always be checked for fit to the parameter values used to create them. This is done by reading the population covariance matrix into a program that specifies the same model as was used to create it. This analysis should result in parameter values that match exactly those used to create the data, a χ^2 value of exactly zero, optimal values of all other fit indices, and values of zero for all residuals.

Here we demonstrate the data generation process using the smallest of the models described in the previous section, and shown on the left-hand side of Figure 15.1. The unstandardized population parameter values for this model were as follows: loadings were all specified as .7, measurement error variances were all set at .51, factor variances were 1.0, and the factor correlation was set to .5.

The program in Figure 15.2 was used to generate data for samples of size 150 for the 8 variable model. Using the MONTECARLO command in Mplus allows for the data to be both generated and analyzed in a single pro-

```
TITLE: Example CFA Simulation
MONTECARLO:
        NAMES = y1-y8;
        NOBSERVATIONS = 150;
        NREPS = 500;
        SEED = 741355;
        GENERATE = y1 - y8 (1);
        CATEGORICAL = y1 - y8;
        REPSAVE = ALL;
        SAVE = M1N150CAT1*.DAT;
        RESULTS = M1N150CAT1RES;
MODEL POPULATION:
        f1 BY y1-y4*1;
        f2 BY y5-y8*1;
        f1-f2*.50;
        f1 WITH f2*.25;
        y1-y8*.50;ᵃ
        [y1$1-y8$1*1.39];
MODEL:
        f1 BY y1-y4;
        f2 BY y5-y8;
        f1-f2*.50;
        f1 WITH f2*.25;
        [y1$1-y8$1*1.39];
OUTPUT: TECH9;
```

Figure 15.2 Input for Mplus MONTECARLO program.

ᵃ Note that, although measurement error variances (y1–y8) must be included when *creating* the categorized simulated data through the *MODEL MONTECARLO* command, measurement error variances should *not* be included when analyzing the data using the *MODEL* command. This is because threshold and measurement error parameters are not simultaneously identified for categorized data, so both cannot be estimated.

gram. The printed output contains a summary of the results for parameter estimates, standard errors, and χ^2 values. The generated data and results can also be saved into files and analyzed using external programs using the SAVE= and RESULTS= commands, respectively.

The MONTECARLO command signals that a Monte Carlo study is to be done and specifies the variable names, the sample size for each replication (NOBSERVATIONS), the number of replications (NREPS), and the random seed (SEED) to be used. The MODEL POPULATION command provides the population model and parameter values on which the samples are to be generated. The symbol "*" is used here to give the value of each population parameter (the "@" symbol can also be used). The MODEL command describes the model to be estimated in the samples. In this case it is the same as the population model, but if the researcher wanted to introduce some type of model misspecification, the misspecified model would be described here.

In this study we were interested in the behavior of the DWLS estimators with categorical data. Because DWLS is the default estimator for categorical data, use of this estimator does not need to be specified in the program. To obtain categorical data the GENERATE and CATEGORICAL options are added to the MONTECARLO command. The GENERATE option indicates the number of thresholds to be used for data generation. In this example, it is specified as (1), indicating that variables y1 – y8 will be generated as categorical variables with 2 categories each. The CATEGORICAL option indicates which variables are to be analyzed as categorical. Although the variables specified in the GENERATE and CATEGORICAL options will typically be the same, having both options allows for model parameters to be generated as categorical but analyzed as continuous. To do this you would specify GENERATE = y1 – y8 (1) as above but leave out the CATEGORICAL option. The distributions of these variables are controlled through the values of the thresholds. By default, variable scores are generated to be normally distributed. These distributions can be changed through judicious use of thresholds, however. For example, for a two category variable specifying the threshold to be 0 would create a uniform distribution. In this example, the threshold for all 8 variables has been set at 1.39 (see the following discussion on the MODEL POPULATION command) which would result in approximate values of skew and kurtosis of 3.0 and 7.0, respectively.

The next three lines have to do with saving data and results. The REPSAVE option indicates the replications for which raw data should be saved. If all of the raw data are to be saved, as in the current example, this is indicated by specifying REPSAVE=ALL. The SAVE option specifies the name of the file(s) into which the data should be saved. If data from more than one replication are to be saved, separate files will be created for each replication. In this situation, use of an asterisk in the name of the data file will cause the files to be given the name specified plus consecutive integers indicating the replication number. In this example, we have used the name "M1N150CAT1" to indicate the data are from the first model with $n = 150$ using the first categorization method. The 500 data files will be named "M1N150CAT11.DAT, M1N150CAT12.DAT,... M1N150CAT1500. DAT." The RESULTS option can be used to save selected output from each replication in an external ASCII file. In this case, results from all replications will be saved in a single file, under the name specified in the option (in this example, all results will be saved in the file "M1N150CAT1.RES."). Results are saved in the following order: replication number, χ^2 value, parameter estimates, standard errors, and fit statistics. The order of parameter estimates, standard errors, and fit statistics is given in the output from the MONTECARLO run. These results are saved in free format, separated by spaces, in a format of E15.8. The format E15.8 indicates that the data are in scientific notation and that each variable consists of fifteen columns, eight of which are to the right of the decimal point.

Under the MODEL POPULATION and MODEL commands, the line "[y1$1–y8$1*1.39]" specifies that the thresholds, designated with "$", should be set to 1.39. For the population data this will result in a fixed threshold value of 1.39. For the sample data, thresholds will have start values of 1.39, but estimated values will vary due to sampling error. This value represents the cutpoint described previously that will yield nonnormally distributed variables with skew and kurtosis of approximately 3.0 and 7.0, respectively. Finally, the option "TECH9" in the OUTPUT command will cause the program to print error messages in the output for replications that fail to converge.

Selected output from this program regarding fit indexes is shown in Figure 15.3. This portion of the output provides information on fit indices and indicates the number of replications for which estimation was successful. In this example, 481 out of 500 replications ran successfully. The 19 failures are probably due to the combination of small sample size, high nonnormality, and small number of categories (2). For DWLS estimation, fit indices given include the χ^2 statistic, RMSEA, and WRMR. WRMR, or the weighted root mean square residual, is a measure of the difference between the sample and model-implied matrices based on the WLS estimation. Yu and Muthén (2002) have suggested an upper cutoff of 1.0 for this index when applied to nonnormally distributed data.

In this section, two tables, labeled "Proportions" and "Percentiles," are shown for each fit index. Under each of these headings there are columns labeled "expected' and "observed." For the χ^2 statistic, the "expected" proportions and percentiles columns contain the probability values and critical values, respectively, from a central χ^2 distribution with degrees of freedom associated with the model being tested. For example, the value of 0.010 in the "expected proportion" column corresponds to the value of 21.666 in the "expected percentile" column and indicates that, for the degrees of freedom associated with this model, a proportion of .01 of the sample χ^2 values would be expected to equal or exceed a critical value of 21.666. Values in the columns labeled "observed" contain the proportion of actual sample values at or beyond the value of χ^2 shown in the corresponding "observed percentile" column. For example, the lowest observed proportion of .091 corresponds to a χ^2 value of 32.199, indicating that .091 or 9.1% of the sample χ^2 values were greater than or equal to 32.199.

Following information on fit statistics, summaries of parameter estimates and standard errors are given under "Model Results." This portion of the output is shown in Figure 15.4, in which the first column contains the popu-

```
Chi-Square Test of Model Fit
        Number of successful computations      481

            Proportions                    Percentiles
    Expected     Observed     Expected        Observed
       0.990       1.000        2.088           3.746
       0.980       0.998        2.532           4.527
       0.950       0.998        3.325           5.194
       0.900       0.996        4.168           6.079
       0.800       0.952        5.380           7.552
       0.700       0.894        6.393           9.018
       0.500       0.773        8.343          11.720
       0.300       0.588       10.656          15.008
       0.200       0.478       12.242          18.162
       0.100       0.331       14.684          21.744
       0.050       0.225       16.919          25.182
       0.020       0.129       19.679          28.709
       0.010       0.091       21.666          32.199


RMSEA (Root Mean Square Error Of Approximation)

    Mean                                     0.062
    Std Dev                                  0.030
    Number of successful computations        481

            Proportions                    Percentiles
    Expected     Observed     Expected        Observed
       0.990       1.000       -0.007           0.005
       0.980       1.000        0.001           0.006
       0.950       0.952        0.013           0.013
       0.900       0.869        0.024           0.020
       0.800       0.792        0.037           0.042
       0.700       0.682        0.046           0.044
       0.500       0.482        0.062           0.056
       0.300       0.314        0.078           0.071
       0.200       0.204        0.087           0.083
       0.100       0.104        0.100           0.095
       0.050       0.052        0.111           0.104
       0.020       0.017        0.123           0.121
       0.010       0.015        0.132           0.133
```

Figure 15.3 Selected Fit Index Output from Mplus MONTECARLO program.

lation values specified in the program. The mean and standard deviation of each parameter estimate across the successful replications are given in columns two and three. The fourth column contains the average standard error across the completed replications, while the fifth column contains the mean squared error (M.S.E.). The latter is the average of the squared deviations of the sample parameter estimates from their population value.

MODEL RESULTS

	Population	ESTIMATES Average	Std. Dev.	S.E. Average	M.S.	95% Cover	% Sig Coeff
F1 BY							
Y1	1.000	1.0000	0.0000	0.0000	0.0000	1.000	0.000
Y2	1.000	1.2701	3.8987	2.3897	15.2415	0.892	0.794
Y3	1.000	1.1199	1.3495	1.0222	1.8317	0.909	0.821
Y4	1.000	1.4511	6.2867	8.0346	39.6442	0.909	0.807
F2 BY							
Y5	1.000	1.0000	0.0000	0.0000	0.0000	1.000	0.000
Y6	1.000	1.2860	2.5821	1.7941	6.7352	0.881	0.815
Y7	1.000	1.2700	1.9479	1.0736	3.8592	0.900	0.821
Y8	1.000	1.1729	1.3715	0.7642	1.9071	0.890	0.805
F1 WITH							
F2	0.250	0.2384	0.1688	0.1134	0.0286	0.721	0.526
Thresholds							
Y1$1	1.390	1.4042	0.1617	0.1505	0.0263	0.940	1.000
Y2$1	1.390	1.4078	0.1552	0.1507	0.0243	0.963	1.000
Y3$1	1.390	1.4057	0.1521	0.1504	0.0233	0.963	1.000
Y4$1	1.390	1.4003	0.1523	0.1499	0.0232	0.956	1.000
Y5$1	1.390	1.4032	0.1596	0.1504	0.0256	0.960	1.000
Y6$1	1.390	1.3932	0.1545	0.1494	0.0238	0.960	1.000
Y7$1	1.390	1.3980	0.1546	0.1498	0.0239	0.958	1.000
Y8$1	1.390	1.4021	0.1542	0.1501	0.0239	0.944	1.000
Variances							
F1	0.500	0.5635	0.3437	0.2469	0.1219	0.877	0.663
F2	0.500	0.5604	0.3435	0.2302	0.1214	0.842	0.640

Figure 15.4 Selected Parameter Estimate Output from Mplus MONTECARLO program.

Error messages from replications that failed to converge are included under "Technical 9 output." These messages indicate the sequence number of the replication and the reason for the unsuccessful run. In this example, 19 replications failed to run successfully.

In some situations, researchers might prefer to generate data in either Mplus or another program and then run another Mplus program to read in and analyze the data. This is referred to as an "external" Monte Carlo study in Mplus. An example of an Mplus program that reads in the data created by the program in Figure 15.2 is included among the online materials for this chapter, available at http://education.ufl.edu/leite/publications/. Although in this example the data to be analyzed were created by Mplus, data created from any software could be used. However, the Mplus Monte Carlo function requires that a list of the names of all data files to be analyzed be read in. This is done automatically when generating data using Mplus Monte Carlo, but such a list could also be prepared easily using a word processing program.

Setting the Random Seed

One important consideration in generating sample data is the choice of a random seed. The seed value is used by the program as a starting point for the random draws that create the samples. One seed is needed for each set of sample data that is generated. Most programs allow two ways of setting the seed. One is to allow the computer's internal clock to derive the seed. Using this method will result in a different random seed, and thus randomly different samples, to be drawn each time the program is run. The second method is to provide a value for the starting seed using a random numbers table. This method is preferable in most situations for two reasons. The first is that, by keeping a record of the random seeds used to create the samples for different cells of the design, the data for a particular cell can be regenerated if the data are lost or damaged in some way. Another reason we prefer this method is that it allows for the seed to be tested. This is done by generating a series of random normal variates, or z-scores, using the seed and applying the various tests of randomness to the data. Such tests include the Runs test, the one-sample Kolmogorov-Smirnov test, or the Sign test (see Siegel & Castellan, 1988, for more details). While this is possible with seeds generated by the computer, it involves extra steps unless raw data are generated instead of covariance matrices. Studies have evaluated some of the random number generators available in commercially available software (e.g., Bang, Schumacker, & Schlieve, 1998; Fishmann & Moore, 1982) and should be consulted when choosing a software package.

One final issue with regard to seed selection is whether the same seed should be used for generating data for different cells of the design. For example, if a researcher is interested in studying the effects of 5, 10, and 15 indicators per variable on the power of the model χ^2 test in SEM, data for the three levels could be generated independently using separate seeds or the same seed could be used to generate data for all three levels. The advantage of the latter method is that it reduces the amount of sampling error across the three levels. However, it also introduces a degree of dependence among them. In terms of data analysis, this dependence could be modeled through the use of a repeated measures or other dependent-samples design. In our view, the most important consideration in choosing between the two methods is the degree to which it corresponds to the real world situation being modeled. For example, in the example involving the number of indicators, use of the same seed to generate data for all three levels would correspond to a situation in which an investigator had access to 15 total items, but might have chosen to use only 5 or 10 of them in a given study. In this case the indicators are considered to be from the same indicator pool and should be generated using the same seed. In contrast, use of separate seeds would correspond to the situation in which different sets of items were used for each of the three levels of number of indicators.

Choosing the Number of Replications

As noted by Harwell et al. (1996) in the context of item response theory Monte Carlo studies, the choice of the number of replications depends on the purpose of the study, the desire to minimize sampling variance, and the need for adequate power. The purpose of the study will influence the number of replications needed because some effects of interest are less stable than others. For example, more replications may be needed to study the behavior of standard errors than parameter estimates because the former generally have a greater sampling variance. However, if the object of the study is to compare the results of different parameterizations of a model, or of different software packages, it is not necessary to obtain an empirical sampling distribution so a large number of replications might not be needed. In some cases, the study design may require researcher judgments to be made. For example, the researcher might base some decisions on several competing criteria, as in the decision to eliminate or retain items in scale development, or the number of factors suggested by a scree plot. Simulating this process would require such a judgment to be made for each replication, thus severely limiting the number of replications that is feasible. Powell and Schafer (2001) reported numbers of replications ranging from 20 to 1000 with a median of 200 in their review of χ^2 robustness studies in SEM.

Robey and Barcikowski (1992) argued that Monte Carlo researchers do not typically have a strong rationale for their choice of the number of replications to be used, resulting either in a lack of power or in excessive power. As in applied research, the sample size (or number of replications in the case of Monte Carlo research) will depend on the desired levels of power and Type I error, and on the sizes of the effects to be detected. One advantage of Monte Carlo studies in determining power levels of individual parameter estimates is that the population parameters are generally known, and various authors have suggested criteria for robustness of Type I error rates (e.g., Bradley, 1978; Serlin, 2000), thus making it much easier to obtain effect sizes. Robey and Barcikowski and Serlin also provided formulae for determining the number of replications necessary to detect departures from the nominal Type I error rate in statistical tests.

We used the procedure outlined by Robey and Barcikowski (1992) to determine the number of replications needed to detect departures of the actual Type I error rate from a nominal rate of .05 with power of .80. Use of this procedure requires the researcher to specify the degree of departure from the nominal rate that is to be detected. Under conditions such as those in our example study, rejection rates for structural equation models have been found to greatly exceed the nominal rate. We therefore felt that Bradley's (1978) "liberal" or "very liberal" (i.e., actual error rate within ±.75α of the nominal rate) criteria were reasonable. For situations in which researchers expect a greater degree of correspondence between the actual and nominal error rates, a more conservative criterion should be used. For the "very liberal" criterion, we found that the number of replications needed was 276, while use of the "liberal" criterion (±.5α of the nominal rate) yielded a value of 574 replications. Guided by these two values, we set the number of replications at 500.

Generating the Sample Data

If the program being used does not generate both the population and sample matrices, the sample data would be generated as the next step in the study. This would be the case if the LISREL program were used to conduct the Monte Carlo study. An example of the necessary SIMPLIS and PRELIS code is in the online materials for this chapter at http://education.ufl.edu/leite/publications/. Unless raw data are needed, covariance matrices are typically generated because they require less storage space. However, some estimators for nonnormally distributed data as well as some missing data treatments require raw data. This is just as easy to generate but requires more storage space.

Once the sample data are generated some final checks are necessary. First, the sample data should be checked to make sure they match with theoretical expectations. For example, if sample size is varied, standard errors of parameter estimates should decrease as sample size increases. If increasing levels of model misspecification are included in the study, values of the fit indices should be examined to confirm that these increase with the level of misspecification. If distributions with varying levels of nonnormality were created, the obtained values of skew and kurtosis should be compared with the target values.

It is also important to verify that the expected amount of output was actually obtained. The primary reason that some cells might not contain the full number of replications is that some samples might result in nonconvergence and/or improper solutions. The issue of whether or not to include nonconvergent solutions is a matter of debate, but in general researchers tend to exclude these samples from further analysis. This seems consistent with practice in applied studies because nonconvergence is usually an indication of a problem with the model or data (or both) and interpreting the results of nonconvergent solutions is therefore generally considered to be poor practice. Improper solutions present more of a problem because these are not typically screened out by computer programs. However, under some conditions it is not unusual to obtain parameter values that are clearly out of range, such as negative error variances or standardized factor loadings exceeding ±1.0. Although Mplus, LISREL, and EQS all generate warning messages when nonconvergent solutions are found, these programs may not always provide warnings for improper solutions. For this reason it is a good practice to screen parameter estimates to determine if they are within acceptable limits (e.g., Enders & Bandalos, 2001). If improper or out-of-range values are obtained in practice, one would hope that applied researchers would recognize them as indicative of a problem with the analysis and not interpret the results. If this is the case, eliminating such samples from the study would yield results that are more representative of the real world, which should be one's goal. Finally, researchers must decide if they will generate other samples to replace those with nonconvergent or improper solutions, or simply use the samples that do converge. Replacing the samples with new ones has the advantage of maintaining a balanced design. However, it should be noted that convergent replications are not a random subset of all possible replications. In fact, convergent replications often differ from nonconvergent solutions in terms of fit, parameter estimate bias, and other outcomes of interest. Results from such replications can therefore only be generalized to the subset of convergent solutions. Whichever method is used, the number or percentage of nonconvergent or improper solutions for each cell of the design should be reported.

AUTOMATING MONTE CARLO STUDIES

As we have demonstrated, SEM packages such as LISREL and Mplus have built-in facilities for conducting Monte Carlo studies. Although these can be quite useful for some work, large Monte Carlo studies often necessitate the creation and analysis of many sets of data based on different models, parameter values, sample sizes, or other features. Because the built-in simulation facilities are only capable of generating one such dataset at a time, they are not necessarily the most efficient choice of software for a large study in which many features are to be manipulated. For instance, in order to conduct our example study using the built-in facilities we have described for either LISREL or Mplus, we would have to run the example programs 288 (8 models × 2 loading magnitudes × 2 sample sizes × 3 nonnormality levels × 3 categorization levels) separate times, changing the appropriate values each time. Under such circumstances, we consider that any time savings that could be achieved through automatizing the process would be well worth the extra effort we would need to expend to set it up.

In the following sections we offer examples of two such methods that can be used to generate data, call up an existing SEM package (Mplus, in our examples) to estimate the model, save the output, and analyze it. The methods we illustrate involve the use of the SAS matrix language IML and of the free software package R. The advantages of such methods are that: (1) data generation, model estimation, and analyses of the resulting parameters can be conducted from within the same program, if desired, and (2) datasets with different parameter values, sample sizes, or other features can be generated by the same program by incorporating loops in which these features are systematically changed. It is not necessary to utilize all of these features, however; the researcher can choose the IML or SAS functions that are most useful in a given simulation. To illustrate, we use SAS IML to create Mplus MONTECARLO files to generate and save data with the desired characteristics, and to create another set of Mplus "external" MONTECARLO programs to read in and analyze the data. Output from the Mplus analyses is then returned to SAS IML, which saves selected values into new data files. Similarly, in our R examples we use R to generate the data, call Mplus to estimate the models, and then return the Mplus output to R for analysis.

Automating Monte Carlo Studies Using SAS IML

The matrix language IML, available as part of the SAS statistical software package, includes many features that are useful in generating and saving data for Monte Carlo studies. It is also possible to use the IML language to call external SEM programs, such as LISREL or Mplus to estimate models based on the generated data. The output from such estimations can then be returned to SAS IML for further analyses, if desired. In this section we provide a small demonstration in which we use SAS IML to create an Mplus MONTECARLO program which generates data with the desired features, analyzes the data according to a specified model, and saves selected results (parameter estimates, standard errors, and fit statistics) into an external file.

Our SAS IML programs make use of a versatile feature known as "macros." At a basic level, macros consist of lines of code that define a procedure or set of procedures. The macro provides an easy way to reference these procedures. An additional feature of SAS IML macros that is particularly useful in simulation studies is their use in loops. Loops allow the researcher to run the same lines of code sequentially, changing one or more features. For example, a loop could be used to generate data for the same model three times, changing the sample size each time. Series of nested loops can be used to change multiple data features in the same program. In our CFA example, we will use two loops to change the sample sizes and nonnormality levels in each pass through the code. Although these loops do not represent the entirety of the design we discussed previously, we have limited the analysis for the sake of simplicity. Interested readers should be able to expand the code to include additional loops with little difficulty. Looping can be done without the use of macros; however, macros are necessary in our program in order to insert the value of the looping variable into our Mplus program. This will become clearer as the reader goes through the example program, but without the use of macros insertion of looping variable values into other parts of the program is not allowed in SAS IML.

The SAS IML code used to create the Mplus MONTECARLO programs to generate and analyze data and save selected results for the 2-factor CFA model is included in the online materials for this chapter at http://education .ufl.edu/leite/publications/. In the interest of saving space, we have annotated the program liberally with comments to explain the different commands rather than providing detailed explanations in the text.

Automating Monte Carlo Studies Using the R Software

In this section, we briefly describe how to perform a Monte Carlo simulation using the R statistical software (R Development Core Team, 2010). R is a free program for data analysis and graphics. In this demonstration of the use of R for Monte Carlo studies, we perform a small study of whether the DWLS estimator is sensitive to model size when dichotomous data are used. We simulate data for the CFA model in Figure 15.1 and use the same population parameters shown in Figure 15.2 for the Mplus-based simulation. First, we show how to define the population covariance matrix of interest using R. Then, we demonstrate how to generate multivariate normal and non-normal data, given the defined covariance matrix. Once data are generated, there are multiple options to fit the models of interest. Within R, we could use the packages *sem* (Fox, 2006), *lavaan* (Rosseel, 2011), or *OpenMx* (Boker et al., 2011) to fit a CFA model. These packages can estimate a variety of SEM models. Another option is to integrate R with Mplus, EQS, or LISREL to fit SEM models. This integration consists of calling the external SEM software from within R to fit a SEM to a specific dataset, then capturing the results and placing them within R objects. Calling an external SEM software can be accomplished with the function *system* of the *base* R package, but the R packages *REQS* (Mair, Wu, & Bentler, 2010) and *MplusAutomation* (Hallquist, 2011) can also be used to facilitate integration with EQS and Mplus, respectively. These packages have been specifically designed to integrate R with existing SEM programs. After the models of interest have been fit to the simulated datasets using any of these software options and the results have been placed within R objects, the data analysis for the Monte Carlo study can be completed entirely within R.

For our demonstration we have opted to develop our own R code for calling Mplus to run the SEM analyses. *REQS* and *MplusAutomation* are well documented in the papers by Mair et al. (2010) and Hallquist (2011), respectively; readers interested in using these approaches can consult those references. Another reason we chose to create a custom function to integrate R and Mplus is that the *MplusAutomation* package does not allow for Mplus programs that analyze multiple datasets at once. We will provide an example of a Monte Carlo study in which we create three R programs. The first program simulates the data, the second calls Mplus to fit the CFA models and organizes and saves the simulation results, and the third calls the first two programs, fully automating the simulation. A Monte Carlo study in R can be implemented with a series of computational steps (i.e., procedural programming) or with object-oriented programming. For the example study, we will demonstrate a procedural strategy in which separate functions are written to execute each step of the simulation and placed in separate files, and then a short program is written to use the previously created functions for each of the conditions of the study.

As the first step to programming a Monte Carlo study in R, we created the *montecarlo* function to generate the desired data. This function takes as arguments (i.e., inputs) the population parameters of a CFA model for dichotomous items (i.e., thresholds, factor loadings, measurement errors,[2] factor variances and covariances), the sample size, and the number of iterations to produce simulated datasets. The *montecarlo* function contains a *for* loop to generate the desired number of datasets for each condition (i.e., 500) and save them as comma-delimited files on the computer. We could handle the data generation and storage step in two different ways:

We could simulate a single dataset at a time, call Mplus to fit the model and capture the results, and then proceed to simulate and analyze the next dataset. This option is economical in terms of disk space, because after each iteration the single simulated dataset is replaced. Alternatively, we could simulate all 500 datasets of a condition, save them to the computer, then call an Mplus program containing the code "DATA: FILE IS filelist.txt; type = montecarlo;" to analyze the 500 datasets at once (See explanation of the Mplus code for such an "external" simulation study in example program in the accompanying on-line materials at http://education.ufl.edu/leite/publications/). We chose the second option because it is much faster than the first. For the second option, it is necessary to create a list of all datasets for a condition that were saved to the computer (i.e., filelist.txt). In other words, the researcher must create a text file in which the names of all the generated datasets are listed on successive lines of the file. The Mplus code for this example shows how to analyze multiple datasets with a single run of the software (also see the Mplus 7.0 manual, p. 430). The *montecarlo* function is shown in the online materials for this chapter at http://education.ufl.edu/leite/publications/.

Our second programming step was to create the *run_sem* function to call Mplus from within R to fit the CFA model to the simulated data and to capture the Mplus results (see online materials at http://education.ufl.edu/leite/publications/). The *run_sem* function also organizes the results from Mplus for further analyses which could be done either within R or in another program.

Our third program for this simulation calls the previously defined *montecarlo* and *run_sem* functions in this sequence for each condition of the study and accumulates the results into a single dataset of condition descriptors, parameter estimates, standard errors, and fit information for all simulated datasets. This dataset can be used to calculate the outcomes of the Monte Carlo simulation study, such as relative bias of parameter estimates and standard errors, and answer the study's research questions. The third program can be viewed at http://education.ufl.edu/leite/publications/.

DATA ANALYSIS FOR MONTE CARLO STUDIES

Because Monte Carlo studies typically include several independent variables with several levels of each, the number of design cells and of replications can quickly become overwhelming. Although the use of descriptive statistics and graphical techniques can help to illustrate the results of Monte Carlo studies, especially when there are complicated interactions present, the complexity of most Monte Carlo studies requires a more powerful method of detecting the degree to which the dependent variables are affected by the independent variables. For this reason, many researchers recommend the use of appropriate inferential statistical methods in analyzing results of these studies (Harwell, 1992; 1997; Hauck & Anderson, 1984; Hoaglin & Andrews, 1975; Skrondal, 2000). Use of inferential statistics is also necessary to quantify and compare the effects of the various independent variables and their interactions. How many researchers would be able to detect three-way or even complex two-way interactions from perusing tables of means? Although this point may seem obvious, Harwell et al. (1996) found that only about 7% of Monte Carlo studies published in the psychometric literature reported the use of inferential statistics.

One common argument against the use of inferential statistics in analyzing the results of Monte Carlo studies is that the large numbers of replications typically used in these studies renders even the smallest of effects statistically significant. Several counter-arguments have been advanced, however (see Harwell, 1997). First, any presentation of the results of significance testing should be accompanied by estimates of effect size such as w^2 for ANOVA designs. While it is the case that the large numbers of replications included in most Monte Carlo studies result in very high levels of power, researchers can use effect sizes to gain some perspective on the practical significance of the results. In some cases a decision may be made to interpret only those effects reaching some pre-specified level of effect size. As a second argument against this criticism, it should be pointed out that guidelines for determining the number of replications to obtain given levels of power have been offered (Robey & Barcikowski, 1992; Serlin, 2000). By following guidelines such as these, researchers can ensure that their studies will have adequate but not excessive levels of power. Another argument for the use of inferential statistics in Monte Carlo studies is that these studies do incorporate sampling error through the inclusion of multiple replications. The use of inferential statistics allows for this sampling error to be taken into account when determining the size of effects. Finally, the use of inferential statistics in Monte Carlo studies allows for study results to be integrated quantitatively through the use of meta-analysis, as illustrated by Harwell (2003).

To summarize, we recommend the use of both inferential and descriptive or graphical methods in analyzing and conveying the results of Monte Carlo studies. While the use of inferential statistics, along with reporting of effect sizes and/or confidence intervals, is invaluable in detecting complicated effects, including interactions, descriptive and graphical techniques are useful in elucidating the nature of these effects.

With regard to the type of inferential statistics to be used, this should be determined by the study design and research questions. If the design factors are primarily categorical, an ANOVA design is more appropriate. This design is preferred by many researchers because it lends itself to the use of tables of means in presenting results. If design factors are primarily continuous, regression analyses may be used. These analyses offer the advantage of allowing for the estimation of values of the dependent variables for various values of the independent variables within the range of those studied. For example, if sample size were varied at 100, 300, and 500, values of the dependent variables for sample sizes of 200 and 400 could also be estimated using the obtained regression equation (assuming linearity). Of course, these values must be interpreted with caution as no data for those conditions were actually observed.

Issues of Data Management

It is perhaps an understatement to say that Monte Carlo studies result in a great deal of output that must somehow be sorted into files, saved, and analyzed. Paxton et al. (2001) provided an excellent discussion of issues involved in the actual execution of Monte Carlo studies and the attendant data management. These matters will be discussed briefly here; readers desiring a more thorough treatment should consult the Paxton et al. article.

What Should Be Saved? The answer to this question will of course depend on the research questions. Most SEM software packages now incorporate Monte Carlo capabilities that allow for parameter estimates, standard errors, and various fit indices to be output to files. In most software packages results from the different replications are concatenated and saved into one file, or the average values across a set of replications may be saved. Information about each run, such as whether estimates converged and were admissible and the replication number, are also commonly included, either in the same file or as a separate file.

There are two basic options for analyzing this output. Both LISREL and Mplus have limited data analysis capabilities and allow for the calculation of basic descriptive statistics for the output data. However, most Monte Carlo researchers prefer to read the output data into statistical software packages such as SPSS, SAS, or R and conduct their data analyses within one of those programs. The advantage of this is that the various measures of bias and variability described previously can be calculated and saved in these programs, and values designating the categorical design factors and their levels can be inserted to allow for inferential statistics to be used.

Because SEM software packages typically save the results of interest, one may question the necessity of saving the original raw data or sample covariance matrices. If the researcher has kept a log of the seeds used to generate the data, they can always be regenerated if any questions arise. However, this may be extremely time-consuming if a complicated design was used or if time-intensive analyses such as bootstrapping or ADF estimation was used.

This decision will therefore depend to some extent on the type of study that was done. However, if the raw data are not stored, provision should be made for regenerating the data should the need arise.

Finally, the issue of documentation should be addressed. As Paxton et al. (2001) argued, it is not possible to have too much documentation in a Monte Carlo study. In addition to saving the raw and summary data, copies of all programs used should be annotated liberally and archived. Annotation of programs is particularly important because it is often difficult to remember the rationale for program code even after a short period of time. Also, if the research is being carried out collaboratively, program annotation provides an easy way to communicate the purpose of each program to other members of the research group. Paxton et al. recommended keeping multiple copies of each program, and based on our experience in Monte Carlo research we would second this recommendation. Programs should also be dated and corrections or changes noted in the program so that there is no confusion over which version of a program is correct. We also like to keep written records of every analysis from the generation of the population matrices to final analysis of the data, with dates to indicate when each was completed. If analyses are rerun or data regenerated, as sometimes proves necessary, it is vital to be able to identify the most current or corrected data and analyses.

SUMMARY

Monte Carlo studies in SEM involve the generation of sample data in order to study issues such as the robustness of statistical estimators to assumption violations, the small sample behavior of asymptotic estimators, or the behavior of ad hoc fit indices. These studies are essentially the same as other research studies in that the research questions should be based on theory and previous research and should be clearly stated. Similarly, the choice of independent variables and their levels and of dependent variables should be based on a thorough review of the available literature and theory. In this chapter we introduced the most commonly used independent and dependent variables in SEM Monte Carlo studies and provided information that we hope will be useful in making choices among these. We also illustrated the use of the Mplus and LISREL computer packages for conducting Monte Carlo studies in SEM, and provided examples of how Monte Carlo studies can be automated using two popular software programs: SAS and R. Finally, we emphasized that the analysis of data from Monte Carlo studies should include inferential statistics chosen to provide answers to the initial research questions as well as tables and/or graphical displays to clarify or elucidate aspects of the results. We ended by discussing the types of output that should be saved and by stressing the importance of clear and thorough documentation of all syntax and data files.

NOTES

1. We use the term *independent variables* here to designate facets of the research design. We use the term *dependent variable* to designate outcome variables. To avoid confusion the terms *exogenous* and *endogenous variables* are used to designate independent and dependent variables within a measured variable or latent variable path model.
2. Although measurement errors and thresholds cannot both be *estimated* for dichotomous data, both should be included in the *generation* of the data (see footnote, Figure 15.2).

REFERENCES

Arnold-Berkovits, I. (2002). *Structural equation modeling with ordered polytomous and continuous variables: A simulation study comparing full-information Bayesian estimation to correlation/covariance methods.* Unpublished doctoral dissertation, University of Maryland, College Park.

Bandalos, D. L., & Enders, C. K. (1996). The effects of nonnormality and number of response categories on reliability. *Applied Measurement in Education, 9,* 151–160.

Bang, J. W., Schumacker, R. E., & Schlieve, P. (1998). Random-number generator validity in simulation studies: An investigation of normality. *Educational and Psychological Measurement, 58,* 430–450.

Boker, S., Neale, M., Maes, H., Wilde, M., Spiegel, M., Brick, T., et al. (2011). Openmx: An open source extended structural equation modeling framework. *Psychometrika, 76,* 306–317.

Bradley, J. V. (1978). Robustness? *British Journal of Mathematical and Statistical Psychology, 31,* 144–152.

Browne, M. W., & Cudeck, R. (1992). Alternative ways of assessing model fit. *Sociological Methods & Research, 21,* 230–258.

Collins, L. M., Schafer, J. L., & Kam, C.-M. (2001). A comparison of inclusive and restrictive strategies in modern missing data procedures. *Psychological Methods, 6,* 330–351.

Cudeck, R., & Henly, S. J. (1991). Model selection in covariance structure analysis and the "problem" of sample size: A clarification. *Psychological Bulletin, 109,* 512–519.

Dolan, C. V. (1994). Factor analysis of variables with 2, 3, 5, and 7 response categories: A comparison of categorical variable estimators using simulated data. *British Journal of Mathematical and Statistical Psychology, 47,* 309–326.

Enders, C. K. (2001). The impact of nonnormality on full information maximum-likelihood estimation for structural equation models with missing data. *Psychological Methods, 6,* 352–370.

Enders, C. K., & Bandalos, D. L. (2001). The relative performance of full information maximum likelihood estimation for missing data in structural equation models. *Structural Equation Modeling: A Multidisciplinary Journal, 8,* 430–457.

Enders, C. E., & Finney, S. J. (2003, April). *Examining the sensitivity of fit indices to model misspecification and complexity when modeling Likert data: An examination of recommended cutoffs.* Paper presented at the annual meeting of the American Educational Research Association, Chicago.

Fan, X., Felsővályi, A., Sivo, S. A., & Keenan, S. C. (2001). *SAS for Monte Carlo Studies: A Guide for Quantitative Researchers.* Cary, NC: SAS Institute.

Fishmann, G. S., & Moore, L. R. (1982). A statistical evaluation of multiplicative congruential random number generators with Modulus $2^{31\text{-}1}$. *Journal of the American Statistical Association, 77,* 129–136.

Flora, D. P., & Curran, P. J. (2004). An empirical evaluation of alternative methods of estimation for confirmatory factor analysis with ordinal data. *Psychological Methods, 9,* 466–491.

Fox, J. (2006). Structural equation modeling with the sem package in r. *Structural Equation Modeling: A Multidisciplinary Journal, 13,* 465–486.

Hallquist, M. (2011). Mplusautomation: Automating Mplus model estimation and interpretation, from http://cran.r-project.org/web/packages/MplusAutomation/

Harwell, M. R. (1992). Summarizing Monte Carlo results in methodological research. *Journal of Educational Statistics, 17,* 297–313.

Harwell, M. R. (1997). Analyzing the results of Monte Carlo studies in Item Response Theory. *Educational and Psychological Measurement, 57,* 266–279.

Harwell, M. R. (2003). Summarizing Monte Carlo results in methodological research: The single-factor, fixed-effects ANCOVA case. *Journal of Educational and Behavioral Statistics, 28,* 45–70.

Harwell, M. R., Stone, C. A., Hsu, T.-C., & Kirisci, L. (1996). Monte Carlo studies in Item Response Theory. *Applied Psychological Measurement, 20,* 101–125.

Hauck, W. W., & Anderson, S. (1984). A survey regarding the reporting of simulation studies. *The American Statistician, 38,* 214–216.

Hoaglin, D.C., & Andrews, D.F. (1975). The reporting of computation-based results in statistics. *The American Statistician, 29,* 122–126.

Hoogland, J. J., & Boomsma, A. (1998). Robustness studies in covariance structure modeling: An overview and meta-analysis. *Sociological Methods & Research, 26,* 329–367.

Hu, L. T., & Bentler, P. M. (1999). Cutoff criteria for fit indexes in covariance structure analysis: Conventional criteria versus new alternatives. *Structural Equation Modeling: A Multidisciplinary Journal, 6,* 1–55.

Hu, L. T., Bentler, P. M., & Kano, Y. (1992). Can test statistics in covariance modeling be trusted? *Psychological Bulletin, 112,* 351–362.

Kaplan, D. (1988). The impact of specification error on the estimation, testing, and improvement of structural equation models. *Multivariate Behavioral Research, 23,* 69–86.

Kaplan, D., & Wenger, R. N. (1993). Asymptotic independence and separability in covariance structure models: Implications for specification error, power, and model identification. *Multivariate Behavioral Research, 28,* 467–482.

Lei, P.-W. (2009). Evaluating estimation methods for ordinal data in structural equation modeling. *Quality and Quantity, 43,* 495–507.

MacCallum, R. (2003).Working with imperfect models. *Multivariate Behavioral Research, 38,* 113–139.

MacCallum, R. C., Browne, M. W., & Sugawara, H. M. (1996). Power analysis and determination of sample size for covariance structure modeling. *Psychological Methods, 1,* 130–149.

Mair, P., Wu, E., & Bentler, P. M. (2010). EQS goes R: Simulations for SEM using the package reqs. *Structural Equation Modeling: A Multidisciplinary Journal, 17,* 333–349.

Mattson, S. (1997). How to generate non-normal data for simulation of structural equation models. *Multivariate Behavioral Research, 32,* 355–373.

Metropolis, N., & Ulam, S. (1949). The Monte Carlo method. *Journal of the American Statistical Association, 44,* 335–341.

Micceri, T. (1989). The unicorn, the normal curve, and other improbably creatures. *Psychological Bulletin, 105,* 156–166.

Muthén, B. O., du Toit, S. H. C., & Spisic, D. (1997). Robust inference using weighted least squares and quadratic estimating equations in latent variable modeling with categorical outcomes. *Unpublished manuscript.*

Muthén, B. O., & Kaplan, D. (1985). A comparison of some methodologies for the factor analysis of non-normal Likert variables. *British Journal of Mathematical and Statistical Psychology, 38,* 171–189.

Muthén, B. O., & Kaplan, D. (1992). A comparison of some methodologies for the factor analysis of non-normal Likert variables: A note on the size of the model. *British Journal of Mathematical and Statistical Psychology, 45,* 19–30.

Muthén, B. O., Kaplan, D., & Hollis, M. (1987). On structural equation modeling with data that are not missing completely at random. *Psychometrika, 52,* 431–462.

Muthén, L. K., & Muthén, B. O. (2002). How to use a Monte Carlo study to decide on sample size and determine power. *Structural Equation Modeling: A Multidisciplinary Journal, 9,* 599–620.

Olsson, U. H., Foss, T., Troye, S. V., & Howell, R. D. (2000). The performance of ML, GLS, and WLS estimation in structural equation modeling under conditions of misspecification and nonnormality. *Structural Equation Modeling: A Multidisciplinary Journal, 7,* 557–595.

Paxton, P., Curran, P. J., Bollen, K. A., Kirby, J., & Chen, F. (2001). Monte Carlo experiments: Design and implementation. *Structural Equation Modeling: A Multidisciplinary Journal, 8,* 287–312.

Powell, D. A., & Schafer, W. D. (2001). The robustness of the likelihood ratio chi-square test for structural equation models: A meta-analysis. *Journal of Educational and Behavioral Statistics, 26,* 105–132.

Preacher, K. J. (2006). Quantifying parsimony in structural equation modeling. *Multivariate Behavioral Research, 41,* 227–259.

R Development Core Team. (2010). R: A language and environment for statistical computing. Vienna, Austria: R Foundation for Statistical Computing. Retrieved from http://www.R-project.org.

Reinartz, W. J., Echambadi, R., & Chinn, W. (2002). Generating non-normal data for simulation of structural equation models using Mattson's method. *Multivariate Behavioral Research, 37,* 227–244.

Robey, R. R., & Barcikowski, R. S. (1992). Type I error and the number of iterations in Monte Carlo studies of robustness. *British Journal of Mathematical and Statistical Psychology, 45,* 283–288.

Rosseel, Y. (2011). Lavaan—Latent variable analysis, from http://lavaan.ugent.be/

Serlin, R. C. (2000). Testing for robustness in Monte Carlo studies. *Psychological Methods, 5,* 230–240.

Siegel, S., & Castellan, N. J., Jr. (1988). *Nonparametric statistics for the behavioral sciences.* New York, NY: McGraw-Hill.

Skrondal, A. (2000). Design and analysis of Monte Carlo experiments: Attacking the conventional wisdom. *Multivariate Behavioral Research, 35,* 137–167.

Vale, C. D., & Maurelli, V. A. (1983). Simulating multivariate non-normal distributions. *Psychometrika, 48,* 465–471.

Yu, C.-Y., & Muthén, B. O. (2002, March). *Evaluation of model fit indices for latent variable models with categorical and continuous outcomes.* Paper presented at the annual meeting of the American Educational Research Association, New Orleans, LA.

Yuan, K.-H., & Chan, W. (2005). On nonequivalence of several procedures of structural equation modeling. *Psychometrika, 70,* 791–798.

ABOUT THE CONTRIBUTORS

Deborah L. Bandalos is Director of the Assessment and Measurement Doctoral Program at James Madison University where she holds the rank of Professor. She is the author of numerous methodological articles and book chapters in the areas of structural equation modeling, exploratory factor analysis, and the conduct of simulation studies, and of a forthcoming book entitled *Measurement Theory and Applications for the Social Sciences.* Debbi is Associate Editor of *Multivariate Behavioral Research* and served previously as Associate Editor of *Structural Equation Modeling: A Multidisciplinary Journal.* She is active in governance for both the American Psychological Association and the American Educational Research Association.

Jaehwa Choi is Associate Professor of educational research methods in the Department of Educational Leadership at The George Washington University. His research interests include structural equation modeling, latent growth models, Markov chain Monte Carlo estimation methods for latent variable models, and formative assessment system applications. He is the chief inventor of *Computer Adaptive Formative Assessment*™ (*CAFA*™ www.eMathTest.com/CAFA) that is the core engine of a web based mathematics learning system, www.eMathTest.com.

Christine DiStefano is Associate Professor of educational research and measurement at the University of South Carolina. She teaches classes in survey research, measurement theory, structural equation modeling, and classroom assessment. Christine's research interests are in the areas of emotional/behavioral assessment for young children, Rasch modeling, classification, and structural equation modeling with ordinal data.

Structural Equation Modeling: A Second Course (2nd ed.), pages 667–673
Copyright © 2013 by Information Age Publishing
667

Craig K. Enders is Associate Professor in the Quantitative Psychology concentration in the Department of Psychology at Arizona State University, where he teaches graduate-level courses in missing data analyses, multilevel modeling, and longitudinal modeling. The majority of his research focuses on analytic issues related to missing data analyses and multilevel modeling. His book, *Applied Missing Data Analysis*, was published with Guilford Press in 2010.

Sara J. Finney has a dual appointment at James Madison University as Associate Professor in the Department of Graduate Psychology and as Associate Assessment Specialist in the Center for Assessment and Research Studies. In addition to teaching multivariate statistics and structural equation modeling for the assessment and measurement Ph.D. program, she is Coordinator of the quantitative psychology concentration within the psychological sciences M.A. program. Much of her research involves the application of structural equation modeling to better understand the functioning of self-report instruments.

Brian F. French is Associate Professor of educational psychology and Director of the Learning and Performance Research Center in the College of Education at Washington State University in Pullman, WA. He teaches courses in measurement, statistics, and advanced quantitative methods. His research is in the area of educational and psychological measurement with an emphasis on test score validity. A sample of topics of interest include: measurement invariance, structural equation modeling, item response theory, factor analysis, and Monte Carlo studies.

Phill Gagné is a federal researcher and earned his doctorate in Measurement, Statistics, and Evaluation from the University of Maryland at College Park in 2004. His published research predominantly pertains to mixture modeling, item response theory, and Monte Carlo simulation methodology. His other research interests include multilevel modeling and single-case research design.

Samuel B. Green is Professor in the T. Denny Sanford School of Social and Family Dynamics at Arizona State University. He is currently on the editorial boards of *Structural Equation Modeling: A Multidisciplinary Journal*, *Psychological Methods*, and *Educational and Psychological Measurement*. He also is a past chair of AERA's SIG/Structural Equation Modeling. Sam conducts research primarily in the areas of structural equation modeling, multivariate analyses of means, and reliability. His focus of research in SEM has been in testing of differences in factor means, exploratory SEM methods, reliability estimation, and analyses of item data.

Gregory R. Hancock is Professor and Program Director of measurement, statistics, and evaluation in the Department of Human Development and

Quantitative Methodology at the University of Maryland, and Director of the Center for Integrated Latent Variable Research (CILVR). His SEM-related research has appeared in such journals as *Psychometrika, Multivariate Behavioral Research, British Journal of Mathematical and Statistical Psychology*, and *Journal of Educational and Behavioral Statistics*. He is Associate Editor of *Structural Equation Modeling: A Multidisciplinary Journal*, and has taught dozens of SEM workshops in the U.S., Canada, and abroad.

Jeffrey R. Harring is Associate Professor of measurement, statistics, and evaluation in the Department of Human Development and Quantitative Methodology at the University of Maryland. His research interests include statistical models and methods for repeated measures data, nonlinear structural equation models, and general statistical computing. His research has appeared in such publication outlets as *Multivariate Behavioral Research, Journal of Educational and Behavioral Statistics, Psychological Assessment, Structural Equation Modeling: A Multidisciplinary Journal, Psychological Methods*, and *Annual Review of Psychology*.

Kit-Tai Hau is Vice-President and Chair Professor of Educational Psychology at The Chinese University of Hong Kong. He is currently the President of the Educational Psychology Division, International Association of Applied Psychology. His research interest includes motivation and quantitative methods, and he has published a very popular Chinese textbook on structural equation modeling, as well as in *American Psychologist, Journal of Educational Psychology*, and *Journal of Educational Measurement*. He has served in various government advisory boards in education. Over the last 15 years, he has conducted more than 80 advanced applied statistics workshops in China, each with hundreds of participants.

Andrew F. Hayes is a faculty member at The Ohio State University. He is the author of *Introduction to Mediation, Moderation, and Conditional Process Analysis* (2013, The Guilford Press), *Statistical Methods for Communication Science* (2005, Lawrence Erlbaum Associates), and has authored or coauthored many journal articles and book chapters in research methods and data analysis. He is one of the founding editors of *Communication Methods and Measures* and serves as Editor-in-Chief through 2015. He teaches research design and data analysis and frequently conducts workshops and short-courses on moderation and mediation analysis throughout the world. He can be located in cyberspace at http://www.afhayes.com/.

Scott L. Hershberger is a manager of biostatistics at United Health Group and formerly a professor of psychology in the Department of Psychology at California State University, Long Beach. He has published extensively in the areas of structural equation modeling, psychometric theory, behavior genetics, and sexual orientation and behavior. He is a fellow of the Royal Statistical Society and International Statistical Institute.

Rex B. Kline is Professor of Psychology at Concordia University in Montréal. Since earning a doctorate in clinical psychology, his areas of research and writing have included the psychometric evaluation of cognitive abilities, cognitive and scholastic assessment of children, structural equation modeling, the training of behavioral science researchers, and usability engineering in computer science. Dr. Kline has published six books and nine chapters in these areas (http://tinyurl.com/rexkline).

Frank R. Lawrence is a senior statistical consultant at Michigan State University. He was a research professor at the University of Alabama, an assistant professor in the College of Health and Human Development at the Pennsylvania State University, senior data modeler at the Department of Defense Manpower Data Center, and biostatistician at Henry Ford Health Systems. Articles he has coauthored have appeared in the literature of several different disciplines including epidemiology, communication disorders, media, psychology, and education.

Walter L. Leite is Associate Professor and Program Coordinator of the Research and Evaluation Methodology Program at the School of Human Development and Organizational Studies in Education at the University of Florida. His primary research interests are structural equation modeling, multilevel modeling, and propensity score methods. He has served as chair and vice-chair of the Structural Equation Modeling and Hierarchical Linear Modeling special interest groups of the American Educational Research Association, and serves as consulting editor for the *Journal of Experimental Education*. He has been teaching statistical analysis and Monte Carlo simulation with the R software package for 8 years.

Roy Levy is Associate Professor of measurement and statistical analysis in the T. Denny Sanford School of Social and Family Dynamics at Arizona State University. His primary research and teaching interests include structural equation modeling, item response theory, Bayesian networks, and Bayesian approaches to inference and modeling. His other interests include assessment design and applications of design and psychometric principles to simulation- and game-based assessments. He is a past chair of the American Educational Research Association's special interest group on Structural Equation Modeling and serves on the editorial board of several methodological and applied journals.

George A. Marcoulides is Professor of research methods and statistics in the Graduate School of Education and in the Interdepartmental Graduate Program in Management (IGPM) in the A. Gary Anderson Graduate School of Management at the University of California, Riverside. He is a fellow of the American Educational Research Association, a fellow of the Royal Statistical Society, and a member of the Society of Multivariate Experimental Psychology. He is currently Editor of the journals *Structural Equation Modeling: A Multidisciplinary Journal* and *Educational and Psychological Measurement*, editor of the Quantitative Methodology Book Series, and on the editorial board of numerous other scholarly journals.

Herbert W. Marsh holds joint appointments at the Centre for Positive Psychology and Education at the University of Western Sydney, at King Saud University in Saudi Arabia, and at Oxford University. He is an *ISI highly cited researcher* (http://isihighlycited.com/) and recently achieved a Google Scholar H-Index of 100 (based on 762 publications and 41,580 citations). He founded and directs the SELF Research Centre that has 450 members and satellite centres at leading Universities around the world, and co-edits the SELF monograph series. He coined the phrase *substantive-methodological research synergy* which underpins his research efforts. His major research/scholarly interests include self-concept and motivational constructs; evaluations of teaching effectiveness; developmental psychology, quantitative analysis; value-added and contextual models; sports psychology; the peer review process; gender differences; peer support and anti-bullying.

Alexandre J. S. Morin is Associate Professor at the Centre for Positive Psychology and Education (http://www.uws.edu.au/cppe), University of Western Sydney, Australia, where he heads the Positive Substantive Methodological Synergy research program, and seeks ways to promote secondary data analyses of multiple rich data bases available locally and internationally. He defines himself as a lifespan developmental psychologist with broad research interests anchored in the exploration of the social determinants (school, workplace) of psychological wellbeing across the lifespan. His research is often anchored into a substantive-methodological synergy framework, representing joint ventures in which new methodological developments are applied to substantively important research issues.

Ralph O. Mueller is Dean of the College of Education, Nursing and Health Professions (ENHP) and founding Director of the Institute for Translational Research at the University of Hartford, CT. He holds faculty rank as Professor of educational leadership in the College of ENHP and as Professor of psychology in the College of Arts and Sciences. Ralph was Chair of AERA's SIG/Structural Equation Modeling and co/taught many national and international SEM workshops for universities, associations, and software companies. He is the author of an introductory SEM text and of other didactic writings about applied statistics.

Benjamin Nagengast is Professor of Educational Psychology at the Center for Educational Science and Psychology, University of Tübingen, Germany. He received his Ph.D. in Psychology from the University of Jena, Germany, and worked as a post-doctoral research fellow at the SELF Research Centre at the University of Oxford. His main research interests include the application of advanced quantitative methods (structural equation modeling, multilevel modeling, causal inference) to research questions in psychology and education. His substantive research interests include educational effectiveness studies, motivation, and academic self-concept.

Dena A. Pastor has a dual appointment at James Madison University as Associate Professor in the Department of Graduate Psychology and as Associate Assessment Specialist in the Center for Assessment and Research Studies. She teaches courses in hierarchical linear modeling, categorical data analysis, and data management. Her research applies statistical and psychometric techniques to the modeling and measurement of college student learning and development. Her publications have appeared in *Contemporary Educational Psychology, Applied Psychological Measurement,* and *Applied Measurement in Education.* She serves on the editorial board for *Educational and Psychological Measurement* and the statistical and methodological advisory board for *Journal of School Psychology.*

Kristopher J. Preacher is a faculty member in the Quantitative Methods Program in the Peabody College at Vanderbilt University. His research concerns the use (and combination) of structural equation modeling and multilevel modeling to model longitudinal and correlational data. Other interests include developing techniques to test mediation and moderation hypotheses, bridging the gap between theory and practice, and studying model evaluation and model selection in the application of multivariate methods to social science questions. He serves on the editorial boards of *Psychological Methods, Communication Methods and Measures,* and *Multivariate Behavioral Research.*

Edward E. Rigdon is Professor of marketing in the Robinson College of Business at Georgia State University in Atlanta, GA. Dr. Rigdon's approach to modeling structural equations with multiple indicators encompasses both factor and composite methods. His work has appeared in many journals, including *Journal of Marketing Research, Journal of Consumer Research, MIS Quarterly, Multivariate Behavioral Research,* and *Structural Equation Modeling: A Multidisciplinary Journal.* He was a co-founder and is currently list owner for SEMNET, an email discussion list devoted to structural equation modeling (https://listserv.ua.edu/archives/semnet.html).

Laura M. Stapleton is Associate Professor in Measurement, Statistics and Evaluation in the Department of Human Development and Quantitative Methodology at the University of Maryland. Her research interests include multivariate analysis of survey data obtained under complex sampling designs and multilevel latent variable models, including tests of mediation within a multilevel framework. She has served as the chair of AERA's SIG/Structural Equation Modeling and has taught structural equation modeling at the University of Texas, the University of Maryland, Baltimore County, and the University of Maryland.

Marilyn S. Thompson is Associate Professor in the T. Denny Sanford School of Social and Family Dynamics at Arizona State University. She is past chair of AERA's Structural Equation Modeling Special Interest Group. Marilyn's structural equation modeling research interests include evaluation of factorial invariance and differences in latent means, approaches to model respecification, and modeling of developmental data. She enjoys writing pedagogical chapters and articles relating to SEM and is also interested in the use and misuse of data to inform educational policy and practice.

Zhonglin Wen is Professor in the Center for Studies of Psychological Application, School of Psychology, South China Normal University. Mainly educated at universities in mainland China and the Chinese University of Hong Kong, he studied as a visiting scholar at Oxford University, the University of Manchester, and the University of Western Sydney. His main research interests include mathematical statistics and advanced quantitative methods in psychology and education, especially structural equation modeling and analyses of mediating and moderating effects. His papers are the most frequently cited in the field of psychology in China over the last three years.

Made in the USA
San Bernardino, CA
11 December 2015